Mechanical Engineering Series

Frederick F. Ling
Series Editor

Springer
New York
Berlin
Heidelberg
Barcelona
Hong Kong
London
Milan
Paris
Singapore
Tokyo

Mechanical Engineering Series

(continued after index)

James F. Doyle

Nonlinear Analysis of Thin-Walled Structures

Statics, Dynamics, and Stability

With 267 Figures

 Springer

James F. Doyle
Department of Aeronautics and Astronautics
Purdue University
West Lafayette, IN 47907, USA
jfdoyle@ecn.purdue.edu

Series Editor
Frederick F. Ling
Ernest F. Gloyna Regents Chair in Engineering
Department of Mechanical Engineering
The University of Texas at Austin
Austin, TX 78712-1063, USA
    and
William Howard Hart Professor Emeritus
Department of Mechanical Engineering,
    Aeronautical Engineering and Mechanics
Rensselaer Polytechnic Institute
Troy, NY 12180-3590, USA

Library of Congress Cataloging-in-Publication Data
Doyle, James F., 1951–
    Nonlinear analysis of thin-walled structures : statics, dynamics,
and stability / James F. Doyle.
        p. cm. — (Mechanical engineering series)
    Includes bibliographical references and index.
    ISBN 0-387-95216-0 (alk. paper)
        1. Thin-walled structures.   2. Structural analysis (Engineering)   3. Nonlinear theories.
    I. Title.   II. Mechanical engineering series (Berlin, Germany)
    TA660.T5 D69 2001
    624.1'7—dc21   00-053774

Printed on acid-free paper.

Production managed by Allan Abrams; manufacturing supervised by Joe Quatela.
Photocomposed copy prepared from the author's LaTeX files.

9 8 7 6 5 4 3 2 1

ISBN 0-387-95216-0   SPIN 10791718

Springer-Verlag   New York Berlin Heidelberg
A member of BertelsmannSpringer Science+Business Media GmbH

To my sister, Marie,
and my brothers, Paddy and Fran;
this one is for you, thanks, Séamus.

Mechanical Engineering Series

Frederick F. Ling   Ward O. Winer
Series Editors

Series Preface

Mechanical engineering, an engineering discipline born of the needs of the Industrial Revolution, is once again asked to do its substantial share in the call for industrial renewal. The general call is urgent as we face the profound issues of productivity and competitiveness that require engineering solutions, among others. The Mechanical Engineering Series is a new series, featuring graduate texts and research monographs, intended to address the need for information in contemporary areas of mechanical engineering.

The series is conceived as a comprehensive one that will cover a broad range of concentrations important to mechanical engineering graduate education and research. We are fortunate to have a distinguished roster of consulting editors, each an expert in one of the areas of concentration. The names of the consulting editors are listed on page vi. The areas of concentration are applied mechanics, biomechanics, computational mechanics, dynamic systems and control, energetics, mechanics of materials, processing, thermal science, and tribology.

We are pleased to present *Nonlinear Analysis of Thin-Walled Structures* by James F. Doyle.

Austin, Texas   Frederick F. Ling

Preface

This book is concerned with the challenging subject of the nonlinear static, dynamic, and stability analyses of thin-walled structures. It carries on from where *Static and Dynamic Analysis of Structures*, published by Kluwer 1991, left off; that book concentrated on frames and linear analysis, while the present book is focused on plated structures, nonlinear analysis, and a greater emphasis on stability analysis.

It is worth restating the justification used for the first volume because it seems even truer today, nearly a decade later. As pointed out, with the widespread availability and use of computers, today's engineers have on their desks an analysis capability undreamt of by previous generations. However, the ever increasing quality and range of capabilities of commercially available software packages have divided the engineering profession into two groups: a small group of specialist program writers that know the details of the coding, algorithms, and solution strategies; and a much larger group of practicing engineers who use the programs. It is possible for this latter group to use this enormous power without really knowing anything of its source. Therein lies the potential danger — the engineer is seduced by the power, the litany of capabilities, the seeming ease of use, and forgets how to perform simple consistency and validation checks. We use, and we should use, commercial packages when they are available. But to make safe, efficient, and intelligent use of them, we need to have some idea of their inner workings as well as the mechanics foundations on which they are built. That is the purpose of this book.

To be an intelligent user of these powerful commercial programs requires some appreciation of the full range of assumptions and procedures on which they are based. Without doubt, an understanding of the mechanics principles is essential, but it is not sufficient, because these principles are transformed in subtle ways when converted into algorithms and code. This situation is exacerbated even more when nonlinear dynamics and stability are involved.

With the foregoing in mind, this book sets as its goal the treatment of nonlinear behavior of thin-walled structures starting with the basic mechanics principles and going all the way to their implementation on digital computers. It is only by studying this in its complete extent do the unique difficulties of computational mechanics manifest themselves. Rather than discuss particular commercial packages, we use the program NonStaD: a complete (but lean) program to perform each of the standard procedures used in commercial programs.

Most topics from that first volume are not repeated but some (such as the finite difference schemes) are revisited since they are affected by the nonlinear case.

No source code is included in this volume, but to encourage readers to try the algorithms, I have posted on my Web homepage the source code to many of the algorithms and problems discussed in the text. The URL is:

`http://aae.www.ecn.purdue.edu/~jfdoyle`

Look under the section on **Source Code**. In a similar vein, I have tried to supplement each chapter with a collection of pertinent problems plus specific references that can form the basis for further studies.

Lafayette, Indiana   James F. Doyle
February, 2001

Contents

Notation

Roman letters:

a	radius, plate width
A	surface area, cross-sectional area
b, b_i	thickness, depth, plate length, body force
c_o	longitudinal wave speed, $\sqrt{EA/\rho A}$
$C, [\,C\,]$	damping, damping matrix
D	plate stiffness, $Eh^3/12(1-\nu^2)$
$\hat{e}_i$	unit vectors
e_{ij}	Eulerian strain tensor
$E, \hat{E}$	Young's modulus, viscoelastic modulus
EI	beam flexural stiffness
E_{ij}	Lagrangian strain tensor
$F, \bar{F}, \bar{F}_o$	member axial force, element nodal force
$\mathcal{F}$	equilibrium path
$g_i(x)$	element shape functions
h	beam or rod height, plate thickness
h_i	area coordinates
i	complex $\sqrt{-1}$, counter
I	second moment of area, $I = bh^3/12$ for rectangle
I_n	modified Bessel functions of the first kind
J^o, J	Jacobian, polar moment of area, $J = \pi d^4/32$ for circle
J_n	Bessel functions of the first kind
k, k_1, k_2	wavenumbers
$K, [\,k\,], [\,K\,]$	stiffness, stiffness matrices
K_n	modified Bessel functions of the second kind
L	length
M, M_x	moment
$M, [\,m\,], [\,M\,]$	mass, mass matrix
N_i	shape functions
$P(t), \hat{P}, \{P\}$	applied force history, vector of nodal loads
$[\,P\,]$	projector matrix
q	distributed load
r, R	radial coordinate, radius
$[\,R\,]$	rotation matrix
t, t_i	time, traction
T	time window, kinetic energy, temperature
$[\,T\,]$	transformation matrix
$u(t)$	response; velocity, strain, etc.
u, v, w	displacements
U	strain energy
V	member shear force, volume, potential, Lyapunov function
W	space transform window
x^o, y^o, z^o	original rectilinear coordinates
x, y, z	deformed rectilinear coordinates
Y_n	Bessel functions of the second kind

Greek letters:

α	coefficient of thermal expansion
β_{ij}	matrix of direction cosines
δ	small quantity, variation
δ_{ij}	Kronecker delta
Δ	determinant, increment
ϵ	small quantity, strain
ϵ_{ijk}	permutation symbol
η	viscosity, damping, principal coordinate
θ	angular coordinate
κ	plate curvature
λ	Lamê constant, eigenvalue
μ	Shear modulus, complex frequency
ν	Poisson's ratio
ξ	variational coordinate
Π	total potential energy
ρ^o, ρ	mass density
σ, ϵ	stress, strain
ϕ, ϕ_x, ϕ_y	Airy stress function, rotation
$[\ \Phi\ ]$	modal matrix
ζ	damping ratio
ω	angular frequency
ω_{ij}	rotation tensor

Special Symbols:

∇^2	differential operator, $\frac{\partial^2}{\partial x^2} + \frac{\partial^2}{\partial y^2}$
$[\quad]$	square matrix
$\{\ \}$	vector

Subscripts:

E, G, T	elastic, geometric, tangent stiffness matrix
i, j, k	tensor components

Superscripts:

K	Kirchhoff stresses
o	original configuration
$*$	complex conjugate
$\bar{\ }$	bar, local coordinates
$\dot{\ }$	dot, time derivative
$\hat{\ }$	hat, frequency dependent, vector
$\prime$	prime, derivative with respect to argument

Abbreviations:

DoF	degree of freedom
CST	constant strain triangle element
DKT	discrete Kirchhoff triangle element
FEM	finite element method
MRT	membrane with rotation triangle element

Introduction

Physical science has two different directions of progress, which have
been called the ascending and the descending scale, the inductive and
the deductive method, the way of analysis and of synthesis. In every
physical science, we must ascend from facts to laws, by the way of
induction and analysis; and we must descend from laws to
consequences, by the deductive and synthetic way.

W.R. HAMILTON [31]

Owing to the necessity to save weight and material in the design of modern
structures, thin-walled reinforced constructions have emerged as a dominant
style. These light-weight structures, however, are more susceptible than their
traditional counterparts to problems originating from large deflections, nonlinear
vibrations, and structural instabilities, and therefore require a greater depth and
breadth of analyses.

This book sets as its goal the treatment of nonlinear behavior of thin-walled
structures starting with the basic mechanics principles and going all the way to
their implementation on computers. It is only by studying this in its complete
extent do the unique difficulties of computational mechanics (as well as the
limitations of the theory) fully manifest themselves. An attempt is made for this
book to be more than just a collection of disparate topics on nonlinear structures;
rather, topics are introduced and developed in such a way that they are given
meaning as part of a coherent whole. The central theme and thread running
through the book is the notion that instability of the equilibrium is synonymous
with motion and large displacements, and therefore requires a fully nonlinear
dynamic analysis capability.

Types of Structures Considered

Structures that can be satisfactorily idealized as a collection of line elements
are called *frame* or *skeletal* structures; Figure I.1 shows a few examples. Usually
their members are assumed to be connected either by frictionless pins or by rigid
joints.

A *rod* member can support only axial loads, whereas a *beam* member supports
bending as well as transverse loads. A *truss* consists of a collection of arbitrarily
oriented rod members that are interconnected at pinned joints. They are loaded
only at their joints and (because the joints cannot transmit bending moments)

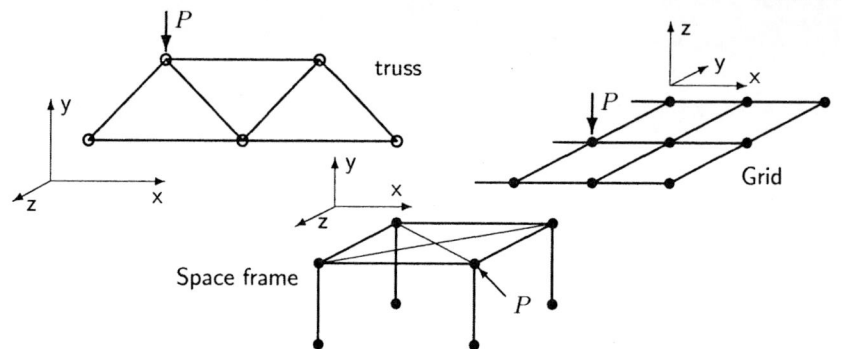

Figure I.1: Some types of skeletal structures.

must be triangulated to avoid collapse. A *frame* structure, on the other hand, is one that consists of beam members that are connected rigidly or by pins at the joints. The members can support bending (in any direction) as well as axial loads, and at the rigid joints the relative positions of the members remain unchanged after deformation. Rigidly jointed frames are often loaded along their members as well as at their joints. Plane frames, like plane trusses, are loaded only in their own plane. In contrast, *grids* (or *grills*) are always loaded normal to the plane of the structure. Space frames can be loaded in any plane. The space frame is the most complicated type of jointed framework — each member can undergo axial deformation, torsional deformation, and flexural deformation (in two planes). Its supports may be fixed, pinned, elastic, or there may be roller supports.

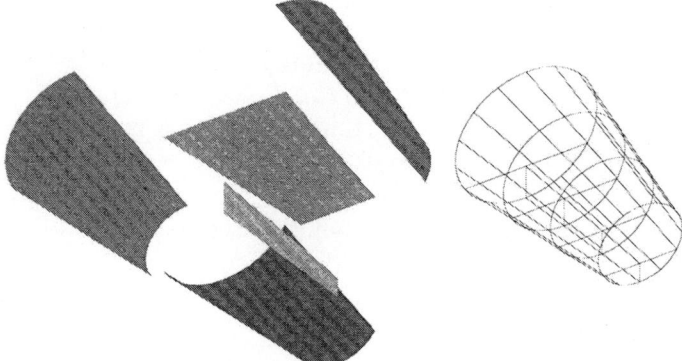

Figure I.2: Exploded view of a complex multicelled thin-walled structure modeled as a collection of flat and curved plates plus frame reinforcers.

A *plate* is an extended body where one of the dimensions is substantially smaller than the other two. Structures that can be satisfactorily idealized as a collection of flat platelets are called *folded plate* structures. These platelets are usually connected by frictionless pins or by rigid joints and undergo in-plane deformations (called the *membrane* action) as well as out-of-plane bending and twisting. When the plates have continuous curvature, they are called *shells*. Folded plate and shell structures are collectively called *thin-walled* structures and, when they are combined with frame members, they are called *reinforced thin-walled* structures; Figure I.2 shows an example.

The total possible displacement components at each point in a structure is known as the *degrees of freedom* (DoF); the degrees of freedom for different structural types is shown in the following table, where u, v, w are translational displacements and ϕ_x, ϕ_y, ϕ_z are rotations about the indicated axes:

Structure	Dimension	u	v	w	ϕ_x	ϕ_y	ϕ_z
Rod	1-D	✓					
Beam	1-D		✓				✓
Shaft	1-D				✓		
Truss	2-D	✓	✓				
Frame/Membrane	2-D	✓	✓				✓
Grill/Plate	2-D			✓	✓	✓	
Truss	3-D	✓	✓	✓			
Frame/FoldedPlate	3-D	✓	✓	✓	✓	✓	✓
GeneralStructure	3-D	✓	✓	✓	✓	✓	✓

This table shows how the frame structure and folded plate structure share common types of degrees of freedom. This choice allows us to conveniently combine them together to form complex reinforced structures.

Sources of Nonlinearity

Nonlinearities can arise in numerous ways; three of the most common in structural applications are material nonlinearity, large deflections, and contact loadings. The plastic forming of a component is an example of the first, an example of the second is the bending vibration of an aircraft wing, which shows a change of stiffness when the skin and stringers are alternatively in tension and compression, while impact loading is an example of the third. In this book we are primarily concerned with nonlinearities arising out of the geometry and loading, and give only a cursory treatment of the other two. The following four examples show more definitely how the nonlinearities can arise.

I: Nonlinear Material Behavior

Consider the load/unload cycle of a simple uniaxial specimen. If there is a one to one relation between the stress and strain and on unloading all the strain is instantaneously recovered, then the material is said to be *elastic*. An elastic solid is characterized by

$$\epsilon = f(\sigma)$$

which, of course, may be nonlinear.

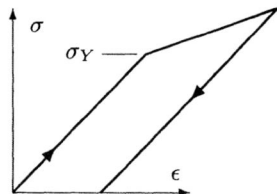

Figure I.3: Stress-strain cycle for elastic-plastic material behavior.

For most structural materials, it is found that beyond a certain stress level (called the yield stress) large deformations (flow) occur for small increments in load; and furthermore, much of the deformation is not recovered when the load is removed. On the load/unload cycle, if $\sigma > \sigma_Y$ (σ_Y = yield stress), this material cannot recover the deformation caused after yielding. This remaining deformation is called the permanent or plastic strain. Structures are designed so as not to have operational stresses that exceed the yield stress, hence we will not devote too much time to this type of nonlinearity. However, discussion of the yield criteria is important.

II: Nonlinearity from Large Deflections

We motivate some of the aspects by considering the simple truss structure shown in Figure I.4.

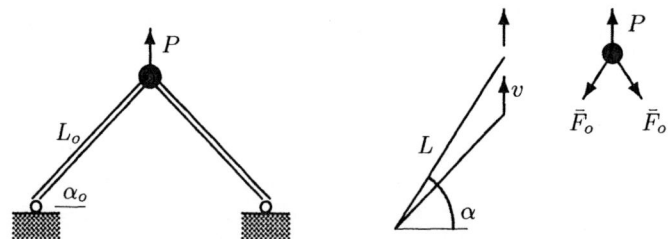

Figure I.4: Pinned truss with concentrated mass.

Ignoring the mass of the truss, dynamic equilibrium of the large concentrated mass gives

$$P - 2\bar{F}_o \sin \alpha = M\ddot{v}$$

where both $\bar{F}_o$ and α are dependent on the deflection v. From geometry considerations, we obtain

$$L \cos \alpha = L_o \cos \alpha_o, \qquad L \sin \alpha = L_o \sin \alpha_o + v$$

where L_o and α_o are the original length and orientation, respectively, and v is the vertical deflection at the load point. Squaring both sides and adding, gives the new length and orientation as

$$L = \sqrt{L_o^2 + 2vL_o \sin \alpha_o + v^2}, \qquad \sin \alpha = \frac{L_o \sin \alpha_o + v}{\sqrt{L_o^2 + 2vL_o \sin \alpha_o + v^2}}$$

The new length of the member is related to the old length by $L = L_o + \bar{u}$, consequently, the axial displacement in the member is

$$\bar{u} = \sqrt{L_o^2 + 2vL_o \sin \alpha_o + v^2} - L_o$$

This gives rise to an axial force of

$$\bar{F}_o = \frac{EA}{L_o} \bar{u}$$

Note that we consider the parameters of the constitutive relation to be unchanged during the large deflection.

Consider the case when the deflections are somewhat small, then

$$\bar{F}_o = \frac{EA}{L_o} \left[\sqrt{1 + 2\frac{v}{L_o} \sin \alpha_o + \left(\frac{v}{L_o}\right)^2} - 1 \right] L_o \approx \frac{EA}{L_o} \left[\sin \alpha_o + \frac{v}{2L_o} \right] v$$

and

$$\sin \alpha = \frac{\sin \alpha_o + \dfrac{v}{L_o}}{\sqrt{1 + 2\dfrac{v}{L_o} \sin \alpha_o + \left(\dfrac{v}{L_o}\right)^2}} \approx \sin \alpha_o + \frac{v}{L_o} \cos^2 \alpha_o$$

We can write the equation of motion in the form of a single degree-of-freedom oscillator as

$$M\ddot{v} + Kv = P(t)$$

but the spring "constant" is actually a function of the deflection

$$K \approx \frac{2EA}{L_o} \left[\sin^2 \alpha_o + \tfrac{3}{2} \sin \alpha_o \frac{v}{L_o} \right] = \frac{2EA}{L_o} \sin^2 \alpha_o \left[1 + \frac{3}{2L_o \sin \alpha_o} v \right]$$

This is an example of a nonlinear system where the nonlinearity comes from the geometry and enters the equations as a nonlinear stiffness term; depending on the direction of the displacement, the stiffness can either increase or decrease.

As another case of geometric nonlinearity, consider the example of a driven pendulum governed by

$$ML\frac{d^2\theta}{dt^2} + C\frac{d\theta}{dt} + Mg\sin\theta = \hat{P}\sin(\omega t)$$

Here θ is the angle off the vertical. The nonlinearity comes from allowing the angle to be large. There is no analytical solution to this problem. Furthermore, this simple looking equation is capable of exhibiting a variety of different dynamical phenomena (including chaos) as demonstrated in Reference [6]. For somewhat small angles, we can replace the sine function with its Taylor series expansion to get

$$ML\frac{d^2\theta}{dt^2} + C\frac{d\theta}{dt} + Mg[\theta - \tfrac{1}{6}\theta^3] = \hat{P}\sin(\omega t)$$

The nonlinearity has a negative stiffness contribution, where $K = Mg[1 - \tfrac{1}{6}\theta^2]$ irrespective of angle direction.

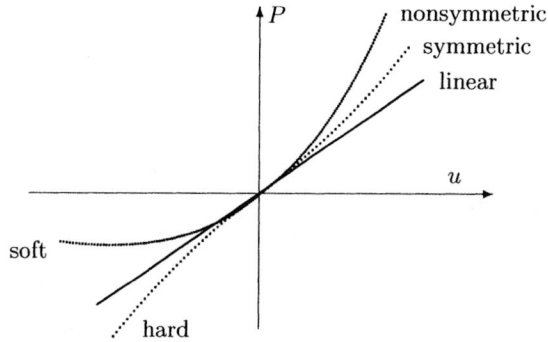

Figure I.5: Symmetric and non-symmetric return force behavior.

These examples highlight the two main stiffness changes we will encounter; we represent these as a symmetric return force and the non-symmetric return force as shown in Figure I.5. Their effects on the stiffness term are given by the static relations

$$P = K(1 + \alpha u^2)u \qquad \text{or} \qquad P = K(1 + \beta u)u$$

The coefficients α and β are constants. When β is negative, the spring can soften in compression and become unstable. This quintessential nonlinear phenomenon is a dominant concern in the later chapters and so we discuss it more later in this introduction.

III: Contact Laws

When one object strikes another, there is a momentum transfer; this occurs through the exertion of a force between the two objects. In order to establish the exact force history during the impact, it is necessary to know the *contact law*, that is, the relation between the contact force and the indentation. While the equations of motion for the two impacting objects could remain linear, invariably the contact relation introduces a nonlinearity into the problem. We illustrate this with the example of two impacting spheres.

The stresses induced when two elastic bodies with curved surfaces are pressed in contact are described by the Hertzian theory of contact stresses [27]. This theory predicts a nonlinear contact law. For two contacting spheres as shown in Figure I.6, for example, the relation is in the form

$$P = K\alpha^{3/2} = K(v_s - v_t)^{3/2}$$

$$K = \frac{4}{3}\sqrt{\frac{R_s R_t}{R_s + R_t}}\left(\frac{k_s k_t}{k_s + k_t}\right), \qquad k_s = \frac{E_s}{1 - \nu_s^2}, \qquad k_t = \frac{E_t}{1 - \nu_t^2}$$

Here the subscripts s and t refer to the striker and target, respectively, E is the Young's modulus, ν is the Poisson's ratio, and R is the radius.

Figure I.6: Hertzian contact.

There are contact forces between the two spheres that are equal and opposite. The equations of motion of both the striker and target are, respectively,

$$m_s \ddot{v}_s = -P = -K(v_s - v_t)^{3/2}$$
$$\chi m_t \ddot{v}_t = +P = +K(v_s - v_t)^{3/2}$$

where χ is a proportionality factor for the amount of target mass put into motion. Introducing the relative indentation $\alpha \equiv v_s - v_t$, the above equations can be rewritten as

$$M\ddot{\alpha} + K\alpha^{3/2} = 0, \qquad \frac{1}{M} = \frac{1}{m_s} + \frac{1}{\chi m_t} \quad \text{or} \quad M = \frac{m_s \chi m_t}{m_s + \chi m_t}$$

Again the nonlinearity appears as a nonlinear stiffness, but its origin arose out of the changing geometry of contact. This is an example of a deformation-dependent load.

It is worth pointing out that if the force $P(t)$ were specified, then the equations of motion of the individual spheres would be linear; however, since the force between the two is specified as a nonlinear function of the deformation, then the coupled problem is nonlinear.

IV: Nonlinear Friction

Our final example of nonlinearity is also that of contact, but it is frictional contact. Examples are break linings, a violin bow touching the strings, the behavior of joints in robots, machining tools, and objects on a conveyor belt.

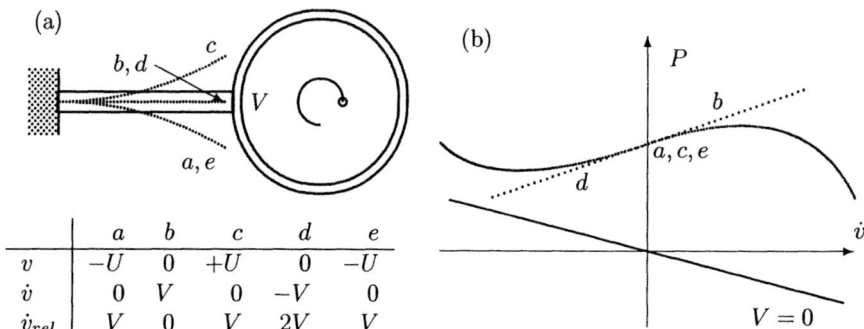

	a	b	c	d	e
v	$-U$	0	$+U$	0	$-U$
$\dot{v}$	0	V	0	$-V$	0
$\dot{v}_{rel}$	V	0	V	$2V$	V

Figure I.7: Friction due to contact. (a) Beam and rotating drum. (b) Nonlinear frictional force against velocity.

As a simple illustration, Figure I.7(a) shows a cantilever beam in contact with a rotating drum. If the drum is at rest, and we set the beam vibrating, then for small amplitudes the frictional resistance can be considered to be proportional to the velocity just as is usually done for viscous damping. This friction is such as to retard the motion.

Let us now set the drum in motion. The friction causes a positive vertical force the value of which depends on the relative velocity of the moving surfaces. Furthermore, experiments show that the coefficient of static friction (when the two surfaces do not move relative to each other) is larger than the coefficient of kinetic friction (when the surfaces do move relative to each other). Therefore, as the beam vibrates while the drum is rotating, there is a changing frictional force that depends on the velocity. The table inset shows that relative velocities (assuming $\dot{v}_{max}$ is the same as the drum) at different stages of the oscillation, and Figure I.7(b) shows the corresponding forces. For very small oscillations, we can approximate the force as

$$P \approx P_o + a\dot{v}$$

with a being positive. The equation of motion can be written in the form of a single degree-of-freedom oscillator as

$$M\ddot{v} + Kv = P(t) = P_o + a\dot{v} \qquad \text{or} \qquad M\ddot{v} - a\dot{v} + Kv = P_o$$

This resembles a simple linear oscillator with viscous damping; however, it has a significant difference in that the "damping" is negative. That is, instead of energy being dissipated during the motion, energy is actually being pumped into the system. As we will see later, the dynamics are greatly affected by this, with the small vibrational motion becoming unstable.

The above linear approximation is useful for very small oscillations; as the vibration velocity increases relative to the drum velocity, some nonlinear effects occur. Clearly, for example, if the vibration velocity is larger than the drum velocity, then the frictional force should become a retarding force (the linear approximation indicates that it would just continue to get larger). On the other hand, if the relative velocity becomes very large (with the vibration velocity being negative), we would expect the friction force to asymptote to a retarding force. An approximation that takes these limits into account is

$$P \approx P_o + \mu[1 - \beta \dot{v}^2]\dot{v}$$

This is an example of a velocity-dependent nonlinearity.

Instability of the Equilibrium

Our large deflection example of Figure I.4 showed the possibility of stiffness softening if the load is applied downward. Let us now follow some implications of this.

Consider the quasi-static case, equilibrium in the deformed configuration gives

$$\tfrac{1}{2}P - \bar{F}_o \sin \alpha = 0$$

Substituting for $\bar{F}_o$ and α leads to the force/deflection relation

$$P = 2EA\left[\sin\alpha_o + \frac{v}{L_o}\right]\left[1 - \frac{1}{\sqrt{1 + 2\dfrac{v}{L_o}\sin\alpha_o(\dfrac{v}{L_o})^2}}\right] \quad \text{or} \quad P - F(v) = 0$$

This determines P uniquely as a function of v; however, v is not uniquely determined as a function of P. Nonuniqueness is a fundamental aspect of nonlinear problems.

Consider this as a displacement-driven problem: that is, determine the load P as a function of the vertical displacement v. The results are shown in Figure I.8. Because it is a one degree-of-freedom system, we can identify the slope of $P(v)$ as a stiffness called the *tangent stiffness*. When v is positive, the structure stiffens, but when v is negative, the stiffness decreases and indeed goes to zero. Note that this occurs at a positive value of apex height. At this stage, the structure is *unstable* in the sense that if under load control, the load is increased, then the next equilibrium point is at B, which is a large displacement away. This phenomenon of taking a large displacement to the next equilibrium position is called *snap-through*. The nonuniqueness of this nonlinear problem is that the

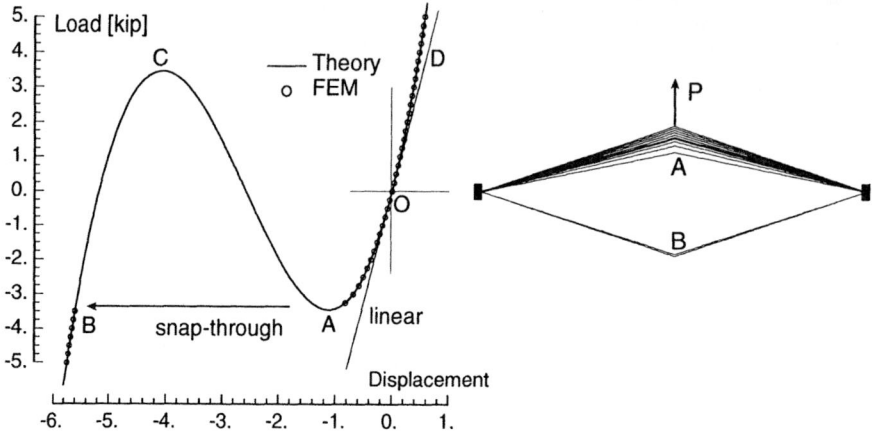

Figure I.8: Deflections and deformed shape at various stages of loading.

load levels at A and B are the same but they correspond to completely different configurations.

On unloading, we can follow an equilibrium path to C at which stage another snap-through occurs over to point D. The region between A and C is an unstable equilibrium path. The load path $OABCDO$ is nonconservative and hence energy is lost. This energy is converted into kinetic energy that propels the dynamic event during the snapping. Thus, dynamics is also an intimate aspect of stability phenomena.

The stability of the equilibrium is a very important aspect of the analysis of structures and therefore we devote Chapters 6 and 7 to considering this issue in greater detail.

Outline of the Book

Although nonlinear equations often look simple, there are no general solutions to them. Indeed their simple form belie the complexity and richness of their solutions. The only general applicable solution methods are numerical and that is the primary focus of this book.

This book takes a synthesis approach to developing the material: the elemental blocks are developed on first principles, these are then combined to model more complicated problems, and from this new principles are learned. Chapter 1 considers the basics of nonlinear deformable body mechanics. The chapter ends with the realization that computer methods are the only viable schemes for general purpose solutions. Chapter 2 looks at the in-plane (membrane) and out-of-plane (flexural) behaviors of thin plated shells. The analysis is developed fully from the governing differential equations all the way to the computational formulation. Our elemental structural building block is a triangular finite element; although there are more sophisticated elements available, this seems the most appropriate vehicle for discussing all the issues of integration and assemblage for complex systems. Chapter 3 then looks at the large (nonlinear) deflection of structures composed of shells and frames. Emphasis is placed on the large deflection and large rotation but small strain situation because that is the predominant situation with thin-walled structures. The corotational scheme emerges as a most appropriate (both conceptually and practically) procedure for describing the nonlinear behavior of these structures.

The second part of the book begins with a summary of the small deflection linear vibrations of structures in Chapter 4. Having recast the inertia properties in discrete form, it then introduces the powerful method of modal analysis for understanding the dynamics of complex systems. Chapter 5 presents the computational formulations for the nonlinear dynamic analysis of 3-D structures. Chapter 6 refines the concept of equilibrium in the process of discussing structural stability. This is done in the context of large deflections so that the relation between buckling analysis and nonlinear stability can be understood. Chapter 7 introduces the stability of the motion and also looks at the very difficult topic of stability of motion in the large.

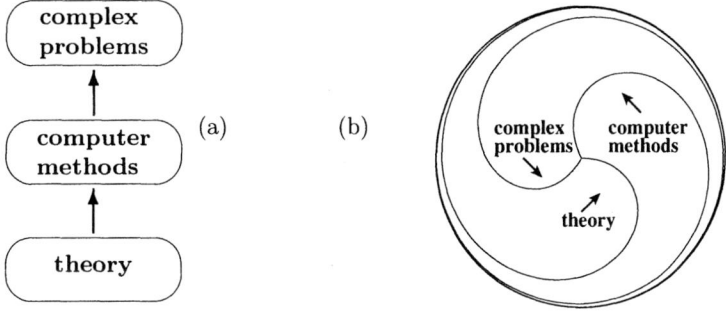

Figure I.9: Two views of the role of theory and computational methods. (a) Traditional view. (b) Interactive view.

Each chapter is divided roughly into three segments that basically correspond to review of mechanics plus some analytical solutions; the computational implementation and application to test problems; and applications to nonsimple problems used for exploring additional aspects of the theory. It is worth contrasting the approach with the traditional approach. The traditional development of these topics would be laid out as in Figure I.9(a), which says that theory (through simple examples) is used to justify the computational methods, and that these computational methods in turn are used to solve the complex problems. Furthermore, it says that theory is separate from the computational methods in the sense that these methods derive from the theory.

But this model does not work in the analysis of nonlinear problems (especially dynamic problems) because there are no "simple cases." A particular example is the richly complex phenomenon of chaos, which can appear even in a single degree-of-freedom system; only through computer methods was this phenomenon amenable to exploration. A more appropriate view, therefore, seems to be the interactive one, as shown in Figure I.9(b). That is, the computational implementations reflect onto the theory to more fully explain and develop the theory. Simultaneously, the few analytical solutions developed are used to help put order in the results for the complex problems. It is the interplay that is the true relation between theory and computer methods. The computer implementations become, in a sense, a laboratory for experimentation to discover new facts about the complex system that are then used to enhance the theory.

With these ideas in mind, the example problems in the book basically fall into three categories. The first illustrate aspects of the theory; an example is how the sequence of vibration mode shapes of a plate changes with membrane loading. The second investigate aspects of the numerical performance; examples are the convergence tests. The third group are the exploratory problems for which there are no simple solutions; examples are the limit cycles achieved by a beam with follower forces and mode jumping in rectangular plates. It is emphasized that it is only because the computational methods were developed could these phenomena be approached in any satisfactory way.

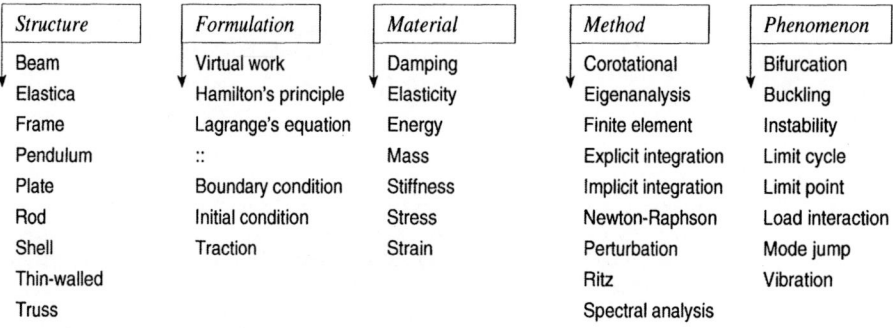

Figure I.10: Subset of key words from the Index.

Finally, a word about the index. Some books lend themselves to indexing according to the first mention of a key word, while others are best done as indexing every occurrence of the key word or phrase. The method chosen here is in the form of threads linking the key ideas listed in Figure I.10. For example, following the thread for plates leads through Hamilton's principle, boundary conditions, spectral analysis, vibrations, stability, and so on. On the other hand, following the thread for virtual work goes through beams, trusses, corotational method, and so on. This approach seems more apt to catch the interwoven nature of the subjects.

1
Mechanics of Solids

This chapter introduces some basic concepts in the mechanics of deformable bodies. We consider how lengths, areas, and volumes change during a deformation, and this leads to the important concept of strain. We then introduce stress and the equations of motion. To complete the mechanics formulation of problems, we also describe the constitutive (or material) behavior. This is a complicated and extensive subject; our interest, however, is elastic behavior because that is the operational regime for most structural materials. Staying within the elastic limit is an important consideration and so some consideration is given to failure theories.

The chapter concludes with a reformulation of the governing equations in terms of a variational principle; that is, equilibrium is seen as the achievement of a stationary value of the total potential energy. Figure 1.1 shows an example of a simple truss and how its total potential (Π) changes as a function of vertical deflection for different values of applied load P; equilibrium corresponds to where the slope is zero. This formulation lends itself well to the approximate computer methods needed to solve nonlinear problems.

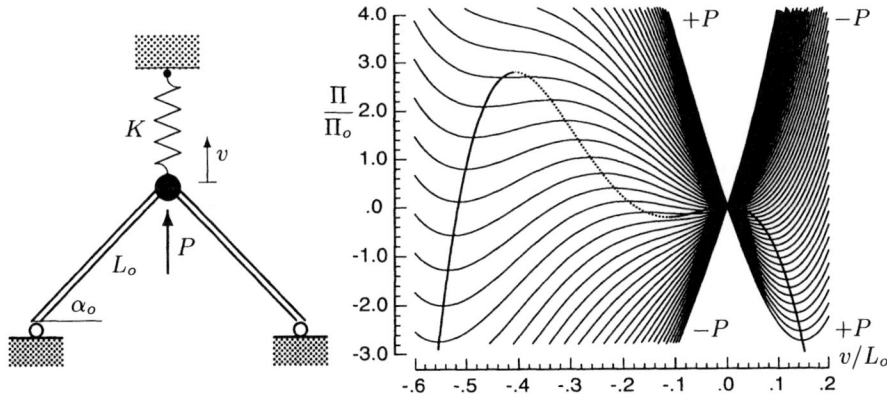

Figure 1.1: Total potential energy as a function of deflection for different values of applied load. The equilibrium positions are where the potential has stationary values; the dashed line indicates unstable equilibrium positions.

1.1   Cartesian Tensors

Solid mechanics deals with groups of things such as u, v, w representing displacements; x, y, z representing coordinates; and $\sigma_{xx}, \sigma_{yy}, \sigma_{xy}, \cdots$, representing stresses. We would like a notation that handles such groups conveniently — we will use the subscript or indicial notation to achieve this. Also, whenever convenient, we will introduce the matrix notation equivalents because these are better suited for computer implementation.

Tensor Fields

Let the symbol z_i with the range $i = 1, \cdots, n$ be used to denote any one of the variables in the set $\{z_1, z_2, \cdots, z_n\}$. The symbol i is called an *index*. Similar notations with multiple indices such as t_{ij}, $i, j = 1, \cdots, n$, are also used to represent individual components in the set of $[n \times n]$ elements $\{t_{11}, t_{12}, \cdots, t_{nn}\}$. For most cases in this chapter, the range will go from 1 to 3.

A number of special symbols have been introduced as a convenience in using the tensor notation. Two especially useful symbols are the Kronecker delta and the Permutation symbol. The Kronecker delta is denoted by δ_{ij} and is defined as

$$\delta_{ij} = \begin{cases} 1 & \text{if } i = j \\ 0 & i \neq j \end{cases} \qquad \text{or} \qquad [\delta_{ij}] = \begin{bmatrix} 1 & 0 & 0 \\ 0 & 1 & 0 \\ 0 & 0 & 1 \end{bmatrix} = \lceil\ I\ \rfloor$$

Note that written in matrix form, it is the same as the identity matrix. The permutation symbol is defined as

$$\epsilon_{ijk} = \begin{cases} 1 & \text{when } ijk \text{ form an even permutation of 123; e.g., 312} \\ -1 & \text{when } ijk \text{ form an odd permutation of 123; e.g., 321} \\ 0 & \text{otherwise; e.g., 122} \end{cases}$$

Consider two Cartesian coordinate systems (x_1, x_2, x_3) and (x'_1, x'_2, x'_3) as shown in Figure 1.2. A *base* vector is a unit vector parallel to a coordinate axis. Let $\hat{e}_1, \hat{e}_2, \hat{e}_3$ be the base vectors for the (x_1, x_2, x_3) coordinate system and $\hat{e}'_1, \hat{e}'_2, \hat{e}'_3$ the base vectors for the (x'_1, x'_2, x'_3) system as also shown in the figure.

Because the coordinate axes are mutually orthogonal, we have for the vector dot and cross products

$$\hat{e}_i \cdot \hat{e}_j = \delta_{ij}, \qquad \hat{e}'_i \cdot \hat{e}'_j = \delta_{ij}, \qquad \hat{e}_i \times \hat{e}_j = \sum_k \epsilon_{ijk}\hat{e}_k, \qquad \hat{e}'_i \times \hat{e}'_j = \sum_k \epsilon_{ijk}\hat{e}_k$$

A vector $\hat{x}$ can be projected onto the two coordinate systems with the result

$$\hat{x} = \sum_j x_j\hat{e}_j = \sum_j x'_j\hat{e}'_j = \hat{x}'$$

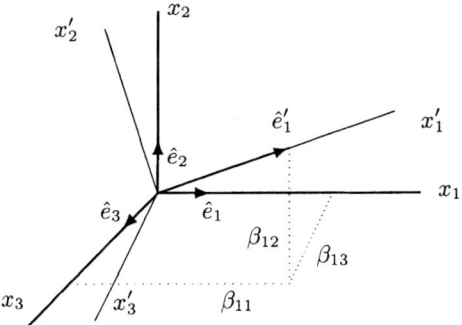

Figure 1.2: Base vectors and rotated coordinate system.

This can be rewritten by first introducing the matrix of *direction cosines*

$$\beta_{ij} \equiv \hat{e}'_i \cdot \hat{e}_j \tag{1.1}$$

and substituting to get

$$x'_i = \sum_j \beta_{ij} x_j , \qquad x_i = \sum_j \beta_{ji} x'_j$$

This gives the relation for transformation of components in one coordinate system into components in another. These two equations taken together yield the conclusion

$$\sum_j \beta_{ij} \beta_{kj} = \delta_{ik} , \qquad [\ \beta\ ][\ \beta\ ]^T = \lceil\ I\ \rfloor$$

Thus, $[\beta_{ij}]$ are orthogonal and the relation is known as the *orthogonality relation*.

A system of quantities is called by different names depending on how the components of the system are defined in the coordinates x_1, x_2, x_3 and how they are transformed when the coordinates are changed to x'_1, x'_2, x'_3. A system is called a *scalar* if it has only a single component ϕ in the variables x_i and a single component ϕ' in the coordinates x'_i, and if ϕ and ϕ' are numerically equal at the corresponding points. That is,

$$\phi(x_1, x_2, x_3) = \phi'(x'_1, x'_2, x'_3)$$

A system is called a *vector* field or a tensor field of order one if it has three components v_i in the coordinates x_i, and three components v'_i in the coordinates x'_i and the components are related by the transformation law

$$v'_i = \sum_k \beta_{ik} v_k$$

The tensor field of order two is a system that has nine components t_{ij} in the coordinates x_i, and nine components t'_{ij} in the coordinates x'_i, and the components are related by the transformation law

$$t'_{ij} = \sum_{m,n} \beta_{im}\beta_{jn}t_{mn}$$

This is easily generalized to a tensor of order n.

Because tensor fields are continuous functions of position, then they are amenable to calculus operations such as differentiation and integration. An important issue is determining the type of tensor resulting from differentiation and integration. Consider a vector v_j with the transformation

$$v'_j(\hat{x}') = \beta_{jk}v_k(\hat{x})$$

Differentiating both sides of the equation, we obtain

$$\frac{\partial v'_j}{\partial x'_i} = \sum_k \beta_{jk}\frac{\partial v_k}{\partial x'_i} = \sum_{k,m} \beta_{jk}\frac{\partial v_k}{\partial x_m}\frac{\partial x_m}{\partial x'_i} = \sum_{k,m} \beta_{jk}\beta_{im}\frac{\partial v_k}{\partial x_m}$$

This says that partial derivatives of any tensor field behave like the components of a Cartesian tensor. (It should be noted that this is not true in curvilinear coordinate systems.) From this it is apparent that the term

$$\frac{\partial T_{ij}}{\partial x_k}$$

is a third-order tensor. In general, differentiation with respect to x_i increases the tensor order of a term by 1. Consider a tensor field $T_{jkm\ldots}(x)$ in a region V bounded by a surface A. Then,

$$\int_V \frac{\partial}{\partial x_i}T_{jkm\ldots}dV = \int_A n_iT_{jkm\ldots}dA \qquad (1.2)$$

where n_i are components of the unit vector $\hat{n}$ along the exterior normal of A. This is known as the *Integral Theorem*.

We use the notation

$$\{v\} = \begin{Bmatrix} v_1 \\ v_2 \\ v_3 \end{Bmatrix}$$

to represent a vector. The notation $\{v\}^T$ then represents the quantities

$$\{v\}^T = \{v_1,\, v_2,\, v_3\}$$

The transformation relations are expressed in matrix notation as

$$\{v_G\} = [\,T\,]\{v_L\}, \qquad [\,T\,] \equiv [\{\hat{e}_1\}\{\hat{e}_2\}\{\hat{e}_3\}]$$

where the subscripts G and L refer to global and local, respectively. We call $[\,T\,]$ a *triad* of the unit vectors $\hat{e}_i$.

Properties of Second-Order Tensors

Because second-order tensors are so prevalent in solid mechanics (they are used to represent stress and strain, for example), it is of value now to summarize some of their major properties. Because second-order tensors can be represented conveniently by matrices, many of the following results can also be established simply from matrix theory.

A tensor S_{ij} is symmetric if $S_{ij} = S_{ji}$, while a tensor A_{ij} is antisymmetric if $A_{ij} = -A_{ji}$. A nonsymmetric tensor can always be represented as the sum of a symmetric and an antisymmetric tensor. The contraction of a symmetric and an antisymmetric tensor is zero, that is, $\sum_{i,j} S_{ij} A_{ij} = 0$.

Consider the relation

$$V_i = \sum_j S_{ij} n_j , \qquad V_i = \text{vector} , \qquad n_i = \text{unit vector}$$

and let S_{ij} be a symmetric tensor. This says a vector V_i is produced by contracting the unit vector with the second-order tensor S_{ij}; in general, $\hat{V}$ is not parallel to $\hat{n}$. An interesting question is: Under what circumstances does the transformation relation produce a $\hat{V}$ that is parallel to $\hat{n}$?

Figure 1.3: Meaning of eigenvectors.

To answer this, assume that there is an $\hat{n}$ such that it is proportional to $\hat{V}$, that is,

$$V_i = \lambda n_i = \sum_j S_{ij} n_j \qquad \text{or} \qquad \sum_j [S_{ij} - \lambda \delta_{ij}] \, n_j = 0$$

This is given in expanded matrix form as

$$\begin{bmatrix} S_{11} - \lambda & S_{12} & S_{13} \\ S_{12} & S_{22} - \lambda & S_{23} \\ S_{13} & S_{23} & S_{33} - \lambda \end{bmatrix} \begin{Bmatrix} n_1 \\ n_2 \\ n_3 \end{Bmatrix} = 0$$

A nontrivial solution for n_i exists only if the determinant of the coefficient matrix vanishes. Expanding the determinantal equation, we obtain the characteristic equation

$$\lambda^3 - I_1 \lambda^2 + I_2 \lambda - I_3 = 0 \tag{1.3}$$

where the invariants I_1, I_2, I_3 are defined as

$$I_1 = \sum_j S_{ii} = S_{11} + S_{22} + S_{33}$$

$$I_2 = \sum_{i,j} \tfrac{1}{2}[S_{ii}S_{jj} - S_{ij}S_{ji}]$$

$$I_3 = \det[T_{ij}] \tag{1.4}$$

The characteristic equation yields three roots or possible values for λ

$$\lambda^{(1)}, \lambda^{(2)}, \lambda^{(3)}$$

These are called *eigenvalues*. For each eigenvalue there is a corresponding solution for $\hat{n}$. The three $\hat{n}$'s

$$\hat{n}^{(1)}, \hat{n}^{(2)}, \hat{n}^{(3)}$$

are called *eigenvectors*. We say that the matrix $[\ S\ ]$ has the eigenpairs $(\lambda, \hat{n})^{(i)}$, $i = 1, 2, 3$. These are obtained by solving the eigenvalue problem

$$[\ S\ ]\{n\} - \lambda\{n\} = 0$$

This is an eigenvalue problem because it is homogeneous and has an unknown parameter (λ). We will frequently encounter eigenvalue problems in the subsequent chapters.

The measures of stress and strain we introduce later will be symmetric second-order tensors; then the $\lambda^{(k)}$ will be identified as principal stresses and strains, respectively. Principal values are extremal values of the tensor.

Example 1.1:   Given two directions of an orthogonal triad, determine the third direction by using the orthogonality conditions.

Let the given vectors be

$$\hat{e}_1' = \tfrac{1}{2}\hat{e}_1 - \tfrac{1}{2}\hat{e}_2 + \tfrac{1}{\sqrt{2}}\hat{e}_3, \qquad \hat{e}_2' = \tfrac{1}{2}\hat{e}_1 - \tfrac{1}{2}\hat{e}_2 - \tfrac{1}{\sqrt{2}}\hat{e}_3$$

We can obtain the third direction from knowledge that

$$\hat{e}_3' = \hat{e}_1' \times \hat{e}_2' = \tfrac{1}{\sqrt{2}}\hat{e}_1 + \tfrac{1}{\sqrt{2}}\hat{e}_2 + 0\hat{e}_3$$

At this stage, the direction cosines are easily established as

$$[\ \beta_{ij}\ ] = \begin{bmatrix} \frac{1}{2} & -\frac{1}{2} & \frac{1}{\sqrt{2}} \\ \frac{1}{2} & -\frac{1}{2} & -\frac{1}{\sqrt{2}} \\ \frac{1}{\sqrt{2}} & \frac{1}{\sqrt{2}} & 0 \end{bmatrix}$$

Note that the rows of $[\ \beta_{ij}\ ]$ are the components of the vectors $\hat{e}_i'$ referred to $\hat{e}_j$.

Example 1.2:   Determine the new components T_{ij} when transformed to a coordinate system defined by its eigenvectors.

Let the original coordinate system be defined by the base vectors

$$\hat{e}_1 = \{1, 0, 0\}, \qquad \hat{e}_2 = \{0, 1, 0\}, \qquad \hat{e}_3 = \{0, 0, 1\}$$

and the transformed system by

$$\hat{e}_1' = \hat{n}^{(1)}, \qquad \hat{e}_2' = \hat{n}^{(2)}, \qquad \hat{e}_3' = \hat{n}^{(3)}$$

Then the direction cosines are given by

$$\beta_{ij} = \hat{e}_i' \cdot \hat{e}_j = \hat{n}^{(i)} \cdot \hat{e}_j = \sum_p n_p^{(i)} \hat{e}_p \cdot \hat{e}_j = \sum_p n_p^{(i)} \delta_{pj} = n_j^{(i)}$$

From the eigenvalue problem, we also have that

$$\sum_j S_{ij} n_j^{(k)} = \lambda^{(k)} n_i^{(k)} \qquad \text{or} \qquad \sum_j S_{ij} \beta_{kj} = \lambda^{(k)} \beta_{ki}$$

Multiply both sides by β_{pi} and recognizing the left-hand side as the transformation of S_{ij} and the right-hand side as the Kronecker delta gives

$$S_{pk}' = \lambda^{(k)} \delta_{kp}$$

or, in expanded matrix form,

$$\begin{bmatrix} S_{11}' & S_{12}' & S_{13}' \\ & S_{22}' & S_{23}' \\ Sym & & S_{33}' \end{bmatrix} = \begin{bmatrix} \lambda^{(1)} & 0 & 0 \\ 0 & \lambda^{(2)} & 0 \\ 0 & 0 & \lambda^{(3)} \end{bmatrix}$$

That is, with respect to the new coordinate system, S_{ij} has a diagonal form.

We can show that $\lambda^{(i)}$'s are the maximums and minimums of the associated quadratic surface.

1.2 Deformation and Rotation

A deformation is a comparison of two states. In the mechanics of deformable bodies, we are particularly interested in the deformation of neighboring points; that they are different is in the nature of deformable bodies.

Deformation Gradient

Set up a common global coordinate system as shown in Figure 1.4 and associate x_i^o with the undeformed configuration and x_i with the deformed configuration. That is,

$$\text{Initial position:} \quad \hat{x}^o = \sum_i x_i^o \hat{e}_i \qquad \qquad \text{Final position:} \quad \hat{x} = \sum_i x_i \hat{e}_i$$

where both vectors are referred to the common set of unit vectors $\hat{e}_i$. The variables x_i^o and x_i are called the *Lagrangian* and *Eulerian* variables, respectively.

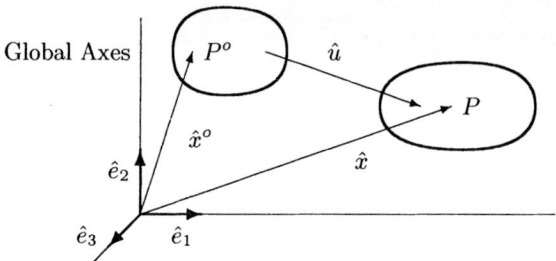

Figure 1.4: Undeformed and deformed configurations.

A displacement is the shortest distance traveled when a particle moves from one location to another, that is,

$$\hat{u} = \hat{r} - \hat{r}^o = \sum_i x_i \hat{e}_i - \sum_i x_i^o \hat{e}_i \qquad \text{or} \qquad u_i = x_i - x_i^o$$

and is illustrated in Figure 1.4.

A motion is expressed in the following form:

$$x_i = x_i(x_1^o, x_2^o, x_3^o, t)$$

In the Lagrangian system, all quantities are expressed in terms of the initial position coordinates and time; in the Eulerian system, the independent variables are x_i and t, where x_i are the position coordinates at the time of interest. Realizing that the description of deformation is essentially geometric, we can understand the Lagrangian description by putting a rectangular grid on the original (undeformed) body and determining what it will look like during the motion. In other words, the Lagrangian grid is always superposed on the same material points and therefore deforms as shown in Figure 1.5. The Eulerian description puts a grid on the currently deforming body.

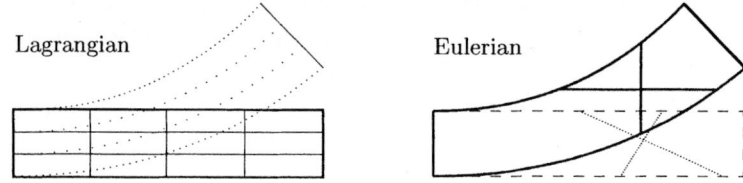

Figure 1.5: Grids illustrating Lagrangian and Eulerian descriptions.

Consider a deformation in the vicinity of the point P; that is, consider two points separated by dx_i^o in the undeformed configuration and by dx_i in the deformed configuration as shown in Figure 1.6. The positions of the two points are related through the Taylor series expansion

$$P: \qquad x_i = x_i(x_i^o)$$

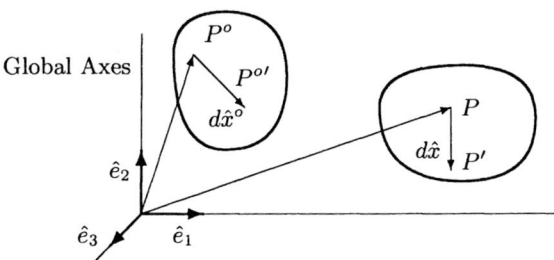

Figure 1.6: Deformation of neighboring points.

P' : $x_i' = x_i + dx_i = x_i(x_i^o + dx_i^o)$

$$\approx x_i(x_i^o) + \sum_j \frac{\partial x_i}{\partial x_j^o} dx_j^o + \tfrac{1}{2} \sum_{j,k} \frac{\partial^2 x_i}{\partial x_j^o \partial x_k^o} dx_j^o dx_k^o + \dots$$

If dx_i^o is small, that is, the neighboring points are very close to each other, then

$$x_i' - x_i = dx_i \approx \sum_j \frac{\partial x_i}{\partial x_j^o} dx_j^o \tag{1.5}$$

This describes how the separation in the deformed configuration is related to the separation in the undeformed configuration. It is expressed in matrix form as

$$
\begin{Bmatrix} dx_1 \\ dx_2 \\ dx_3 \end{Bmatrix} =
\begin{bmatrix}
\dfrac{\partial x_1}{\partial x_1^o} & \dfrac{\partial x_1}{\partial x_2^o} & \dfrac{\partial x_1}{\partial x_3^o} \\[2mm]
\dfrac{\partial x_2}{\partial x_1^o} & \dfrac{\partial x_2}{\partial x_2^o} & \dfrac{\partial x_2}{\partial x_3^o} \\[2mm]
\dfrac{\partial x_3}{\partial x_1^o} & \dfrac{\partial x_3}{\partial x_2^o} & \dfrac{\partial x_3}{\partial x_3^o}
\end{bmatrix}
\begin{Bmatrix} dx_1^o \\ dx_2^o \\ dx_3^o \end{Bmatrix}
$$

The quantities $\dfrac{\partial x_i}{\partial x_j^o}$ are called the *deformation gradients* and form the basis of the description of any deformation. Deformation gradients relate the behavior of neighboring points.

The above relation uniquely specifies dx_1, dx_2, dx_3 in terms of dx_1^o, dx_2^o, dx_3^o. On the assumption that the deformation is continuous, then we should be able to write the inverse; but this is true only if the determinant is non-zero, that is,

$$\det[\frac{\partial x_i}{\partial x_j^o}] \neq 0$$

Define the Jacobian, referenced to the undeformed configuration, as

$$J^o \equiv \det[\frac{\partial x_i}{\partial x_j^o}] \quad = \quad \sum_{i,j,k} \epsilon_{ijk} \frac{\partial x_1}{\partial x_i^o} \frac{\partial x_2}{\partial x_j^o} \frac{\partial x_3}{\partial x_k^o}$$

Note that the Jacobian is a scalar quantity. We will impose the restriction on any deformation that no region of finite volume is deformed into a region of zero or infinite volume. That is, we restrict the values of the Jacobian as

$$0 < J^o < \infty$$

It is necessary to always check this to see if the deformation is physically possible.

The deformation gradient can be decomposed into its symmetric and anti-symmetric parts as

$$\frac{\partial x_i}{\partial x_j^o} = \delta_{ij} + \frac{\partial u_i}{\partial x_j^o} = \delta_{ij} + \bar{\epsilon}_{ij} + \bar{\omega}_{ij}$$

where

$$\bar{\epsilon}_{ij} \equiv \tfrac{1}{2}\left(\frac{\partial u_j}{\partial x_i^o} + \frac{\partial u_i}{\partial x_j^o}\right), \qquad \bar{\omega}_{ij} \equiv \tfrac{1}{2}\left(\frac{\partial u_j}{\partial x_i^o} - \frac{\partial u_i}{\partial x_j^o}\right)$$

The symmetric tensor $\bar{\epsilon}_{ij}$ is often referred to as the small or infinitesimal strain tensor. The tensor $\bar{\omega}_{ij}$ is called the (Lagrangian) rotation tensor. The reasons for these names will become clearer later in the chapter. Similarly, the deformation gradient with respect to the deformed configuration can be written as

$$\frac{\partial x_i^o}{\partial x_j} = \delta_{ij} + \frac{\partial u_i}{\partial x_j} = \delta_{ij} + \epsilon_{ij} + \omega_{ij}, \quad \epsilon_{ij} \equiv \tfrac{1}{2}\left(\frac{\partial u_j}{\partial x_i} + \frac{\partial u_i}{\partial x_j}\right), \quad \omega_{ij} \equiv \tfrac{1}{2}\left(\frac{\partial u_j}{\partial x_i} - \frac{\partial u_i}{\partial x_j}\right)$$

The symmetric tensor ϵ_{ij} is also often referred to as the small or infinitesimal strain tensor. The tensor ω_{ij} is called the (Eulerian) rotation tensor.

Deformation of Lines, Areas, and Volumes

The descriptions of a line segment before and after deformation are

$$d\hat{x}^o = \sum_i dx_i^o \hat{e}_i, \qquad d\hat{x} = \sum_i dx_i \hat{e}_i$$

A straightforward substitution of the deformation gradient gives

$$dx_i = \sum_j \frac{\partial x_i}{\partial x_j^o} dx_j^o \tag{1.6}$$

Even if $d\hat{x}^o$ is only horizontal, $d\hat{x}$ has all non-zero components, in general.

The area of a parallelogram region can be calculated by considering the vector cross product of lines that bound it. That is, if the region is defined by two vectors $d\hat{x}^a$ and $d\hat{x}^b$, we have

$$\text{Area} = d\hat{x}^a \times d\hat{x}^b = \text{vector}$$

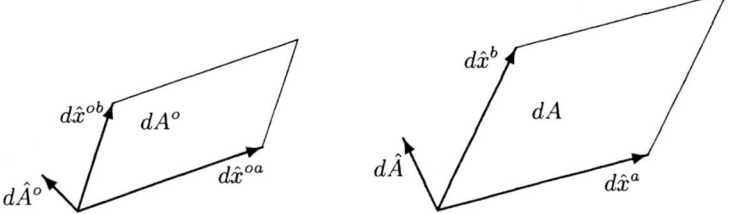

Figure 1.7: Deformation of areas.

Note that we consider the area to be a vector; it has a direction as well as a magnitude, and the direction is given by a normal to the surface. We can show [21] that the components of area are related through

$$dA_i = \sum_k J^o \frac{\partial x_k^o}{\partial x_i} dA_k^o \qquad (1.7)$$

This elegant formula for the deformation of areas is somewhat similar to the corresponding one for line segments; note, however, that it is the Eulerian form of the deformation gradient that is used.

Consider the parallelepiped of sides $d\hat{x}^{oa}$, $d\hat{x}^{ob}$, $d\hat{x}^{oc}$, which deforms into $d\hat{x}^{a}$, $d\hat{x}^{b}$, $d\hat{x}^{c}$. The volume before deformation is $dV^o = (d\hat{x}^{oa} \times d\hat{x}^{ob}) \cdot d\hat{x}^{oc}$. Expand dV using the deformation gradient and recognizing the collection of gradients as J^o, rearrange to get

$$dV = J^o dV^o$$

Mass is conserved during the deformation of a solid giving

$$\rho dV = \rho^o dV^o \qquad \text{or} \qquad \rho = \rho^o / J^o$$

That is, the density changes during the deformation.

Example 1.3:   Consider the following plane inhomogeneous deformation

$$x_1 = [R - x_2^o] \sin(x_1^o/R) , \qquad x_2 = R - [R - x_2^o] \cos(x_1^o/R) , \qquad x_3 = x_3^o$$

where R is a parameter. What (if any) are the restrictions for this to be a valid deformation? Draw the deformed shape of material initially lying between $-h < x_2^o < h$, and calculate the orientation and magnitude of the deformed areas.

This deformation is shown in Figure 1.8. Note that initially horizontal lines become arcs of concentric circles, while initially vertical lines become radial lines emanating from a common point. This deformation resembles that of bending.

The deformation gradient is given by

$$\left[\frac{\partial x_i}{\partial x_j^o} \right] = \begin{bmatrix} (R - x_2^o)\cos(x_1^o/R)/R & -\sin(x_1^o/R) & 0 \\ (R - x_2^o)\sin(x_1^o/R)/R & \cos(x_1^o/R) & 0 \\ 0 & 0 & 1 \end{bmatrix}$$

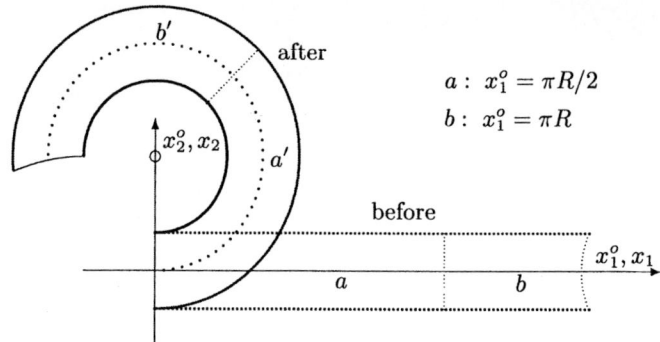

$$a: x_1^o = \pi R/2$$
$$b: x_1^o = \pi R$$

Figure 1.8: Shape before and after deformation.

The Jacobian is the determinant of the deformation gradient matrix and on multiplication simplifies to

$$J^o = 1 - \frac{x_2^o}{R}$$

We note that as long as $x_2^o < R$ that the volume remains positive.

The areas are related through the inverse of the above deformation gradient. Since it is a 2-D problem, we get

$$\left\{ \begin{array}{c} dA_1 \\ dA_2 \end{array} \right\} = \left[\begin{array}{cc} \cos(x_1^o/R) & \sin(x_1^o/R) \\ -J^o \sin(x_1^o/R) & J^o \cos(x_1^o/R) \end{array} \right] \left\{ \begin{array}{c} dA_1^o \\ dA_2^o \end{array} \right\}$$

Areas that were initially vertical and facing the 1-direction are preserved in size; this can be checked by comparing dA_2 for $x_1^o = \pi R/2$ to dA_1 for $x_1^o = \pi R$. Areas that were initially horizontal either contract ($x_2^o > 0$) or expand ($x_2^o < 0$). This is the hallmark of a beam or plate in bending.

In the limit as $x_2^o/R \ll 1$, then $J^o \approx 1$ and areas are preserved. This is the situation that will be prevalent when we consider the bending of thin-walled structures.

Rotation at a Point

A general deformation can be conceived as a straining action plus a rotation. The large rotations of rigid bodies is taken up in Chapter 3; what we are interested in here is the description of the rotations of deformable bodies. This is not so straightforward primarily because different lines through a given point in the body can have different rotations. We will find it necessary to introduce the idea of an "average" or mean rotation.

Consider the rotation of a general line element OP that deforms to $O'P'$ as shown in Figure 1.9. For convenience, let the two lines occupy the same position at the origin so that $O = O'$.

Focus on a line segment that is initially lying in the 1-2 plane; that is, look at the lines OP^* and OP'^* (the latter is the projection of the deformed line onto the 1-2 plane) rotating about the x_3-axis. The rotation is obtained as a change

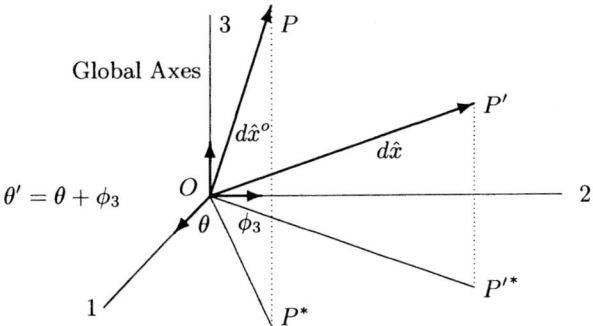

Figure 1.9: Rotation of a line element.

in orientation as follows:

$$\text{orientation of } OP^*: \quad \tan\theta \;=\; \frac{dx_2^o}{dx_1^o}$$

$$\text{orientation of } OP'^*: \quad \tan\theta' \;=\; \frac{dx_2}{dx_1} = \frac{\dfrac{\partial x_2}{\partial x_1^o}dx_1^o + \dfrac{\partial x_2}{\partial x_2^o}dx_2^o}{\dfrac{\partial x_1}{\partial x_1^o}dx_1^o + \dfrac{\partial x_1}{\partial x_2^o}dx_2^o}$$

The rotation of the line projection is $\phi_3 = \theta' - \theta$ or

$$
\tan\phi_3 \;=\; \tan(\theta'-\theta) = \frac{\tan\theta' - \tan\theta}{1 + \tan\theta'\tan\theta}
$$

$$
= \frac{\left(\dfrac{\partial x_2}{\partial x_1^o} - \dfrac{\partial x_1}{\partial x_2^o}\right) + \left(\dfrac{\partial x_2}{\partial x_1^o} + \dfrac{\partial x_1}{\partial x_2^o}\right)\cos 2\theta + \left(\dfrac{\partial x_2}{\partial x_2^o} - \dfrac{\partial x_1}{\partial x_1^o}\right)\sin 2\theta}{\left(\dfrac{\partial x_1}{\partial x_1^o} + \dfrac{\partial x_2}{\partial x_2^o}\right) + \left(\dfrac{\partial x_1}{\partial x_1^o} - \dfrac{\partial x_2}{\partial x_2^o}\right)\cos 2\theta + \left(\dfrac{\partial x_1}{\partial x_2^o} + \dfrac{\partial x_2}{\partial x_1^o}\right)\sin 2\theta}
$$

where

$$
\cos\theta = dx_1^o/\sqrt{(dx_1^o)^2 + (dx_2^o)^2}\,, \qquad \sin\theta = dx_2^o/\sqrt{(dx_1^o)^2 + (dx_2^o)^2}
$$

was used. From this, it is clear that different line elements will have different amounts of rotation (since θ will be different). In fact, some will be positive, some negative. To characterize the rotation at a point, it is necessary to remove the angle dependence. This will be done by averaging.

A measure of average rotation is given as [53]

$$
\tan\bar\phi_3 \equiv \frac{1}{2\pi}\int_o^{2\pi} \tan\phi_3\, d\theta = \frac{1}{2\pi}\int \frac{A + B\cos 2\theta - C\sin 2\theta}{D + C\cos 2\theta + B\sin 2\theta}\, d\theta
$$

where the coefficients

$$
A = \frac{\partial x_2}{\partial x_1^o} - \frac{\partial x_1}{\partial x_2^o}, \quad B = \frac{\partial x_2}{\partial x_1^o} + \frac{\partial x_1}{\partial x_2^o}, \quad C = \frac{\partial x_1}{\partial x_1^o} - \frac{\partial x_2}{\partial x_2^o}, \quad D = \frac{\partial x_1}{\partial x_1^o} + \frac{\partial x_2}{\partial x_2^o}
$$

are independent of θ. Let the denominator be written as

$$f = D + C \cos 2\theta + B \sin 2\theta, \qquad df = (0 - 2C \sin 2\theta + 2B \cos 2\theta)d\theta$$

Therefore, the integral can be rewritten as the sum of two parts:

$$\tan \bar{\phi}_3 = \frac{1}{2\pi} \int_o^{2\pi} \frac{\frac{1}{2}df}{f} + \frac{A}{2\pi} \int_o^{2\pi} \frac{d\theta}{D + C \cos 2\theta + B \sin 2\theta} = I_1 + I_2$$

The first integral is simply

$$I_1 = \frac{1}{4\pi} \ln[\ f\ ]_o^{2\pi} = 0$$

The denominator of the second integral can be rearranged as (using the sine of the sum of two angles)

$$
\begin{aligned}
D + C \cos 2\theta + B \sin 2\theta &= D + \sqrt{C^2 + B^2}(\sin \beta \cos 2\theta + \cos \beta \sin 2\theta) \\
&= D + \sqrt{C^2 + B^2} \sin(\beta + 2\theta)
\end{aligned}
$$

with $\tan \beta \equiv C/B$. Hence, the integral becomes on using $\sigma \equiv \beta + 2\theta$

$$I_2 = \frac{A}{2\pi} \int_o^{2\pi} \frac{d\theta}{D + \sqrt{C^2 + B^2} \sin(\beta + 2\theta)} = \frac{A}{4\pi} \int_\beta^{\beta+4\pi} \frac{d\sigma}{D + \sqrt{C^2 + B^2} \sin \sigma}$$

Provided that $D^2 > C^2 + B^2$, this gives

$$I_2 = \frac{2A}{4\pi} \frac{1}{\sqrt{D^2 - (C^2 + B^2)}} \tan^{-1} \left\{ \frac{D \tan(\sigma/2) + \sqrt{C^2 + B^2}}{\sqrt{D^2 - (C^2 + B^2)}} \right\} \Big|_\beta^{\beta+4\pi}$$

The $\tan^{-1}(\ )$ term is zero or multiples of 2π. Because for small deformations, it is required that

$$A = \frac{\partial x_2}{\partial x_1^o} - \frac{\partial x_1}{\partial x_2^o} = \frac{\partial u_2}{\partial x_1^o} - \frac{\partial u_1}{\partial x_2^o}$$

be a measure of rotation, then let $\tan^{-1}(\ )$ be 2π. Hence

$$I_2 = \frac{\frac{1}{2}A}{\sqrt{D^2 - (C^2 + B^2)}}$$

Consequently, the average rotation becomes (after substituting for the coefficients)

$$\tan \bar{\phi}_3 = \frac{1}{2}\left(\frac{\partial x_2}{\partial x_1^o} - \frac{\partial x_1}{\partial x_2^o}\right) \Big/ \sqrt{\left(\frac{\partial x_1}{\partial x_1^o}\right)\left(\frac{\partial x_2}{\partial x_2^o}\right) - \frac{1}{4}\left(\frac{\partial x_1}{\partial x_2^o} + \frac{\partial x_2}{\partial x_1^o}\right)^2} \qquad (1.8)$$

Similarly, for lines initially in the other planes, we have

$$\tan \bar{\phi}_2 = \tfrac{1}{2}(\frac{\partial x_1}{\partial x_3^o} - \frac{\partial x_3}{\partial x_1^o})\Big/ \sqrt{(\frac{\partial x_1}{\partial x_1^o})(\frac{\partial x_3}{\partial x_3^o}) - \tfrac{1}{4}(\frac{\partial x_1}{\partial x_3^o} + \frac{\partial x_3}{\partial x_1^o})^2}$$

$$\tan \bar{\phi}_1 = \tfrac{1}{2}(\frac{\partial x_3}{\partial x_2^o} - \frac{\partial x_2}{\partial x_3^o})\Big/ \sqrt{(\frac{\partial x_2}{\partial x_2^o})(\frac{\partial x_3}{\partial x_3^o}) - \tfrac{1}{4}(\frac{\partial x_2}{\partial x_3^o} + \frac{\partial x_3}{\partial x_2^o})^2}$$

In the limit of small deformations, these three angles are simply related to the anti-symmetric component of the deformation gradient. That is

$$\bar{\phi}_k = \tfrac{1}{2}(\frac{\partial x_j}{\partial x_i^o} - \frac{\partial x_i}{\partial x_j^o}) = \omega_{ij}$$

using cyclic permutation on i, j, k.

Example 1.4: Consider a simple shear deformation parallel to the $x_1^o - x_2^o$ plane and given mathematically by

$$x_1 = x_1^o + k\,x_2^o, \qquad x_2 = x_2^o, \qquad x_3 = x_3^o$$

Determine the average rotation of a point.

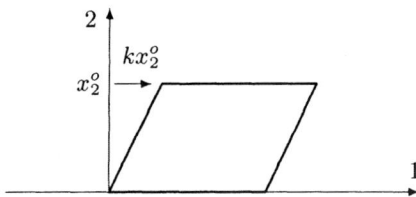

Figure 1.10: Simple shear deformation.

Substituting for $x_i(x_i^o)$ into the formula for the average rotation, we get

$$\tan \bar{\phi}_1 = 0, \qquad \tan \bar{\phi}_2 = 0, \qquad \tan \bar{\phi}_3 = \frac{\tfrac{1}{2}(0 - k)}{\sqrt{(1)(1) - \tfrac{1}{4}(k+0)^2}} = \frac{-\tfrac{1}{2}k}{\sqrt{1 - \tfrac{1}{4}k^2}}$$

If the deformation is small, this gives approximately

$$\bar{\phi}_3 \approx -\tfrac{1}{2}k$$

In looking at Figure 1.10, we see that the vertical and horizontal lines rotate angles of

$$\tan \phi_{90} = -k, \qquad \tan \phi_0 = 0$$

respectively. For small deformations, the average rotation is the average of the rotations of these two mutually perpendicular lines.

The constraint $D^2 > C^2 + B^2$ for this rotation about the 3-axis becomes

$$4(\frac{\partial x_1}{\partial x_1^o})(\frac{\partial x_2}{\partial x_2^o}) > (\frac{\partial x_1}{\partial x_2^o} + \frac{\partial x_2}{\partial x_1^o})^2$$

Specifically, for the simple shear problem, this is equivalent to

$$4 > k^2 \quad \text{or} \quad k < 2$$

When $D^2 < C^2 + B^2$, that is, when the deformation is larger, then the integration must be performed differently.

1.3   Strain

Strain is a measure of the "stretching" of the material points within a body; it is a measure of the relative displacement without rigid body motion and is an essential ingredient for the description of the constitutive behavior of materials. There are many measures of strain in existence, so it is worthwhile to first review some of the more common ones so as to put into perspective the measures we will actually introduce.

Strain Measures

Assume that a line segment of original length L_o is changed to length L. Some of the common measures of strain are:

$$\text{Engineering:} \quad \epsilon = \frac{\text{change in length}}{\text{original length}} = \frac{\Delta L}{L_o}$$

$$\text{True:} \quad \epsilon^T = \frac{\text{change in length}}{\text{final (current) length}} = \frac{\Delta L}{L} = \frac{\Delta L}{L_o + \Delta L}$$

$$\text{Logarithmic:} \quad \epsilon^N = \int_{L_o}^{L} \text{true strain} = \int_{L_o}^{L} \frac{dL}{L} = \log_n\left(\frac{L}{L_o}\right)$$

The relations among the measures are

$$\epsilon^T = \frac{\epsilon}{1 + \epsilon}, \qquad \epsilon^N = \log_n(1 + \epsilon)$$

An essential requirement of a strain measure is to allow the final length to be calculated knowing the original length. This is true of each of the above since

$$\text{Engineering:} \quad L = L_o + \Delta L = L_o + L_o\epsilon = L_o(1 + \epsilon)$$

$$\text{True:} \quad L = L_o + \Delta L = L_o + \frac{\epsilon^T L_o}{(1 - \epsilon^T)} = \frac{L_o}{(1 - \epsilon^T)}$$

$$\text{Logarithmic:} \quad L = L_o \exp(\epsilon^N)$$

The measures give different numerical values for the strain but all are equivalent in that they allow ΔL (or L) to be calculated knowing L_o. Because the measures are equivalent, then it is a matter of convenience as to which measure is to be chosen in an analysis.

The difficulty with these strain measures is that they do not have convenient transformation properties. This poses a problem in developing our three-dimensional theory because the quantities involved should transform as tensors of the appropriate order.

As a body deforms, various points will translate and rotate. The easiest way to distinguish between deformation and the local rigid-body motion is to consider the change in distance between two neighboring material particles. We will use this to establish our strain measures.

Let two material points before deformation have coordinates (x_i^o) and $(x_i^o + dx_i^o)$; and after deformation have the coordinates (x_i) and $(x_i + dx_i)$. The initial and final distances between these neighboring points are given by

$$dS_o^2 = \sum_i dx_i^o dx_i^o = (dx_1^o)^2 + (dx_2^o)^2 + (dx_3^o)^2$$

and

$$dS^2 = \sum_i dx_i dx_i = \sum_{i,j,k} \frac{\partial x_m}{\partial x_i^o} \frac{\partial x_m}{\partial x_j^o} dx_i^o dx_j^o$$

respectively. Only in the event of stretching or straining is dS^2 different from dS_o^2. That is,

$$dS^2 - dS_o^2 = dS^2 - \sum_i dx_i^o dx_i^o = \sum_{i,j,m} \left(\frac{\partial x_m}{\partial x_i^o} \frac{\partial x_m}{\partial x_j^o} - \delta_{ij} \right) dx_i^o dx_j^o$$

is a measure of the relative displacements. It is insensitive to rotation as can be easily demonstrated by considering a rigid-body motion. These equations can be written as

$$dS^2 - dS_o^2 = \sum_{i,j} 2E_{ij} dx_i^o dx_j^o \tag{1.9}$$

by introducing the strain measure

$$E_{ij} \equiv \frac{1}{2} \sum_m \left(\frac{\partial x_m}{\partial x_i^o} \frac{\partial x_m}{\partial x_j^o} - \delta_{ij} \right) \tag{1.10}$$

It is easy to observe that E_{ij} is a symmetric tensor of the second order. It is called the Lagrangian strain tensor.

In a similar manner, using Eulerian variables, we can introduce the Eulerian strain tensor through

$$dS^2 - dS_o^2 = dx_i dx_i - dS_o^2 = \sum_{i,j} \left(\delta_{ij} - \frac{\partial x_m^o}{\partial x_i} \frac{\partial x_m^o}{\partial x_j} \right) dx_i dx_j \equiv \sum_{i,j} 2e_{ij} dx_i dx_j$$

In the subsequent developments, however, we will mostly use the Lagrangian strain tensor.

Physical Interpretation of Normal and Shear Strains

To relate the strain tensor to the strain quantities with which we are familiar, consider the line element

$$dx_1^o = dS_o, \qquad dx_2^o = dx_3^o = 0$$

at the initial state as shown in Figure 1.11. After deformation, the line element is given by dx_i with magnitude dS.

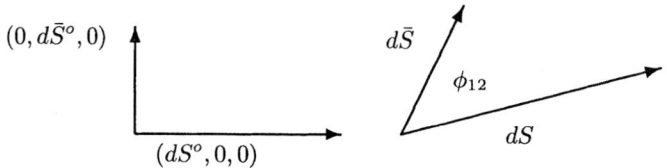

Figure 1.11: Deformation of two initially perpendicular line elements.

Let E_1 be the extension per unit original length of the element, that is,

$$E_1 = \frac{dS - dS_o}{dS_o} \qquad \text{or} \qquad dS = (1 + E_1)dS_o$$

For this line element we also have

$$dS^2 - dS_o^2 = 2E_{11}dS_o^2$$

and combining with the above yields

$$E_{11} = E_1 + \tfrac{1}{2}E_1^2 \qquad \text{or} \qquad E_1 = \sqrt{1 + 2E_{11}} - 1$$

There are similar relations for line elements originally in the x_2^o and x_3^o directions. The components E_{11}, E_{22}, and E_{33} are called the normal components of strain.

Note that there is a certain asymmetry (as regards the stretching direction) in the meaning of the normal components. For example, as the line is stretched, E_1 increases possibly without limit resulting in E_{11} doing the same. If, however, the line is shrunk so that E_1 is negative, then there is a definite limit given by $E_1 = -1$ which corresponds to $\Delta L = -L_o$, meaning that the line has shrunk to zero length. That is, we have limits on the strains of

$$-1 < E_1 < \infty, \qquad -0.5 < E_{11} < \infty$$

This asymmetry between the stretching and shrinking directions is important when considering the constitutive behavior.

Consider now a line segment at an arbitrary angle θ in the undeformed configuration, then its length after deformation is obtained from

$$dS^2 - dS_o^2 = 2\left[E_{11}dx_1^o dx_1^o + E_{21}dx_1^o dx_2^o + E_{12}dx_2^o dx_1^o + E_{22}dx_2^o dx_2^o\right]$$

Realizing that $dx_1^o = dS_o \cos\theta$ and so on, and that the strain of this arbitrary line is

$$dS^2 - dS_o^2 = 2E_{\theta\theta}dS_o^2$$

leads to

$$E_{\theta\theta} = E_{11}\cos^2\theta + 2E_{12}\cos\theta\sin\theta + E_{22}\cos^2\theta \qquad (1.11)$$

where use is made of $E_{12} = E_{21}$. This gives us the transformation rule for the components of strain — they transform as second-order tensors.

A deformation can also exhibit distortion (change in relative angles) in the configuration. Consider, in the initial state, two line elements parallel to x_1^o and x_2^o, respectively, as shown in Figure 1.11. The two line elements are denoted by dx_i^o and $d\bar{x}_i^o$, respectively, with

$$dx_1^o = dS_o, \qquad\qquad dx_2^o = dx_3^o = 0$$
$$d\bar{x}_2^o = d\bar{S}_o, \qquad\qquad dx_1^o = dx_3^o = 0$$

These two elements are perpendicular to each other initially. After deformation, dx_i^o is deformed into dx_i, and $d\bar{x}_i^o$ into $d\bar{x}_i$.

Denoting the angle between dx_i and $d\bar{x}_i$ by ϕ_{12} and taking the dot product of these two vectors, we obtain

$$dS\,d\bar{S}\cos\phi_{12} = \sum_i dx_i\,d\bar{x}_i = \sum_{i,k,m}\frac{\partial x_i}{\partial x_k^o}\frac{\partial x_i}{\partial x_m^o}dx_k^o\,d\bar{x}_m^o = \sum_i \frac{\partial x_i}{\partial x_1^o}\frac{\partial x_i}{\partial x_2^o}dx_1^o dx_2^o$$

which can readily be rewritten in terms of the Lagrangian strain tensor as

$$dS\,d\bar{S}\cos\phi_{12} = 2E_{12}dS_o d\bar{S}_o$$

By substituting for dS and $d\bar{S}$ in terms of the extensions, we get

$$dS = (1 + E_1)dS_o, \qquad d\bar{S} = (1 + E_2)d\bar{S}_o$$

thus leading to

$$\cos\phi_{12} = \frac{2E_{12}}{(1 + E_1)(1 + E_2)}$$

Denoting the change in angle by

$$\alpha_{12} \equiv \frac{\pi}{2} - \phi_{12}$$

and using the expressions for the elongations in terms of the strain components, we finally obtain

$$\sin\alpha_{12} = \frac{2E_{12}}{\sqrt{1 + 2E_{11}}\sqrt{1 + 2E_{22}}}$$

All the Lagrangian strain components E_{11}, E_{22}, and E_{12} contribute to the change of angle. However, it is only when $E_{12} = 0$ that the angle between the two line elements would be preserved. The component E_{12} therefore seems a good measure of the "shearing" of perpendicular line segments.

Because the term $\sin\alpha_{12}$ must lie in the range ± 1, we then have the limits on E_{12} of

$$-\tfrac{1}{2}\pi < \alpha_{12} < \tfrac{1}{2}\pi \quad \text{or} \quad -\sqrt{1+2E_{11}}\sqrt{1+2E_{22}} < 2E_{12} < \sqrt{1+2E_{11}}\sqrt{1+2E_{22}}$$

The limits on E_{12} are a combination of those on α_{12} and on the stretches.

Consider now two perpendicular line segments oriented at an arbitrary angle θ in the undeformed configuration, the change of angle after deformation is obtained from

$$dS\,d\bar{S}\sin\alpha'_{12} = \sum_i dx_i\,d\bar{x}_i = \sum_{i,k,m}\frac{\partial x_i}{\partial x_k^o}\frac{\partial x_i}{\partial x_m^o}dx_k^o\,d\bar{x}_m^o = \sum_{k,m}[2E_{km}+\delta_{km}]dx_k^o\,d\bar{x}_m^o$$

Expanding this gives

$$dS\,d\bar{S}\sin\alpha'_{12} = (2E_{11}+1)dx_1^o d\bar{x}_1^o + 2E_{21}dx_1^o d\bar{x}_2^o + 2E_{12}dx_2^o d\bar{x}_1^o + (2E_{22}+1)dx_2^o d\bar{x}_2^o$$

Realizing that the undeformed segment lengths are given by

$$dx_i^o = dS_o\{\cos\theta,\ \sin\theta,\ 0\}\,, \qquad d\bar{x}_i^o = d\bar{S}_o\{-\sin\theta,\ \cos\theta,\ 0\}$$

and that the shear strain of these arbitrary perpendicular lines is

$$dS\,d\bar{S}\sin\alpha'_{12} = 2E'_{12}dS_o d\bar{S}_o$$

leads to

$$E'_{12} = -(E_{11}-E_{22})\cos\theta\sin\theta + E_{12}(\cos^2\theta - \sin^2\theta) \qquad (1.12)$$

This gives us the transformation rule for the components of shear strain — they, too, transform as components of a second-order tensors.

Example 1.5: Express the components of the Lagrangian strain tensor in terms of the displacement components.

Sometimes it is convenient to deal with displacements and displacement gradients instead of the deformation gradient. These are obtained by using the relations

$$x_m = x_m^o + u_m\,, \qquad \frac{\partial x_m}{\partial x_i^o} = \frac{\partial u_m}{\partial x_i^o} + \delta_{im}$$

The Lagrangian strain tensor E_{ij} can be written in terms of the displacement by

$$\begin{aligned} E_{ij} &= \sum_m \tfrac{1}{2}\left[\left(\frac{\partial u_m}{\partial x_i^o}+\delta_{im}\right)\left(\frac{\partial u_m}{\partial x_j^o}+\delta_{jm}\right)-\delta_{ij}\right] \\ &= \tfrac{1}{2}\left[\frac{\partial u_i}{\partial x_j^o}+\frac{\partial u_j}{\partial x_i^o}+\sum_m\frac{\partial u_m}{\partial x_i^o}\frac{\partial u_m}{\partial x_j^o}\right] \end{aligned}$$

Typical expressions for E_{ij} in unabridged notations are

$$E_{11} = \frac{\partial u_1}{\partial x_1^o} + \frac{1}{2}\left[\left(\frac{\partial u_1}{\partial x_1^o}\right)^2 + \left(\frac{\partial u_2}{\partial x_1^o}\right)^2 + \left(\frac{\partial u_3}{\partial x_1^o}\right)^2\right]$$

$$E_{22} = \frac{\partial u_2}{\partial x_2^o} + \frac{1}{2}\left[\left(\frac{\partial u_2}{\partial x_2^o}\right)^2 + \left(\frac{\partial u_2}{\partial x_2^o}\right)^2 + \left(\frac{\partial u_3}{\partial x_2^o}\right)^2\right]$$

$$E_{12} = \frac{1}{2}\left(\frac{\partial u_1}{\partial x_2^o} + \frac{\partial u_2}{\partial x_1^o}\right) + \frac{1}{2}\left[\frac{\partial u_1}{\partial x_1^o}\frac{\partial u_1}{\partial x_2^o} + \frac{\partial u_2}{\partial x_1^o}\frac{\partial u_2}{\partial x_2^o} + \frac{\partial u_3}{\partial x_1^o}\frac{\partial u_3}{\partial x_2^o}\right] \qquad (1.13)$$

Note the presence of the nonlinear terms.

Example 1.6: Show that the Lagrangian strain tensor is zero for a rigid-body motion.

In a rigid-body rotation all the points are given a displacement but the relative distance between points is unchanged. Consider the two-dimensional case described by

$$x_1 = x_1^o \cos\phi - x_2^o \sin\phi, \qquad x_2 = x_1^o \sin\phi + x_2^o \cos\phi, \qquad x_3 = x_3^o$$

where ϕ is the angle of rotation about the 3-axis. The corresponding displacements are

$$u_1 = x_1^o(\cos\phi - 1) - x_1^o \sin\phi, \qquad u_2 = x_1^o \sin\phi + x_2^o(\cos\phi - 1), \qquad u_3 = 0$$

from which the displacement gradients are determined to be

$$\left[\frac{\partial u_i}{\partial x_j^o}\right] = \left[\begin{array}{ccc} (\cos\phi - 1) & -\sin\phi & 0 \\ \sin\phi & (\cos\phi - 1) & 0 \\ 0 & 0 & 0 \end{array}\right]$$

It is now straightforward to show that all Lagrangian strain components are zero. For example, from Equation (1.13)

$$E_{11} = (C - 1) + \frac{1}{2}\left[(C - 1)^2 + (-S)^2 + (0)^2\right] = C - 1 + \frac{1}{2}C - C + \frac{1}{2} + \frac{1}{2}S = 0$$

where we used $C \equiv \cos\phi$ and $S \equiv \sin\phi$. We obtain a similar result for the other components.

The infinitesimal strain tensor has the components

$$[\,\epsilon_{ij}\,] = \frac{1}{2}\left[\frac{\partial u_j}{\partial x_i^o} + \frac{\partial u_i}{\partial x_j^o}\right] = \left[\begin{array}{ccc} (\cos\phi - 1) & 0 & 0 \\ 0 & (\cos\phi - 1) & 0 \\ 0 & 0 & 0 \end{array}\right] \approx \frac{1}{2}\left[\begin{array}{ccc} \phi^2 & 0 & 0 \\ 0 & \phi^2 & 0 \\ 0 & 0 & 0 \end{array}\right]$$

It is only when ϕ is very small is this strain tensor nearly zero.

Example 1.7: Determine the strain for the deforming body of Figure 1.8.

The displacement gradient is

$$\left[\frac{\partial u_i}{\partial x_j^o}\right] = \left[\begin{array}{ccc} J^o \cos(x_1^o/R) - 1 & -\sin(x_1^o/R) & 0 \\ J^o \sin(x_1^o/R) & \cos(x_1^o/R) - 1 & 0 \\ 0 & 0 & 1 \end{array}\right]$$

with $J^o = 1 - x_2^o/R$. The strains are

$$
\begin{aligned}
E_{11} &= J^o C - 1 + \tfrac{1}{2}[(J^o C - 1)^2 + (J^o S)^2 + (0)^2] = -\frac{x_2^o}{R} + \frac{1}{2}(\frac{x_2^o}{R})^2 \\
E_{22} &= C - 1 + \tfrac{1}{2}[(C - 1)^2 + (-S)^2 + (0)^2] = 0 \\
E_{12} &= \tfrac{1}{2}[-S + J^o S] + \tfrac{1}{2}[(J^o C - 1)(-S) + (J^o S)(C - 1) + (0)^2] = 0
\end{aligned}
$$

where we used $C \equiv \cos(x_1^o/R)$ and $S \equiv \sin(x_1^o/R)$. Only line segments initially in the x_1-direction are strained. There is a line, $x_2^o = 0$, which is not strained; other lines are strained in proportion to their distance from this line. In the limit of a very thin body ($x_2^o/R << 1$), the strain distribution is linear

$$
E_{11} \approx -\frac{x_2^o}{R} \approx -x_2^o \frac{dv}{dx_1^o}
$$

where v is the u_2 displacement of the $x_2^o = 0$ line. These are the strain characteristics of a beam or plate in bending.

Example 1.8:   Establish the relation between the incremental components of the Lagrangian strain tensor and the incremental components of the infinitesimal strain tensor.

The change in length of a line element can be written in terms of the Lagrangian strain as

$$
2\sum_{i,j} E_{ij} dx_i^o dx_j^o = S^2 - S_o^2 = \sum_i dx_i dx_i - \sum_i dx_i^o dx_i^o
$$

The increments in strains are obtained from this as (noting that dx_i^o is not changed)

$$
2\sum_{i,j} \Delta E_{ij} dx_i^o dx_j^o = 2\sum_i d\Delta x_i dx_i - 0
$$

But we also have that the change of the new positions can be rewritten as

$$
d\Delta x_i = d\Delta(x_i^o + u_i) = d\Delta u_i = \sum_k \frac{\partial \Delta u_i}{\partial x_k} dx_k = \sum_k \Delta(\frac{\partial u_i}{\partial x_k}) dx_k = \sum_k \Delta(\epsilon_{ik} + \omega_{ik}) dx_k
$$

Therefore

$$
\sum_{i,j} \Delta E_{ij} dx_i^o dx_j^o = \sum_{i,j} \Delta \epsilon_{ij} dx_i dx_j + \sum_{i,j} \Delta \omega_{ij} dx_i dx_j = \sum_{i,j} \Delta \epsilon_{ij} dx_i dx_j
$$

where the anti-symmetry of ω_{ij} and symmetry of $dx_i dx_j$ was used to set the product to zero. Substituting for dx_i in terms of dx_i^o now gives

$$
\Delta E_{mn} = \sum_{i,j} \frac{\partial x_i}{\partial x_m^o} \frac{\partial x_j}{\partial x_n^o} \Delta \epsilon_{ij} \qquad \text{and} \qquad \Delta \epsilon_{mn} = \sum_{i,j} \frac{\partial x_i^o}{\partial x_m} \frac{\partial x_j^o}{\partial x_n} \Delta E_{ij}
$$

This surprising result shows that although ΔE_{ij} and $\Delta \epsilon_{ij}$ are small, they are not equal. The main reason for this is because they are referred to different configurations.

We will utilize this relation when we consider small variations of the strain field.

Infinitesimal Strain and Rotation

The full nonlinear analysis of problems is quite difficult and so simplifications are often sought. Three situations are shown in Figure 1.12.

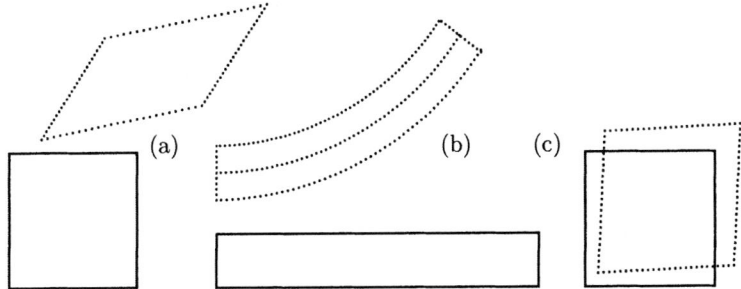

Figure 1.12: Combinations of displacements and strains. (a) Large displacements, rotations, and strains. (b) Large displacements and rotations, small strains. (c) Small displacements, rotations, and strains.

The general case is that of large displacements, large rotations, and large strains. In the chapters dealing with the linear theory, where both the displacements and strains are small, Case (c) prevails. Our nonlinear analysis of thin-walled structures will be primarily restricted to Case (b), where the deflections and rotations can be large but the strains are small. This is a reasonable approximation because structural materials do not exhibit large strains without yielding and structures are designed to operate without yielding.

If the displacement gradients are small, that is,

$$\left| \frac{\partial u_i}{\partial x_j^o} \right| << 1$$

then the product terms in the Lagrangian strain tensor E_{ij} can be neglected. The result is

$$E_{ij} \simeq \epsilon_{ij} = \tfrac{1}{2} \left(\frac{\partial u_i}{\partial x_j^o} + \frac{\partial u_j}{\partial x_i^o} \right)$$

where ϵ_{ij} is the infinitesimal strain tensor. This assumption also leads to the conclusion that the components E_{ij} are small as compared with unity. Thus, the infinitesimal strain components have direct interpretations as extensions or change of angles.

If, in addition to the above, the following condition exists

$$\left| \frac{u_i}{L} \right| << 1$$

where L is the smallest dimension of the body, then

$$x_i \simeq x_i^o$$

and the distinction between the Lagrangian and Eulerian variables vanishes. As a result, the functional forms of the displacement components u_i in these two variables become identical. Henceforth, when we use the small-strain approximations, ϵ_{ij} will be used to denote both the infinitesimal Lagrangian and Eulerian strain tensors, ω_{ij} will denote both the infinitesimal Lagrangian and Eulerian rotation, and x_i will denote both Lagrangian and Eulerian variables.

1.4   Cauchy Stress Principle

The kinetics of rigid bodies are described in terms of forces; the equivalent concept for continuous media is stress (loosely defined as force over unit area). Actions can be exerted on a continuum through either contact forces or forces contained in the mass. The contact force is often referred to as a surface force or traction as its action occurs on a surface. We are primarily concerned with contact forces.

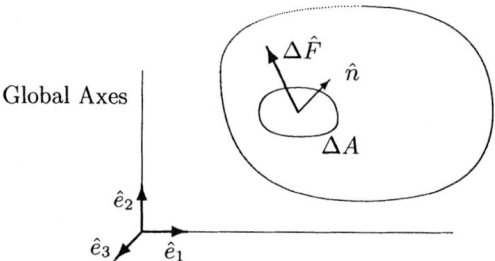

Figure 1.13: Exposed forces on an arbitrary section cut.

Traction Vector

Consider a small surface element of area ΔA on our imagined exposed surface A in the deformed configuration. There must be resultant forces and moments acting on ΔA to make it equipollent to the effect of the rest of the material. That is, when the pieces are put back together, these forces cancel each other. Let these forces be thought of as contact forces and so give rise to contact stresses (even though they are inside the body). Cauchy formalized this by introducing his concept of traction vector.

Let $\hat{n}$ be the unit vector that is perpendicular to the surface element ΔA and let $\Delta \hat{F}$ be the resultant force exerted from the other part of the surface element with the negative normal vector. We assume that as ΔA becomes vanishingly small, the ratio $\Delta \hat{F}/\Delta A$ approaches a definite limit $d\hat{F}/dA$. The vector obtained in the limiting process

$$\lim_{\Delta A \to 0} \frac{\Delta \hat{F}}{\Delta A} = \frac{d\hat{F}}{dA} \equiv \hat{t}^{(\hat{n})}$$

is called the *traction vector*. This vector represents the force per unit area acting on the surface and its limit exists because the material is assumed continuous. The superscript $\hat{n}$ is a reminder that the traction is dependent on the orientation of the area.

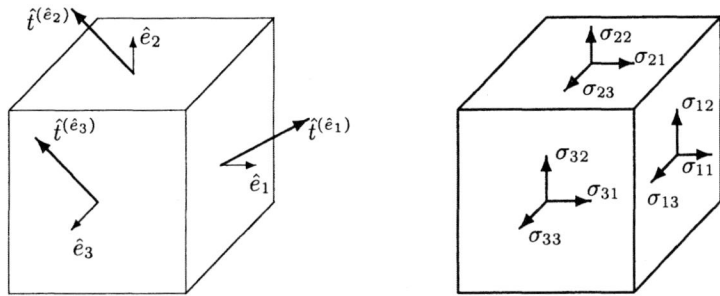

Figure 1.14: Stressed cube.

To give explicit representation of the traction vector, consider its components on the three faces of a cube as shown in Figure 1.14. The traction on the 1-face is

$$\hat{n} = \hat{e}_1 : \qquad \hat{t}^{(\hat{n})} = \sum_i t_i^{(\hat{e}_1)} \hat{e}_i = t_1^{(\hat{e}_1)} \hat{e}_1 + t_2^{(\hat{e}_1)} \hat{e}_2 + t_3^{(\hat{e}_1)} \hat{e}_3$$

while on the 2-face

$$\hat{n} = \hat{e}_2 : \qquad \hat{t}^{(\hat{n})} = \sum_i t_i^{(\hat{e}_2)} \hat{e}_i = t_1^{(\hat{e}_2)} \hat{e}_1 + t_2^{(\hat{e}_2)} \hat{e}_2 + t_3^{(\hat{e}_2)} \hat{e}_3$$

Because this description is somewhat cumbersome, we simplify the notation by introducing

$$\sigma_{ij} \equiv t_j^{(\hat{e}_i)}$$

where i refers to the face and j to the component. More specifically,

$$\sigma_{11} \equiv t_1^{(\hat{e}_1)} , \qquad \sigma_{13} \equiv t_3^{(\hat{e}_1)} , \qquad \sigma_{31} \equiv t_1^{(\hat{e}_3)} , \qquad \dots$$

The normal projections of $\hat{t}^{(\hat{n})}$ on these special faces are the normal stress components $\sigma_{11}, \sigma_{22}, \sigma_{33}$, while projections perpendicular to $\hat{n}$ are shear stress components $\sigma_{12}, \sigma_{13}; \sigma_{21}, \sigma_{23}; \sigma_{31}, \sigma_{32}$.

It is important to realize that while $\hat{t}$ resembles the elementary idea of stress (force over area) it is not stress; $\hat{t}$ transforms as a vector and has only three components. The tensor σ_{ij} is our definition of stress; it has nine components with units of force over area, but at this stage we do not know how these components transform.

Relation between t_i and n_j

We know that the traction vector $\hat{t}^{(\hat{n})}$ acting on an area $dA\hat{n}$ depends on the normal $\hat{n}$ of the area. The particular relation can be obtained by considering a traction on an arbitrary surface of the tetrahedron shown in Figure 1.15. On the three faces perpendicular to the coordinate directions the components of the three traction vectors are denoted by σ_{ij}. The vector acting on the inclined surface ABC is $\hat{t}$ and the unit normal vector $\hat{n}$. The equilibrium of the tetrahedron requires that the resultant force acting on it must vanish.

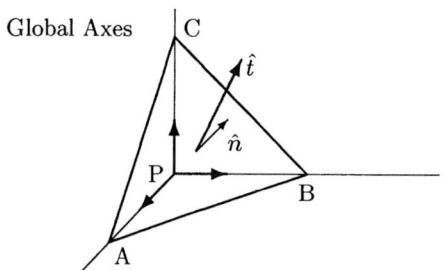

Figure 1.15: Tetrahedron.

The equation for the balance of forces in the x_1-direction for the tetrahedron is given by

$$t_1 dA - \sigma_{11} dA_1 - \sigma_{21} dA_2 - \sigma_{31} dA_3 + b_1 \rho dV = 0$$

where b_1 is the x_1-component of the body force $\hat{b}$ (which may also contain inertia terms), t_1 is the x_1-component of the traction vector, dA_i is the area of the face perpendicular to the x_i axis, dA is the area of the inclined surface, and

$$dV = \tfrac{1}{3} h dA$$

is the volume of the tetrahedron. In this, h is the smallest distance from point P to the inclined surface ABC. Noting that the normal to the area has the components

$$\hat{n} = n_1 \hat{e}_1 + n_2 \hat{e}_2 + n_3 \hat{e}_3$$

we conclude that the components of area are

$$
\begin{aligned}
\text{area of face 1:} \quad dA_1 &= n_1 dA \\
\text{area of face 2:} \quad dA_2 &= n_2 dA \\
\text{area of face 3:} \quad dA_3 &= n_3 dA
\end{aligned}
$$

Now divide through by dA in the equilibrium relation, and letting $h \to 0$, we obtain

$$t_1 = \sigma_{11} n_1 + \sigma_{21} n_2 + \sigma_{31} n_3 = \sum_j \sigma_{j1} n_j$$

Similar equations can be derived from the consideration of the balance of forces in the x_2- and x_3-directions. These three equations can be written in the indicial notation as

$$t_i = \sum_j \sigma_{ji} n_j \tag{1.14}$$

This compact relation says that we need only know nine numbers $[\sigma_{ij}]$ to be able to determine the traction vector on any area passing through a point. These elements are called the Cauchy stress components and form the Cauchy stress tensor. It is a second-order tensor because t_i and n_j transform as first-order tensors. Later, we will establish that it is symmetric.

Kirchhoff Stresses

We have chosen to use the Lagrangian variables for the description of a body with finite deformation. For consistency, we need to introduce a measure of stress referred to the undeformed configuration. Because true loading exists only in the deformed state, the corresponding loading and stress in the body at the initial (undeformed) state could be considered as fictitious. To appreciate the motivation in introducing the new definitions of stress, it is worthwhile to keep the following in mind:

- The traction vector is first defined in terms of a force divided by area.
- The stress tensor is then defined according to a transformation relation for the traction and area normal.

To refer our description of tractions to the surface before deformation, we must define a traction vector $\hat{t}^o$ acting on an area dA^o as indicated in Figure 1.16. The introduction of such a vector is somewhat arbitrary, so we first reconsider the Cauchy stress so as to motivate the developments.

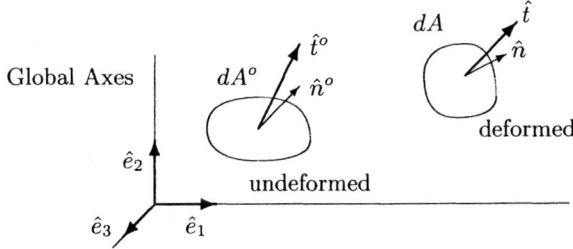

Figure 1.16: Traction vectors in the undeformed and deformed configurations.

In the deformed state, on every plane surface passing through a point, there is a traction vector $\hat{t}_i$ defined in terms of the deformed surface area. That is, letting the traction vector be $\hat{t}$ and the total resultant force acting on dA be $d\hat{F}$, then

$$\hat{t} \equiv \frac{d\hat{F}}{dA} \qquad \text{or} \qquad t_i \equiv \frac{dF_i}{dA}$$

Let all the traction vectors and unit normals in the deformed body form two respective vector spaces. Then the Cauchy stress tensor σ_{ij} was shown to be the transformation between these two vector spaces, that is,

$$t_i = \sum_j \sigma_{ji} n_j$$

The tensorial property of the Cauchy stress tensor can be established from the quotient rule. Defined in this manner, the Cauchy stress tensor is an abstract quantity; however, on special plane surfaces such as those with unit normals parallel to $\hat{e}_1, \hat{e}_2$, and $\hat{e}_3$, respectively, the nine components of $[\sigma_{ij}]$ can be related to the traction vector and thus have physical meaning.

Thus, the meaning of σ_{ij} are the components of stress derived from the force vector dF_i divided by the deformed area. This, in elementary terms, is called *true stress*.

We will now do a parallel development for the undeformed configuration. Let the resultant force $d\hat{F}^o$, referred to the undeformed configuration, be given by a transformation of the force $d\hat{F}$ acting on the deformed area. One possibility is to take $dF_i^o = dF_i$, and this gives rise to the so-called Lagrange stress tensor, which in simple terms would correspond to "force divided by original area." Instead, let

$$dF_i^o = \sum_j \frac{\partial x_i^o}{\partial x_j} dF_j$$

which follows the analogous rule for the deformation of line segments. The reason for this choice will become apparent later when we consider the equations of motion. It is important to realize that this is not a rotation transformation but that the force components are being "deformed." The Kirchhoff traction vector is defined as

$$t_i^o \equiv \frac{dF_i^o}{dA^o} = \sum_j \frac{\partial x_i^o}{\partial x_j} \frac{dF_j}{dA^o}$$

This leads to the definition of the Kirchhoff stress tensor σ_{ij}^K:

$$t_i^o = \sum_j \sigma_{ji}^K n_j^o$$

The meaning of σ_{pq}^K are components of stress derived from the transformed components of the force vector, divided by the original area. There is no elementary equivalent to this stress. This stress is usually referred to as the second Piola-Kirchhoff stress tensor; we will abbreviate it simply as the Kirchhoff stress tensor.

Example 1.9:   Establish the relation between the Cauchy and Kirchhoff stress tensors.

From the definition of the Kirchhoff traction vector, we have

$$t_i^o dA^o = dF_i^o = \sum_j \frac{\partial x_i^o}{\partial x_j} dF_j = \sum_j \frac{\partial x_i^o}{\partial x_j} t_j dA$$

Replacing the tractions with their respective stress tensors gives

$$\sum_j \sigma_{ji}^K n_j^o dA^o = \sum_{j,p} \frac{\partial x_i^o}{\partial x_j} \sigma_{pj} n_p dA$$

Because we also have the relation between the areas

$$n_p \, dA = \sum_k J^o \frac{\partial x_k^o}{\partial x_p} n_k^o \, dA^o$$

then

$$\sum_j \sigma_{ji}^K n_j^o dA^o = \sum_{j,k,p} J^o \frac{\partial x_i^o}{\partial x_j} \sigma_{pj} \frac{\partial x_k^o}{\partial x_p} n_k^o dA^o$$

With $J^o = \rho^o/\rho$, the relation between Kirchhoff and Cauchy stress tensors becomes

$$\sigma_{ji}^K = \sum_{m,n} \frac{\rho^o}{\rho} \frac{\partial x_i^o}{\partial x_m} \frac{\partial x_j^o}{\partial x_n} \sigma_{mn} \,, \qquad \sigma_{ji} = \sum_{m,n} \frac{\rho}{\rho^o} \frac{\partial x_i}{\partial x_m^o} \frac{\partial x_j}{\partial x_n^o} \sigma_{mn}^K \qquad (1.15)$$

We will show that the Cauchy stress tensor is symmetric, hence these relations show that the Kirchhoff stress tensor is also a symmetric tensor.

Example 1.10: A unit cubic solid is subjected to the applied load as shown in the Figure 1.17. Determine the Cauchy and Kirchhoff stresses.

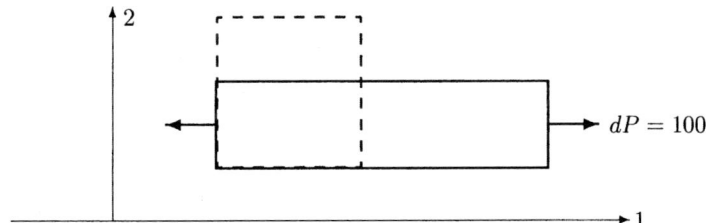

Figure 1.17: Cube with uniaxial load.

The deformation is given by

$$x_1 = \lambda_1 x_1^o \,, \qquad x_2 = \lambda_2 x_2^o \,, \qquad x_3 = \lambda_3 x_2^o$$

where the stretches are $\lambda_1 = 2$, $\lambda_2 = \lambda_3 = 0.5$. For this problem, the basic information is given in terms of the forces and so the stresses will be established by using the connection between them, the tractions, and the stresses.

The deformation gradients are given by

$$\left[\frac{\partial x_i}{\partial x_p^o} \right] = \begin{bmatrix} \lambda_1 & 0 & 0 \\ 0 & \lambda_2 & 0 \\ 0 & 0 & \lambda_3 \end{bmatrix}, \qquad \left[\frac{\partial x_p^o}{\partial x_i} \right] = \begin{bmatrix} 1/\lambda_1 & 0 & 0 \\ 0 & 1/\lambda_2 & 0 \\ 0 & 0 & 1/\lambda_3 \end{bmatrix}$$

The Jacobian is therefore

$$J^o = \lambda_1\lambda_2\lambda_3 = 2(0.5 \times 0.5) = 0.5$$

which shows that there is a volume change. The Cauchy stress tensor is obtained from information about the traction vectors

$$t_i = \sum_j \sigma_{ji}n_j \equiv \frac{dF_i}{dA} = \sigma_{1i}n_1 + \sigma_{2i}n_2 + \sigma_{3i}n_3$$

On the x_1-face $\hat{n} = \hat{e}_1$, giving $t_i = \sigma_{1i}$, and the components of force and area are therefore

$$dF_i \;=\; \{dP, 0, 0\} = \{100, 0, 0\}$$

$$dA_1 \;=\; J^o \frac{\partial x_1^o}{\partial x_1} dA_1^o = \lambda_2\lambda_3\, dA_1^o = 0.5 \times 0.5 = 0.25\,, \quad dA_2 = 0, \quad dA_3 = 0$$

On the x_2-face and x_3-face, we have $t_i = \sigma_{2i}$, and $t_i = \sigma_{3i}$, respectively, and in both cases

$$dF_i = \{0, 0, 0\}$$

Thus, for the respective faces, we have

$$\hat{n}^{(1)} = \{1,0,0\}: \quad \sigma_{11} = \frac{dF_1}{dA_1} = \frac{dP}{dA_1} = \frac{1}{\lambda_2\lambda_3}\frac{dP}{dA_1^o} = \frac{100}{.25} = 400, \quad \sigma_{12} = \sigma_{13} = 0$$

$$\hat{n}^{(2)} = \{0,1,0\}: \quad \sigma_{21} = \sigma_{23} = \sigma_{22} = 0$$

$$\hat{n}^{(3)} = \{0,0,1\}: \quad \sigma_{31} = \sigma_{32} = \sigma_{33} = 0$$

In summary, the components of the stress tensor are

$$[\,\sigma_{ij}\,] = \frac{1}{\lambda_2\lambda_3}\frac{dP}{dA_1^o}\begin{bmatrix} 1 & 0 & 0 \\ 0 & 0 & 0 \\ 0 & 0 & 0 \end{bmatrix} = \begin{bmatrix} 400 & 0 & 0 \\ 0 & 0 & 0 \\ 0 & 0 & 0 \end{bmatrix}$$

Because the deformed area is $dA = 0.25$, the Cauchy stress has the interpretation of force divided by deformed area.

The components of the Kirchhoff stress will be obtained by using this and the deformation gradients. Convert the Cauchy stress to Kirchhoff stress by

$$
\begin{aligned}
\sigma_{pq}^K \;=\;& \sum_{i,j} \frac{\rho^o}{\rho}\sigma_{ij}\frac{\partial x_p^o}{\partial x_i}\frac{\partial x_q^o}{\partial x_j} \\[2mm]
=\;& J^o\Bigg[\sigma_{11}\frac{\partial x_p^o}{\partial x_1}\frac{\partial x_q^o}{\partial x_1} + \sigma_{12}\frac{\partial x_p^o}{\partial x_1}\frac{\partial x_q^o}{\partial x_2} + \sigma_{13}\frac{\partial x_p^o}{\partial x_1}\frac{\partial x_q^o}{\partial x_3} \\[2mm]
&\; + \sigma_{21}\frac{\partial x_p^o}{\partial x_2}\frac{\partial x_q^o}{\partial x_1} + \sigma_{22}\frac{\partial x_p^o}{\partial x_2}\frac{\partial x_q^o}{\partial x_2} + \sigma_{23}\frac{\partial x_p^o}{\partial x_2}\frac{\partial x_q^o}{\partial x_3} \\[2mm]
&\; + \sigma_{31}\frac{\partial x_p^o}{\partial x_3}\frac{\partial x_q^o}{\partial x_1} + \sigma_{32}\frac{\partial x_p^o}{\partial x_3}\frac{\partial x_q^o}{\partial x_2} + \sigma_{33}\frac{\partial x_p^o}{\partial x_3}\frac{\partial x_q^o}{\partial x_3}\Bigg]
\end{aligned}
$$

Because only $\sigma_{11} \neq 0$, we have simply

$$\sigma_{pq}^K = J^o\sigma_{11}\frac{\partial x_p^o}{\partial x_1}\frac{\partial x_q^o}{\partial x_1}$$

and this gives, for instance,

$$\sigma_{11}^K = (\lambda_1\lambda_2\lambda_3)(\frac{1}{\lambda_2\lambda_3}\frac{dP}{dA_1^o})(\frac{1}{\lambda_1})(\frac{1}{\lambda_1}) = \frac{1}{\lambda_1}\frac{dP}{dA_1^o} = \frac{1}{2}(400)(\frac{1}{2})^2 = 50$$

$$\sigma_{22}^K = \frac{1}{2}(400)(0)^2 = 0$$

$$\sigma_{33}^K = \frac{1}{2}(400)(0)^2 = 0$$

In summary,

$$[\sigma_{pq}^K] = \frac{1}{\lambda_1}\frac{dP}{dA_1^o}\begin{bmatrix} 1 & 0 & 0 \\ 0 & 0 & 0 \\ 0 & 0 & 0 \end{bmatrix} = \begin{bmatrix} 50 & 0 & 0 \\ 0 & 0 & 0 \\ 0 & 0 & 0 \end{bmatrix}$$

Because the original area is $dA^o = 1$, the Kirchhoff stress does not have a simple interpretation such as force divided original area.

1.5   Governing Equations of Motion

Recall that Newton's laws for the equation of motion of a rigid body can be written as

$$\sum \hat{F} = m\ddot{\hat{u}}$$

$$\sum \hat{M} = m\,\hat{x} \times \ddot{\hat{u}}$$

where $\ddot{\hat{u}}$ is the acceleration and m is the mass. These equations will now be used to establish the equations of motion of a deformable body. It will turn out, however, that they are not the most suitable form, and we look at other formulations. In particular, we look at the forms arising from the principle of virtual work and leading to stationary principles such as Hamilton's principle and Lagrange's equation.

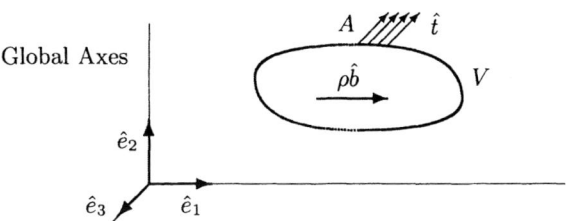

Figure 1.18: Arbitrary small volume.

Equations of Motion in Terms of Stress

Consider an arbitrary volume V taken from the deformed body as shown in Figure 1.18; it has tractions $\hat{t}$ on the boundary surface A, and body force per unit mass $\hat{b}$. Newton's laws of motion become, respectively,

$$\int_A \hat{t}\, dA + \int_V \rho\hat{b}\, dV = \int_V \rho\hat{\ddot{u}}\, dV$$

$$\int_A \hat{x} \times \hat{t}\, dA + \int_V \hat{x} \times \hat{b}\rho\, dV = \int_V \hat{x} \times \hat{\ddot{u}}\rho\, dV$$

These are written in the indicial notation as

$$\int_A t_i dA + \int_V \rho b_i dV = \int_V \rho\ddot{u}_i dV$$

$$\sum_{j,k} \epsilon_{ijk}[\int_A x_j t_k dA + \int_V x_j b_k \rho dV = \int_V x_j \ddot{u}_k \rho dV]$$

These are the equations of motion in terms of t_i. We now obtain the equations of motion in terms of the stress. In doing this, there is a choice between the deformed state and the undeformed state.

Using $t_i = \sum_p \sigma_{pi} n_p$ and by the integral theorem of Equation (1.2), we get

$$\int_A t_i\, dA = \int_A \sum_p \sigma_{pi} n_p\, dA = \int_V \sum_p \frac{\partial \sigma_{pi}}{\partial x_p} dV$$

The equations of motion become (after simplification and noting that, because the volume V is arbitrary, the integrands must vanish)

$$\sum_p \frac{\partial \sigma_{pi}}{\partial x_p} + \rho b_i = \rho\ddot{u}_i$$

$$\sum_{j,k} \epsilon_{ijk}\sigma_{jk} = 0 \qquad (1.16)$$

The second equation shows that σ_{ij} is a symmetric tensor, because the contraction of a symmetric tensor with an antisymmetric tensor is zero. Hence the two equations of motion become, in expanded notation,

$$\frac{\partial \sigma_{11}}{\partial x_1} + \frac{\partial \sigma_{21}}{\partial x_2} + \frac{\partial \sigma_{31}}{\partial x_3} + \rho b_1 = \rho\ddot{u}_1$$

$$\frac{\partial \sigma_{12}}{\partial x_1} + \frac{\partial \sigma_{22}}{\partial x_2} + \frac{\partial \sigma_{32}}{\partial x_3} + \rho b_2 = \rho\ddot{u}_2$$

$$\frac{\partial \sigma_{13}}{\partial x_1} + \frac{\partial \sigma_{23}}{\partial x_2} + \frac{\partial \sigma_{33}}{\partial x_3} + \rho b_3 = \rho\ddot{u}_3 \qquad (1.17)$$

It is worth repeating that due to the symmetry property of the stress tensor, only six components are independent. That is, the number of independent stress components in the above are reduced because $\sigma_{12} = \sigma_{21}, \sigma_{13} = \sigma_{31}$ and $\sigma_{23} = \sigma_{32}$.

To complete our duality of treatments of stress and deformation, we need to consider the equations of motion with respect to the undeformed configuration. Specifically, the Cauchy equations of motion are in terms of the spatial partial derivatives and these must be changed to derivatives with respect to the undeformed state.

We begin with the body force $\hat{b}$, which is the body force per unit mass in the deformed configuration. Define the body force per unit mass in the undeformed state as $\hat{b}^o$ such that

$$b_i^o \rho^o dV^o \equiv b_i \rho dV$$

In view of mass conservation, $\rho^o dV^o = \rho dV$, we obtain

$$b_i^o = b_i$$

This body force relation is also valid for inertial forces. Note that the result would be different if the body force were defined per unit volume instead of per unit mass.

We change the spatial derivatives to material derivatives as follows:

$$\sum_j \frac{\partial \sigma_{ji}}{\partial x_j} = \sum_{j,p} \frac{\partial \sigma_{ji}}{\partial x_p^o}\frac{\partial x_p^o}{\partial x_j} = \sum_{j,p} \frac{\partial}{\partial x_p^o}\left(\sigma_{ji}\frac{\partial x_p^o}{\partial x_j}\right) - \sum_{j,p} \sigma_{ji}\frac{\partial^2 x_p^o}{\partial x_p^o \partial x_j}$$

$$= \sum_{j,p} \frac{\partial}{\partial x_p^o}\left(\sigma_{ji}\frac{\partial x_p^o}{\partial x_j}\right)$$

Noting that $\rho/\rho^o = J$ and $\partial J/\partial x_p^o = 0$, we obtain

$$\sum_{j,p} \frac{\partial}{\partial x_p^o}\left(\frac{\rho^o}{\rho}\sigma_{ji}\frac{\partial x_p^o}{\partial x_j}\right) + \rho^o b_i = \rho^o \ddot{u}_i$$

It remains now to replace the term in parentheses with a quantity that has meaning in the undeformed configuration. We would prefer to have a symmetric stress tensor, to that end let us replace the term in parentheses with

$$\sum_j \frac{\partial x_i}{\partial x_j^o}\sigma_{kj}^K \equiv \sum_j \frac{\rho^o}{\rho}\sigma_{ji}\frac{\partial x_p^o}{\partial x_j}$$

to give the equations of motion

$$\sum_{j,k} \frac{\partial}{\partial x_k^o}\left[\frac{\partial x_i}{\partial x_j^o}\sigma_{kj}^K\right] + \rho^o b_i^o = \rho^o \ddot{u}_i, \qquad \sigma_{ij}^K = \sigma_{ji}^K \tag{1.18}$$

These equations of motion are slightly more complicated than those using the Cauchy stress because they explicitly include the deformed state. Note also that each one of the equations of motion contains all the stress components and therefore are too cumbersome to write explicitly.

Arrived at this way, the stress σ_{pi}^K is just a convenience variable, which, because of the role it plays, can be interpreted as a stress. However, because of our previous development, we can give it physical interpretation in terms of tractions and forces acting on undeformed areas.

Virtual Work Formulation of Equilibrium

Let $u_i(x_i^o)$ be the displacement field which satisfies the equilibrium equations in V. On the surface A, the surface traction t_i is prescribed on A_t and the displacement on A_u. Consider a variation of displacement δu_i (we will sometimes call this the virtual displacement), then

$$\bar{u}_i = u_i + \delta u_i$$

where u_i satisfy the equilibrium equations and the given boundary conditions. Thus, δu_i must vanish over A_u but be arbitrary over A_t. Let δW_e be the virtual work done by the body force b_i and traction t_i; that is,

$$\delta W_e = \sum_i \int_V \rho b_i \delta u_i \delta V + \sum_i \int_{A_t} t_i \delta u_i \, dA + \sum_i \int_{A_u} t_i \delta u_i \, dA \qquad (1.19)$$

The last term is zero. We can also express this virtual work as

$$
\begin{aligned}
\delta W_e &= \sum_i \int_V \rho b_i \delta u_i dV + \sum_i \int_A \sigma_{ji} \delta u_i n_j \, dA \\
&= \sum_i \int_V \rho b_i \delta u_i dV + \sum_{i,j} \int_V \frac{\partial}{\partial x_j}(\sigma_{ji}\delta u_i)dV \\
&= \sum_{i,j} \int_V \left(\rho b_i + \frac{\partial \sigma_{ji}}{\partial x_j}\right)\delta u_i dV + \sum_{i,j} \int_V \sigma_{ji}\, \delta\!\left(\frac{\partial u_i}{\partial x_j}\right)dV \\
&= \sum_{i,j} \int_V \left(\rho b_i + \frac{\partial \sigma_{ji}}{\partial x_j}\right)\delta u_i dV + \sum_{i,j} \int_V \sigma_{ji}\delta\epsilon_{ji}dV
\end{aligned}
$$

where the last term was reduced using the decomposition of the deformation gradient into $\epsilon_{ij} + \omega_{ij}$ and noting that the contraction of the antisymmetric rotation with the symmetric stress is zero. Define the total virtual work as

$$\delta W = \delta W_e - \sum_{i,j} \int_V \sigma_{ji}\delta\epsilon_{ji}dV = \sum_{i,j} \int_V \left(\rho b_i + \frac{\partial \sigma_{ji}}{\partial x_j}\right)\delta u_i dV$$

These developments actually paralleled what was done in deriving the Cauchy stress equations of motion, therefore, we can look at it in one of two ways. First, because the term in parentheses is zero due to equilibrium, then we conclude that the total virtual work is zero. That is,

$$\delta W = \delta W_e - \sum_{i,j} \int_V \sigma_{ji}\delta\epsilon_{ji}dV = 0$$

On the other hand, if we say that the total virtual work is zero for any arbitrary virtual displacement δu_i, then we conclude that the term in parentheses is zero. That is, we obtain the equilibrium equations in terms of the Cauchy stress. The principle of virtual work states that a deformable body is in equilibrium if the total virtual work is zero for every independent kinematically admissible virtual displacement. We will interpret the symbol δ as meaning a *variation* and the above equation as a variational principle.

We would also like to write the virtual work expression in terms of the undeformed configuration. Following developments similar to the example in Section 1.3, we get the relation

$$\delta\epsilon_{mn} = \sum_{i,j} \frac{\partial x_i^o}{\partial x_m} \frac{\partial x_j^o}{\partial x_n} \delta E_{ij}$$

The relation between Cauchy and Kirchhoff stress is

$$\sigma_{mn} = \sum_{i,j} \frac{\rho}{\rho^o} \frac{\partial x_m}{\partial x_i^o} \frac{\partial x_n}{\partial x_j^o} \sigma_{ij}^K$$

and recalling that the deformed and undeformed volumes are related by $dV = dV^o \rho^o / \rho$, the internal virtual work term becomes

$$\sum_{m,n} \sigma_{mn} \delta\epsilon_{mn}\, dV = \sum_{p,q} \sigma_{pq}^K \delta E_{pq}\, dV^o$$

Hence the Cauchy stress / Eulerian (small) strain combination is energetically equivalent to the Kirchhoff stress / Lagrangian strain combination.

We are now in a position to write the virtual work form of equilibrium as

$$\delta W = \delta W_e - \sum_{m,n} \int_V \sigma_{mn} \delta\epsilon_{mn} dV = \delta W_e - \sum_{p,q} \int_{V^o} \sigma_{pq}^K \delta E_{pq} dV^o = 0 \qquad (1.20)$$

In contrast to the differential equations of motion, there are no added complications using the undeformed state as reference. It is useful to realize that, during a deformation, the reference state $t = 0$ could be any one of the previous equilibrium positions and not necessarily the original stress-free state. We will make use of this in our incremental formulation for the computer.

Stationary Principles

The virtual work form is completely general, but there are further developments that are more convenient to use in some circumstances. We now look at some of these developments.

I: Stationary Potential Energy

The internal virtual work is associated with the straining of the body and therefore we will use the representation

$$\delta U = \sum_{i,j} \int_V \sigma_{ij} \delta \epsilon_{ij} dV = \sum_{i,j} \int_{V^o} \sigma_{ij}^K \delta E_{ij} dV^o$$

and call U the strain energy of the body.

A system is *conservative* if the work done in moving the system around a closed path is zero. We say that the external force system is conservative if it can be obtained from a potential function. For example, for a set of discrete force, we have

$$P_i = -\frac{\partial V}{\partial u_i} \qquad \text{or} \qquad V = -\sum_i P_i u_i$$

where u_i is the displacement associated with the load P_i. The negative sign in the definition of V is arbitrary, but choosing it so gives us the interpretation of V as the capacity (or potential) to do work. The external work term now becomes

$$\delta W_e = \sum_i P_i \delta u_i = -\sum_i \frac{\partial V}{\partial u_i} \delta u_i = -\delta V$$

We get almost identical representations for conservative body forces and conservative traction distributions. The principle of virtual work can be rewritten as

$$\delta U + \delta V = 0 \qquad \text{or} \qquad \delta \Pi \equiv \delta[U + V] = 0 \qquad (1.21)$$

The term inside the brackets is called the *total potential energy*. This relation is called the *principle of stationary potential energy*. We may now restate the principle of virtual work as: *For a conservative system to be in equilibrium, the first-order variation in the total potential energy must vanish for every independent admissible virtual displacement.* Another way of stating this is that among all the displacement states of a conservative system that satisfy compatibility and the boundary constraints, those that also satisfy equilibrium make the potential energy stationary. In comparison to the conservation of energy theorem, this is much richer, because instead of one equation it leads to as many equations are there are degrees of freedom (independent displacements).

Example 1.11: Determine the equilibrium conditions for the nonlinear system shown in Figure 1.19.

Identify u, the resulting displacement at the point of application of the load, as the independent admissible displacement. The response of the nonlinear spring is shown in Figure 1.19(a): under tension it stiffens, under compression it shows softening. The virtual work for this spring is

$$\delta W = F \delta u = K[1 + \alpha u] u \delta u$$

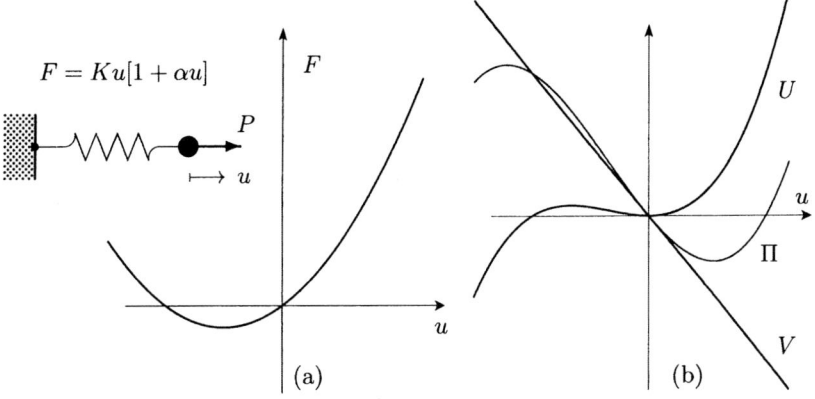

$$F = Ku[1 + \alpha u]$$

Figure 1.19: Equilibrium of a nonlinear spring. (a) Force/deflection relation. (b) Potential energy.

This is also the virtual strain energy δU. Integrating then gives

$$U = \tfrac{1}{2}Ku^2[1 + \tfrac{1}{3}\alpha u]$$

The potential of the applied force is

$$V = -Pu$$

The total potential energy of the system is, therefore,

$$\Pi = \tfrac{1}{2}Ku^2[1 + \tfrac{1}{3}\alpha u] - Pu$$

These terms are shown plotted in Figure 1.19(b) for different values of displacement u. It is apparent that Π can achieve two stationary values — a valley and a peak. The principle indicates that both occur at equilibrium positions.

A stationary potential energy requires that

$$\frac{\partial \Pi}{\partial u} = 0 \qquad \Rightarrow \qquad Ku[1 + \alpha u] - P = 0$$

We recognize this as the equilibrium balance between the external applied load P and the internal force F of the spring. If the spring were linear ($\alpha = 0$), it would reduce to the single equilibrium equation

$$Ku = P \qquad \text{or} \qquad u = P/K$$

In the nonlinear case, however, we have two possible positions

$$u = \frac{-1 \pm \sqrt{1 + 4\alpha P/K}}{2\alpha} \approx \frac{P}{K}, \frac{-1}{\alpha}$$

(The approximation is for slight nonlinearity when α is small.) The first is close to the linear equilibrium position, but what is the meaning of the second position? Furthermore, this second position corresponds to a negative displacement, which surely cannot happen because the load is positive. A hallmark of nonlinear systems is the possibility of multiple equilibrium positions. Indeed, looking at

Figure 1.19(a), we see that even at zero load ($F = 0$) there is the equal possibility of two deflections.

It is hard to imagine an "ordinary" material or spring behaving in this way. However, engineering structures do behave this way, and we will cover many such cases in the later chapters. Many of the structured materials (an example is corrugated cardboard) also behave this way.

II: Hamilton's Principle

To apply the idea of virtual work to dynamic problems, we need to account for the presence of inertia forces, and the fact that all quantities are functions of time. We will take care of the former by use of D'Alembert's principle and the latter by time averaging.

D'Alembert's principle converts a dynamic problem into an equivalent problem of static equilibrium by treating the inertia as a body force. That is, the total body force is comprised of $\rho b_i \rightarrow (\rho b_i - \rho \ddot{u}_i)$ where $-\rho \ddot{u}_i dV$ is the inertia force of an infinitesimal volume. The virtual work of the body force is

$$\delta W^b = \sum_i \int_V \rho b_i \delta u_i \, dV - \sum_i \int_V \rho \ddot{u}_i \delta u_i \, dV$$

In writing this relation, we suppose that the performance of the virtual displacement consumes no time; that is, the real motion of the system is stopped while the virtual displacement is performed. Consequently, the time variable is conceived to remain constant while the virtual displacement is executed.

We will concentrate on the inertia term in the above virtual work expression. Noting that

$$\sum_i \frac{d}{dt}(\dot{u}_i \, \delta u_i) = \sum_i \ddot{u}_i \, \delta u_i + \sum_i \dot{u}_i \, \delta \dot{u}_i$$

the inertia term can be written as

$$\sum_i \int \rho \ddot{u}_i \, \delta u_i dV = \sum_i \int \rho \frac{d}{dt}(\dot{u}_i \, \delta u_i) dV - \sum_i \int \rho \dot{u}_i \, \delta \dot{u}_i dV$$

Introducing the concept of *kinetic energy*, defined as

$$T \equiv \tfrac{1}{2} \sum_i \int_V \rho \dot{u}_i \dot{u}_i dV \qquad \text{such that} \qquad \delta T \equiv \sum_i \int_V \rho \dot{u}_i \, \delta \dot{u}_i \, dV$$

the principle of virtual work becomes

$$\delta W = \delta W^s + \delta W^b - \delta U + \delta T - \frac{d}{dt} \sum_i \int_V (\dot{u}_i \delta u_i) \rho dV = 0$$

It remains now to treat the last integral term.

Hamilton refined the concept that a motion can be viewed as a path in configuration space; he showed that, for a system with given configurations at times

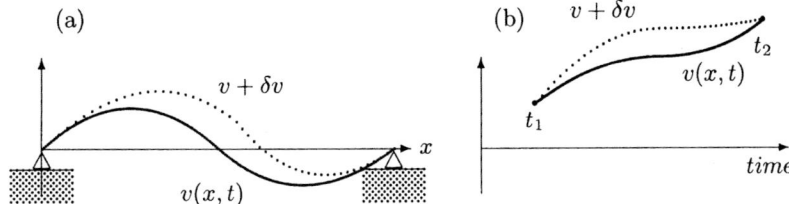

Figure 1.20: Hamilton's configuration space for a pinned/pinned beam. (a) Varied path in space. (b) Varied path in time.

t_1 and t_2, of all the possible configurations between these two times the actual that occurs satisfies a stationary principle. This essentially geometric idea is illustrated in Figure 1.20, where a varied path in both space and time are shown for a beam. At a particular instant in time, t, the beam has the deformed shape of the solid line shown in (a). We can imagine a varied deformation shape shown as the dotted line in (a), but the end constraints at $x = 0$ and L are not varied. Now consider a particular point on the beam and plot its position over time; this gives the solid line of (b) and it represents the "Newtonian path" of the point. The addition of the virtual displacement $\delta v(x,t)$ gives a path that may look like the dotted line in (b); again there are no variations at the extreme times t_1 and t_2.

Hamilton disposed of the last term in the virtual work relation by integrating the equation over time between the limits of the 1 and 2 configurations. The last term is a time derivative and so may be integrated explicitly to give

$$-\sum_i \int \dot{u}_i \delta u_i \, \rho dV \Big|_{t_1}^{t_2}$$

By stipulation, the configuration has no variations at the extreme times and hence the term is zero. Consequently, the virtual work relation becomes

$$\int_{t_1}^{t_2} [\delta W^s + \delta W^b + \delta T - \delta U]dt = 0 \qquad (1.22)$$

This equation is generally known as the *extended Hamilton's principle*. An important feature of this principle is that it is formulated without reference to any particular system of coordinates. That is, it holds true for constrained as well as generalized coordinates.

In the special case when the applied loads, both body forces and surface tractions, can be derived from a scalar potential function V, the variations become complete variations and we can write

$$\delta \int_{t_1}^{t_2} [T - (U + V)]dt = 0 \qquad (1.23)$$

This equation is Hamilton's principle. Hamilton's principle is a variational principle and it is as general as Newton's second law. When we apply this principle, we need to identify two classes of boundary conditions, called *essential* and *natural* boundary conditions, respectively. The essential boundary conditions are also called *geometric* boundary conditions because they correspond to prescribed displacements and rotations; the geometric conditions must be rigorously imposed. The natural boundary conditions are associated with the applied loads and are implicitly contained in the variational principle.

Example 1.12: Consider a particle of mass M subjected to a force P. Show that Newton's law governing the motion of the particle can be recovered from Hamilton's principle.

The kinetic energy, strain energy, and potential of the applied force are given by, respectively,
$$T = \tfrac{1}{2}M\dot{u}^2, \qquad U = 0, \qquad V = -Pu$$
Hamilton's principle for the particle is
$$\delta \int_{t_1}^{t_2} [T - (U + V)]dt = \delta \int_{t_1}^{t_2} [\tfrac{1}{2}M\dot{u}^2 - (0 - Pu)]dt = \int_{t_1}^{t_2} [m\dot{u}\,\delta\dot{u} + P\delta u]dt = 0$$
Noting that
$$\dot{u}\,\delta\dot{u}\,dt = \dot{u}\frac{d(\delta u)}{dt}dt = \dot{u}\,d(\delta u)$$
then we can integrate the first term in the integral by parts to give
$$M\dot{u}\delta u\Big|_{t_1}^{t_2} + \int_{t_1}^{t_2} [-M\ddot{u} + P]\delta u\,dt = 0$$

By stipulation, the variation δu at the times t_1 and t_2 are zero, then the first term is also zero. Since the time limits of integration are arbitrary, and since the variations between these limits can be arbitrary, then we conclude that the integrand must be zero. This gives
$$-M\ddot{u} + P = 0 \qquad \text{or} \qquad P = M\ddot{u}$$

This is Newton's second law.

Example 1.13: Use Hamilton's principle to derive the equations of motion of a rod taking the lateral contraction into account. Assume the material behavior is linear.

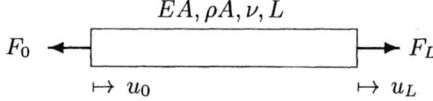

Figure 1.21: Rod with end loads.

As the rod deforms longitudinally, it also contracts due to the Poisson ratio effect. Thus, each particle of the rod also has a transverse component of velocity. We will now add this to the kinetic energy term in order to have a more accurate accounting of the energy. The transverse strain is related to the axial strain by $\epsilon_t = -\nu\epsilon$, therefore, the transverse velocity is given by

$$\dot{u}_t = r\dot{\epsilon}_t = -\nu r\dot{\epsilon} = -\nu r\frac{\partial\dot{u}}{\partial x}$$

In this, we have assumed that the velocity is proportional to the distance r from the centroid of the cross-section. The total kinetic energy of the rod is readily found to be

$$T = \int_V \tfrac{1}{2}\rho[\dot{u}(x,t)^2 + \dot{u}_t(x,t)^2]dV = \tfrac{1}{2}\int_0^L\int_A \rho[\dot{u}^2 + \nu^2 r^2(\frac{\partial\dot{u}}{\partial x})^2]dA\,dx$$

Because $\dot{u}$ is a function only of x (and time), then we can perform the integration with respect to the cross-section to give

$$T = \tfrac{1}{2}\int_0^L [\rho A\dot{u}^2 + \nu^2\rho J(\frac{\partial\dot{u}}{\partial x})^2]dx\,, \qquad J \equiv \int_A r^2\,dA$$

where J is the polar moment of area. The total strain energy is given by

$$U = \tfrac{1}{2}\int_V \sigma\epsilon\,dV = \tfrac{1}{2}\int_V E\epsilon^2\,dV = \int_0^L EA(\frac{\partial u}{\partial x})^2 dx$$

The final term we need is the potential of the applied forces. Assume there are no distributed loads — only end loads as shown in Figure 1.21 — then we have

$$V = -(-F_0 u_0 + F_L u_L) = F_0 u_0 - F_L u_L$$

Substitute these energies into Hamilton's principle to get

$$\delta\int_{t_1}^{t_2}\left[\tfrac{1}{2}\int_0^L [\rho A\dot{u}^2 + \nu^2\rho J(\frac{\partial\dot{u}}{\partial x})^2]dx - \tfrac{1}{2}\int_0^L EA(\frac{\partial u}{\partial x})^2 dx - F_0 u_0 + F_L u_L\right]dt = 0$$

Take the variation inside

$$\int_{t_1}^{t_2}\left[\int_0^L \left\{\rho A\dot{u}\delta\dot{u} + \nu^2\rho J(\frac{\partial\dot{u}}{\partial x})(\frac{\partial\delta\dot{u}}{\partial x}) - EA(\frac{\partial u}{\partial x})(\frac{\partial\delta u}{\partial x})\right\}dx - F_0\delta u_0 + F_L\delta u_L\right]dt = 0$$

We now use integration by parts in order to have all terms multiplied by a common variation δu. For example, the time integration of the first term can be rewritten as

$$\int_{t_1}^{t_2}\rho A\dot{u}\,\delta\dot{u}\,dt = \int_{t_1}^{t_2}\rho A\dot{u}\,d(\delta u) = \rho A\dot{u}\,\delta u\Big|_{t_1}^{t_2} - \int_{t_1}^{t_2}\rho A\ddot{u}\,\delta u\,dt$$

By stipulation, there is no variation of the configuration at times t_1 and t_2, hence the term evaluated at these limits is zero. For the third term, we have for the space integration

$$\int_0^L EA(\frac{\partial u}{\partial x})(\frac{\partial\delta u}{\partial x})dx = \int_0^L EA(\frac{\partial u}{\partial x})d(\delta u) = EA\frac{\partial u}{\partial x}\,\delta u\Big|_0^L - \int_0^L EA\frac{\partial^2 u}{\partial x^2}dx\,\delta u$$

The middle term requires both space and time integration and this is done as a combination of the previous two terms. The result is

$$\int_{t_1}^{t_2} \int_0^L \nu^2 \rho J (\frac{\partial \dot{u}}{\partial x})(\frac{\partial \delta \dot{u}}{\partial x}) dx\, dt = - \int_{t_1}^{t_2} \nu^2 \rho J \frac{\partial \ddot{u}}{\partial x}\Big|_0^L \delta u\, dt + \int_{t_1}^{t_2} \int_0^L \nu^2 \rho J \frac{\partial^2 \ddot{u}}{\partial x^2} \delta u\, dx\, dt$$

Add all these terms together to get

$$\int_{t_1}^{t_2} \Big[\int_0^L \Big[EA\frac{\partial^2 u}{\partial x^2} + \nu^2 \rho J \frac{\partial^2 \ddot{u}}{\partial x^2} - \rho A \ddot{u} \Big] \delta u\, dx - \Big[EA\frac{\partial u}{\partial x} + \nu^2 \rho J \frac{\partial \ddot{u}}{\partial x} - F \Big] \delta u \Big|_0^L \Big] dt = 0$$

Because the time limits and space limits in the integrations are arbitrary, then the first integrand is zero, giving the governing differential equation as

$$EA\frac{\partial^2 u}{\partial x^2} + \nu^2 \rho J \frac{\partial^2 \ddot{u}}{\partial x^2} - \rho A \frac{\partial^2 u}{\partial t^2} = 0 \tag{1.24}$$

If either Poisson's ratio or the polar moment of area is negligibly small, then we recover the elementary rod theory [22].

The remaining terms must also be zero and thereby specify the boundary conditions; at either end of the rod, we specify

$$u \qquad \text{or} \qquad F = EA\frac{\partial u}{\partial x} + \nu^2 \rho J \frac{\partial \ddot{u}}{\partial x} \tag{1.25}$$

The natural boundary condition is a rather surprising result. We may recognize the first term of it as arising from the linear elastic assumption of

$$\sigma = \frac{F}{A} = E\epsilon = E\frac{\partial u}{\partial x} \qquad \text{or} \qquad F = EA\frac{\partial u}{\partial x}$$

Does the presence of the second term mean that this relation is no longer valid? A very important point in the variational approach to problems is that both the differential equation and the associated boundary conditions are implied in the potential. Because we started with an approximation for the potential function, we derived a governing differential equation and a set of boundary conditions most consistent with that approximation. We can imagine, therefore, proposing a different potential and having a natural boundary condition that is actually the same as the axial force relation. In fact, such a situation arises in the higher-order rod theory referred to as *Mindlin-Herrmann rod theory* [23]. The modified rod theory just developed is referred to as *Love's rod theory*.

III: Lagrange's Equation

Hamilton's principle provides a complete formulation of a dynamical problem; however, to obtain solutions to some problems, the Hamilton integral formulation must be converted into one or more differential equations of motion in a manner as just shown in the examples. For computer solution, these must be further reduced to equations using discrete unknowns. That is, we introduce some generalized coordinates (or degrees of freedom with the constrained degrees removed). At present, we will not be explicit about which coordinates we are considering but accept that we can write a function as

$$u = u(u_1, u_2, \ldots, u_N)$$

where u_i are the generalized coordinates. We get these generalized coordinates by the imposition of *holonomic* constraints — the constraints are geometric of the form $f_i(u_1, u_2, \ldots, u_N, t) = 0$ and do not depend on the velocities.

The time derivative of such a function is

$$\dot{u} = \sum_{j=1}^{N} \frac{\partial u}{\partial u_j} \dot{u}_j$$

Consequently, we see that the kinetic energy is a function of the following form:

$$T = T(u_1, u_2, \ldots, u_N; \dot{u}_1, \dot{u}_2, \ldots, \dot{u}_N)$$

The variation in the kinetic energy is given by

$$\delta T = \sum_{j=1}^{N} \frac{\partial T}{\partial u_j} \delta u_j + \sum_{j=1}^{N} \frac{\partial T}{\partial \dot{u}_j} \delta \dot{u}_j$$

We can use integration by parts on the second term to obtain

$$\int_{t_1}^{t_2} \delta T \, dt = \int_{t_1}^{t_2} \sum_{j=1}^{N} \left\{ \frac{\partial T}{\partial u_j} - \frac{d}{dt} \frac{\partial T}{\partial \dot{u}_j} \right\} \delta u_j \, dt$$

where we used the fact that the variations at the extreme times are zero.

The total potential of the conservative forces is a function of the form

$$U + V = \Pi = \Pi(u_1, u_2, \ldots, u_N)$$

and its variation is given by

$$\delta \Pi = \sum_{j=1}^{N} \frac{\partial \Pi}{\partial u_j} \delta u_j$$

Additionally, we have that the virtual work of the non-conservative forces is given by

$$\delta W^d = \sum_{j=1}^{N} Q_j \delta u_j$$

Hamilton's extended principle now takes the form

$$\int_{t_1}^{t_2} \sum_{j=1}^{N} \left\{ -\frac{d}{dt} \left(\frac{\partial T}{\partial \dot{u}_j} \right) + \frac{\partial T}{\partial u_j} - \frac{\partial (U+V)}{\partial u_j} + Q_j \right\} \delta u_j dt = 0$$

Because the virtual displacements δu_j are independent and arbitrary, and because the time limits are arbitrary, then each integrand is zero. This leads to the *Lagrange's equation of motion*:

$$\mathcal{F}_i \equiv \frac{d}{dt} \left(\frac{\partial T}{\partial \dot{u}_i} \right) - \frac{\partial T}{\partial u_i} + \frac{\partial}{\partial u_i} \left(U + V \right) - Q_i = 0 \qquad (1.26)$$

for $i = 1, 2, \ldots, N$. The expression, $\mathcal{F}_i = 0$, is our statement of (dynamic) equilibrium. It is apparent from the Lagrange's equation that, if the system is not in motion, then we recover the principle of stationary potential energy expressed in terms of generalized coordinates.

We emphasize that the transition from Hamilton's principle to Lagrange's equation was possible only by identifying u_i as generalized coordinates. That is, Hamilton's principle holds true for constrained as well as generalized coordinates but Lagrange's equation is valid only for the latter. A nice historical discussion of Hamilton's principle and Lagrange's equation is given in Reference [84].

Example 1.14:   A rigid pendulum is constrained by a linear spring as shown in Figure 1.22. Determine the equations of motion.

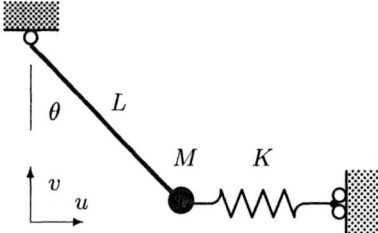

Figure 1.22: Elastically restrained pendulum.

The spring has its natural length when the mass is at its lowest position, hence the strain energy is

$$U = \tfrac{1}{2}Ku^2 = \tfrac{1}{2}K[L\sin\theta]^2$$

Note that the massless spring moves so as to be always horizontal. The mass has a velocity component $L\dot{\theta}$ about the pendulum axis, hence the kinetic energy is

$$T = \tfrac{1}{2}M[\dot{u}^2 + \dot{v}^2] = \tfrac{1}{2}M[L\dot{\theta}]^2$$

Additionally, gravity acts as a conservative force opposite to the coordinate direction v and hence has the potential

$$V = Mgv = MgL[1 - \cos\theta]$$

The total potential is given by

$$\Pi = U + V = \tfrac{1}{2}K[L\sin\theta]^2 + MgL[1 - \cos\theta]$$

Substituting these into Lagrange's equation

$$\mathcal{F}_\theta = \frac{d}{dt}\{\frac{\partial T}{\partial\dot{\theta}}\} - \{\frac{\partial T}{\partial\theta}\} + \{\frac{\partial(U+V)}{\partial\theta}\} - Q_\theta = 0$$

(where we have identified θ as the generalized coordinate) leads to

$$\frac{d}{dt}\{ML^2\dot{\theta}\} - 0 + KL^2\cos\theta\sin\theta + MgL\sin\theta - Q_\theta = 0$$

We can write this as

$$ML\ddot{\theta} + KL\cos\theta\sin\theta + Mg\sin\theta = \frac{1}{L}Q_\theta$$

Even though the spring is linear, the resulting equation is highly nonlinear because the rotational angle can be large. This is typical of the nonlinear systems we will be considering in the next chapters.

Example 1.15: Specialize Lagrange's equation to the case when the motions are small.

Consider small motions about an equilibrium position defined by $u_i = 0$ for all i. Perform a Taylor series expansion on the strain energy function to get

$$U(u_1, u_2, \ldots) = U(0) + \sum_i \left.\frac{\partial U}{\partial u_i}\right|_0 u_i + \tfrac{1}{2}\sum_i\sum_j \left.\frac{\partial^2 U}{\partial u_i \partial u_j}\right|_0 u_i u_j + \ldots$$

The first term in this expansion is irrelevant and the second term is zero since, by assumption, the origin is an equilibrium position. We therefore have the representation of the strain energy as

$$U(u_1, u_2, \ldots) \approx \tfrac{1}{2}\sum_i\sum_j K_{ij}u_i u_j, \qquad K_{ij} \equiv \left.\frac{\partial^2 U}{\partial u_i \partial u_j}\right|_0$$

We can do a similar expansion for the kinetic energy; in this case, however, we also assume that the system is linear in such a way that T is a function only of the velocities $\dot{u}_j$. We get

$$T(\dot{u}_1, \dot{u}_2, \ldots) \approx \tfrac{1}{2}\sum_i\sum_j M_{ij}\dot{u}_i\dot{u}_j, \qquad M_{ij} \equiv \left.\frac{\partial^2 T}{\partial \dot{u}_i \partial \dot{u}_j}\right|_0$$

The potential of the conservative forces also has an expansion similar to that for U, but we retain only the linear terms in u_j such that

$$V = -\sum_j P_j u_j, \qquad P_j \equiv -\left.\frac{\partial V}{\partial u_j}\right|_0$$

Finally, assume that the non-conservative forces are of the viscous type such that the virtual work is

$$\delta W^d = Q^d \delta u = -c\dot{u}\,\delta u$$

This suggests the introduction of a function analogous to the potential for the conservative forces

$$Q_j^d = -\frac{\partial D}{\partial \dot{u}_j} \qquad \text{where} \qquad D = D(\dot{u}_1, \dot{u}_2, \ldots, \dot{u}_N)$$

For small motions, we get

$$D(\dot{u}_1, \dot{u}_2, \ldots, \dot{u}_N) \approx \tfrac{1}{2}\sum_i\sum_j C_{ij}\dot{u}_i\dot{u}_j, \qquad C_{ij} \equiv \left.\frac{\partial^2 D}{\partial \dot{u}_i \partial \dot{u}_j}\right|_0$$

The function D is called the *Rayleigh dissipation function.*

Substitute these forms for U, V, T, and D into Lagrange's equation to get

$$\sum_j \{K_{ij}u_j + C_{ij}\dot{u}_j + M_{ij}\ddot{u}_j\} = P_i, \qquad i = 1, 2, \ldots, N$$

This is put in the familiar matrix form as

$$[\, K \,]\{u\} + [\, C \,]\{\dot{u}\} + [\, M \,]\{\ddot{u}\} = \{P\} \qquad (1.27)$$

By comparison with the one-degree-of-freedom case, we have the meaning of $[\, K \,]$, $[\, M \,]$, and $[\, C \,]$ as the (generalized) structural stiffness, mass, and damping matrices, respectively. As yet, we have not said how the actual coefficients can be obtained or the actual meaning of the generalized coordinates; this is the subject of a later section, and the next few chapters.

Discussion

In the subsequent chapters when we need to derive governing equations, we will use Hamilton's principle for continuous systems and Lagrange's equation for discrete systems. Sometimes, as in element formulations, we will deal directly with the principle of virtual work.

Keep in mind that these governing equations are actually just different forms of (dynamic) equilibrium and are not sufficient in themselves to solve problems. To complete the formulation of a problem, we must also make the material (or constitutive) behavior explicit as was done in the spring and rod examples. This is the topic of the next two sections. In stating the various forms of the governing equations, we often referred to conservative systems or systems with a potential. We will also make this explicit in regards to the material behavior.

1.6   Material Behavior

The concepts of stress, on the one hand, and strain, on the other hand, were developed independently of each other and apart from the assumption of a continuum, the development placed no restrictions on the material. That is, the concepts developed so far apply whether the material is elastic or plastic, isotropic or anisotropic. Indeed, they apply even if the material is a fluid. This section makes the material behavior explicit.

Types of Materials

Similarly shaped bodies with similarly applied loads may have different deformation responses. This is due to the internal constitution of their matter. A constitutive equation is an experimentally determined relation between the applied loads and the deformation response for a particular material. There is a wide range of materials available, but engineering structures are made from relatively few. (Reference [28] gives an enjoyable account of a wide variety of structures and the types of materials used for their construction.) Furthermore, most structures are designed to sustain only elastic loads and are fabricated from isotropic and homogeneous materials (steel and aluminum, for instance); the literature is abundant with data on their material constants. Analysis of structural components fabricated with composite materials, on the other hand, requires the use of anisotropic elasticity theory. Analysis of problems in metal-forming and ductile fracture are based on the inelastic and plastic responses of materials, particularly those under large deformation. Polymeric materials require knowledge of their time-dependent stress relaxation and creep properties.

To simplify the relations in this section, only small deformations will be considered. One further restraint is that the material is assumed homogeneous, that is, irrespective of specimen size, the specimen will have the same material response. This will not preclude study of inhomogeneous structures — it is only the local material (or small-scale) behavior that is assumed homogeneous.

In broad terms, failure refers to any action that leads to an inability on the part of a structure to function as intended. Common modes of failure include permanent deformation (yielding), fracture, buckling, creep, and fatigue. The successful use of a material in any application requires assurance that it will function safely. Therefore, the design process must involve steps where the predicted in-service stresses, strains, and deformations are limited to appropriate levels using failure criteria based on experimental data. At present, formulation of failure theories for particular materials is an area of widespread research.

While the behavior of a real material is very complicated, nevertheless, most structural materials can be divided into certain classes and four of the main classes will be considered here. In characterizing materials, it is not the force and the displacement that are used but the stress and the strain; this is reasonable since both of these concepts are local in nature. Historically, then, it was common to obtain "Stress/Strain diagrams," but with more specialized materials being used in structures, this approach must be extended to include time dependency. Therefore, in plotting the responses to be shown next, it is assumed that a load/unload cycle is imposed; other cycles could also be used.

I: Elastic Material

If there is a one to one relation between the stress and strain and, on unloading, all the strain is instantaneously recovered, then the material is said to be elastic. Most structural materials in common use are adequately described by this type of material. We write the relation as

$$\epsilon = f(\sigma)$$

which, of course, may be nonlinear. In linear elasticity, the deformation is a linear function of the stress

$$\sigma = E\epsilon$$

where E is called the Young's modulus.

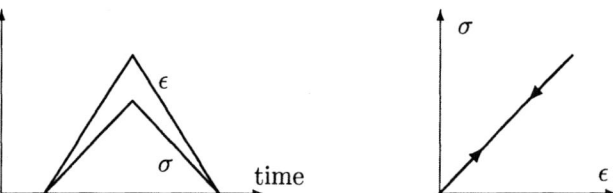

Figure 1.23: Elastic behavior.

II: Elastic/Plastic Material

For some materials, it is found that beyond a certain stress level large deformations occur for small increments in load, and furthermore, much of the deformation is not recovered when the load is removed. On the load/unload cycle, if $\sigma > \sigma_Y$ (σ_Y = yield stress), this material cannot recover the deformation caused after yielding. This remaining deformation is called the permanent or plastic strain. The amount occuring depends on the geometry of the problem.

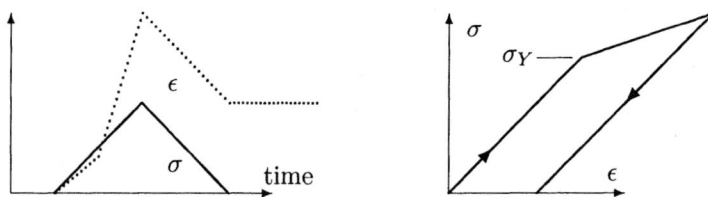

Figure 1.24: Stress/strain cycle for elastic-plastic material behaviors.

The total strain is considered as composed of elastic and plastic parts

$$\epsilon = \epsilon^e + \epsilon^p$$

and constitutive relations are written for the separate parts. Since plastic flow is an incremental process, then the constitutive relations must be written in incremental form

$$Elastic: \qquad d\epsilon = \frac{1}{E}d\sigma$$

$$Plastic: \qquad d\epsilon^p = \lambda\sigma$$

This is the Levy-Mises theory, and it says that the plastic strain increments depend on the current stress state, and also that the principal directions of the increments coincide with those of the stress. The parameter λ is determined experimentally, but unlike E, say, it is not constant but depends on the level of stress.

In the solution of an actual problem, the increments of deformation are obtained for each increment of load, then the total deformation is obtained by summing all the increments over the loading history.

The maximum stress, referred to as the yield stress, beyond which the loading and unloading curves differ, is the elastic limit of the material. The unloading curve is usually parallel to the elastic loading curve. The difference in yield stresses between the virgin and plastically deformed material is caused by its strain-hardening response.

III: Viscoelastic Material

The mechanical properties of all solids are affected to varying degrees by the temperature and rate of deformation. Although such effects are not measurable at low temperature, they become noticeable at high temperatures relative to the glassy transition temperature for polymers or the melting temperature for metals. Above the glassy transition temperature, many amorphous polymers flow like a Newtonian fluid and are referred to as viscoelastic materials.

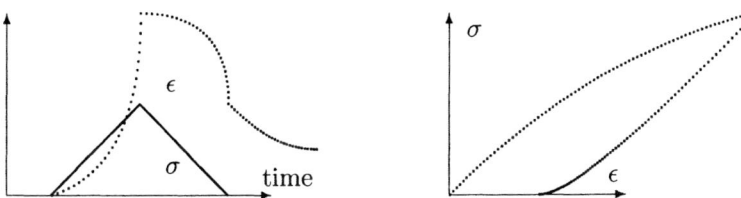

Figure 1.25: Viscoelastic behavior.

That is, they exhibit a combination of solid and fluid effects. If the solid is elastic, then the deformations are proportional to the stress, and if the fluid is Newtonian, then the deformation rate is proportional to the stress. Two very simple viscoelastic materials can be represented by a combination of a spring and dashpot. The Maxwell model has the spring in series with the dashpot, while the Kelvin model has the spring in parallel with the dashpot. These materials are time dependent and the relation must be written in terms of the time derivatives; the standard linear solid, for example, is described by

$$\frac{d\epsilon}{dt} + \frac{E_2}{\eta}\epsilon = \frac{1}{E}\frac{d\sigma}{dt} + \frac{1}{\eta}[1 + \frac{E_2}{E_1}]\sigma$$

In addition to the time dependence, there is also a dependence on more material coefficients.

This material exhibits a residual strain after the load is removed, but over time it recovers this strain to leave the material in its original state.

IV: Structural Materials

The structural properties of some base materials can be improved by combining them with other materials to form composite materials. A historical survey is given in Reference [28]. For a sandwich material such as plywood, the cross-grain weakness of the wood is improved by alternating directions of the lamina. Concrete is an example of particulate materials with cement as the bonding agent. Reinforced concrete uses steel bars to improve the tensile strength of the base concrete. Sheets of glass are stiff but prone to brittle failure because of small defects. Glass-fiber-reinforced composites combine the stiffness of glass in the form of fibers with the bonding of a matrix material such as epoxy; in this way, a defect in individual fibers does affect the overall behavior.

It is possible to analyze these materials through their constituent behavior, but for structural analyses this is very rarely done because it is computationally prohibitive. Furthermore, as in the case of concrete, the structural properties depend significantly on the particulars of the bonding behavior and this is rather difficult to predict. It is more usual to treat these materials as homogeneous materials and establish average properties based on a representative volume. This is not an area we will pursue here; suffice it to say, that the variety of behaviors from these structured materials is much richer than from traditional homogeneous materials and the analytical description must be adequate for the purpose. For example, these materials are more likely to be anisotropic (have different properties in different directions).

Failure Criteria

The basic assumption underlying all yield criteria is that failure is predicted to occur at a particular point in a material only when the value of a certain measure of stress reaches a critical value. The critical level of the selected measure is obtained experimentally, usually by a uniaxial test. The most important, and widely applied, failure criteria have been combined stress yield theories for isotropic metals and we discuss some of the more well-known below. Such criteria predict yielding in multiaxial states of stress using uniaxial yield stress as the only input parameter.

We can characterize the stress state of a body in terms of the principal stresses because they are the extremum values. In the following, we assume the principal stresses are ordered according to $\sigma_3 < \sigma_2 < \sigma_1$.

I: Maximum Principal Stress Theory [Rankine]

The maximum principal stress theory predicts that failure will occur at a point in a material when the maximum principal normal stress becomes equal to or exceeds the uniaxial failure stress for that material. That is, this criterion predicts failure to occur at a point when

$$\sigma_1 \geq \sigma_f \qquad \text{or} \qquad \sigma_3 \leq -\sigma_f$$

where σ_f is the magnitude of the uniaxial failure stress in tension and an equal but opposite failure stress is assumed in compression.

This theory provides a generally poor prediction of yield onset for most metals and is not typically used for materials that behave in a ductile fashion. It has, however, been applied successfully to predict fracture of some brittle materials in multiaxial stress states.

II: Maximum Shear Stress Theory [Coulomb, Tresca, Guest]

The maximum shear stress theory states that yield is predicted to occur at a point in a material when the absolute maximum shear stress at that point becomes equal to or exceeds the magnitude of the maximum shear stress at yield in a uniaxial tensile test of the same material. Based on the stress transformation equations, this critical yielding value of the maximum shear stress during a uniaxial test is $\tau_Y = \frac{1}{2}\sigma_Y$. Thus, yield failure is predicted to occur in a multiaxial state of stress if

$$\tau_{max} \geq \tfrac{1}{2}\sigma_Y$$

In terms of the principal stresses, this can be recast into the form

$$\sigma_1 - \sigma_3 \geq \sigma_Y$$

This criterion is shown in Figure 1.26.

This criterion will not predict yielding to occur under any level of applied hydrostatic loading (because the shear is zero). Experimental evidence for other states of stress attest that the maximum shear stress criterion is a good theory for predicting yield failure of ductile metals.

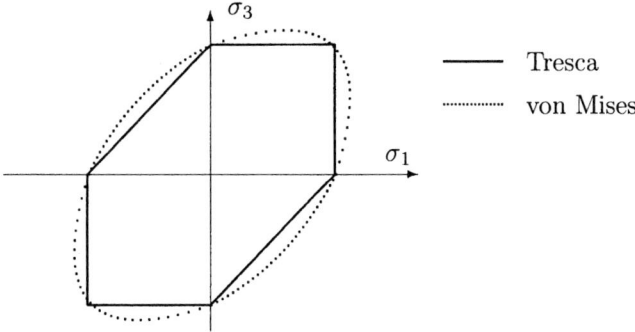

Figure 1.26: Tresca and von Mises yield criteria.

III: Energy of Distortion Theory [Huber, von Mises, Henchy]

The total strain energy of an isotropic linear elastic material is often divided into two parts: the dilatation energy, which is associated with change of volume under a mean hydrostatic pressure, and the distortion energy, which is associated with change in shape. It can be shown that the distortion energy takes the following form in terms of the principal stresses:

$$U_d = \frac{1+\nu}{6E}[(\sigma_1 - \sigma_2)^2 + (\sigma_2 - \sigma_3)^2 + (\sigma_3 - \sigma_1)^2]$$

The energy of distortion theory states that yield is predicted to occur at a point in a material when the distortion energy at that point becomes equal to or exceeds the magnitude of the distortion energy at yield in a uniaxial tensile test of the same material. The critical yielding value of the distortion energy in a uniaxial test is

$$U_d = \frac{1+\nu}{6E}[2\sigma_Y^2]$$

Combining these results, the statement of the energy of distortion theory is that yield will occur if

$$(\sigma_1 - \sigma)^2 + (\sigma_2 - \sigma_3)^2 + (\sigma_3 - \sigma_1)^2 \geq 2\sigma_Y^2$$

The behavior of this is shown in Figure 1.26. The predictions are close to those of the maximum shear stress criterion, but it has the slight advantage of using a single function for any state of stress.

1.7 Elastic Constitutive Relations

Because of our interest in thin-walled structures, the elastic material is the one of most relevance to us. Furthermore, the situations that arise are usually of the type of large rotations but small strains and we utilize this to make approximations.

Hyperelastic Materials

Consider a small volume of material under the action of applied loads on its surface. One assumption about elastic behavior is that: *The work done by the applied forces is transformed completely into potential energy. Furthermore, the potential energy is stored entirely in the form of strain energy.* That the work is transformed into potential energy means that it is completely recoverable and the material system is conservative.

We will use Lagrangian variables. The increment of work done on the small volume is

$$dW_e = \int_{V^o} \left[\sum_{i,j} \sigma_{ij}^K dE_{ij}\right] dV^o$$

The potential is comprised entirely of the strain energy U; the increment of strain energy is

$$dU = dU(E_{ij}) = \int_{V^o} \Big[\sum_{i,j} \frac{\partial U}{\partial E_{ij}} dE_{ij} \Big] dV^o$$

From the hypothesis, we can equate dW_e and dU to give

$$\int_{V^o} \Big[\sum_{i,j} \sigma_{ij}^k dE_{ij} \Big] dV^o = \int_{V^o} \Big[\sum_{i,j} \frac{\partial U}{\partial E_{ij}} dE_{ij} \Big] dV^o$$

Because the volume is arbitrary, the integrands must be equal, hence we have

$$\sigma_{ij}^K = \frac{\partial U}{\partial E_{ij}} \tag{1.28}$$

A material described by this relation is called *hyperelastic*. Note that it is valid for large deformations and for anisotropic materials; rather than develop this general case, we will look at each of these separately.

Nonlinear Isotropic Materials

Many structural materials (steel and aluminum, for example) are essentially isotropic in that the stiffness of a sheet is about the same in all directions. We will use this to simplify the elastic relation.

Because the material is isotropic, the strain energy is a function of the strain invariants only; that is, $U = U(I_1, I_2, I_3)$, where the definition of the invariants from Equation (1.4) is

$$I_1 = \sum_k E_{kk}\,, \qquad I_2 = \tfrac{1}{2} I_1^2 - \tfrac{1}{2} \sum_k E_{ik} E_{ik}\,, \qquad I_3 = \det[E_{ij}]$$

The stress/strain relation becomes

$$\sigma_{ij}^K = \frac{\partial U}{\partial I_1} \frac{\partial I_1}{\partial E_{ij}} + \frac{\partial U}{\partial I_2} \frac{\partial I_2}{\partial E_{ij}} + \frac{\partial U}{\partial I_3} \frac{\partial I_3}{\partial E_{ij}}$$

The various derivatives of the invariants with respect to strain are

$$\frac{\partial I_1}{\partial E_{ij}} = \delta_{ij}\,, \qquad \frac{\partial I_2}{\partial E_{ij}} = I_1 \delta_{ij} - E_{ij}\,, \qquad \frac{\partial I_3}{\partial E_{ij}} = I_2 \delta_{ij} - I_1 E_{ij} + \sum_k E_{ik} E_{kj}$$

On substituting these into the above constitutive relation, and rearranging, we get

$$\sigma_{ij}^K = \beta_o \delta_{ij} + \beta_1 E_{ij} + \beta_2 \sum_p E_{ip} E_{pj} \tag{1.29}$$

which is a nice compact relation. The coefficients have the explicit representation

$$
\beta_o = \frac{\partial U}{\partial I_1} + \frac{\partial U}{\partial I_2}I_1 + \frac{\partial U}{\partial I_3}I_2
$$

$$
\beta_1 = -\frac{\partial U}{\partial I_2} - \frac{\partial U}{\partial I_3}I_1
$$

$$
\beta_2 = \frac{\partial U}{\partial I_3}
$$

If $U(I_1, I_2, I_3)$ is now considered to be expanded as a polynomial in the invariants, it is seen how this form gives an elasticity description with many material coefficients.

Linear Elastic Anisotropic Materials

Reinforced materials are likely to have directional properties and are therefore anisotropic. They are also more likely to have small operational strains, and we take advantage of this to effect another set of material approximations.

Because the strains are assumed small, we can take the Taylor series expansion of the strain energy density function

$$
U(E_{ij}) = U(0) + \sum_{i,j} \left[\frac{\partial U}{\partial E_{ij}}\right]_0 E_{ij} + \frac{1}{2} \sum_{i,j,p,q} \left[\frac{\partial^2 U}{\partial E_{ij}\partial E_{pq}}\right]_0 E_{ij}E_{pq} + \cdots
$$

By using the Lagrangian strain tensor, the expansion is valid for large deflections and rotations but for small strains. Noting that $U(0) = 0$ and

$$
\frac{\partial E_{ij}}{\partial E_{pq}} = \delta_{ip}\delta_{jq}
$$

then get

$$
\sigma_{pq}^K = \frac{\partial U}{\partial E_{pq}} \approx \left[\frac{\partial U}{\partial E_{pq}}\right]_0 + \sum_{r,s} \left[\frac{\partial^2 U}{\partial E_{pq}\partial E_{rs}}\right]_0 E_{rs} + \cdots
$$

$$
= \sigma_{pq}^o + \sum_{r,s} C_{pqrs}E_{rs} + \cdots
$$

Let the stress be zero when the strains are zero, then for small strains

$$
\sigma_{pq}^K = \sum_{r,s} C_{pqrs}E_{rs}
$$

Because of symmetry in σ_{ij}^K and E_{ij}, C_{pqrs} reduces to 36 coefficients. But because of the explicit form of C_{pqrs} in terms of derivatives, we have the further restriction

$$
C_{pqrs} = \frac{\partial^2 U}{\partial E_{pq}\partial E_{rs}}\bigg|_0 = \frac{\partial^2 U}{\partial E_{rs}\partial E_{pq}}\bigg|_0 = C_{rspq}
$$

This additional symmetry reduces the elastic tensor to 21 constants. This is usually considered to be the most general linearly elastic material.

We can write this relation in the matrix form

$$\{\sigma\} = [\,D\,]\{E\}, \qquad \{\sigma\} \equiv \{\sigma_{11}, \sigma_{22}, \cdots\}, \qquad \{\epsilon\} \equiv \{\epsilon_{11}, \epsilon_{22}, \cdots\}$$

where $[\,D\,]$ is of size $[6 \times 6]$. Because of the symmetry of both the stress and strain, we have $[\,D\,]^T = [\,D\,]$. Special materials are reduced forms of this relation. For example, an *orthotropic* material has three planes of symmetry and this reduces the number of material coefficients to nine and the elastic matrix is given by

$$
\begin{bmatrix}
d_{11} & d_{12} & d_{13} & 0 & 0 & 0 \\
d_{12} & d_{22} & d_{23} & 0 & 0 & 0 \\
d_{13} & d_{23} & d_{33} & 0 & 0 & 0 \\
0 & 0 & 0 & \frac{1}{2}(d_{11} - d_{12}) & 0 & 0 \\
0 & 0 & 0 & 0 & \frac{1}{2}(d_{11} - d_{12}) & 0 \\
0 & 0 & 0 & 0 & 0 & \frac{1}{2}(d_{11} - d_{12})
\end{bmatrix}
$$

For a transversely isotropic material this reduces to five coefficients because $d_{55} = d_{44}$ and $d_{66} = (d_{11} - d_{12})/2$. A thin fiber-reinforced composite sheet is usually considered to be transversely isotropic [37].

For the isotropic case, every plane is a plane of symmetry and every axis is an axis of symmetry. It turns out that there are only two independent elastic constants, and the elastic matrix is given as above but with

$$d_{11} = d_{22} = d_{33} = \lambda + 2\mu, \qquad d_{12} = d_{23} = d_{13} = \lambda$$

The constants λ and μ are called the Lamé constants. The stress/strain relations for isotropic materials are usually expressed in the form

$$\sigma_{ij}^K = 2\mu E_{ij} + \lambda \delta_{ij} \sum_k E_{kk}, \qquad 2\mu E_{ij} = \sigma_{ij}^K - \frac{\lambda}{3\lambda + 2\mu} \delta_{ij} \sum_k \sigma_{kk}^K \qquad (1.30)$$

This is called *Hooke's law* and is the linearized version of Equation (1.29). The expanded form of the Hooke's law for strains in terms of stresses is

$$
\begin{aligned}
E_{xx} &= \frac{1}{E}\left[\sigma_{xx}^K - \nu(\sigma_{yy}^K + \sigma_{zz}^K)\right] \\[2mm]
E_{yy} &= \frac{1}{E}\left[\sigma_{yy}^K - \nu(\sigma_{zz}^K + \sigma_{xx}^K)\right] \\[2mm]
E_{zz} &= \frac{1}{E}\left[\sigma_{zz}^K - \nu(\sigma_{xx}^K + \sigma_{yy}^K)\right] \qquad (1.31) \\[2mm]
2E_{xy} &= \frac{2(1+\nu)}{E}\sigma_{xy}^K, \quad 2E_{yz} = \frac{2(1+\nu)}{E}\sigma_{yz}^K, \quad 2E_{xz} = \frac{2(1+\nu)}{E}\sigma_{xz}^K
\end{aligned}
$$

and for stresses in terms of strains

$$\sigma_{xx}^K = \frac{E}{(1+\nu)(1-2\nu)}\left[(1-\nu)E_{xx} + \nu(E_{yy} + E_{zz})\right]$$

$$\sigma_{yy}^K = \frac{E}{(1+\nu)(1-2\nu)}\left[(1-\nu)E_{yy} + \nu(E_{zz} + E_{xx})\right]$$

$$\sigma_{zz}^K = \frac{E}{(1+\nu)(1-2\nu)}\left[(1-\nu)E_{zz} + \nu(E_{xx} + E_{yy})\right] \qquad (1.32)$$

$$\sigma_{xy}^K = \frac{E}{2(1+\nu)}2E_{xy}, \quad \sigma_{yz}^K = \frac{E}{2(1+\nu)}2E_{yz}, \quad \sigma_{xz}^K = \frac{E}{2(1+\nu)}2E_{xz}$$

where E is the Young's modulus and ν is the Poisson's ratio related to the Lamé coefficients by

$$E = \frac{\mu(3\lambda + 2\mu)}{\lambda + \mu}, \quad \nu = \frac{\lambda}{2(\lambda + \mu)}, \quad \lambda = \frac{\nu E}{(1-2\nu)(1+\nu)}, \quad \mu = G = \frac{E}{2(1+\nu)}$$

The coefficient $\mu = G$ is called the shear modulus.

Viewing the relation between the normal components of stress and strain as forming a $[3 \times 3]$ matrix, then it can be inverted only if the determinant is positive. The determinant is

$$\det = (1 - 2\nu)(1 + \nu)^2 > 0$$

Hence we conclude that

$$-1 < \nu < 0.5$$

A negative Poisson's ratio would indicate a material that, under uniaxial tension, would expand in the transverse direction. This is possible for some of the structured materials.

A temperature change can affect the constitutive behavior in two ways: first it can change the values of the material coefficients; and second, it causes a volumetric expansion. We are only concerned with the later effect — this is called *thermoelasticity*. Because the temperature change only causes a volume change, then only the normal strain components are affected and the Hooke's law of Equation (1.31) is modified to

$$E_{xx} = \frac{1}{E}\left[\sigma_{xx}^K - \nu(\sigma_{yy}^K + \sigma_{zz}^K)\right] + \alpha\Delta T$$

$$E_{yy} = \frac{1}{E}\left[\sigma_{yy}^K - \nu(\sigma_{zz}^K + \sigma_{xx}^K)\right] + \alpha\Delta T$$

$$E_{zz} = \frac{1}{E}\left[\sigma_{zz}^K - \nu(\sigma_{xx}^K + \sigma_{yy}^K)\right] + \alpha\Delta T \qquad (1.33)$$

where α is the coefficient of thermal expansion and ΔT is the temperature change.

A special case that arises in the analysis of thin-walled structures is that of *plane stress*. Here, the stress through the thickness of the plate is approximately zero such that $\sigma_{zz}^K \approx 0$, $\sigma_{xz}^K \approx 0$, and $\sigma_{yz}^K \approx 0$. This leads to

$$E_{zz} = \frac{-\nu}{E}[\sigma_{xx}^K + \sigma_{yy}^K] = \frac{-\nu}{1-\nu}[E_{xx} + E_{yy}]$$

Substituting this into the 3-D Hooke's law then gives

$$E_{xx} = \frac{1}{E}[\sigma_{xx}^K - \nu\sigma_{yy}^K], \qquad \sigma_{xx}^K = \frac{E}{(1-\nu^2)}[E_{xx} + \nu E_{yy}]$$

$$E_{yy} = \frac{1}{E}[\sigma_{yy}^K - \nu\sigma_{xx}^K], \qquad \sigma_{yy}^K = \frac{E}{(1-\nu^2)}[E_{yy} + \nu E_{xx}]$$

The shear relation is unaffected.

A final point to note is that, except for the isotropic material, the material coefficients are given with respect to a particular coordinate system. Hence, we must transform the coefficients into the new coordinate system when the axes are changed.

Example 1.16: The following quantities were recorded during the large deformation testing of a uniaxial specimen: P, the applied force; $\epsilon \equiv \Delta L/L_o$, the unit change of axial length; $\epsilon_t \equiv \Delta W/W_o$, the unit change of transverse width. Establish the relationships necessary to convert this information to stress and strain.

Following from the examples of Section 1.4, we have that the stretches are

$$\lambda_1 = 1 + \epsilon, \qquad \lambda_2 = 1 + \epsilon_t = 1 - \nu\epsilon, \qquad \lambda_3 = 1 + \epsilon_t = 1 - \nu\epsilon = \lambda_2$$

where we have introduced $\nu \equiv -\epsilon_t/\epsilon$ as the ratio of the axial straining to the transverse straining. The Lagrangian and Eulerian strains in the axial direction are

$$E_{11} = \epsilon + \tfrac{1}{2}\epsilon^2, \qquad e_{11} = e - \tfrac{1}{2}e^2 = \frac{\epsilon}{(1+\epsilon)^2}\left[1 + \tfrac{1}{2}\epsilon\right]$$

The stresses are

$$\sigma_{11}^K = \frac{1}{\lambda_1}\frac{dP}{dA_1^o} = \frac{\sigma_o}{(1+\epsilon)}, \qquad \sigma_{11} = \frac{1}{\lambda_2\lambda_3}\frac{dP}{dA_1^o} = \frac{\sigma_o}{(1-\nu\epsilon)^2}, \qquad \sigma_o \equiv \frac{dP}{dA_1^o}$$

The stress σ_o can be thought of as the "force over original area," although here it is introduced solely as a normalizing factor.

As shown in Figure 1.27, there are three possibilities for the behavior of $\sigma_o = P/A_o$ against $\epsilon \equiv \Delta L/L_o$: it can be concave up indicating hardening, be concave down indicating softening, or be linear. The corresponding stress/strain curves are also shown in Figure 1.27. Note that for the range of nonlinear behaviors shown, all the Kirchhoff stress/Lagrangian strain relations show softening, whereas the Cauchy stress/Eulerian strain show hardening. Therefore, whether a material is physically linear or nonlinear, softening or hardening, is not a definite concept but depends on the measures used for the stress and strain. Of course, the mechanical

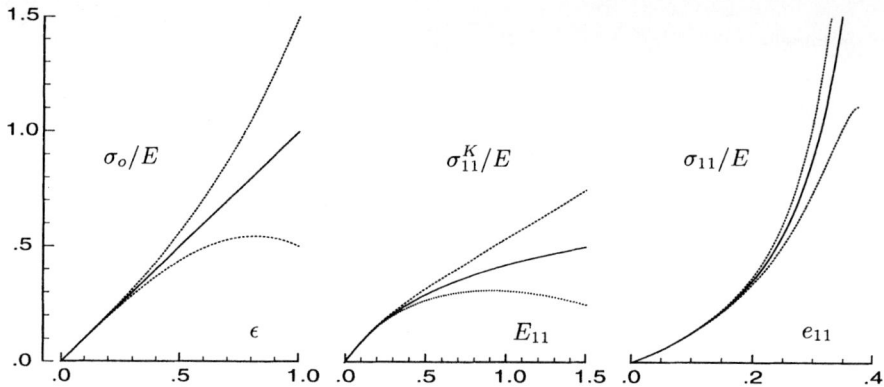

Figure 1.27: The forms of constitutive behavior for the experimental observed behavior.

problem can be objectively nonlinear even though the constitutive relation is linear because the description of the geometry can be nonlinear.

Example 1.17:   Contrast the physical response of materials described by linear constitutive relations.

Consider the uniaxial constitutive relations

$$\text{material:}\quad \sigma_{11}^{K} = E E_{11}\,, \qquad \text{spatial:}\quad \sigma_{11} = E e_{11}$$

where, for simplicity, we let the modulus of both materials be the same. Substituting the respective expressions for stress and strain leads to

$$\text{material:}\quad \sigma_o = E\epsilon(1+\epsilon)(1+\tfrac{1}{2}\epsilon)\,, \qquad \text{spatial:}\quad \sigma_o = E\epsilon\Big(\frac{1-\nu\epsilon}{1+\epsilon}\Big)^2\big[1+\tfrac{1}{2}\epsilon\big]$$

These are shown plotted in Figure 1.28.

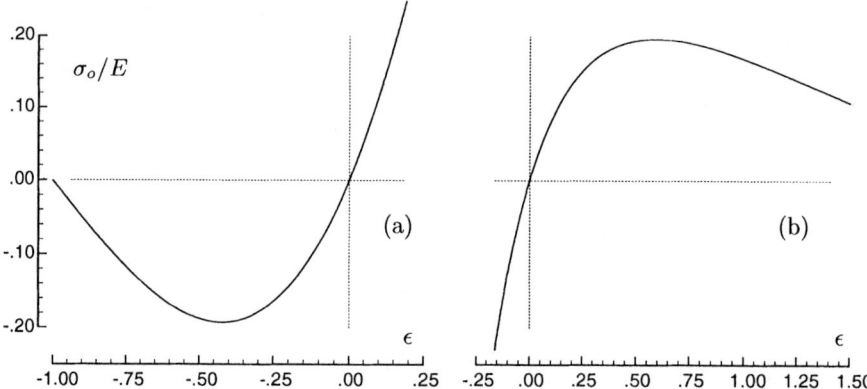

Figure 1.28: Physical responses for linear constitutive relations. (a) Material. (b) Spatial.

There are two obvious implications from Figure 1.28: linear constitutive relations imply highly nonlinear physical behaviors, and the two descriptions are completely different. In addition, both descriptions exhibit instabilities. Consider the spatial description, for example: as the load is increased, a point is reached where further load increments cannot be sustained and large deformations ensue. This is an example of a *limit point* instability and will be discussed in greater detail (within a structural context) in Chapters 6 and 7. It is worth noting that the instability occurs while the cross-sectional area is still sizable. The material description exhibits an instability in compression.

The form of the spatial behavior is that of a *cohesive strength* model. This has been used successfully [83] to model the fast propagation of cracks in brittle materials.

In subsequent chapters, we will be dealing with large displacements and rotations, but relatively small strains. In those cases, we will use a linear constitutive relation and restrict ourselves to strain levels such that $\epsilon < 0.20$; this avoids both instabilities, and all relations can be reasonably approximated as linear. For structural materials such as aluminum, these strain levels would have been associated with gross plastic yielding.

Example 1.18: Continuing the bending problem from Section 1.2, let the constitutive behavior be

$$\sigma_{ij}^K = 2G\, E_{ij} + \lambda \delta_{ij} E_{kk}$$

Determine the Cauchy stress.

From the previously obtained deformation gradient, we can determine the Lagrangian strains to be

$$E_{11} = -\frac{x_2^o}{R} + \frac{1}{2}\left(\frac{x_2^o}{R}\right)^2, \qquad \text{others:} \quad E_{ij} = 0$$

The Kirchhoff stresses are therefore

$$\sigma_{11}^K = (2G + \lambda)E_{11}, \qquad \sigma_{22}^K = \lambda E_{11}, \qquad \text{others:} \quad \sigma_{ij}^K = 0$$

We can think of these stresses as acting on the undeformed configuration. The tractions on a surface $x_1^o = $ constant are parabolic and independent of the position x_1^o; in the limit of small x_2^o, however, they have the familiar linear distribution of a beam in bending. This situation therefore resembles a beam in pure bending. The stress σ_{22}^K would give rise to a normal traction on the lateral surface; what this implies is that the given deformation could be achieved only with the aid of additional tractions on the lateral surfaces.

Since σ_{11}^K and σ_{22}^K are the only non-zero stresses, we have for the Cauchy stresses

$$\sigma_{ji} = \frac{\rho}{\rho_o}\left[\frac{\partial x_i}{\partial x_1^o}\frac{\partial x_j}{\partial x_1^o}\sigma_{11}^K + \frac{\partial x_i}{\partial x_2^o}\frac{\partial x_j}{\partial x_2^o}\sigma_{22}^K\right]$$

This leads to three non-zero components of the Cauchy stress

$$\sigma_{11} = \frac{\rho}{\rho_o}\left[\frac{1}{R^2}(R - x_2^o)^2 C^2 \sigma_{11}^K + S^2 \sigma_{22}^K\right] = J^o C^2 \sigma_{11}^K + \frac{1}{J^o}S^2 \sigma_{22}^K$$

$$\sigma_{22} = \frac{\rho}{\rho_o}\left[\frac{1}{R^2}(R - x_2^o)^2 S^2 \sigma_{11}^K + C^2 \sigma_{22}^K\right] = J^o S^2 \sigma_{11}^K + \frac{1}{J^o}C^2 \sigma_{22}^K$$

$$\sigma_{12} = \frac{\rho}{\rho_o}\left[\frac{1}{R^2}(R - x_2^o)^2 SC \sigma_{11}^K - CS \sigma_{22}^K\right] = J^o SC \sigma_{11}^K - \frac{1}{J^o}CS \sigma_{22}^K$$

where $C \equiv \cos(x_1^o/R)$ and $S \equiv \sin(x_1^o/R)$. These stresses exhibit a rather complex dependence on both x_1^o and x_2^o.

The presence of the non-zero shear stress is, perhaps, a bit surprising. Keep in mind, however, as x_1^o is changed, that the components σ_{ij} are not necessarily oriented with respect to the deformed lateral surface. It is instructive, therefore, to consider the components of the Cauchy stress with respect to the deformed lateral surface. Consider a line that is initially horizontal, then after deformation it has the orientation

$$n_1 = \cos(x_1^o/R) = C\,, \qquad n_2 = \sin(x_1^o/R) = S$$

We now transform the stress components to get

$$
\begin{aligned}
\sigma_{nn} &= \sigma_{11}C^2 + \sigma_{22}S^2 + 2\sigma_{12}CS = J^o\sigma_{11}^K \\
\sigma_{tt} &= \sigma_{11}S^2 + \sigma_{22}C^2 - 2\sigma_{12}CS = \frac{1}{J^o}\sigma_{22}^K \\
\sigma_{tn} &= -(\sigma_{11} - \sigma_{22})CS + \sigma_{12}(C^2 - S^2) = 0
\end{aligned}
$$

Thus, the Cauchy stress components with respect to line-preserving orientations show a close connection to the Kirchhoff stress. Indeed, if we consider the case when $x_2^o \ll R$ (that is, it is like a very narrow beam) but we still allow the large deflections, then we get

$$J^o = \frac{1}{R}(R - x_2^o) \approx 1$$

leading to

$$\sigma_{nn} \approx \sigma_{11}^K\,, \qquad \sigma_{tt} \approx \sigma_{22}^K$$

In some of the later chapters, we will consider situations where the deflections are large but the strains small, we will then find it useful to invoke this approximate relation between the Kirchhoff and transformed Cauchy stresses.

Example 1.19: Show how an initial stress state affects the current relation between an increment of stress and an increment of strain for an isotropic material.

Let the initial stress state σ_{ij}^o be associated with the displacement field u_i^o. Furthermore, let the current displacement u_i be represented as

$$u_i = u_i^o + \xi_i$$

where ξ_i is the (small) increment of displacement from the current value of u_i^o. Using this in the strain/displacement relation allows the total strain to be decomposed as

$$
\begin{aligned}
2E_{ij} &= \frac{\partial u_i}{\partial x_j^o} + \frac{\partial u_j}{\partial x_i^o} + \sum_k \frac{\partial u_k}{\partial x_i^o}\frac{\partial u_k}{\partial x_j^o} \\
&= \left(\frac{\partial u_i^o}{\partial x_j^o} + \frac{\partial u_j^o}{\partial x_i^o} + \sum_k \frac{\partial u_k^o}{\partial x_i^o}\frac{\partial u_k^o}{\partial x_j^o}\right) \\
&\quad + \left(\frac{\partial \xi_i}{\partial x_j^o} + \frac{\partial \xi_j}{\partial x_i^o} + \sum_k \frac{\partial \xi_k}{\partial x_i^o}\frac{\partial \xi_k}{\partial x_j^o}\right) + \left(\sum_k \frac{\partial u_k^o}{\partial x_i^o}\frac{\partial \xi_k}{\partial x_j^o} + \sum_k \frac{\partial \xi_k}{\partial x_i^o}\frac{\partial u_k^o}{\partial x_j^o}\right)
\end{aligned}
$$

The various collections of terms in parentheses are labeled as follows

$$E_{ij} = E_{ij}^o + \epsilon_{ij} + \eta_{ij}$$

Note that ϵ_{ij} is an increment of strain from the current configuration but referenced to the zero configuration. The interaction term η_{ij} contains components of both u_i^o and ξ_i; this is the term we are especially interested in.

Let the constitutive relation be

$$\sigma_{ij}^K = 2\mu E_{ij} + \lambda \delta_{ij} \sum_k E_{kk}, \qquad 2\mu E_{ij} = \sigma_{ij}^K - \frac{\lambda}{2\mu + 3\lambda} \delta_{ij} \sum_k \sigma_{kk}^K$$

Then, after substituting for the strains, the stresses are

$$\sigma_{ij}^K = \sigma_{ij}^o + 2\mu\epsilon_{ij} + \lambda\delta_{ij} \sum_k \epsilon_{kk} + 2\mu\eta_{ij} + \lambda\delta_{ij} \sum_k \eta_{kk}$$

We are interested in taking derivatives of this stress with respect to ϵ_{pq}. Since

$$\frac{\partial \xi_k}{\partial x_j^o} \approx \epsilon_{ij} + \omega_{ij}$$

then, for the purpose of differentiation, we can replace the gradient of ξ_i with ϵ_{ij}. We now get for the σ_{11}^K stress, for example,

$$\sigma_{11}^K = \sigma_{11}^o + 2\mu\epsilon_{11} + \lambda[\epsilon_{11} + \epsilon_{22} + \epsilon_{33}] + 2\mu \sum_k \frac{\partial u_k^o}{\partial x_1^o} \epsilon_{k1} + \lambda \sum_{k,p} \frac{\partial u_k^o}{\partial x_p^o} \epsilon_{kp}$$

with similar expressions for the other components. Let us define the current tangent moduli as

$$
\begin{aligned}
E_{T11} \equiv \frac{\partial \sigma_{11}^K}{\partial \epsilon_{11}} &= (2\mu + \lambda) + (2\mu + \lambda)\frac{\partial u_1^o}{\partial x_1^o} \approx (2\mu + \lambda)\left[1 + E_{11}^o\right] \\
&= (2\mu + \lambda)\left[1 + \frac{1}{2\mu}\left\{\sigma_{11}^o - \frac{\lambda}{2\mu + 3\lambda}(\sigma_{11}^o + \sigma_{22}^o + \sigma_{33}^o)\right\}\right]
\end{aligned}
$$

The derivatives are taken such that the other strains are kept constant. The effect of the initial stress is to change the tangent modulus — an increase in stress causes an increase in modulus. This is the same phenomenon as observed when tuning a violin string, say. Suppose the initial stress is uniaxial such that only $\sigma_{11}^o \neq 0$, then two of the moduli are

$$
\begin{aligned}
E_{T11} &= (2\mu + \lambda)\left[1 + \frac{1}{2\mu}\left\{1 - \frac{\lambda}{2\mu + 3\lambda}\right\}\sigma_{11}^o\right] \\
E_{T22} &= (2\mu + \lambda)\left[1 + \frac{1}{2\mu}\left\{ - \frac{\lambda}{2\mu + 3\lambda}\right\}\sigma_{11}^o\right]
\end{aligned}
$$

We see that the material becomes anisotropic due to the stress σ_{11}^o.

This is our first encounter with a tangent modulus (or stiffness), and it will play a central role in the study of nonlinear deformations. A way to visualize the above results is to consider a block of material that is under a quasi-static load state. Now superpose a stress wave disturbance of small amplitude. The current tangent modulus relates the small increments (due to the stress wave) of strain to the small increments of stress. Although the material is isotropic, the stress wave experiences the material as being anisotropic. As an aside, residual stresses can be detected by monitoring the small changes in wave speed caused by the small changes in tangent moduli [30].

This is also our first encounter with *load interactions* — the situation where the application of one load causes a change in the load/deformation response for another load. This nonlinear phenomenon, in fact, becomes one of the subthemes for our analysis of stability problems in Chapters 6 and 7.

Strain Energy for Some Linear Elastic Structures

Consider a local coordinate system in which the rotation of the structural member is negligible.

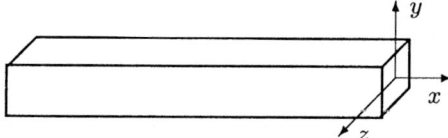

Figure 1.29: Cross-section of structural member in local coordinates.

When the strains are small, we need not distinguish between the undeformed and deformed configurations. Under this circumstance, let the material obey Hooke's law and be summarized in the matrix forms

$$\{\epsilon\} = [\,\mathcal{C}\,]\{\sigma\}, \qquad \{\sigma\} = [\,\mathcal{D}\,]\{\epsilon\}, \qquad [\,\mathcal{C}\,] = [\,\mathcal{D}\,]^{-1}$$

The general expression for the strain energy is

$$U = \tfrac{1}{2} \int_V [\sigma_{xx}\epsilon_{xx} + \sigma_{yy}\epsilon_{yy} + \sigma_{xy}\gamma_{xy} + \cdots]dV = \tfrac{1}{2}\int_V \{\sigma\}^T\{\epsilon\}dV$$

Using Hooke's law, this can be put in the alternate forms

$$U = \tfrac{1}{2}\int_V \{\epsilon\}^T[\,\mathcal{D}\,]\{\epsilon\}dV = \tfrac{1}{2}\int_V \{\sigma\}^T[\,\mathcal{C}\,][\,\sigma\,]dV \qquad (1.34)$$

The above relations will now be particularized to some structural systems of interest by writing the distributions of stress and strain in terms of resultants.

For the rod member, there is only an axial stress present and it is uniformly distributed on the cross-section. Let F be the resultant force; then $\sigma = F/A = E\epsilon$ and

$$\text{axial:} \qquad U = \tfrac{1}{2}\int_0^L \frac{F^2}{EA}dx = \tfrac{1}{2}\int_0^L EA\left(\frac{du}{dx}\right)^2 dx$$

For the beam member in bending, there is only an axial stress, but it is distributed linearly on the cross-section in such a way that there is no resultant axial force. Let M be the resultant moment; then $\sigma = -My/I = E\epsilon$ and

$$\text{bending:} \qquad U = \tfrac{1}{2} \int_0^L \frac{M^2}{EI} dx = \tfrac{1}{2} \int_0^L EI \left(\frac{d^2v}{dx^2}\right)^2 dx$$

The shear forces in a beam can also do some work. Let the shear stress be assumed to be uniformly distributed on the cross-section, and the resultant shear force be V, then $\tau = V/A = G\gamma$ and

$$\text{shear:} \qquad U = \tfrac{1}{2} \int_0^L \frac{V^2}{GA} dx = \tfrac{1}{2} \int_0^L GA \left(\frac{dv}{dx}\right)^2 dx$$

For a circular shaft in torsion, there is only a shear stress and it is linearly distributed on the radius. Let T be the resultant torque, then $\tau = Tr/J = G\gamma$ and

$$\text{torsion:} \qquad U = \tfrac{1}{2} \int_0^L \frac{T^2}{GJ} dx = \tfrac{1}{2} \int_0^L GJ \left(\frac{d\phi}{dx}\right)^2 dx$$

where ϕ is the twist per unit length and J is the polar moment of area.

For the four cases considered above, there are the resultant loads

$$F, \quad M, \quad V, \quad T$$

the corresponding deformations

$$\frac{du}{dx}, \quad \frac{d^2v}{dx^2}, \quad \frac{dv}{dx}, \quad \frac{d\phi}{dx}$$

and the associated stiffnesses

$$EA, \quad EI, \quad GA, \quad GJ$$

and in each case, the energy expression is of the form

$$\text{energy} = \tfrac{1}{2} \int_0^L \frac{[\text{load}]^2}{(\text{stiffness})} dx = \tfrac{1}{2} \int_0^L (\text{stiffness})[\text{deformation}]^2 dx$$

Note that even the general expression, Equation (1.34), follows this form. There are other types of structures, and an energy expression can be set up for these also. They will all have a similar form.

Example 1.20: Use the principle of stationary potential energy to establish the equilibrium condition for the simple truss shown in Figure 1.1.

From geometry, we can establish that the axial displacement of the member is

$$\bar{u} = \sqrt{L_o^2 + 2vL_o \sin \alpha_o + v^2} - L_o$$

The axial strain is $\bar{\epsilon} = \bar{u}/L_o$, hence the strain energies for the spring and two truss members are

$$U = \tfrac{1}{2}Kv^2 + 2\tfrac{1}{2}EAL_o\bar{\epsilon}^2 = \tfrac{1}{2}Kv^2 + EAL_o\left[\sqrt{1 + 2\frac{v}{L_o}\sin\alpha_o + (\frac{v}{L_o})^2} - 1\right]^2$$

The applied load is acting in the coordinate direction, hence the potential of this load is

$$V = -Pv$$

The total potential for the problem is therefore

$$\Pi = \tfrac{1}{2}Kv^2 + EAL_o\left[\sqrt{1 + 2\frac{v}{L_o}\sin\alpha_o + (\frac{v}{L_o})^2} - 1\right]^2 - Pv$$

This is shown plotted in Figure 1.1 for different values of load P. The equilibrium path corresponds to where Π has an extremum — these are shown connected in the figure. We see for large positive or large negative P that there is only one equilibrium position. However, for small negative P there are three sets of equilibrium points: two correspond to minima, while the third corresponds to a maximum. As we will see in Chapters 6 and 7, the maximum corresponds to an unstable equilibrium position and these are indicated with the dashed line. The critical load for the structure is when the stable and unstable paths converge.

The equilibrium path (load/deflection curve) is

$$\mathcal{F}_v = \frac{\partial\Pi}{\partial v} = Kv + 2EA\left[\sin\alpha_o + \frac{v}{L_o}\right]\left[1 - \frac{1}{\sqrt{1 + 2\frac{v}{L_o}\sin\alpha_o + (\frac{v}{L_o})^2}}\right] - P = 0$$

which could be written $P = F(v)$. It is worth emphasizing again that, although the material behavior is linear, the structural behavior can still be highly nonlinear. We will develop this example further in the later chapters.

1.8 Approximate Weak Form of Problems

The examples of the previous sections show that we have two alternative ways of stating our problem. The first is by a set of differential equations plus a set of associated boundary conditions; this is known as the *strong* form or *classical* form of the problem. The alternate way is by extremizing a functional; this is known as the *weak* form or *variational* form of the problem. They are both equivalent (as shown by the examples) but lend themselves to approximation in different ways. What we wish to pursue in the following is approximation arising from the weak form; specifically, we will approximate the functional itself and use the variational principle to obviate consideration of the natural boundary conditions. This is called the Ritz (or Rayleigh-Ritz) method. The computational implementation will be in the form of the finite element method.

Ritz Method

In general, a continuously distributed deformable body consists of an infinity of
material points and therefore has infinitely many degrees of freedom. The Ritz
method is an approximate procedure by which continuous systems are reduced
to systems with finite degrees of freedom. The fundamental characteristic of the
method is that we operate on the functional corresponding to the problem. To
fix ideas, consider the static case, where we are looking for the solution of $\delta \Pi = 0$
with prescribed boundary conditions on u. Let

$$u(x, y, z) = \sum_{i=1}^{\infty} a_i \phi_i(x, y, z)$$

where ϕ_i are independent *trial functions*, and the a_i are multipliers to be determined in the solution. The trial functions satisfy the essential (geometric)
boundary conditions but not necessarily the natural boundary conditions. The
variational problem states that

$$\Pi(u) = \Pi(a_1 \phi_1 + a_2 \phi_2 + \cdots) = \text{stationary}$$

Thus, $\Pi(a_1 \phi_1 + a_2 \phi_2 + \cdots)$ can be regarded as a function of the variables
$a_1, a_2, \cdots$. To satisfy $\Pi = \text{stationary}$, we require that

$$\mathcal{F}_1 = \frac{\partial \Pi}{\partial a_1} = 0, \qquad \mathcal{F}_2 = \frac{\partial \Pi}{\partial a_2} = 0, \qquad \cdots$$

These equations are then used to determine the coefficients a_i. Normally, we
only include a finite number of terms in the expansion.

An important consideration is the selection of the trial functions ϕ_i. Select-
ing efficient admissible functions may not be easy; fortunately, many problems
closely resemble other problems that have been solved before, and the literature is full of examples that can serve as a guide. It must also be kept in mind
that these functions need only satisfy the essential boundary conditions and not
(necessarily) the natural boundary conditions. For practical analyses, this is a
significant point and largely accounts for the effectiveness of the displacement-
based finite element analysis procedure.

For convenience in satisfying the boundary conditions on u, we usually set

$$u = u_o + \sum_n a_n \phi_n$$

where u_o conforms to the non-homogeneous (nonzero) boundary conditions. For
homogeneous displacement boundary conditions, we set $u_o = 0$.

Example 1.21: Consider an inhomogeneous rod fixed at one end and subjected to an axial concentrated force at the other end, as shown in Figure 1.30.

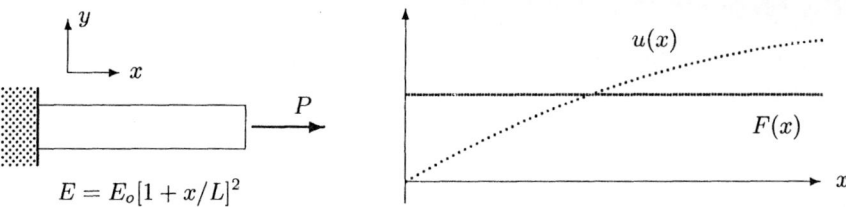

Figure 1.30: Rod with variable modulus.

The variation of Young's modulus is $E(x) = E_o[1 + x/L]^2$. Determine a Ritz approximate solution.

The boundary conditions for this problem are:

$$\text{essential: } u\big|_{x=0} = 0, \qquad \text{natural: } EA\frac{du}{dx}\big|_{x=L} = P$$

The exact solution is easily calculated to give

$$u(x) = \int_0^x \epsilon(x)dx = \int_0^x \frac{P}{EA}dx = \frac{PL}{E_oA}\frac{x/L}{(1+x/L)}, \qquad F(x) = EA\frac{du}{dx} = P$$

The member force distribution $F(x)$ is constant. Both the displacement distribution and force distribution are shown plotted in Figure 1.30. We will use these results to evaluate the quality of the Ritz approximate solutions. Specifically, we will investigate the use of different trial functions.

Because the deformation is one-dimensional, then the strain is

$$\epsilon_{xx} = \frac{\partial u}{\partial x} = \frac{du}{dx}$$

and the total potential energy of the body is

$$\Pi = U + V = \frac{1}{2}\int_0^L EA\left(\frac{du}{dx}\right)^2 dx - Pu_L$$

The integration over the cross-section has already been performed. We will calculate the displacement and force distributions using the following assumed form for the displacement:

$$u(x) = a_0 + a_1x + a_2x^2$$

This must satisfy the essential boundary condition, hence $a_0 = 0$. Note that the remaining polynomial does not necessarily satisfy the natural boundary condition. Substituting the assumed displacements into the total potential energy expression, we obtain

$$\Pi = \frac{1}{2}\int_0^L E_oA[1 + x/L]^2[a_1 + 2a_2x]^2 dx - P[a_1L + a_2L^2]$$

Invoking the stationarity of Π with respect to the coefficients a_1, a_2, we obtain the following equilibrium equations for determining a_1 and a_2:

$$\mathcal{F}_1 = \frac{\partial\Pi}{\partial a_1} = \int_0^L E_oA[1 + x/L]^2[a_1 + 2a_2x]dx - PL = 0$$

x/L	$u(x)$ exact	1-term	2-term	bi-lin	$F(x)$ exact	1-term	2-term	bi-lin
0.0	0.0	0.0	0.0	0.0	1.0	0.428	0.804	0.632
0.5	0.3333	.2143	.3247	.3158	1.0	0.96	1.113	1.421
0.5	0.3333	.2143	.3247	.3158	1.0	0.96	1.113	0.729
1.0	0.5	.4285	.4948	.4779	1.0	1.714	0.740	1.297

Table 1.1: Displacement and force results for the non-uniform rod.

$$\mathcal{F}_2 = \frac{\partial \Pi}{\partial a_2} = \int_0^L E_o A [1 + x/L]^2 [a_1 + 2a_2 x] 2x \, dx - PL^2 = 0$$

Performing the required integrations gives

$$\frac{E_o A}{30} \begin{bmatrix} 70L & 85L^2 \\ 85L^2 & 124L^3 \end{bmatrix} \left\{ \begin{array}{c} a_1 \\ a_2 \end{array} \right\} = \left\{ \begin{array}{c} PL \\ PL^2 \end{array} \right\}$$

Solving this system gives for the two coefficients

$$a_1 = \frac{78}{97} \frac{P}{E_o A}, \qquad a_2 = \frac{-30}{97} \frac{P}{E_o A L}$$

This Ritz analysis, therefore, yields the approximate solution

$$u(x) = \frac{78P}{97 E_o A} [x - \frac{10}{26L} x^2]$$

and the force distribution is

$$F(x) = EA \frac{du}{dx} = \frac{78P}{97} [1 - \frac{10}{13L} x][1 + x/L]^2$$

These results are shown in Table 1.1 as the 2-term columns. The most striking aspect of these results is the accuracy of the displacements and yet the axial force is not constant and equal to P.

The other terms in the table correspond to using a_1 only (1-term) and using two domains (bi-linear) as discussed presently.

Completeness and Convergence

A number of observations can now be made about the use of stationary principles. First, if the functional contains derivatives up to order m, then there must be continuity of displacement derivative up to $m - 1$, and the order of the highest derivative that is present in the governing differential equation is then $2m$. For example, in a beam-bending problem where the strain is $d^2 v/dx^2$, $m = 2$ because the highest derivative in the functional is of order 2, and there must be continuity of v and dv/dx. The reason for obtaining a derivative of order $2m = 4$ in the governing differential equation is that integration by parts is employed $m = 2$ times.

A second observation is that through the stationarity condition, we obtain the governing differential equations *and* the appropriate boundary conditions. Hence, the effect of the natural boundary conditions are implicitly contained in the expression for the potential Π. (Note that the essential boundary conditions must be stated separately.)

Some of the specific characteristics of the Ritz method are:

- Usually, the accuracy of the assessed displacement is increased with an increase in the number of trial functions.
- While fairly accurate expressions for the displacements are obtained, the corresponding forces may differ significantly from the exact values.
- Equilibrium is satisfied in an average sense through minimization of the total potential energy. Therefore, forces (computed on the basis of the displacements) do not, in general, satisfy the equilibrium equations of the original problem.
- The approximate system is stiffer than the actual system and therefore buckling loads and vibration resonances are overestimated.

A question arises as to what are the appropriate additional terms to be used if more terms are to be included so as to achieve a converged accurate solution. The sequence of terms should be *complete*. For example, for a 1-D problem the simple polynomial sequence

$$1 \quad x \quad x^2 \quad x^3 \quad \ldots$$

is complete. The trigonometric sequence

$$1 \quad \sin x \quad \cos x \quad \sin 2x \quad \cos 2x \quad \sin 3x \quad \cos 3x \quad \ldots$$

is also complete. Note, however, that the cosines on their own could be used to represent a symmetric distribution and therefore would be complete. As we go to higher dimensions, the question of completeness gets a little more involved. Clearly, for a complicated domain, finding a complete set of appropriate functions is very difficult. Therefore, we now take a different approach altogether — instead of representing the response as a collection of (complicated) functions over the complete domain, we represent the domain as a collection of many subdomains over which the Ritz functions are relatively simple. This is the essence of the finite element method.

The Finite Element Discretization

One disadvantage of the conventional Ritz analysis is that the trial functions are defined over the whole region. This causes a particular difficulty in the selection of appropriate functions; in order to solve accurately for large stress gradients, say, we may need many functions. However, these functions are also defined

over the regions in which the stresses vary rather slowly and where not many
functions are required. Another difficulty arises when the total region is made
up of subregions with different kinds of strain distributions. As an example,
consider a building modeled by plates for the floors and beams for the vertical
frame. In this situation, the trial functions used for one region (e.g., the floor) are
not appropriate for the other region (e.g., the frame), and special displacement
continuity conditions and boundary relations must be introduced. We conclude
that the conventional Ritz analysis is, in general, not particularly computer-
oriented.

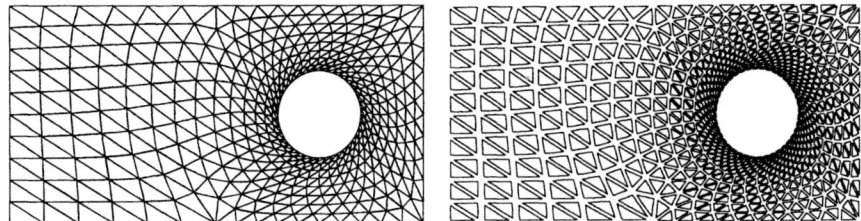

Figure 1.31: Continuous domain discretized as finite elements. Right figure has shrunk
elements for easier viewing.

We can view the finite element method as an application of the Ritz method
where, instead of the trial functions spanning the complete domain, the individual functions span only subdomains (the finite elements) of the complete region.
Figure 1.31 shows an example of a bar with a hole modeled as a collection of
many triangular regions. The use of relatively many functions in regions of high
strain gradients is made possible simply by using many elements as shown around
the hole in the figure. The combination of domains with different kinds of strain
distributions (e.g., a frame member connected to a plate) may be achieved by
using different kinds of elements to idealize the domains.

In order that a finite element solution be a Ritz analysis, it must satisfy
the essential boundary conditions. However, in the selection of the displacement
functions, no special attention need be given to the natural boundary conditions,
because these conditions are imposed with the load vector and are satisfied approximately in the Ritz solution. The accuracy with which these natural boundary conditions are satisfied depends on the specific trial functions employed and
on the number of elements used to model the problem. This idea is demonstrated
in the convergence studies of the next few chapters.

Example 1.22: As another Ritz solution to the inhomogeneous problem of
Figure 1.30, assume that the displacements are given in a piecewise linear form as

$$u(x) = \frac{2x}{L}u_2 \qquad\qquad\qquad\qquad 0 \le x \le L/2$$

$$u(x) = \left[\frac{2L - 2x}{L}\right]u_2 + \left[\frac{2x - L}{L}\right]u_3 \qquad L/2 \le x \le L$$

where u_2 and u_3 are the displacements at points mid-way and the end of the rod. This displacement distribution satisfies the essential boundary condition at $x = 0$, and also the continuity of displacement condition at $x = L/2$. There is no continuity of the first derivative du/dx at $x = L/2$, but that is permissible because the highest derivative *in the potential* is du/dx. We will refer to such piecewise simple displacement distributions as *interpolation* or *shape* functions.

Using these trial functions in the potential energy gives

$$\Pi = \tfrac{1}{2}\int_0^{L/2} E_o A[1+x/L]^2 \left[\frac{2u_2}{L}\right]^2 dx + \tfrac{1}{2}\int_{L/2}^{L} E_o A[1+x/L]^2 \left[-\frac{2u_2}{L}+\frac{2u_3}{L}\right]^2 dx - Pu_3$$

In this case, the displacements u_2 and u_3 are the generalized coordinates or degrees of freedom. Invoking that Π is stationary with respect to these, we obtain the equilibrium equations

$$\mathcal{F}_2 = \frac{\partial \Pi}{\partial u_2} = \frac{E_o A}{6L}[56u_2 - 37u_3] = 0$$

$$\mathcal{F}_3 = \frac{\partial \Pi}{\partial u_3} = \frac{E_o A}{6L}[-37u_2 + 37u_3] - P = 0$$

These can be arranged in matrix form as

$$\frac{E_o A}{6L}\begin{bmatrix} 56 & -37 \\ -37 & 37 \end{bmatrix} \begin{Bmatrix} u_2 \\ u_3 \end{Bmatrix} = \begin{Bmatrix} 0 \\ P \end{Bmatrix} \qquad \text{or} \qquad [\,K\,]\{u\} = \{P\}$$

where $[\,K\,]$ is called the *structural stiffness matrix*. Solving the system of simultaneous equations, we get for the degrees of freedom

$$u_2 = \frac{P}{E_o A}\frac{222}{703}, \qquad u_3 = \frac{P}{E_o A}\frac{336}{703}$$

The displacement distributions are piecewise linear. The member forces vary with x and are given by

$$F(x) = P\frac{444}{703}[1+x/L]^2 \qquad 0 \le x \le L/2$$

$$F(x) = P\frac{228}{703}[1+x/L]^2 \qquad L/2 \le x \le L$$

These results are shown in Table 1.1 as the bi-linear columns. Note again that the displacements are quite accurate, but the forces are significantly off. Indeed, there is not even equilibrium at the joint. The main idea here, however, is that accuracy is improved by increasing the number of subdomains.

Example 1.23:   Illustrate the process of element assemblage.

We begin with the decomposition of the total potential according to subregions

$$\Pi = \Pi_1(u_2) + \Pi_1(u_2, u_3) - P_2 u_2 - P_3 u_3$$

where

$$\Pi_1 = \tfrac{1}{2}\int_0^{L/2} E_o A[1+x/L]^2 \left[\frac{2u_2}{L}\right]^2 dx = \frac{E_o A}{12L}[u_2]^2$$

$$\Pi_2 = \tfrac{1}{2}\int_{L/2}^{L} E_o A[1+x/L]^2 \left[-\frac{2u_2}{L}+\frac{2u_3}{L}\right]^2 dx = \frac{E_o A}{12L}[-u_2+u_3]^2$$

and, to generalize, we have put applied forces at locations 2 and 3. The partial derivatives for each region are

$$\mathcal{F}_{12} = \frac{\partial \Pi_1}{\partial u_2} = \frac{E_o A}{6L} 19[u_2]$$

$$\mathcal{F}_{22} = \frac{\partial \Pi_2}{\partial u_2} = \frac{E_o A}{6L} 37[+u_2 - u_3]$$

$$\mathcal{F}_{23} = \frac{\partial \Pi_2}{\partial u_3} = \frac{E_o A}{6L} 37[-u_2 + u_3]$$

These are not zero since we are not considering the total potential. In fact, we refer to these as *element nodal forces* because they are computed for each element (subregion) and are associated with each node (point where the degree-of-freedom is monitored). We will use the notation $\{F\}$ to refer to the vector of nodal forces.

Returning to the total potential as an assemblage of potentials for each region, one interpretation of assemblage is simply the summing of all the element nodal forces in conjunction with the applied loads. That is,

$$\mathcal{F}_2 = \frac{\partial \Pi}{\partial u_2} = \mathcal{F}_{12} + \mathcal{F}_{22} - P_2 = 0$$

$$\mathcal{F}_3 = \frac{\partial \Pi}{\partial u_3} = \mathcal{F}_{13} + \mathcal{F}_{23} - P_3 = 0$$

Arrange these equations in the matrix form

$$\begin{Bmatrix} \mathcal{F}_2 \\ \mathcal{F}_3 \end{Bmatrix} = \begin{Bmatrix} \mathcal{F}_{12} \\ \mathcal{F}_{13} \end{Bmatrix} + \begin{Bmatrix} \mathcal{F}_{22} \\ \mathcal{F}_{23} \end{Bmatrix} - \begin{Bmatrix} P_2 \\ P_3 \end{Bmatrix} = \begin{Bmatrix} 0 \\ 0 \end{Bmatrix} \qquad \text{or} \qquad \{F\} - \{P\} = 0$$

This form of the assemblage is valid even for nonlinear problems.

We can arrange the nodal force relations in the form of stiffness relations

$$\{\mathcal{F}_{12}\} = \frac{E_o A}{6L} 19[\; 1 \;]\{u_2\}, \qquad \begin{Bmatrix} \mathcal{F}_{22} \\ \mathcal{F}_{23} \end{Bmatrix} = \frac{E_o A}{6L} 37 \begin{bmatrix} 1 & -1 \\ -1 & 1 \end{bmatrix} \begin{Bmatrix} u_2 \\ u_3 \end{Bmatrix}$$

Both are of the form $\{F\} = [\; k \;]\{u\}$, where $[\; k \;]$ is called the *element stiffness matrix*. This representation is possible only because the problem is linear. Note that the first relation could be expanded to $[2 \times 2]$ by including the degree-of-freedom $u_1 = 0$. We get another interpretation of assemblage by augmenting the stiffness relation to full system size and adding. Thus

$$\begin{Bmatrix} \mathcal{F}_2 \\ \mathcal{F}_3 \end{Bmatrix} = \begin{Bmatrix} \mathcal{F}_{12} \\ \mathcal{F}_{13} \end{Bmatrix} + \begin{Bmatrix} \mathcal{F}_{22} \\ \mathcal{F}_{23} \end{Bmatrix} - \begin{Bmatrix} P_2 \\ P_3 \end{Bmatrix}$$

$$= \frac{E_o A}{6L} 19 \begin{bmatrix} 1 & 0 \\ 0 & 0 \end{bmatrix} \begin{Bmatrix} u_2 \\ u_3 \end{Bmatrix} + \frac{E_o A}{6L} 37 \begin{bmatrix} 1 & -1 \\ -1 & 1 \end{bmatrix} \begin{Bmatrix} u_2 \\ u_3 \end{Bmatrix} - \begin{Bmatrix} P_2 \\ P_3 \end{Bmatrix} = \begin{Bmatrix} 0 \\ 0 \end{Bmatrix}$$

Performing the addition of the stiffnesses gives

$$\frac{E_o A}{6L} \begin{bmatrix} 56 & -37 \\ -37 & 37 \end{bmatrix} \begin{Bmatrix} u_2 \\ u_3 \end{Bmatrix} = \begin{Bmatrix} P_2 \\ P_3 \end{Bmatrix} \qquad \text{or} \qquad [\; K \;]\{u\} = \{P\}$$

where $[\; K \;]$ is called the *structural stiffness matrix*. Therefore, the assemblage process can also be thought of as adding the element stiffnesses suitably augmented (with zeros) to full system size. It is important to realize, however, that this interpretation is only valid for linear systems.

To summarize, we have for the assemblage process

$$\{F\} = \sum_m \{F\}^{(m)} = \{P\}$$

where $\{P\}$ is the collection of external loads made up of applied loads and inertia loads. This relation must be satisfied throughout the loading history. For linear problems, this can be put in the more familiar form using stiffness matrices. On an element level, the nodal forces are written in terms of the element stiffness matrix as

$$\{F\}^{(m)} = [\ k\ ]^{(m)}\{u\}$$

and the assemblage process simply becomes the addition of element stiffness matrices. That is,

$$\{F\} = \sum_m \{F\}^{(m)} = \sum_m [\ k\ ]^{(m)}\{u\} = [\ K\ ]\{u\} = \{P\}$$

where the element stiffness matrices are suitably augmented to the size of the structural stiffness matrix $[\ K\ ]$.

Example 1.24:   Establish the element stiffness relations for a frame member.

Consider the axial stretching and rotation of the member shown Figure 1.32. We establish a local coordinate system at the first node with the $\bar{x}$ axis directed along the member. Descriptions of quantities in the local coordinates will have an overhead bar.

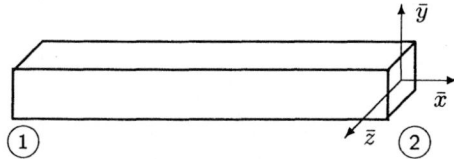

Figure 1.32: Cross-section of truss and frame members in local coordinates.

In the local description, the member has the deformation system

$$\{\bar{u}\}^T = \{\bar{u}_1,\ \bar{v}_1,\ \bar{w}_1,\ \bar{\phi}_{1x},\ \bar{\phi}_{1y},\ \bar{\phi}_{1z};\ \bar{u}_2,\ \bar{v}_2,\ \bar{w}_2,\ \bar{\phi}_{2x},\ \bar{\phi}_{2y},\ \bar{\phi}_{2z}\}$$

We assume the member is long and slender and hence there is only 1-D axial stress and a torsional shear stress. The strain is a combination of the axial stretching, the bending action in two planes, and a twisting action about the axis. The axial and shear strains are written as

$$\bar{\epsilon}(\bar{x},\bar{y},\bar{z}) = \frac{\partial \bar{u}}{\partial \bar{x}} - \bar{y}\frac{\partial^2 \bar{v}}{\partial \bar{x}^2} - \bar{z}\frac{\partial^2 \bar{w}}{\partial \bar{x}^2}, \qquad \bar{\gamma}(\bar{x},\bar{y},\bar{z}) = \bar{r}\frac{\partial \bar{\phi}}{\partial \bar{x}}, \qquad \bar{r} = \sqrt{\bar{y}^2 + \bar{z}^2}$$

At the local level, we assume small strains and hence each of the actions are uncoupled. That is, we can analyze them separately and at the end sum them together. Additionally, we need not distinguish between the undeformed and deformed configurations.

We begin with the axial behavior. Let the axial displacement have a linear distribution

$$\bar{u}(\bar{x}) = \left[1 - \frac{\bar{x}}{L}\right]\bar{u}_1 + \left[\frac{\bar{x}}{L}\right]\bar{u}_2 = f_1(\bar{x})\bar{u}_1 + f_2(\bar{x})\bar{u}_2$$

There are two shape functions because there are two degrees of freedom. Substitute this into the expression for the strain to get

$$\bar{\epsilon}(\bar{x}, \bar{y}, \bar{z}) = \frac{1}{L}(-\bar{u}_1 + \bar{u}_2)$$

Write this in matrix form as

$$\bar{\epsilon}(\bar{x}, \bar{y}, \bar{z}) = \{c(\bar{x}, \bar{y}, \bar{z})\}^T \{\bar{u}\}$$

where $\{c\}$ is of size $[2 \times 1]$. The material behavior is assumed to be linearly elastic, hence the axial stress is given by

$$\bar{\sigma} = E\bar{\epsilon} = E\{\bar{c}\}^T \{\bar{u}\}$$

The principle of virtual work can be used to determine a set of element nodal forces consistent with the internal stress as

$$\{\bar{F}\}^T \{\delta\bar{u}\} = \int \sigma\delta\bar{\epsilon}\, dV = \int E\{\bar{u}\}^T \{\bar{c}\}\{\bar{c}\}^T \{\delta\bar{u}\}\, dV$$

The integration is performed with respect to $dV = d\bar{x}\, d\bar{y}\, d\bar{z} = d\bar{x}\, dA$ and leads to

$$\{\bar{F}\}^T \{\delta\bar{u}\} = \bar{F}_1\delta\bar{u}_1 + \bar{F}_2\delta\bar{u}_2 = \frac{EA}{L}[\bar{u}_1 - \bar{u}_2]\delta\bar{u}_1 + \frac{EA}{L}[-\bar{u}_1 + \bar{u}_2]\delta\bar{u}_2$$

Because the virtual displacements are arbitrary, we can equate corresponding coefficients of the variations to get expressions for the nodal forces. Consequently,

$$\left\{\begin{matrix} \bar{F}_1 \\ \bar{F}_2 \end{matrix}\right\} = \frac{EA}{L}\begin{bmatrix} 1 & -1 \\ -1 & 1 \end{bmatrix}\left\{\begin{matrix} \hat{u}_1 \\ \hat{u}_2 \end{matrix}\right\}$$

This is the element stiffness for a rod.

Let the transverse deflection have a cubic distribution

$$\bar{v}(\bar{x}) = \left[1 - 3(\frac{\bar{x}}{L})^2 + 2(\frac{\bar{x}}{L})^3\right]v_1 + (\frac{\bar{x}}{L})\left[1 - 2(\frac{\bar{x}}{L}) + (\frac{\bar{x}}{L})^2\right]L\phi_{1z} + (\frac{\bar{x}}{L})^2\left[3 - 2(\frac{\bar{x}}{L})\right]v_2$$

$$+ (\frac{\bar{x}}{L})^2\left[-1 + (\frac{\bar{x}}{L})\right]L\phi_{2z} = g_1(\bar{x})v_1 + g_2(\bar{x})\phi_{1z} + g_3(\bar{x})v_2 + g_4(\bar{x})\phi_{2z}$$

There are four shape functions because there are four degrees of freedom. Substitute these into the expression for the strain to get

$$\bar{\epsilon}(\bar{x}, \bar{y}, \bar{z}) = -\bar{y}\left\{\left[-\frac{6}{L^2} + \frac{12\bar{x}}{L^3}\right]\bar{v}_1 + \left[-\frac{4}{L} + \frac{6\bar{x}}{L^2}\right]\phi_{1z} + \left[-\frac{6}{L^2} - \frac{12\bar{x}}{L^3}\right]\bar{v}_2 + \cdots\right\}$$

Write this in matrix form as $\bar{\epsilon}(\bar{x}, \bar{y}, \bar{z}) = \{c(\bar{x}, \bar{y}, \bar{z})\}^T \{\bar{u}\}$ where now $\{c\}$ is of size $[4 \times 1]$. The material behavior is assumed to be linearly elastic, and as for the

rod, the principle of virtual work can be used to determine a set of element nodal
forces consistent with the internal stress as

$$\{\bar{F}\}^T\{\delta\bar{u}\} = \int \bar{\sigma}\delta\bar{\epsilon}\,dV = \int E\{\bar{u}\}^T\{\bar{c}\}\{\bar{c}\}^T\{\delta\bar{u}\}\,dV$$

The integrations on the cross-section will give rise to the terms

$$\int \bar{y}^2 dA = I_{zz}\,, \quad \int \bar{y}\,dA = 0$$

The last term is zero because we assume that the local axes are principal axes.
Performing the integrations leads to

$$\begin{aligned}
\{\bar{F}\}^T\{\delta\bar{u}\} &= \bar{V}_1\delta\bar{v}_1 + \bar{M}_{1z}\delta\bar{\phi}_{1z} + \bar{V}_2\delta\bar{v}_2 + \bar{M}_{2z}\delta\bar{\phi}_{1z} \\
&= EI_{zz}\int_L \left(\left[-\frac{6}{L^2}+\frac{12\bar{x}}{L^3}\right]\bar{v}_1 + \left[-\frac{4}{L}+\frac{6\bar{x}}{L^2}\right]\phi_{1z} + \cdots\right) \\
&\qquad \left(\left[-\frac{6}{L^2}+\frac{12\bar{x}}{L^3}\right]\delta\bar{v}_1 + \left[-\frac{4}{L}+\frac{6\bar{x}}{L^2}\right]\delta\phi_{1z} + \cdots\right)\,d\bar{x} \\
&= \frac{EI_{zz}}{L^3}[12\bar{v}_1 + 6L\bar{\phi}_1 - 12\bar{v}_2 + 6L\bar{\phi}_2]\delta\bar{v}_1 + \cdots
\end{aligned}$$

Because the virtual displacements are arbitrary, we can equate corresponding
coefficients of the variations to get expressions for the nodal forces. This results
in

$$\begin{Bmatrix} \bar{V}_1 \\ \bar{M}_{1z} \\ \bar{V}_2 \\ \bar{M}_{2z} \end{Bmatrix} = \frac{EI_{zz}}{L^3}\begin{bmatrix} 12 & 6L & -12 & 6L \\ 6L & 4L^2 & -6L & 2L^2 \\ -12 & -6L & 12 & -6L \\ 6L & 2L^2 & -6L & 4L^2 \end{bmatrix}\begin{Bmatrix} \bar{v}_1 \\ \bar{\phi}_{1z} \\ \bar{v}_2 \\ \bar{\phi}_{2z} \end{Bmatrix}$$

which is the element stiffness for a beam. The bending action in the $\bar{x} - \bar{z}$ plane
is the same as above except that I_{zz} is replaced with I_{yy}.

For the torsion action, for simplicity, assume the cross-section is circular and let
the relative axial twist be linearly distributed, then we have for the shear strain

$$\bar{\gamma}(\bar{x},\bar{y},\bar{z}) = \bar{r}\frac{\partial\bar{\phi}}{\partial\bar{x}} = \frac{\partial}{\partial\bar{x}}\left[1-(\frac{\bar{x}}{L})\right]\bar{\phi}_{1x} + \frac{\partial}{\partial\bar{x}}\left[(\frac{\bar{x}}{L})\right]\bar{\phi}_{2x} = \bar{r}\frac{1}{L}(-\bar{\phi}_{1x}+\bar{\phi}_{2x})$$

where $\bar{r} = \sqrt{\bar{y}^2 + \bar{z}^2}$. The shear stress is then

$$\bar{\tau} = G\bar{\gamma} = G\bar{r}\frac{1}{L}(-\bar{\phi}_{1x}+\bar{\phi}_{2x})$$

Again, the principle of virtual work can be used to determine a set of element
nodal forces consistent with this internal stress as

$$\{\bar{F}\}^T\{\delta\bar{u}\} = \int \bar{\tau}\delta\bar{\gamma}\,dV = \int G\bar{r}(\frac{1}{L^2})(-\bar{\phi}_1+\bar{\phi}_{2x})(-\delta\bar{\phi}_1+\delta\bar{\phi}_{2x})\,dV$$

The integration on the cross-section will give rise to the term $\int \bar{r}^2 d\bar{y}\,d\bar{z} = I_{xx}$ and
we get

$$\{\bar{F}\}^T\{\delta\bar{u}\} = \bar{T}_1\delta\bar{\phi}_{1x} + \bar{T}_2\delta\bar{\phi}_{2x} = \frac{GI_{xx}}{L}\left[(-\bar{\phi}_{1x}+\bar{\phi}_{2x})(-\delta\bar{\phi}_{1x}+\delta\bar{\phi}_{2x})\right]$$

If the cross-section in noncircular, we just substitute an appropriate value for GI_{xx}. Because the virtual displacements are arbitrary, we get the nodal torques as

$$\left\{ \begin{array}{c} \bar{T}_1 \\ \bar{T}_2 \end{array} \right\} = \frac{GI_{xx}}{L} \begin{bmatrix} 1 & -1 \\ -1 & 1 \end{bmatrix} \left\{ \begin{array}{c} \bar{\phi}_{1x} \\ \bar{\phi}_{2x} \end{array} \right\}$$

This is the element stiffness for a rod in torsion.

A 2-D frame deforming in the $x-y$ plane has a combination of axial and flexural behaviors. The element stiffness in local coordinates is then

$$[\bar{k}_E] = \frac{EA}{L} \begin{bmatrix} 1 & 0 & 0 & -1 & 0 & 0 \\ 0 & 0 & 0 & 0 & 0 & 0 \\ 0 & 0 & 0 & 0 & 0 & 0 \\ -1 & 0 & 0 & 1 & 0 & 0 \\ 0 & 0 & 0 & 0 & 0 & 0 \\ 0 & 0 & 0 & 0 & 0 & 0 \end{bmatrix} + \frac{EI_{zz}}{L^3} \begin{bmatrix} 0 & 0 & 0 & 0 & 0 & 0 \\ 0 & 12 & 6L & 0 & -12 & 6L \\ 0 & 6L & 4L^2 & 0 & -6L & 2L^2 \\ 0 & 0 & 0 & 0 & 0 & 0 \\ 0 & -12 & -6L & 0 & 12 & -6L \\ 0 & 6L & 2L^2 & 0 & -6L & 4L^2 \end{bmatrix}$$

which relates $\{\bar{F}_1, \bar{V}_1, \bar{M}_{1z}; \bar{F}_2, \bar{V}_2, \bar{M}_{2z}\}^T$ to $\{\bar{u}_1, \bar{v}_1, \bar{\phi}_{1z}; \bar{u}_2, \bar{v}_2, \bar{\phi}_{2z}\}^T$. A 3-D frame member has an axial load, two bending actions, and an axial twist; these properties are described by EA, EI_{zz}, EI_{yy}, and GI_{xx}, respectively.

Discussion

The preceding sections have laid out various aspects of the mechanics of deformable bodies. We now wish to draw some implications for further developments.

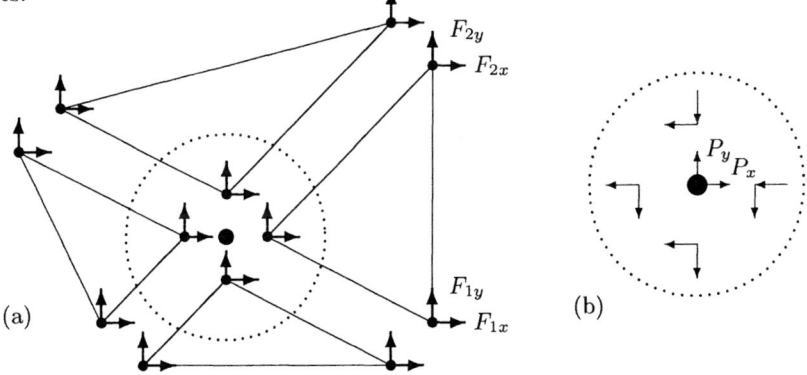

Figure 1.33: Equilibrium of a discretized region. (a) Element nodal forces. (b) Structural nodal forces.

In the discretization process, we impose continuity between the subregions. The assemblage process is then the satisfaction of the equilibrium conditions between these subregions. Thus, in reference to Figure 1.33, equilibrium in terms of the nodal forces is pertinent on two levels:

- The element is in equilibrium with itself even though differential equilibrium is not (necessarily) satisfied at all points.

- At any structural node, the sum of the element nodal point forces is in equilibrium with the externally applied nodal loads (including body, surface, initial, inertial, damping, and reaction loads).

The first of these is relevant in the formulation of the elements, while the second is relevant for the assemblage. As already shown, we have for the assemblage process

$$\{F\} = \sum_m \{F\}^{(m)} = \{P\}$$

where $\{P\}$ is the collection of external loads made up of applied loads and inertia loads. This relation must be satisfied throughout the loading history.

From this preliminary discussion, we can identify a number of ingredients necessary for a complete algorithmic presentation for solving nonlinear plate and shell problems:

- Formulation for thin-plated structures.
- Means to discretize distributed bodies.
- Geometry description for large displacements and rotations.
- Scheme for time/load stepping.
- Nonlinear equation solver.

We consider each of these ingredients in detail in the next few chapters.

Problems

1.1 A block rotates an angle θ about the x_3-axis.
- Write down its deformation and obtain the deformation gradient.
- Show that the volume change is zero.

1.2 Consider the following deformation

$$x_1 = 3x_1^o + k\,x_2^o, \qquad x_2 = 2x_1^o + 4x_2^o, \qquad x_3 = x_3^o$$

- What are the restrictions on k for this to be a valid deformation?
- Draw the deformed shape. Show by measurement the consistency of the physical interpretation of the Lagrangian strains with their connection to the deformation gradient.
- Show that the formulas describing the deformation of areas are in agreement with the geometric construction.
- Determine the principal strains.
- Draw the before and after positions of the principal element.

1.3 The Lagrangian strain tensor at a point is

$$[E_{ij}] = \begin{bmatrix} 2 & -1 & \sqrt{2} \\ -1 & 3 & -\sqrt{2} \\ \sqrt{2} & -\sqrt{2} & 4 \end{bmatrix}$$

- What is the strain of a line element initially oriented as $\hat{n} = \frac{1}{2}\hat{e}_1 - \frac{1}{2}\hat{e}_2 + \frac{1}{\sqrt{2}}\hat{e}_3$.
- What is the shear strain between two line elements initially oriented as $\hat{n}^a = \frac{1}{2}\hat{e}_1 - \frac{1}{2}\hat{e}_2 + \frac{1}{\sqrt{2}}\hat{e}_3$ and $\hat{n}^b = -\frac{1}{2}\hat{e}_1 + \frac{1}{2}\hat{e}_2 + \frac{1}{\sqrt{2}}\hat{e}_3$.

1.4 Consider the deformation of a square such that the corners move as

$$(0,0) \Rightarrow (0,0) \qquad (1,0) \Rightarrow (1,1.5) \qquad (0,1) \Rightarrow (-1,2)$$

- Describe the deformation mathematically.
- Determine the Lagrangian strain tensor.
- What can be said about the deformation given by

$$(0,0) \Rightarrow (0,0) \qquad (0,1) \Rightarrow (1,1.5) \qquad (1,0) \Rightarrow (-1,2)$$

1.5 Consider the following components of a stress tensor

$$[\sigma_{ij}] = \begin{bmatrix} 1 & 2 & 0 \\ 2 & 3 & 0 \\ 0 & 0 & 0 \end{bmatrix}$$

- Determine the components of the traction vector with respect to an area rotated θ about the x_3-axis.
- Determine the components of stress transformed an angle θ about the same axis.
- How do the above compare or are they related ?

1.6 A stress distribution field is described by

$$\sigma_{11} = 3x_1 + k_1 x_2^2, \qquad \sigma_{22} = 2x_1 + 4x_2, \qquad \sigma_{12} = a + bx_1 + cx_1^2 dx_2 + ex_2^2 + fx_1x_2$$

- Under what circumstances (if any) is the symmetric stress field in static equilibrium ?

1.7 Consider the simple shear deformation

$$x_1 = x_1^o + kx_2^o, \qquad x_2 = x_2^o, \qquad x_3 = x_3^o$$

and the constitutive behavior

$$\sigma_{ij}^K = 2G\, E_{ij} + \lambda \delta_{ij} E_{kk}$$

- Determine the Lagrangian stress and Cauchy stress.
- Investigate the forces and the areas they act on.

1.8 A rigid block has a Cauchy stress σ_{11} only acting on it. The block is given a rigid body rotation about the x_3-axis such that σ_{11} moves with it.
- What are the new Cauchy stresses?
- Determine the components of the Kirchhoff stress before and after the rotation.
- Show that the Lagrangian strain tensor is also invariant to the same rigid body rotation.

1.9 Consider a cantilever beam, fixed at the end $x = 0$, and subjected to a concentrated lateral applied force at the other.
- Using the Ritz method, show that the displacement $v(x) = a_0 + a_1 x + a_2 x^2 + a_3 x^3$, leads to the exact solution.
- Show that the addition of extra terms have zero contributions.

2
Thin Plates and Shells

The distribution of displacement and stress fields throughout a generally loaded structure is very complicated and only computational methods can give effective solutions. The key to the finite element method is discretizing the structure into a collection of small regions that are easier to handle. Figure 2.1 shows some examples of thin-walled structures modeled as a collection of many triangular subregions. This chapter considers the formulation of these triangular elements.

A plate is an extended body where one of the dimensions is substantially smaller than the other two. The plates in 3-D thin-walled structures can support both in-plane and out-of-plane loading. Furthermore, because the plates are thin, they lend themselves to approximation — while the structure may be three-dimensional, the local behavior is two-dimensional under plane stress. We take advantage of this to formulate an effective solution. Plates in flexure are the two-dimensional equivalent of beams and *classical plate theory* is its equivalent of the Bernoulli-Euler beam theory; whereas the in-plane or *membrane* behavior of plates is analogous to that of rods. So as to concentrate on the essentials, in this chapter we limit the deflections and strains to being small.

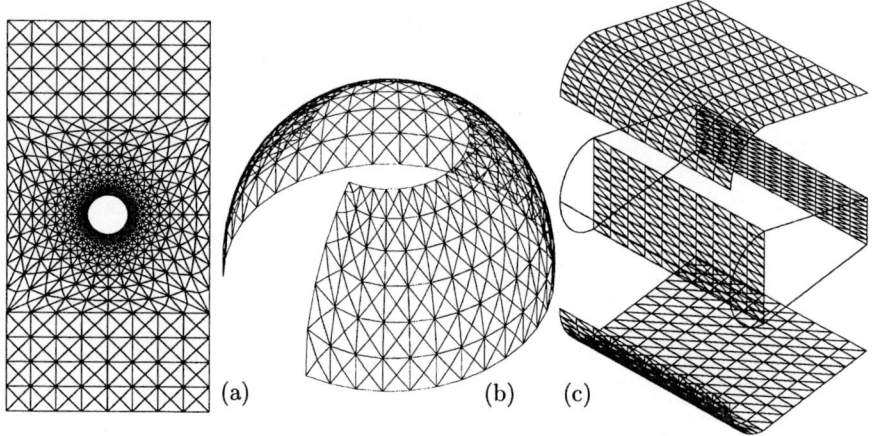

Figure 2.1: Thin-walled structures discretized as collections of triangular finite elements. (a) Plate with hole. (b) Segmented dome. (c) Exploded view of a wing section.

2.1   Flat Plate Theory

Fundamentally, plate theory is an approximate structural theory and therefore it is best to approach it by way of a variational principle. We will begin by developing a plate theory (called *Mindlin plate theory*) that takes the shear deformation into account — this is the plate equivalent of the Timoshenko beam. The transition to achieve the classical or thin-plate theory is then more transparent.

Equations of Motion

Consider a rectangular plate of thickness h as shown in Figure 2.2. The plate lies in the x-y plane and is subjected to both in-plane and transverse loads. The mid-plane of the plate is taken at $z = 0$.

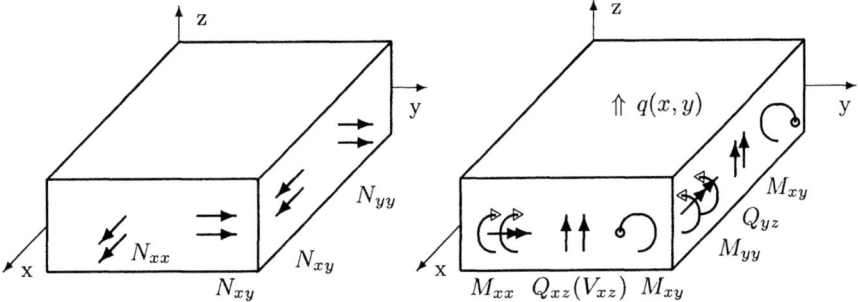

Figure 2.2: Element of stressed plate.

Because the plate is thin, we begin by expanding the displacements in a Taylor series (in terms of z) about the mid-plane values as

$$\begin{aligned}
\bar{u}(x,y,z) &\approx u(x,y) - z\psi_x(x,y) \\
\bar{v}(x,y,z) &\approx v(x,y) - z\psi_y(x,y) \\
\bar{w}(x,y,z) &\approx w(x,y)
\end{aligned} \tag{2.1}$$

where ψ_x and ψ_y are rotations of the subscripted faces in the directions of the curvatures. These say that the deformation is governed by five independent functions: $u(x,y)$, $v(x,y)$ are the in-plane displacements; $w(x,y)$ is the out-of-plane displacement; and $\psi_x(x,y)$, $\psi_y(x,y)$ are the rotations of the mid-plane. It is understood that all variables are also functions of time, but we make the assumption that the above kinematic representations do not change under dynamic conditions.

The normal and shear strains corresponding to the above deformations are

$$\bar{\epsilon}_{xx} \quad = \quad \frac{\partial \bar{u}}{\partial x} = \frac{\partial u}{\partial x} - z \frac{\partial \psi_x}{\partial x}$$

$$\bar{\epsilon}_{yy} = \frac{\partial \bar{v}}{\partial y} = \frac{\partial v}{\partial y} - z\frac{\partial \psi_y}{\partial y}$$

$$\bar{\gamma}_{xy} = \frac{\partial \bar{u}}{\partial y} + \frac{\partial \bar{v}}{\partial x} = \left(\frac{\partial u}{\partial y} + \frac{\partial v}{\partial x}\right) - z\left(\frac{\partial \psi_x}{\partial y} + \frac{\partial \psi_y}{\partial x}\right)$$

$$\bar{\gamma}_{xz} = \frac{\partial \bar{u}}{\partial z} + \frac{\partial \bar{w}}{\partial x} = \left(-\psi_x + \frac{\partial w}{\partial x}\right)$$

$$\bar{\gamma}_{yz} = \frac{\partial \bar{v}}{\partial z} + \frac{\partial \bar{w}}{\partial y} = \left(-\psi_y + \frac{\partial w}{\partial y}\right)$$

Because the plate is thin, the stress in the z direction cannot be very large. We therefore assume that it is approximately zero; that is, we assume a state of plane stress with $\sigma_{zz} = 0$. Substituting for the strains in the Hooke's law for plane stress then leads to

$$\bar{\sigma}_{xx} = \frac{E}{1-\nu^2}[\bar{\epsilon}_{xx} + \nu\bar{\epsilon}_{yy}] = \frac{E}{1-\nu^2}\left[\left(\frac{\partial u}{\partial x} + \nu\frac{\partial v}{\partial y}\right) - z\left(\frac{\partial \psi_x}{\partial x} + \nu\frac{\partial \psi_y}{\partial y}\right)\right]$$

$$\bar{\sigma}_{yy} = \frac{E}{1-\nu^2}[\bar{\epsilon}_{yy} + \nu\bar{\epsilon}_{xx}] = \frac{E}{1-\nu^2}\left[\left(\frac{\partial v}{\partial y} + \nu\frac{\partial u}{\partial x}\right) - z\left(\frac{\partial \psi_y}{\partial y} + \nu\frac{\partial \psi_x}{\partial x}\right)\right]$$

$$\bar{\sigma}_{zz} = 0 \qquad\qquad\qquad\qquad\qquad\qquad\qquad\qquad\qquad (2.2)$$

$$\bar{\sigma}_{xy} = G\gamma_{xy} = G\left[\left(\frac{\partial u}{\partial y} + \frac{\partial v}{\partial x}\right) - z\left(\frac{\partial \psi_x}{\partial y} + \frac{\partial \psi_y}{\partial x}\right)\right]$$

$$\bar{\sigma}_{xz} = G\gamma_{xz} = G\left[-\psi_x + \frac{\partial w}{\partial x}\right], \qquad \bar{\sigma}_{yz} = G\gamma_{yz} = G\left[-\psi_y + \frac{\partial w}{\partial y}\right]$$

Although the plate is treated as being in plane stress, we still retain the $\bar{\sigma}_{xz}$ and $\bar{\sigma}_{yz}$ shear stresses.

The strain energy for the plate is

$$U = \frac{1}{2}\int_V [\bar{\sigma}_{xx}\bar{\epsilon}_{xx} + \bar{\sigma}_{yy}\bar{\epsilon}_{yy} + \bar{\sigma}_{xy}\bar{\gamma}_{xy} + \bar{\sigma}_{xz}\bar{\gamma}_{xz} + \bar{\sigma}_{yz}\bar{\gamma}_{yz}]\, dV$$

Substitute for the stresses and strains and integrate with respect to the thickness to get the total strain energy as

$$U = \frac{1}{2}\int_A D\left[\left(\frac{\partial \psi_x}{\partial x} + \frac{\partial \psi_y}{\partial y}\right)^2 - \frac{1}{2}(1-\nu)\left[4\frac{\partial \psi_x}{\partial x}\frac{\partial \psi_y}{\partial y} - \left(\frac{\partial \psi_y}{\partial x} + \frac{\partial \psi_x}{\partial y}\right)^2\right]\right] dx dy$$

$$+ \frac{1}{2}\int_A Gh\left[\left(\psi_x - \frac{\partial w}{\partial x}\right)^2 + \left(\psi_y - \frac{\partial w}{\partial y}\right)^2\right] dx\, dy \qquad (2.3)$$

$$+ \frac{1}{2}\int_A \left[E^*h\left[\left(\frac{\partial u}{\partial x}\right)^2 + \left(\frac{\partial v}{\partial y}\right)^2 + 2\nu\frac{\partial u}{\partial x}\frac{\partial v}{\partial y}\right] + Gh\left(\frac{\partial u}{\partial y} + \frac{\partial v}{\partial x}\right)^2\right] dx\, dy$$

where $D \equiv Eh^3/12(1-\nu^2)$ is called the plate bending stiffness and $E^* \equiv E/(1-\nu^2)$. The total kinetic energy is

$$T = \frac{1}{2}\int_V \rho[\dot{\bar{u}}(x,y,z,t)^2 + \dot{\bar{v}}(x,y,z,t)^2 + \dot{\bar{w}}(x,y,z,t)^2]dV$$

$$= \frac{1}{2}\int_A \left[\rho h[\dot{u}^2 + \dot{v}^2 + \dot{w}^2] + \frac{1}{12}\rho h^3[\dot{\psi}_x^2 + \dot{\psi}_y^2]\right] dx\, dy \qquad (2.4)$$

If the applied surface tractions and loads on the plate are as shown in Figure 2.2, then the potential of these loads is

$$V = -\int q(x,y)w\, dx\, dy - N_{xx}u - N_{xy}v - M_{xx}\psi_x - M_{xy}\psi_y - V_{xz}w + \cdots$$

where the edge loads can be on each face. The energies de-couple into in-plane (u and v) and out-of-plane (w, ψ_x and ψ_y) sets; hence, we now find it convenient to treat them separately.

In-Plane Membrane Behavior

The energies for the in-plane behavior are

$$U = \tfrac{1}{2}\int_A \left[E^*h\left[\left(\frac{\partial u}{\partial x}\right)^2 + \left(\frac{\partial v}{\partial y}\right)^2 + 2\nu \frac{\partial u}{\partial x}\frac{\partial v}{\partial y}\right] + Gh\left(\frac{\partial u}{\partial y} + \frac{\partial v}{\partial x}\right)^2 \right] dx\, dy$$

$$T = \tfrac{1}{2}\int_A \rho h[\dot{u}^2 + \dot{v}^2]\, dx\, dy, \qquad V = -N_{xx}u - N_{xy}v + \cdots \tag{2.5}$$

Application of Hamilton's principle (as illustrated in Chapter 1) with variations in δu and δv leads to two differential equations

$$\frac{Eh}{1-\nu^2}\left[\nabla^2 u - \tfrac{1}{2}(1+\nu)\left(\frac{\partial^2 u}{\partial y^2} - \frac{\partial^2 v}{\partial x \partial y}\right)\right] = \rho h \frac{\partial^2 u}{\partial t^2} \tag{2.6}$$

$$\frac{Eh}{1-\nu^2}\left[\nabla^2 v - \tfrac{1}{2}(1+\nu)\left(\frac{\partial^2 v}{\partial x^2} - \frac{\partial^2 u}{\partial x \partial y}\right)\right] = \rho h \frac{\partial^2 v}{\partial t^2}, \qquad \nabla^2 \equiv \frac{\partial^2}{\partial x^2} + \frac{\partial^2}{\partial y^2}$$

These are the *Navier's equations*. Damping is easily incorporated into the equations by modifying the inertia terms; that is,

$$\rho h \frac{\partial^2 u_i}{\partial t^2} \quad \longrightarrow \quad \rho h \frac{\partial^2 u_i}{\partial t^2} + \eta h \frac{\partial u_i}{\partial t}$$

where η is the damping (or viscous) coefficient.

For the associated boundary conditions, we specify one condition from either set:

$$\left\{ u \quad \text{or} \quad N_{xx} = \frac{Eh}{1-\nu^2}\left[\frac{\partial u}{\partial x} + \nu \frac{\partial v}{\partial y}\right] \right\}, \quad \left\{ v \quad \text{or} \quad N_{xy} = Gh\left[\frac{\partial u}{\partial y} + \frac{\partial v}{\partial x}\right] \right\}$$

We can give interpretation of the boundary conditions in terms of resultants on the cross-section. For example, the resultants of the normal and shear stresses are defined on the cross-section as

$$N_{xx}(x,y) \equiv \int \bar{\sigma}_{xx}(x,y,z)\, dz, \qquad N_{xy}(x,y) \equiv \int \bar{\sigma}_{xy}(x,y,z)\, dz$$

and leads to

$$N_{xx} = \frac{Eh}{(1-\nu^2)}\left[\frac{\partial u}{\partial x} + \nu\frac{\partial v}{\partial y}\right] = \sigma_{xx}h$$

$$N_{yy} = \frac{Eh}{(1-\nu^2)}\left[\frac{\partial v}{\partial y} + \nu\frac{\partial u}{\partial x}\right] = \sigma_{yy}h$$

$$N_{xy} = \frac{Eh}{2(1+\nu)}\left[\frac{\partial u}{\partial y} + \frac{\partial v}{\partial x}\right] = \sigma_{xy}h \tag{2.7}$$

That is, N_{xx} and so on, are the resultant forces per unit length due to the stresses acting on the edge faces.

Example 2.1: Specialize the Navier's equation to the case where there is only an x dependence.

There are no derivatives with respect to y. The first of the two Navier's equations becomes

$$\frac{Eh}{1-\nu^2}\frac{\partial^2 u}{\partial x^2} = \rho h\frac{\partial^2 u}{\partial t^2} \qquad \text{or} \qquad E^* h\frac{\partial^2 u}{\partial x^2} = \rho h\frac{\partial^2 u}{\partial t^2}$$

which is the one-dimensional wave equation for longitudinal disturbances propagating in a rod [23]. The second of the Navier's equations becomes

$$\frac{Eh}{1-\nu^2}\left[\frac{\partial^2 v}{\partial x^2} - \tfrac{1}{2}(1+\nu)\left(\frac{\partial^2 v}{\partial x^2}\right)\right] = \rho h\frac{\partial^2 v}{\partial t^2} \qquad \text{or} \qquad Gh\frac{\partial^2 v}{\partial x^2} = \rho h\frac{\partial^2 v}{\partial t^2}$$

This is also a one-dimensional wave equation but it is for shear disturbances propagating in a rod [23]. This is not the flexural shear behavior.

For the associated boundary conditions, we specify one condition from either set:

$$\left\{u \quad \text{or} \quad N_{xx} = \frac{Eh}{1-\nu^2}\frac{\partial u}{\partial x}\right\}, \quad \left\{v \quad \text{or} \quad N_{xy} = Gh\frac{\partial v}{\partial x}\right\}$$

Out-of-Plane Flexural Behavior

The energies associated with the out-of-plane behavior are

$$U = \tfrac{1}{2}\int_A D\left[\left(\frac{\partial\psi_x}{\partial x} + \frac{\partial\psi_y}{\partial y}\right)^2 - \tfrac{1}{2}(1-\nu)\left[4\frac{\partial\psi_x}{\partial x}\frac{\partial\psi_y}{\partial y} - \left(\frac{\partial\psi_y}{\partial x} + \frac{\partial\psi_x}{\partial y}\right)^2\right]\right]dxdy$$

$$\qquad + \tfrac{1}{2}\int_A Gh\left[\left(\psi_x - \frac{\partial w}{\partial x}\right)^2 + \left(\psi_y - \frac{\partial w}{\partial y}\right)^2\right]dx\,dy$$

$$T = \tfrac{1}{2}\int_A \left[\rho h[\dot{w}^2] + \tfrac{1}{12}\rho h^3[\dot{\psi}_x{}^2 + \dot{\psi}_y{}^2]\right]dx\,dy$$

$$V = -\int q(x,y)w\,dx\,dy - M_{xx}\psi_x - M_{xy}\psi_y - V_{xz}w + \cdots \tag{2.8}$$

An application of Hamilton's principle with the variations of δw, $\delta\psi_x$ and $\delta\psi_y$, leads to, respectively,

$$q + Gh\frac{\partial}{\partial x}\left[\frac{\partial w}{\partial x} - \psi_x\right] + Gh\frac{\partial}{\partial y}\left[\frac{\partial w}{\partial y} - \psi_y\right] = \rho h\ddot{w}$$

$$\tfrac{1}{2}D\left[(1-\nu)\nabla^2\psi_x + (1+\nu)\frac{\partial}{\partial x}\left(\frac{\partial\psi_x}{\partial x} + \frac{\partial\psi_y}{\partial y}\right)\right] + Gh\left[\frac{\partial w}{\partial x} - \psi_x\right] = \rho I_p\ddot{\psi}_x$$

$$\tfrac{1}{2}D\left[(1-\nu)\nabla^2\psi_y + (1+\nu)\frac{\partial}{\partial y}\left(\frac{\partial\psi_x}{\partial x} + \frac{\partial\psi_y}{\partial y}\right)\right] + Gh\left[\frac{\partial w}{\partial y} - \psi_y\right] = \rho I_p\ddot{\psi}_y$$

$$(2.9)$$

where $I_p \equiv h^3/12$. These are the equations of motion for the *Mindlin plate*; this theory accounts for the shear deformation as well as the rotational inertia. The associated boundary conditions (on each edge face of the plate) are specified in terms of any three conditions selected from the following groups:

$$\left\{ w \quad \text{or} \quad V_{xz} = Gh\left[\frac{\partial w}{\partial x} - \psi_x\right] \right\}$$

$$\left\{ \psi_x \quad \text{or} \quad M_{xx} = D\left[\frac{\partial\psi_x}{\partial x} + \nu\frac{\partial\psi_y}{\partial y}\right] \right\}$$

$$\left\{ \psi_y \quad \text{or} \quad M_{xy} = \tfrac{1}{2}(1-\nu)D\left[\frac{\partial\psi_x}{\partial y} + \frac{\partial\psi_y}{\partial x}\right] \right\}$$

These are specified for an x-face, the other faces are similar.

We can give interpretation to the boundary conditions in terms of resultants of the stresses on the cross section. For example, taking resultants for the shear stress defined as

$$Q_{xz}(x,y) \equiv \int \bar{\sigma}_{xz}(x,y,z)\,dz = \int G\left[-\psi_x + \frac{\partial w}{\partial x}\right]dz$$

leads to

$$Q_{xz} = Gh\left[-\psi_x + \frac{\partial w}{\partial x}\right] = V_{xz}, \quad Q_{yz} = Gh\left[-\psi_y + \frac{\partial w}{\partial y}\right] = V_{yz} \quad (2.10)$$

We can also take a moment due to the stresses acting on the edge faces. For example,

$$M_{xx} \equiv -\int \sigma_{xx}z\,dz = \frac{Eh^3}{12(1-\nu^2)}\left[\frac{\partial\psi_x}{\partial x} + \nu\frac{\partial\psi_y}{\partial y}\right]$$

and all resultants can be written as

$$M_{xx} = D\left[\frac{\partial\psi_x}{\partial x} + \nu\frac{\partial\psi_y}{\partial y}\right]$$

$$M_{yy} = D\left[\frac{\partial\psi_y}{\partial y} + \nu\frac{\partial\psi_x}{\partial x}\right]$$

$$2M_{xy} = D\left[\frac{\partial\psi_y}{\partial x} + \frac{\partial\psi_x}{\partial y}\right](1-\nu) \quad (2.11)$$

These resultants are related only to the rotations.

In order to account for the truncation error of the expansions $\bar{u}$ and $\bar{v}$, we could add correction coefficients to the energies as is usually done with the Timoshenko beam theory [23]. We will not pursue this here because our interest is to develop a theory for thin plates.

Flexural Behavior of Thin Plates

The plate theory derived here (called classical plate theory) is the 2-D equivalent of the Bernoulli-Euler beam theory. Rather than go directly to the governing equations, we will retrace the developments of the Mindlin plate, but with the assumptions of the classical theory.

We modify the Mindlin equations to the thin-plate theory in two steps. First, we assume that the transverse shear deformation is negligible; this is equivalent to saying that the shear stiffness in the transverse direction is infinite. This leads to

$$\frac{\partial w}{\partial x} - \psi_x = 0, \qquad \frac{\partial w}{\partial y} - \psi_y = 0$$

It is important to realize that while these combinations are zero, their product with Gh is nonzero (because it is related to the transverse shear resultant). The displacements for the flexural motion are approximated as

$$\bar{u}(x,y,z) \approx -z\frac{\partial w}{\partial x}(x,y), \quad \bar{v}(x,y,z) \approx -z\frac{\partial w}{\partial y}(x,y), \quad \bar{w}(x,y,z) \approx w(x,y)$$

The normal and shear strains corresponding to these deformations are

$$\bar{\epsilon}_{xx} = \frac{\partial \bar{u}}{\partial x} = -z\frac{\partial^2 w}{\partial x^2}, \qquad \bar{\epsilon}_{yy} = \frac{\partial \bar{v}}{\partial y} = -z\frac{\partial^2 w}{\partial y^2}$$

$$\bar{\gamma}_{xy} = \frac{\partial \bar{u}}{\partial y} + \frac{\partial \bar{v}}{\partial x} = -2z\frac{\partial^2 w}{\partial x \partial y}$$

$$\bar{\gamma}_{xz} = \frac{\partial \bar{u}}{\partial z} + \frac{\partial \bar{w}}{\partial x} = 0, \qquad \bar{\gamma}_{yz} = \frac{\partial \bar{v}}{\partial z} + \frac{\partial \bar{w}}{\partial y} = 0$$

We reiterate that, although the transverse shear strains are zero, the transverse shear forces are nonzero. Also note that there is an in-plane shear that depends on the distance from the midplane — there is no comparable quantity in beam theories. Substituting these strains into the Hooke's law for plane stress gives

$$\bar{\sigma}_{xx} = \frac{-Ez}{1-\nu^2}\left[\frac{\partial^2 w}{\partial x^2} + \nu\frac{\partial^2 w}{\partial y^2}\right]$$

$$\bar{\sigma}_{yy} = \frac{-Ez}{1-\nu^2}\left[\frac{\partial^2 w}{\partial y^2} + \nu\frac{\partial^2 w}{\partial x^2}\right]$$

$$\bar{\sigma}_{xy} = -2Gz\left[\frac{\partial^2 w}{\partial x \partial y}\right] \qquad (2.12)$$

The strain energy for a plate in plane stress is

$$U = \tfrac{1}{2} \int_V \left[\bar{\sigma}_{xx}\bar{\epsilon}_{xx} + \bar{\sigma}_{yy}\bar{\epsilon}_{yy} + \bar{\sigma}_{xy}\bar{\gamma}_{xy} \right] dV$$

Substitute for the stresses and strains and integrate with respect to the thickness to get the total strain energy as

$$
\begin{aligned}
U &= \tfrac{1}{2} \int D\left[(\nabla^2 w)^2 + 2(1-\nu)\left[(\frac{\partial^2 w}{\partial x \partial y})^2 - \frac{\partial^2 w}{\partial x^2}\frac{\partial^2 w}{\partial y^2} \right] \right] dx\, dy \\
T &= \tfrac{1}{2} \int_A \rho h \left[\dot{w}^2 \right] dx\, dy
\end{aligned}
\tag{2.13}
$$

where we have made our second assumption that the rotational inertia is negligible. The potential of the applied loads is

$$V = -\int q(x,y)w\, dx\, dy - M_{xx}\frac{\partial w}{\partial x} - V_{xz}w + \cdots$$

where the edge loads are on each face. Using Hamilton's principle with the variation of only δw then leads to the governing equation

$$D\nabla^2\nabla^2 w + \rho h \frac{\partial^2 w}{\partial t^2} = q \tag{2.14}$$

Again, the effect of damping is easily incorporated into the equations by modifying the inertia terms; that is,

$$\rho h \frac{\partial^2 w}{\partial t^2} \quad \longrightarrow \quad \rho h \frac{\partial^2 w}{\partial t^2} + \eta h \frac{\partial w}{\partial t}$$

where η is the damping (or viscous) coefficient.

Performing the integration by parts required to get the boundary conditions is rather involved for an arbitrary boundary — a detailed description is given in Reference [63]. The associated boundary conditions are found to be

$$
\begin{aligned}
&\left\{ w \quad \text{or} \quad V_{xz} = -D\left[\frac{\partial^3 w}{\partial x^3} + (2-\nu)\frac{\partial^3 w}{\partial x \partial y^2} \right] \right\} \\
&\left\{ \frac{\partial w}{\partial x} \quad \text{or} \quad M_{xx} = D\left[\frac{\partial^2 w}{\partial x^2} + \nu\frac{\partial^2 w}{\partial y^2} \right] \right\}
\end{aligned}
\tag{2.15}
$$

The shear to be specified is called the *Kirchhoff shear*. This shear is not the resultant Q_{xz} but is actually given by

$$\text{Kirchhoff shear:} \quad V_{xz} = Q_{xz} - \frac{\partial M_{xy}}{\partial y}$$

This can be understood physically by realizing that the shear moment M_{xy} can be interpreted as a couple comprised of vertical forces a small distance apart.

Then, because the moment is distributed, so too are the vertical forces, which consequently at any given location will have an imbalance in the vertical forces. Alternatively, the classical plate theory has restrictive degrees of freedom, where the shear strains γ_{xz} and γ_{yz} are zero. That is, the shear resultants Q_{xz} and Q_{yz} do not have a relationship to the corresponding deformation. While this can be rationalized in the constitutive relation by saying that the shear modulus in the transverse direction is very large, it means that the resultant force is associated with higher-order derivatives of the deformation.

The resultants can be written as

$$M_{xx} = D\left[\kappa_{xx} + \nu\kappa_{yy}\right] = D\left[\frac{\partial^2 w}{\partial x^2} + \nu\frac{\partial^2 w}{\partial y^2}\right]$$

$$M_{yy} = D\left[\kappa_{yy} + \nu\kappa_{xx}\right] = D\left[\frac{\partial^2 w}{\partial y^2} + \nu\frac{\partial^2 w}{\partial x^2}\right]$$

$$M_{xy} = M_{yx} = D(1-\nu)\kappa_{xy} = D(1-\nu)\frac{\partial^2 w}{\partial x\partial y} \qquad (2.16)$$

These resultants are related only to the out-of-plane deflection. The stresses are obtained from equations such as

$$\sigma_{xx} = -\frac{M_{xx}z}{I_p}$$

with $I_p \equiv h^3/12$.

In later sections, we will look at boundaries that are straight; let the boundary be located at $x = $ constant, then to summarize, the type of boundary conditions to be satisfied are to be chosen from

$$\text{Displacement}: \quad w = w(x, y, t)$$

$$\text{Slope}: \quad \psi_x = \frac{\partial w}{\partial x}$$

$$\text{Moment}: \quad M_{xx} = +D\left[\frac{\partial^2 w}{\partial x^2} + \nu\frac{\partial^2 w}{\partial y^2}\right]$$

$$\text{Shear}: \quad V_{xz} = -D\left[\frac{\partial^3 w}{\partial x^3} + (2-\nu)\frac{\partial^3 w}{\partial x\partial y^2}\right]$$

$$\text{Loading}: \quad q = D\nabla^2\nabla^2 w + \rho h\frac{\partial^2 w}{\partial t^2} + \eta h\frac{\partial w}{\partial t} \qquad (2.17)$$

The corresponding expressions for the y face are obtained by permuting x and y. Note that Poisson's ratio ν enters the moment and shear relations and acts to couple the gradients in x to those in y.

Example 2.2: Specialize the thin-plate flexural equations when there is no y dependence.

There are no derivatives with respect to y, and the summary of plate equations becomes

$$
\begin{aligned}
\text{Displacement}: & \quad w = w(x, t) \\
\text{Slope}: & \quad \psi_x = \frac{\partial w}{\partial x} \\
\text{Moment}: & \quad M_{xx} = +D\frac{\partial^2 w}{\partial x^2} \\
\text{Shear}: & \quad V_{xz} = -D\frac{\partial^3 w}{\partial x^3} \\
\text{Loading}: & \quad q = D\frac{\partial^4 w}{\partial x^4} + \rho h\frac{\partial^2 w}{\partial t^2} + \eta h\frac{\partial w}{\partial t}
\end{aligned}
\tag{2.18}
$$

These are the equations for a beam if we make the associations

$$
D \iff EI, \qquad \rho h \iff \rho A
$$

A plate deforming as assumed here is called cylindrical bending.

2.2   Membrane Problems

Perhaps the most popular solution method for plane elastostatic problems is via the Airy stress function. We summarize the approach for Cartesian and cylindrical coordinates.

Compatibility of Strains

The Navier's equations can be rewritten in terms of resultants as

$$
\begin{aligned}
\frac{\partial N_{xx}}{\partial x} + \frac{\partial N_{xy}}{\partial y} &= \rho h\frac{\partial^2 u}{\partial t^2} + \eta h\frac{\partial u}{\partial t} + b_x \\
\frac{\partial N_{xy}}{\partial x} + \frac{\partial N_{yy}}{\partial y} &= \rho h\frac{\partial^2 v}{\partial t^2} + \eta h\frac{\partial v}{\partial t} + b_y
\end{aligned}
$$

where b_x and b_y are the body forces. Thus, at each point in the body there are three unknown functions: N_{xx}, N_{xy}, and N_{yy}. In the static case, these obviously must satisfy equilibrium. However, there are only two equilibrium equations, hence, further restrictions must be imposed. These restrictions come from the requirement that the strains associated with the stresses must be *compatible*. Suppose a stress field is proposed and it is equilibrated. The use of Hooke's law converts it to a strain field. Suppose now it is desired to obtain the displacements. This can be done by integrating the strain/displacement relations

$$
\frac{\partial u}{\partial x} = \epsilon_{xx}, \qquad \frac{\partial v}{\partial y} = \epsilon_{yy}, \qquad \frac{\partial u}{\partial y} + \frac{\partial v}{\partial x} = 2\epsilon_{xy}
$$

These can be viewed as a system of three independent partial differential equations for two displacements u, v with ϵ_{ij} prescribed. For arbitrary values of ϵ_{ij}, there may not exist a unique solution for the displacement field. For a unique solution in u_i, some restrictions must be placed on the strains ϵ_{ij}. By differentiating the above, we obtain, for instance,

$$2\frac{\partial^2 \epsilon_{xy}}{\partial x \partial y} = \frac{\partial^2 \epsilon_{xx}}{\partial y^2} + \frac{\partial^2 \epsilon_{yy}}{\partial x^2}$$

This equation is known as the *compatibility equation*, first obtained by St. Venant in 1860.

To obtain compatibility in terms of stress, use Hooke's law to replace the strains in the compatibility equations with stresses and simplify this by utilizing the equilibrium equations to get

$$\sum_{\beta} \nabla^2 (\sigma_{\beta\beta}) + \frac{4\rho}{(1+\kappa)} \sum_{\beta} \frac{\partial b_{\beta}}{\partial x_{\beta}} = 0$$

where the indices range 1, 2. The stress field must satisfy this equation and the equilibrium equations in order to be admissible. The boundary conditions to be satisfied are

$$\text{on } A_t : \qquad \sum_{j} \sigma_{ij} n_j = t_i = \text{ given}$$

$$\text{on } A_u : \qquad u_i = \text{ given}$$

Note that the second set of boundary conditions are obtained by integrating the strain/displacement relations in conjunction with the stress/strain relations.

Airy Stress Function Formulation

Suppose the body forces can be derived from a potential $V(x,y)$ as

$$\rho b_x = -\frac{\partial V}{\partial x}, \qquad \rho b_y = -\frac{\partial V}{\partial y}$$

For example, gravity loading in the y-direction is described by $V = \rho g y$, then $\rho b_x = 0$, $\rho b_y = -\rho g$. Furthermore, let the stresses be obtained from a stress function $\phi(x,y)$ as

$$\sigma_{xx} = \frac{\partial^2 \phi}{\partial y^2} + V, \qquad \sigma_{yy} = \frac{\partial^2 \phi}{\partial x^2} + V, \qquad \sigma_{xy} = -\frac{\partial^2 \phi}{\partial x \partial y} \qquad (2.19)$$

It can be verified directly by substitution that stresses obtained in this manner will automatically satisfy equilibrium. The function ϕ is called the *Airy stress function*.

But the stresses must also satisfy compatibility, that is,

$$\nabla^2(\sigma_{xx} + \sigma_{yy}) = \frac{-4\rho}{(1+\kappa)}\left[\frac{\partial b_x}{\partial x} + \frac{\partial b_y}{\partial y}\right] = \frac{4}{(1+\kappa)}\nabla^2 V$$

On substituting for the stresses in terms of the stress function, this becomes

$$\nabla^2\nabla^2\phi = -\frac{2(\kappa-1)}{(1+\kappa)}\nabla^2 V \qquad \text{where} \qquad \nabla^2\nabla^2 = \frac{\partial^4}{\partial x^4} + 2\frac{\partial^4}{\partial x^2\partial y^2} + \frac{\partial^4}{\partial y^4}$$

The general solution to the above equation can be put in the form

$$\phi = \phi_c + \phi_p$$

where the functions ϕ_c, ϕ_p are the complementary and particular solutions, respectively. They satisfy

$$\nabla^2\nabla^2\phi_c = 0$$
$$\nabla^2\nabla^2\phi_p = -\frac{2(\kappa-1)}{(1+\kappa)}\nabla^2 V$$

Thus, ϕ_c is a bi-harmonic function, while ϕ_p depends on the body force field and is not necessarily bi-harmonic.

A quick way to obtain harmonic functions in Cartesian coordinates is to extract separately the real and imaginary parts of an analytic function. For example, if

$$\phi = \phi_R + i\phi_I = (x+iy)^n, \qquad i \equiv \sqrt{-1}$$

then

n	ϕ_R	ϕ_I
1	x	y
2	$x^2 - y^2$	$2xy$
3	$x^3 - 3xy^2$	$3x^2 y - y^3$
4	$x^4 - 6x^2 y^2 + y^4$	$4x^3 y - 4xy^3$
5	$x^5 - 10x^3 y^2 + 5xy^4$	$5x^4 y - 10x^2 y^3 + y^5$

Each of these is a harmonic function.

If $\phi(x,y)$ is harmonic, then the product functions $x\phi$ and $y\phi$ are bi-harmonic because

$$\left(\frac{\partial^2}{\partial x^2} + \frac{\partial^2}{\partial y^2}\right)[x\phi] = 2\frac{\partial\phi}{\partial x} + x\frac{\partial^2\phi}{\partial x\partial x} + x\frac{\partial^2\phi}{\partial y\partial y} = 2\frac{\partial\phi}{\partial x} + x\nabla^2\phi = 2\frac{\partial\phi}{\partial x}$$

Therefore

$$\nabla^2\nabla^2[x\phi] = 2\frac{\partial}{\partial x}[\nabla^2\phi] = 0$$

Similarly for the y product. This gives a quick scheme for obtaining bi-harmonic functions. For example,

$x\phi$			$y\phi$		
x^2	:	xy	xy	:	y^2
$x^3 - xy^2$	:	$2x^2y$	$x^2y - y^3$	:	$2xy^2$
$x^4 - 3x^2y^2$	:	$3x^3y - xy^3$	$xy - 3xy^3$	:	$3x^2y^2 - y^4$

is a collection of bi-harmonic functions obtained from the table of harmonic functions above.

This can be generalized to the statement: Let ϕ_o, ϕ_1, and ϕ_2 be any harmonic functions, then a representation of a bi-harmonic function can be formed by the linear combination

$$\phi(x,y) = \phi_o(x,y) + x\phi_1(x,y) + y\phi_2(x,y)$$

A bi-harmonic stress function is always the exact solution to some problem — the art of solving practical problems is finding the right combination of these functions to satisfy the given boundary conditions.

Example 2.3:   Show that the stress function

$$\phi(x,y) = Axy + Bxy^3 + Cy^3$$

can be used to solve the problem of a deep cantilever beam with a parabolic shear traction distribution on the end.

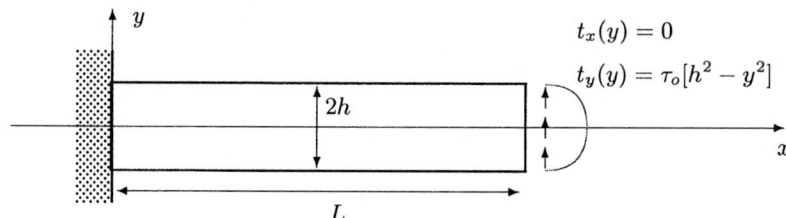

Figure 2.3: Cantilever beam with end shear traction.

First, it is clear that ϕ is bi-harmonic because the highest power in the polynomial is three. The stresses are

$$\sigma_{xx} = \frac{\partial^2 \phi}{\partial y^2} = 6Bxy + 6Cy$$

$$\sigma_{yy} = \frac{\partial^2 \phi}{\partial x^2} = 0$$

$$\sigma_{xy} = -\frac{\partial^2 \phi}{\partial x \partial y} = -A - 3By^2$$

Consider the tractions on the horizontal planes $y = \pm h$ such that $n_x = 0$, $n_y = 1$. That is,

$$t_x = 0 = \sigma_{xy} = -A - 3Bh^2, \qquad t_y = 0 = \sigma_{yy} = 0$$

Note that the normal traction condition is automatically satisfied. In fact, we only get one equation from the four traction conditions and this leads to $A = -3Bh^2$.

Now look at the face at $x = L$. The tractions are

$$t_x = 0 = \sigma_{xx} = 6BLy + 6Cy, \qquad t_y = \tau_0[h^2 - y^2] = \sigma_{xy} = -A - 3By^2$$

These two conditions lead to three equations

$$6BL + 6C = 0, \qquad \tau_0 h^2 = -A, \qquad \tau_0 = -3B$$

which give

$$2A = -\tau_0 h^2, \qquad B = \tau_0/3, \qquad C = -BL$$

Thus the stress solution is

$$
\begin{aligned}
\sigma_{xx} &= 2\tau_0[x - L]y \\
\sigma_{yy} &= 0 \\
\sigma_{xy} &= \tau_0[h^2 - y^2]
\end{aligned}
$$

At this stage, we have a stress field that satisfies the tractions on three sides of the body. In order to guarantee that this is indeed the solution, we must also satisfy the boundary conditions along the face at $x = 0$. But what are the traction conditions? These were not specified as part of the problem.

Example 2.4: Obtain the displacement fields corresponding to the previous example.

To obtain the displacements, we must integrate the strain/displacement relations. Thus, from the normal strains

$$Eu(x,y) = 2\tau_0[x^2 y/2 - xyL] + f_1(y), \qquad v(x,y) = f_2(x)$$

where f_1 and f_2 are functions of integrations. The displacements must also satisfy the shear strain/displacement relation, hence substitute and regroup in terms of only x and y. The separate groups must be equal to a constant (λ, say), therefore integration gives the separate functions $f_1(y)$ and $f_2(x)$. We finally get for the displacements

$$
\begin{aligned}
Eu(x,y) &= 2\tau_0[x^2/2 - xL + h^2 - y^2/3]y - \lambda y + c_1 \\
Ev(x,y) &= \tau_0[L - x/3]x^2 + \lambda x + c_2
\end{aligned}
$$

where λ, c_1, c_2 are unknowns. These contribute a rigid body motion.

Look at the displacements at $x = 0$, we have

$$Eu(0,y) = 2\tau_0[h^2 - y^2/3]y - \lambda y + c_1, \qquad Ev(0,y) = c_2$$

The horizontal displacement is non-zero; not what we wanted for the fixed end condition. The above solution is not the exact solution for the fixed cantilever beam problem; the simple stress function polynomial is not capable of representing the singular stress behavior at the fixed end where $y = \pm h$. The solution, however, is the exact solution if the tractions at $x = 0$ were specified as

$$t_x = +2\tau_o Ly, \qquad t_y = -\tau_o[h^2 - y^2]$$

Note that if these tractions were specified otherwise, then global equilibrium is probably violated.

The above solution gives a good approximation to the cantilever beam problem because it satisfies the exact traction conditions top and bottom, and as can be verified, satisfies an approximate version of the tractions in the form of resultants on the ends. In fact, this is a very useful approach to obtaining practical solutions: satisfy some of the traction conditions exactly, and the others approximately in the form of resultants. If the region of interest is remote from these latter boundaries, then the solution will be quite insensitive to the specific distributions of the applied tractions. This known as *St. Venant's Principle*.

Plane Problems in Cylindrical Coordinates

One of the main difficulties in solving boundary value problems is in satisfying the boundary conditions. This is further exacerbated if the functional form of the tractions are not "similar" to the functional form of the boundary geometry. We illustrate how sometimes a change of coordinate system can lead to effective solutions.

With a change of coordinate system, some quantities follow the usual transformation law. For example, in the cylindrical coordinates (r, θ, z), the strain components may be designated $\epsilon_{rr}, \epsilon_{\theta\theta}, \epsilon_{zz}, \epsilon_{rz}, \epsilon_{r\theta}, \epsilon_{z\theta}$ and they are related to the rectangular components $\epsilon_{xx}, \epsilon_{yy}, \epsilon_{zz}, \epsilon_{xy}, \epsilon_{yz}, \epsilon_{zx}$ by the usual tensor transformation law. That is, the stress and strain components can be referred to a local rectangular frame of reference oriented in the direction of the curvilinear coordinates. However, if displacement vectors are resolved into components in the directions of the curvilinear coordinates, the strain/displacement relationship involves derivatives of the displacement components and, therefore, is influenced by the curvature of the coordinate system. The strain/displacement relations may appear quite different from the corresponding formulas in rectangular coordinates.

We start with the relations between the cylindrical coordinates (r, θ, z) and the rectangular coordinates (x, y, z) given by

$$x = r \cos\theta, \qquad y = r \sin\theta, \qquad z = z$$

and

$$r^2 = x^2 + y^2, \qquad \theta = \tan^{-1}\left(\frac{y}{x}\right), \qquad z = z$$

By using the chain rule, it follows that any derivatives with respect to x and y in the Cartesian equations may be transformed into derivatives with respect to

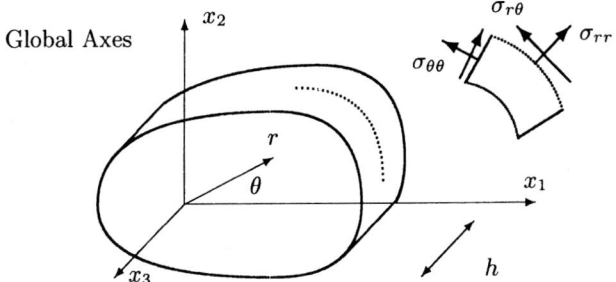

Figure 2.4: Cylindrical coordinates.

r and θ as

$$\frac{\partial}{\partial x} = \frac{\partial r}{\partial x}\frac{\partial}{\partial r} + \frac{\partial \theta}{\partial x}\frac{\partial}{\partial \theta} = \cos\theta\frac{\partial}{\partial r} - \frac{\sin\theta}{r}\frac{\partial}{\partial \theta}$$

$$\frac{\partial}{\partial y} = \frac{\partial r}{\partial y}\frac{\partial}{\partial r} + \frac{\partial \theta}{\partial y}\frac{\partial}{\partial \theta} = \sin\theta\frac{\partial}{\partial r} + \frac{\cos\theta}{r}\frac{\partial}{\partial \theta}$$

We will use these repeatedly to transform our governing equations.

In the cylindrical coordinate system, the components of the displacement vector are denoted by u_r, u_θ. Components of the same vector resolved in the directions of the rectangular coordinates are u_x, u_y. These components of displacement are related by

$$
\begin{aligned}
u_x &= u_r \cos\theta - u_\theta \sin\theta \\
u_y &= u_r \sin\theta + u_\theta \cos\theta
\end{aligned}
$$

Set up a local Cartesian system $(\hat{e}_r, \hat{e}_\theta, \hat{e}_z)$ at point (r, θ, z) in which $\hat{e}_r, \hat{e}_\theta$, and $\hat{e}_z$ are the unit base vectors in the r, θ, and z direction, respectively. Substituting the strain/displacement relation in Cartesian coordinates into the above, we obtain

$$
\begin{aligned}
\epsilon_{rr} &= \frac{\partial u_r}{\partial r} \\
\epsilon_{\theta\theta} &= \frac{u_r}{r} + \frac{1}{r}\frac{\partial u_\theta}{\partial \theta} \\
2\epsilon_{r\theta} &= \frac{1}{r}\frac{\partial u_r}{\partial \theta} + \frac{\partial u_\theta}{\partial r} - \frac{u_\theta}{r}
\end{aligned}
$$

The three components of the stress tensor at a point (r, θ, z) are denoted by $(\sigma_{rr}, \sigma_{\theta\theta}, \sigma_{r\theta})$. The derivation of the equilibrium equations in the cylindrical coordinate system is a straightforward exercise following closely to that of the strain/displacement. We get

$$
\begin{aligned}
\frac{\partial \sigma_{rr}}{\partial r} + \frac{1}{r}\frac{\partial \sigma_{r\theta}}{\partial \theta} + \frac{\sigma_{rr} - \sigma_{\theta\theta}}{r} + \rho b_r &= \rho \ddot{u}_r \\
\frac{\partial \sigma_{r\theta}}{\partial r} + \frac{1}{r}\frac{\partial \sigma_{\theta\theta}}{\partial \theta} + \frac{2}{r}\sigma_{r\theta} + \rho b_\theta &= \rho \ddot{u}_\theta
\end{aligned}
$$

where b_r and b_θ are the components of the body force vector $\hat{b}$ in the r and θ directions, respectively.

The stresses are related to the Airy stress function by

$$
\begin{aligned}
\sigma_{rr} &= \frac{1}{r}\frac{\partial \phi}{\partial r} + \frac{1}{r^2}\frac{\partial^2 \phi}{\partial \theta^2} + V \\[6pt]
\sigma_{\theta\theta} &= \frac{\partial^2 \phi}{\partial r^2} + V \\[6pt]
\sigma_{r\theta} &= -\frac{\partial}{\partial r}\left(\frac{1}{r}\frac{\partial \phi}{\partial \theta}\right)
\end{aligned}
$$

The radial and hoop components of the body force are given by

$$
\rho b_r = -\frac{\partial V}{\partial r}, \qquad \rho b_\theta = -\frac{1}{r}\frac{\partial V}{\partial \theta}
$$

The Airy stress function still satisfies the bi-harmonic equation

$$
\nabla^4 \phi = \nabla^2 \nabla^2 \phi = -2\frac{\kappa - 1}{\kappa + 1}\nabla^2 V, \qquad \nabla^2 \equiv \frac{\partial^2}{\partial r^2} + \frac{1}{r}\frac{\partial}{\partial r} + \frac{1}{r^2}\frac{\partial^2}{\partial \theta^2}
$$

The only difference (in comparison to the Cartesian form) is that the Laplace operator is written in cylindrical coordinates.

The general solution for the homogeneous bi-harmonic equation was obtained by J.H. Michell (1899) by direct substitution of $\phi = f(r)e^{\alpha\theta}$. The solutions are summarized as

$$
\begin{aligned}
\phi(r,\theta) =\ & A_o + B_o\theta + A\log_n r + Br^2\log_n r + Cr^2 + Dr^2\theta \\[4pt]
+\ & \left(A_1 r + B_1 r^3 + C_1\frac{1}{r} + D_1 r\log_n r + E_1 r\theta\right)\begin{Bmatrix} \sin\theta \\ \cos\theta \end{Bmatrix} \\[4pt]
+\ & \sum_{n=2}^{\infty}\left(A_n r^n + B_n r^{n+2} + C_n\frac{1}{r^n} + D_n\frac{r^2}{r^n}\right)\begin{Bmatrix} \sin n\theta \\ \cos n\theta \end{Bmatrix}
\end{aligned}
$$

The braces indicates that either term can be used. Stresses and displacements obtained from these can be found in the charts of Table 1 and Table 2, respectively, of Reference [21]. The constant term A_0 does not yield any nontrivial stresses and is therefore usually omitted. The term A_1 gives zero stresses but is retained because it is associated with rigid body motions.

Example 2.5: A thin annulus, rigidly constrained on its outside, is subjected to a uniform temperature change. Determine the stress distributions.

The geometry and loading are axisymmetric, hence the permissible stress functions are

$$
\phi(r,\theta) = A\log_n r + Br^2\log_n r + Cr^2
$$

While not obvious, but can be easily demonstrated [21], the B term gives rise to a θ-dependent displacement; hence we discard this term to give the stresses as

$$
\sigma_{rr} = \frac{A}{r^2} + 2C, \qquad \sigma_{\theta\theta} = -\frac{A}{r^2} + 2C
$$

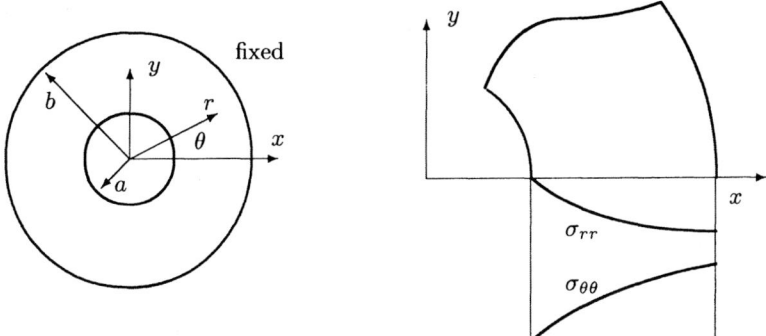

Figure 2.5: Annulus with uniform temperature changes.

This is known as Lamé's solution and can be used to solve a variety of thick cylinder problems. The inner radius $r = a$ is traction free

$$\sigma_{rr} = \frac{A}{a^2} + 2C = 0 \quad \Rightarrow \quad 2C = -\frac{A}{a^2}$$

giving the stress distributions

$$\sigma_{rr} = \frac{A}{a^2}\left[\frac{a^2}{r^2} - 1\right], \qquad \sigma_{\theta\theta} = -\frac{A}{a^2}\left[\frac{a^2}{r^2} + 1\right]$$

It remains now to determine the coefficient A.

The temperature change causes an expansion of the annulus; however, at the outer boundary the total strain is constrained to be zero. Converting Hooke's law of Equation (1.33) to cylindrical coordinates results in

$$\epsilon_{\theta\theta} = \frac{1}{E}[\sigma_{\theta\theta} - \nu\sigma_{rr}] + \alpha\Delta T = 0$$

Substituting for the stresses gives

$$\frac{A}{a^2} = E\alpha\Delta T / [(1 + \nu)\frac{a^2}{b^2} + (1 - \nu)]$$

The stress distributions are shown in Figure 2.5. The hoop stress is the larger stress with the maximum occurring at the inner radius.

If the inner radius is zero, the stress distribution is uniform with the value

$$\sigma_{rr} = \sigma_{\theta\theta} = -\frac{E\alpha\Delta T}{1 - \nu}$$

We will revisit this solution in Chapter 6 when we consider how a temperature change may cause the buckling of a plate.

Example 2.6: Determine the state of stress in a large plate with a small hole, uniformly loaded in the y-direction remote from the hole.

The basic strategy is to add two stress systems together. The first gives the correct applied tractions at infinity, while the second enforces the zero tractions around the edge of the hole without affecting the stresses at infinity.

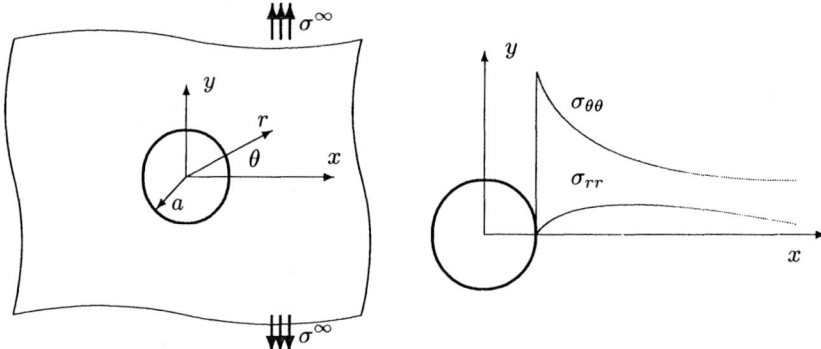

Figure 2.6: Hole in an infinite sheet and the stress distributions.

Initially, neglect the hole and obtain a stress function for the remote stress. That is, knowing

$$\sigma_{xx} = \frac{\partial^2 \phi_o}{\partial y^2} = 0, \qquad \sigma_{yy} = \frac{\partial^2 \phi_o}{\partial x^2} = \sigma^\infty, \qquad \sigma_{xy} = -\frac{\partial^2 \phi_o}{\partial x \partial y} = 0$$

leads us to choose the stress function as

$$\phi_o = \tfrac{1}{2}\sigma^\infty x^2$$

In the vicinity of the hole, we will need to use cylindrical coordinates when satisfying the boundary conditions, hence rewrite ϕ_o as

$$\phi_o = \tfrac{1}{2}\sigma^\infty r^2 \cos^2 \theta = \tfrac{1}{2}\sigma^\infty r^2 (\tfrac{1}{2} + \tfrac{1}{2}\cos 2\theta) = \tfrac{1}{4}\sigma^\infty r^2 + \tfrac{1}{4}\sigma^\infty r^2 \cos 2\theta$$

Our plan is to add to this a stress function that will satisfy the boundary conditions at $r = a$. Whatever form it takes, the stresses must be consistent with this at $r \to \infty$ and therefore they must go to zero at $r \to \infty$.

Although ϕ_o obtained above satisfies the stress condition at $r \to \infty$, it does not satisfy the boundary condition at $r = a$ of

$$0 = t_r = \sigma_{rr}n_r + \sigma_{r\theta}n_\theta = -\sigma_{rr}$$
$$0 = t_\theta = -\sigma_{r\theta}$$

The stress function ϕ_o yields the following stresses at $r = a$

$$\sigma_{rr} = \tfrac{1}{2}\sigma^\infty - \tfrac{1}{2}\sigma^\infty \cos 2\theta$$
$$\sigma_{r\theta} = \tfrac{1}{2}\sigma^\infty \sin 2\theta$$

Additional bi-harmonic functions must be added to ϕ_o in order to clear these tractions without disturbing the stress condition at $r \to \infty$ which are already satisfied by ϕ_o

Using the above-mentioned boundary conditions as a guide, the added bi-harmonic functions must produce stresses that are either independent of θ or dependent on $\cos 2\theta$ (for σ_{rr}) and $\sin 2\theta$ (for $\sigma_{r\theta}$). Meanwhile, the additional stresses must vanish as $r \to \infty$. From the bi-harmonic function table, the suitable candidate stress functions are

$$\log_n r, \quad \frac{1}{r} \cos 2\theta, \quad \cos 2\theta$$

The general stress function that satisfies the remote conditions is therefore

$$\phi = A \log_n r + \frac{\sigma^\infty}{4} r^2 + [\frac{\sigma^\infty}{4} r^2 + C_2 \frac{1}{r^2} + D_2] \cos 2\theta$$

giving the stresses

$$\sigma_{rr} = \frac{A}{r^2} + \frac{\sigma^\infty}{2} - \left[\frac{\sigma^\infty}{2} + \frac{6C_2}{r^4} + \frac{4D_2}{r^2}\right] \cos 2\theta$$

$$\sigma_{\theta\theta} = -\frac{A}{r^2} + \frac{\sigma^\infty}{2} + \left[\frac{\sigma^\infty}{2} + \frac{6C_2}{r^4}\right] \cos 2\theta$$

$$\sigma_{r\theta} = 0 - \left[\frac{-\sigma^\infty}{2} + \frac{6C_2}{r^4} + \frac{2D_2}{r^2}\right] \sin 2\theta$$

There are three constants A, C_2, D_2 to be determined by the boundary conditions at $r = a$. Note that as r becomes very large, the additional terms do indeed vanish.

The boundary conditions at the edge of the hole are that the tractions are zero, that is,

$$t_r = -\sigma_{rr} = 0 = \frac{A}{a^2} + \frac{\sigma^\infty}{2} - \left[\frac{\sigma^\infty}{2} + \frac{6C_2}{a^4} + \frac{4D_2}{a^2}\right] \sin 2\theta$$

$$t_{r\theta} = -\sigma_{r\theta} = 0 = -\left[-\frac{\sigma^\infty}{2} + \frac{6C_2}{a^4} + \frac{2D_2}{a^2}\right] \cos 2\theta$$

Because this must be true for any θ, then

$$\frac{A}{a^2} + \frac{\sigma^\infty}{2} = 0$$

$$\frac{\sigma^\infty}{2} + \frac{6C_2}{a^4} + \frac{4D_2}{a^2} = 0$$

$$-\frac{\sigma^\infty}{2} + \frac{6C_2}{a^4} + \frac{2D_2}{a^2} = 0$$

Solving these simultaneously gives the coefficients as

$$A = -\frac{\sigma^\infty}{2} a^2, \quad C_2 = \frac{\sigma^\infty}{4} a^4, \quad D_2 = -\frac{\sigma^\infty}{2} a^2$$

The stresses are, finally,

$$\sigma_{rr} = \frac{1}{2}\sigma^\infty \left\{1 - \frac{a^2}{r^2} - \left(1 - \frac{4a^2}{r^2} + \frac{3a^4}{r^4}\right) \cos 2\theta\right\}$$

$$\sigma_{\theta\theta} = \frac{1}{2}\sigma^\infty \left\{1 + \frac{a^2}{r^2} + \left(1 + \frac{3a^4}{r^4}\right) \cos 2\theta\right\}$$

$$\sigma_{r\theta} = \frac{1}{2}\sigma^\infty \left\{1 + \frac{2a^2}{r^2} - \frac{3a^4}{r^4}\right\} \sin 2\theta$$

This is known as the Kirsch solution. Figure 2.6 shows the distribution of the hoop stress along the x-axis. Note the high stress gradient close to the edge of the hole. This is an example of a stress concentration where a change in geometry can cause a local increase in stress.

The hoop stress around the edge of the hole is

$$\sigma_{\theta\theta} = \sigma^{\infty} \left\{ 1 + 2\cos 2\theta \right\}$$

showing that at $\theta = 0$, the maximum stress is three times the remote stress. Also note that at $\theta = \pi/2$, $\sigma_{\theta\theta} = -\sigma^{\infty}$.

2.3   Flexural Problems

It is quite difficult to integrate the governing equations for the deflections of plates and have them satisfy arbitrary boundary conditions. We therefore introduce a Fourier analysis method that is based on superposition of particular solutions; these solutions, however, are restricted to having simply supported boundary conditions on at least two opposite edges. Nonetheless, we can effect some useful solutions. As we will see, the technique will also be useful when we consider the dynamics and stability of plates.

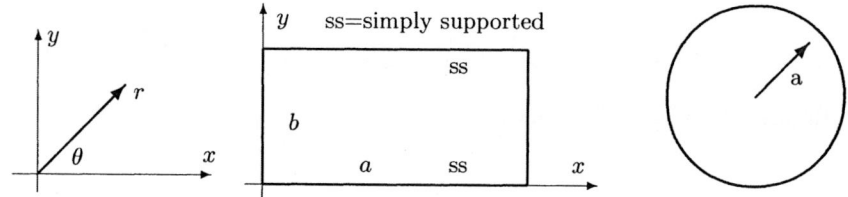

Figure 2.7: Coordinate system for rectangular and circular plates.

Deflection of Rectangular Plates

Consider the bending of rectangular plates. The governing differential equation is

$$D\nabla^2\nabla^2 w(x,y) + Kw(x,y) = q(x,y)$$

where K is the stiffness of an elastic foundation. Let the solutions be represented in the form

$$w(x,y) = \sum_m \tilde{w}_m(x)e^{i\xi_m y}, \qquad \xi_m = \frac{m\pi}{b}$$

where b is the width of the plate. Also let the distributed load be represented in a similar form

$$q(x,y) = \sum_m \tilde{q}_m(x)e^{i\xi_m y}$$

We determine the coefficients $\tilde{w}_m$ by substituting these representations for w and q into the governing differential equation and requiring that it be satisfied for each m. The differential equation for $\tilde{w}$ becomes

$$(\frac{d^2}{dx^2} - \xi^2)(\frac{d^2}{dx^2} - \xi^2)\tilde{w}_m - \beta^4 \tilde{w}_m = \frac{\tilde{q}_m}{D}, \qquad \beta^4 \equiv \frac{(-K)}{D}$$

The definition for β makes the solution structure similar to that for the vibration of plates. The general solution to this equation is comprised of the homogeneous solution and the particular solution. We can see by inspection that the particular solution is given by

$$\tilde{w}_{pm} = \frac{\tilde{q}_m}{D\xi_m^4 + K} \qquad \text{or} \qquad w_p(x,y) = \sum_m \frac{\tilde{q}_m}{D\xi_m^4 + K} e^{i\xi_m y}$$

The homogeneous differential equation has constant coefficients, hence e^{-ikx} is a kernel solution. The characteristic equation for k is

$$k^4 + 2k^2\xi^2 + \xi^4 - \beta^4 = 0$$

The roots of this equation (which we will refer to as *spectrum relations* because of their intimate connection to the spectral analysis method developed in Chapter 4) appear as $\pm$ pairs

$$k_{1,3}(\omega) = \pm \left[\beta^2 - \xi^2\right]^{1/2} \equiv \pm\alpha_m$$

and

$$k_{2,4}(\omega) = \pm i \left[\beta^2 + \xi^2\right]^{1/2} \equiv \pm i\bar{\alpha}_m$$

Thus, the general homogeneous solution is represented by

$$w_h(x,y) = \sum_m \left[\mathbf{A}e^{-i\alpha_m x} + \mathbf{B}e^{-\bar{\alpha}_m x} + \mathbf{C}e^{+i\alpha_m x} + \mathbf{D}e^{+\bar{\alpha}_m x}\right] e^{i\xi_m y} \qquad (2.20)$$

Sometimes, we will find it more convenient to use the solution in the form

$$w_h(x,y) = \sum_m \left[c_1 \cos(\alpha_m x) + c_2 \sin(\alpha_m x) + c_3 \cosh(\bar{\alpha}_m x) + c_4 \sinh(\bar{\alpha}_m x)\right] f(\xi_m y)$$

where $f(\xi_m y) = \cos(\xi_m y)$ or $\sin(\xi_m y)$. Looking at the boundary conditions at $y = 0$ and $y = b$, we see that

$$w = 0, \qquad \frac{\partial^2 w}{\partial x^2} = 0$$

always. That is, this is true for each m term and implies that this particular solution can solve only those problems with simply supported lateral sides.

For $m = 0$ we have beam-like behavior. For $m > 0$, we also have beam-like behavior with a more complicated variation in y, and a more complicated spectrum relation. This association to beam theory will help in specifying the boundary conditions.

Note that, if the spring constant is zero, then the roots are repeated and we have the solution in the form [78]

$$w_h(x,y) = \sum_m \left[\mathbf{A}e^{-i\alpha_m x} + \mathbf{B}xe^{-\alpha_m x} + \mathbf{C}e^{+i\alpha_m x} + \mathbf{D}xe^{+\alpha_m x} \right] e^{i\xi_m y} \quad (2.21)$$

Because of our interest in plate buckling and vibration problems, it is preferable to retain a small K even when it should be zero. The utility of this is shown in the next examples.

Example 2.7:   Determine the deflections of a rectangular plate of size $[a \times b]$, simply supported on all sides with a uniform pressure applied.

We will choose only the $\sin(\xi_m y)$ terms and use the solution form

$$w(x,y) = \sum_m \left[c_1 \cos(\alpha_m x) + c_2 \sin(\alpha_m x) + c_3 \cosh(\bar\alpha_m x) + c_4 \sinh(\bar\alpha_m x) \right.$$
$$\left. + \frac{\tilde q}{D\xi_m^4 + K} \right] \sin(\xi_m y)$$

Only the boundary conditions at $x = 0$ and $x = a$ need be considered, since the lateral boundary conditions are automatically satisfied. We will impose the condition of zero deflection at $x = 0$ and $x = a$. Note that the zero moment condition reduces to

$$\frac{\partial^2 w}{\partial x^2} = 0$$

at both edges.

Imposing the boundary conditions gives

$$\begin{aligned}
\text{at} \quad x = 0 \quad & \tilde w = 0 & = & \quad c_1 + c_3 + Q \\
& \frac{\partial^2 \tilde w}{\partial x^2} = 0 & = & \quad -\alpha^2 c_1 + \bar\alpha^2 c_3 \\
\text{at} \quad x = a \quad & \tilde w = 0 & = & \quad [c_1 C + c_2 S] + [c_3 C_h + c_4 S_h] + Q \\
& \frac{\partial^2 \tilde w}{\partial x^2} = 0 & = & \quad -\alpha^2 [c_1 C + c_2 S] + \bar\alpha^2 [c_3 C_h + c_4 S_h]
\end{aligned}$$

where $C \equiv \cos(\alpha a)$, $C_h \equiv \cosh(\bar\alpha a)$, and so on, and $Q = \tilde q_m / (D\xi_m^4 + K)$. Solving for the coefficients, we get

$$c_1 = -\frac{\bar\alpha^2 Q}{\alpha^2 + \bar\alpha^2}, \quad c_2 = -\frac{\bar\alpha^2 Q}{\alpha^2 + \bar\alpha^2}[\frac{1-C}{S}], \quad c_3 = -\frac{\alpha^2 Q}{\alpha^2 + \bar\alpha^2}, \quad c_4 = -\frac{\alpha^2 Q}{\alpha^2 + \bar\alpha^2}[\frac{1-C_h}{S_h}]$$

The deflected shape is

$$w(x,y) = \sum_{m=1,3,..} \frac{\tilde q_m}{D\xi_m^4 + K} \left[1 - \frac{1}{(\alpha^2 + \bar\alpha^2)} \left[\bar\alpha^2 \cos(\alpha_m x) + \bar\alpha^2 [\frac{1-C}{S}] \sin(\alpha_m x) \right. \right.$$
$$\left. \left. + \alpha^2 \cosh(\bar\alpha_m x) + \alpha^2 [\frac{1-C_h}{S_h}] \sinh(\bar\alpha_m x) \right] \right] \sin(\xi_m y)$$

Because both α and $\bar{\alpha}$ can be complex, the functions $\cos(\alpha x)$, $\cosh(\bar{\alpha},)$, and so on, can exhibit a large dynamic range. To avoid numerical difficulties, the solution is rearranged as

$$w(x,y) = \sum_{m=1,3,..} \frac{\tilde{q}_m}{D\xi_m^4 + K} \left[1 - \bar{\alpha}^2 \frac{\sin \alpha_m (L-x) + \sin \alpha_m x}{(\alpha^2 + \bar{\alpha}^2)S} \right.$$

$$\left. - \alpha^2 \frac{\sinh \bar{\alpha}_m (L-x) + \sinh \bar{\alpha}_m x}{(\alpha^2 + \bar{\alpha}^2)S_h} \right] \sin(\xi_m y)$$

It remains now to impose a particular form of loading. Let the uniform pressure be described by $q(x,y) = q_o = constant$, then the Fourier coefficients are obtained from

$$\tilde{q}_m = \int_0^b \tilde{q}_m(x) \sin(\xi_m y) \, dy = q_o \int_0^b \sin(\xi_m y) \, dy = -\frac{2q_o}{\xi_m}, \qquad m = 1, 3, 5, \ldots$$

All the coefficients for even powers of m are zero.

Example 2.8: Show how the solution for a simply supported plate is affected by the choice of stiffness K and the number of terms in the summation.

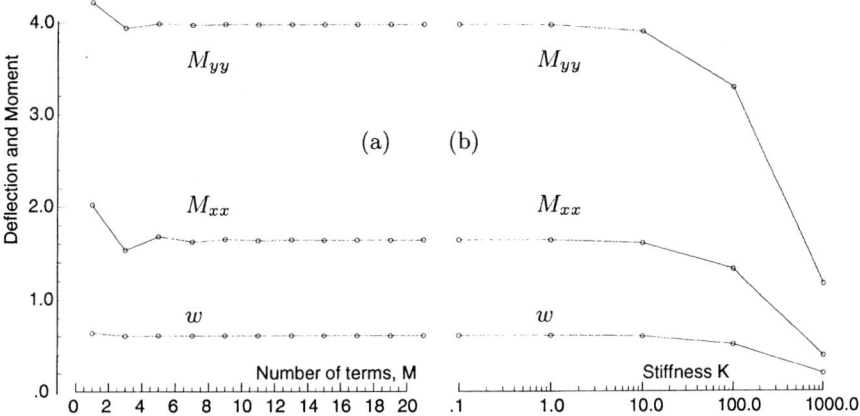

Figure 2.8: Deflection and moment convergence test. (a) Effect of number of terms in the summation. (b) Effect of elastic foundation.

The results we will present are normalized with respect to

$$w_o = \frac{16q_o b^4}{D\pi^6}, \qquad M_o = \frac{q_o b^2}{4\pi^2}$$

These normalizations are taken from Reference [78].

The computed displacements and (stress) moments at the center of a plate with $b = 2a$ are shown in Figure 2.8. Only about three terms ($M = 5$) are needed to give a converged solution.

The effect of the stiffness K is to diminish the deflections. However, in the limit of small stiffness we recover the free-plate solution, thus justifying our use of the solution with non-zero K.

Deflection of Plates in Cylindrical Coordinates

To increase the range of available analytical solutions, we now look at plates with circular boundaries. The coordinate system is in terms of (r, θ) as shown in the Figure 2.7. In some respects, these are simpler that the rectangular plates because they have only a single boundary along $r =$ constant, whereas the rectangular plates have boundaries along two coordinate directions.

The governing differential equation is

$$D\nabla^2\nabla^2 w(r,\theta) = q(r,\theta), \qquad \nabla^2 = \frac{\partial^2}{\partial r^2} + \frac{1}{r}\frac{\partial}{\partial r} + \frac{1}{r^2}\frac{\partial^2}{\partial \theta^2}$$

As with the rectangular plate, we can consider the variation with respect to one of the coordinates to be represented as a Fourier series. For example, consider the forms

$$w(r,\theta) = \sum R_m(r)e^{im\theta}, \qquad q(r,\theta) = \sum q_m(r)e^{im\theta}$$

The differential equation for w becomes

$$\sum_m \left[\frac{d^2}{dr^2} + \frac{1}{r}\frac{d}{dr} - \frac{m^2}{r^2}\right]\left[\frac{d^2 R_m}{dr^2} + \frac{1}{r}\frac{dR_m}{dr} - \frac{m^2}{r^2}R_m\right]e^{im\theta} = \sum q_m(r)e^{im\theta}$$

Setting this to be true for all components m leads to a differential equation for $R_m(r)$ as

$$\left[\frac{d^2}{dr^2} + \frac{1}{r}\frac{d}{dr} - \frac{m^2}{r^2}\right]\left[\frac{d^2 R_m}{dr^2} + \frac{1}{r}\frac{dR_m}{dr} - \frac{m^2}{r^2}R_m\right] = q_m$$

THere are three special cases for the homogeneous solutions

$$m = 0: \qquad R_0(r) = A_0 + B_0 r^2 + C_0 \log r + D_0 r^2 \log r$$
$$m = 1: \qquad R_1(r) = A_1 r + B_1 r^3 + C_1^{-1} + D_1 r \log r$$
$$m > 1: \qquad R_m(r) = A_m r^m + B_m r^{m+2} + C_m^{-m} + D_m r^{-m+2}$$

These solutions can be used to solve a variety of plate problems including those with an inner circular hole as we will show shortly. Note that when $m = 0$, we get the axisymmetric solutions.

The equations for the moments and shears are

$$M_{rr} = D\left[\frac{\partial^2 w}{\partial r^2} + \frac{\nu}{r}\frac{\partial w}{\partial r} + \frac{\nu}{r^2}\frac{\partial^2 w}{\partial \theta^2}\right]$$

$$M_{\theta\theta} = D\left[\frac{\partial^2 w}{\partial \theta^2} + \frac{1}{r}\frac{\partial w}{\partial r} + \nu\frac{\partial^2 w}{\partial r^2}\right]$$

$$M_{r\theta} = (1-\nu)D\left[\frac{1}{r}\frac{\partial^2 w}{\partial r\partial \theta} - \frac{1}{r}\frac{\partial w}{\partial \theta}\right]$$

$$Q_r = -D\frac{\partial}{\partial r}\left[\frac{\partial^2 w}{\partial r^2} + \frac{1}{r}\frac{\partial w}{\partial r} + \frac{1}{r^2}\frac{\partial^2 w}{\partial \theta^2}\right]$$

$$V_r = Q_r - \frac{1}{r}\frac{\partial}{\partial \theta}[M_{r\theta}] \qquad (2.22)$$

The last equation is for the Kirchhoff shear, which we need when we impose the traction free boundary conditions.

Example 2.9: Determine the deflections and moments in a uniformly loaded circular plate clamped on the edge.

For the special case of axisymmetric loading and geometry, the governing equation reduces to

$$D[\frac{d^2}{dr^2} + \frac{1}{r}\frac{d^2}{dr^2}][\frac{d^2}{dr^2} + \frac{1}{r}\frac{d^2}{dr^2}]w = \frac{1}{r}\frac{d}{dr}\left[r\frac{d}{dr}[\frac{1}{r}\frac{d}{dr}[r\frac{d}{dr}]]\right]w = q(r)$$

The equations for the moments are

$$M_{rr} = D\left[\frac{\partial^2 w}{\partial r^2} + \frac{\nu}{r}\frac{\partial w}{\partial r}\right]$$

$$M_{\theta\theta} = D\left[\frac{1}{r}\frac{\partial w}{\partial r} + \nu\frac{\partial^2 w}{\partial r^2}\right]$$

$$M_{r\theta} = 0$$

$$V_r = Q_r = -D\frac{\partial}{\partial r}\left[\frac{\partial^2 w}{\partial r^2}\right]$$

Consider the special case when the distributed load is a constant $q(r) = q_o$. The displacements are obtained by integration as

$$Dw(r) = \frac{1}{64}q_o r^4 + \frac{1}{4}c_1[r^2\log r - r^2] + \frac{1}{4}c_2 r^2 + c_3\log r + c_4$$

The constants of integration c_1 to c_4 are obtained by imposing the boundary conditions.

This solution gives a singularity at $r = 0$, hence we must set $c_1 = 0$ and $c_3 = 0$; these terms would be retained if the inner boundary is not at $r = 0$. The remaining two coefficients are obtained from the boundary conditions at $r = a$, which are that

$$\text{at } r = a: \qquad w = 0, \qquad \frac{\partial w}{\partial r} = 0$$

This leads to the solution

$$w(r) = \frac{q_o a^4}{64D}\left[1 - (\frac{r}{a})^2\right]^2$$

This is shown plotted in Figure 2.9 for $(q_o a^4)/(64D) = 175.8$. The comparisons are with a finite element solution, which we discuss later.

The moment distributions are

$$M_{rr}(r) = \frac{q_o a^2}{16}\left[(1+\nu) - (\frac{r}{a})^2(3+\nu)\right], \qquad M_{\theta\theta}(r) = \frac{q_o a^2}{16}\left[(1+\nu) - (\frac{r}{a})^2(1+3\nu)\right]$$

The maximum stress is the M_{rr} stress and it occurs on the boundary.

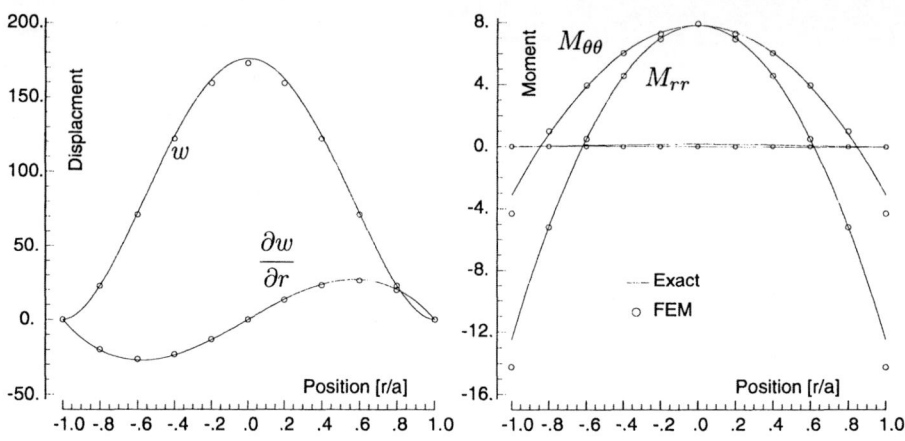

Figure 2.9: Deflection and moment distributions.

Example 2.10: Determine the state of stress in a large plate with a small hole, uniformly loaded with bending moments remote from the hole as shown in Figure 2.10.

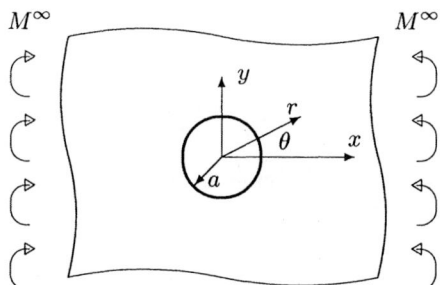

Figure 2.10: Hole in an infinite sheet with remote bending moment.

This problem is the bending equivalent of the Kirsch problem solved earlier and therefore our approach will have many similarities. Let the very large plate have the uniform bending moments

$$M_{xx} = M_o , \qquad M_{yy} = 0$$

After substituting these into the moment/displacement relations and integrating, we get the displacement field

$$w = \frac{M_o}{2D(1 - \nu^2)}[x^2 - \nu y^2] = \frac{M_o}{4D(1 - \nu^2)}[1 - \nu + (1 + \nu)\cos 2\theta]$$

This displacement gives rise to the moments and shears of

$$
\begin{aligned}
M_{rr} &= \tfrac{1}{2}M_o(1 + \cos 2\theta) \\
M_{\theta\theta} &= \tfrac{1}{2}M_o(1 - \cos 2\theta) \\
M_{r\theta} &= -\tfrac{1}{2}M_o \sin 2\theta \\
Q_r &= 0 , \qquad V_{rr} = \frac{1}{r}M_o \cos 2\theta
\end{aligned}
$$

Consider now a hole of radius a in the sheet; the above displacement field gives rise to non-zero stresses along the edge of the hole and we must replace these initial stresses by the action of additional external couples. That is, on the initial state of stress we superpose an additional state of stress that will cancel the couples

$$M_{rr}|_a = \frac{1}{2}M_o(1 + \cos 2\theta), \qquad V_{rr}|_a = \frac{1}{a}M_o \cos 2\theta$$

around the edge of the hole. Furthermore, it must vanish at infinity.

This is achieved by choosing a deflection of

$$w = A \log r + (B + C\frac{a^2}{r^2})\cos 2\theta$$

which also satisfies the governing field equation. On differentiation, it gives the resultants

$$M_{rr} = D\left[-(1-\nu)A\frac{1}{r^2} - (4\nu B\frac{1}{r^2} - 6(1-\nu)C\frac{a^2}{r^4})\cos 2\theta\right]$$

$$M_{\theta\theta} = D\left[(1-\nu)A\frac{1}{r^2} - (4B\frac{1}{r^2} + 6(1-\nu)C\frac{a^2}{r^4})\cos 2\theta\right]$$

$$M_{r\theta} = D(1-\nu)(2B\frac{1}{r^2} + 6C\frac{a^2}{r^4})\sin 2\theta$$

$$V_{rr} = -D\left[4(3-\nu)B\frac{1}{r^3} + 12(1-\nu)C\frac{a^2}{r^5}\right]\cos 2\theta$$

At the edge of the hole $r = a$, we must have that the resulting M_{rr} and V_r be zero, that is,

$$\frac{1}{2}M_o + D\left[-(1-\nu)A\frac{1}{a^2} - (4\nu B\frac{1}{a^2} - 6(1-\nu)C\frac{1}{a^2})\cos 2\theta\right] = 0$$

$$\frac{1}{a}M_o - D\left[4(3-\nu)B\frac{1}{a^3} + 12(1-\nu)C\frac{1}{a^3}\right]\cos 2\theta = 0$$

This gives rise to three equations

$$M_o a^2 - D(1-\nu)2A = 0$$
$$M_o a^2 - D8\nu B + D12(1-\nu)C = 0$$
$$M_o a^2 - D4(3-\nu)B - D12(1-\nu)C = 0$$

Solving this for the three coefficients leads to

$$A = \frac{M_o a^2}{2D(1-\nu)}, \qquad B = \frac{M_o a^2}{2D(3+\nu)}, \qquad C = \frac{-M_o a^2}{4D(3+\nu)}$$

Combining this solution with the original uniform field leads to the moments

$$M_{rr} = \frac{1}{2}M_o\left[1 - \frac{a^2}{r^2} + \left\{1 - \frac{4\nu}{3+\nu}\frac{a^2}{r^2} - \frac{3(1-\nu)}{3+\nu}\frac{a^4}{r^4}\right\}\cos 2\theta\right]$$

$$M_{\theta\theta} = \frac{1}{2}M_o\left[1 + \frac{a^2}{r^2} - \left\{1 + \frac{4}{3+\nu}\frac{a^2}{r^2} - \frac{3(1-\nu)}{3+\nu}\frac{a^4}{r^4}\right\}\cos 2\theta\right]$$

$$M_{r\theta} = \frac{1}{2}M_o\left[\left\{-1 + \frac{2(1-\nu)}{3+\nu}\frac{a^2}{r^2} - \frac{3(1-\nu)}{3+\nu}\frac{a^4}{r^4}\right\}\sin 2\theta\right]$$

$$V_{rr} = -\frac{M_o}{(1-\nu)a}[(6-2\nu)A + -6(1-\nu)C]\cos 2\theta$$

The hoop moment around the edge of the hole is

$$M_{\theta\theta} = \left[1 - \frac{2(1+\nu)}{3+\nu}\cos 2\theta\right]$$

This has a maximum at $\theta = 90\,\mathrm{deg}$ and is

$$M_{rr} = \frac{(5+3\nu)}{(3+\nu)}M_o$$

For typical values of Poisson's ratio this gives a concentration effect of about 1.8, which is about two-thirds of the concentration effect of the equivalent membrane problem.

The behavior of the hoop shear around the edge of the hole is worth noting. The maximum value is

$$Q_\theta = \frac{4}{(3+\nu)a}M_o$$

which shows a dependence on the size of the hole. Thus, as a is made very small, this shear is made very large. All the developments in this chapter are predicated on the transverse shear stress being small; we therefore conclude that applications of the above equations to problems involving very small radii (notches and cracks, for example) must be somewhat suspect.

2.4 Curved Plates and Shells

There is considerable interest in curved plates because of such structural applications as shells, arches, containment vessels, and fuselages to name a few. We now look at some aspects of curved plates — to simplify matters, we consider only circular uniform cylinder segments. More-detailed analysis than what will follow can be found in References [46, 47].

There are a variety of ways to derive the shell equations; we find it most expedient to specify the deformation in the cylindrical coordinates of Figure 2.11(a), obtain the strains, convert to the coordinates of Figure 2.11(b), obtain the energies, and then use Hamilton's principle to derive the equations of motion and the boundary conditions.

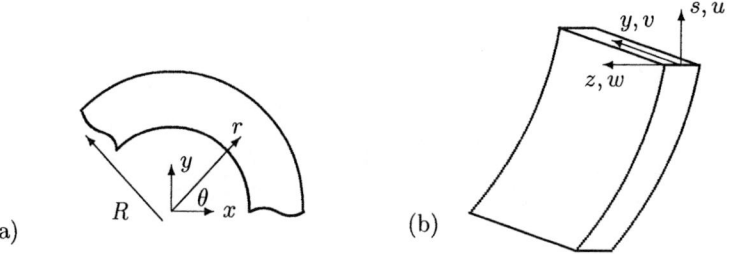

Figure 2.11: Coordinates for curved shell segment. (a) Cylindrical coordinates. (b) Curved shell coordinates.

Deformation of Cylindrical Shells

In the cylindrical coordinate system, (r, θ, z), the components of the displacement vector in the plane are denoted by $\bar{u}_r$ and $\bar{u}_\theta$. The strains are related to these displacements by

$$\bar{\epsilon}_{rr} = \frac{\partial \bar{u}_r}{\partial r}, \qquad \bar{\epsilon}_{\theta\theta} = \frac{\bar{u}_r}{r} + \frac{1}{r}\frac{\partial \bar{u}_\theta}{\partial \theta}$$

$$2\bar{\epsilon}_{r\theta} = \frac{1}{r}\frac{\partial \bar{u}_r}{\partial \theta} + \frac{\partial \bar{u}_\theta}{\partial r} - \frac{\bar{u}_\theta}{r}$$

For thick cross-sections, the strain $\bar{\epsilon}_{\theta\theta}$ is nonlinearly distributed. We will now replace these with an approximate set based on the assumption that the dimension in the r direction is small and this will give a linear distribution.

We begin by expanding the displacements in a Taylor series about the midplane $(r = R)$ using the variable $\xi \equiv r - R$. That is, we approximate the deformation of the shell in cylindrical coordinates as

$$\bar{u}_r(r, \theta, z) \approx u_r(\theta, z)$$

$$\bar{u}_\theta(r, \theta, z) \approx u_\theta(\theta, z) - \xi\left(\frac{1}{R}\frac{\partial u_r}{\partial \theta} - \frac{u_\theta}{R}\right)$$

$$\bar{u}_z(r, \theta, z) \approx u_z(\theta, z) - \xi\frac{\partial u_r}{\partial z} \qquad (2.23)$$

with $\xi \equiv (r - R)$ and where the third equation allows for bending about the z axis. These give the nonzero strains as

$$\bar{\epsilon}_{\theta\theta} = \frac{u_r}{R} + \frac{1}{R}\frac{\partial u_\theta}{\partial \theta} - \frac{\xi}{R^2}\left(\frac{\partial^2 u_r}{\partial \theta^2} - \frac{\partial u_\theta}{\partial \theta}\right)$$

$$\bar{\epsilon}_{zz} = \frac{\partial u_z}{\partial z} - \xi\frac{\partial^2 u_r}{\partial z^2} \qquad (2.24)$$

$$2\bar{\epsilon}_{\theta z} = \frac{1}{R}\frac{\partial u_z}{\partial \theta} + \frac{\partial u_\theta}{\partial z} - \frac{\xi}{R}\left(2\frac{\partial^2 u_r}{\partial z\partial \theta} - \frac{\partial u_\theta}{\partial z}\right)$$

At this stage, it is advantageous to convert the above to a more usual form of notation. It is typical in shell analysis to have a hoop coordinate s, an axial coordinate y, and a transverse coordinate z pointed toward the origin of the circle. That is, we have

$$R\theta \longrightarrow s, \qquad z \longrightarrow -y, \qquad r \longrightarrow -z$$

giving for the corresponding displacements

$$u_\theta \longrightarrow u, \qquad u_z \longrightarrow -v, \qquad u_r \longrightarrow -w$$

Our approximate deformation relations are now

$$\bar{u}(s, y, z) \approx u(s, y) - z\left(\frac{\partial w}{\partial s} + \frac{u}{R}\right)$$

$$\bar{v}(s, y, z) \approx v(s, y) - z\frac{\partial w}{\partial y} \tag{2.25}$$

$$\bar{w}(s, y, z) \approx w(s, y)$$

The nonzero strains are then

$$\epsilon_{ss} = -\frac{w}{R} + \frac{\partial u}{\partial s} - z\left(\frac{\partial^2 w}{\partial s^2} + \frac{1}{R}\frac{\partial u}{\partial s}\right)$$

$$\epsilon_{yy} = \frac{\partial v}{\partial y} - z\frac{\partial^2 w}{\partial y^2} \tag{2.26}$$

$$2\epsilon_{sy} = \frac{\partial v}{\partial s} + \frac{\partial u}{\partial y} - z\left(2\frac{\partial^2 w}{\partial s\partial y} + \frac{1}{R}\frac{\partial u}{\partial y}\right)$$

Other shell theories have slightly different expressions for these strains; the present theory is closest to that of Reissner [51, 58]. An excellent survey of the different theories are given in References [46, 47]. The theory developed here is the shell equivalent of the classical plate theory and the Bernoulli-Euler beam theory.

Equations of Motion

While it is possible to derive the equations of motion based on a free body diagram, it is much more advantageous to use Hamilton's principle, because we then get the set of boundary conditions consistent with these equations.

The strain energy for a small segment of shell in plane stress is

$$U = \tfrac{1}{2}\int_V \left[E^*(\epsilon_{ss}^2 + \epsilon_{yy}^2 + 2\nu\epsilon_{ss}\epsilon_{yy}) + G\gamma_{sy}^2\right] dV$$

where $E^* \equiv E/(1 - \nu^2)$. Substitute for the strains and integrate with respect to the thickness to get the total strain energy as

$$U = \tfrac{1}{2}\int_s \int_y [E^*U_1 + DU_2]\, ds\, dy$$

where

$$U_1 = \left(\frac{\partial u}{\partial s} + \frac{\partial v}{\partial y} - \frac{w}{R}\right)^2 + \tfrac{1}{2}(1 - \nu)\left[-4\left(\frac{\partial u}{\partial s} - \frac{w}{R}\right)\frac{\partial v}{\partial y} + \left(\frac{\partial u}{\partial y} + \frac{\partial v}{\partial s}\right)^2\right]$$

$$U_2 = \left(\frac{\partial^2 w}{\partial s^2} + \frac{\partial^2 w}{\partial y^2} + \frac{1}{R}\frac{\partial u}{\partial s}\right)^2 \tag{2.27}$$

$$-2(1 - \nu)\frac{\partial^2 w}{\partial y^2}\left(\frac{\partial^2 w}{\partial s^2} + \frac{1}{R}\frac{\partial u}{\partial s}\right) + \tfrac{1}{2}(1 - \nu)\left(2\frac{\partial^2 w}{\partial s\partial y} + \frac{1}{R}\frac{\partial u}{\partial y}\right)^2$$

As with flat plates, $D \equiv Eh^3/12(1 - \nu^2)$. In the above, U_1 represents the strain energy due to the membrane strains while U_2 represents the contribution from the bending strains. Retaining only the first two energy contributions gives the so-called *Donnell Shell equations*; retaining all terms will give the *Reissner Shell equations*. The latter are slightly different from the *Timoshenko-Love equations*. The total kinetic energy is

$$T = \tfrac{1}{2} \int_V \rho[\dot{u}(x,y,z,t)^2 + \dot{v}(x,y,z,t)^2 + \dot{w}(x,y,z,t)^2]dV$$

$$= \tfrac{1}{2} \int_s \int_y \rho h \left[\dot{u}^2 + \dot{v}^2 + \dot{w}^2\right] ds\, dy \qquad (2.28)$$

where we have neglected the rotational inertia. Let the potential of the applied loads be

$$V = -Q_u u - Q_v v - Q_w w - Q_\psi \psi, \qquad \psi \equiv \frac{\partial w}{\partial s}$$

Application of Hamilton's principle, taking the variations with respect to δu, δv, and δw, leads to three governing equations

$$E^* \left[\frac{\partial^2 u}{\partial s^2} + \tfrac{1}{2}(1-\nu)\frac{\partial^2 u}{\partial y^2} + \tfrac{1}{2}(1+\nu)\frac{\partial^2 v}{\partial s \partial y} - \frac{1}{R}\frac{\partial w}{\partial s} \right]$$

$$+ \frac{D}{R^2} \left[\frac{\partial^2 u}{\partial s^2} + \tfrac{1}{2}(1-\nu)\frac{\partial^2 u}{\partial y^2} + R\frac{\partial^3 w}{\partial y^2 \partial s} + R\frac{\partial^3 w}{\partial s^3} \right] = \rho h \frac{\partial^2 u}{\partial t^2}$$

$$E^* \left[\tfrac{1}{2}(1+\nu)\frac{\partial^2 u}{\partial s \partial y} + \frac{\partial^2 v}{\partial y^2} + \tfrac{1}{2}(1-\nu)\frac{\partial^2 v}{\partial s^2} - \frac{\nu}{R}\frac{\partial w}{\partial y} \right] = \rho h \frac{\partial^2 v}{\partial t^2}$$

$$-E^* \left[\frac{1}{R}\frac{\partial u}{\partial s} + \frac{\nu}{R}\frac{\partial v}{\partial y} - \frac{w}{R^2} \right] + D \left[\frac{\partial^4 w}{\partial s^4} + 2\frac{\partial^4 w}{\partial s^2 \partial y^2} + \frac{\partial^4 w}{\partial y^4} \right] \qquad (2.29)$$

$$+ \frac{D}{R} \left[\frac{\partial^3 u}{\partial y^2 \partial s} + \frac{\partial^3 u}{\partial s^3} \right] = -\rho h \frac{\partial^2 w}{\partial t^2}$$

These equations are grouped in terms of membrane $E^*[\cdots]$ and flexural $D[\cdots]$ contributions. This rather complicated collection of equations is a combination of the flat membrane, flat plate, and curved beam equations.

The associated boundary conditions on the side $s = $ constant are specified in terms of one each of the following pairs:

$$u \quad \text{or} \quad Q_u = \bar{E}\left[\frac{\partial u}{\partial s} - \frac{w}{R} + \nu\frac{\partial v}{\partial y}\right] + \frac{D}{R}\left[\frac{\partial^2 w}{\partial s^2} + \frac{1}{R}\frac{\partial u}{\partial s} + \nu\frac{\partial^2 w}{\partial y^2}\right]$$

$$v \quad \text{or} \quad Q_v = \tfrac{1}{2}(1-\nu)\bar{E}\left[\frac{\partial v}{\partial s} + \frac{\partial u}{\partial y}\right]$$

$$w \quad \text{or} \quad Q_w = -\frac{D}{R}\left[\frac{\partial^2 u}{\partial s^2} + (1-\nu)\frac{\partial^2 u}{\partial y^2}\right] - D\left[\frac{\partial^3 w}{\partial s^3} + (2-\nu)\frac{\partial^3 w}{\partial s \partial y^2}\right]$$

$$\frac{\partial w}{\partial s} \quad \text{or} \quad Q_m = D\left[\frac{\partial^2 w}{\partial s^2} + \nu\frac{\partial^2 w}{\partial y^2} + \frac{1}{R}\frac{\partial u}{\partial s}\right] \qquad (2.30)$$

It remains now to interpret the resultants and relate them to these boundary conditions.

Referring to Figure 2.11(b), we can form the resultants per unit length as

$$N_{ss} \equiv \int \sigma_{ss}\, dz\,, \qquad N_{sy} \equiv \int \sigma_{sy}\, dz$$

Substituting for the stresses and strains in terms of our approximations leads to

$$N_{ss} = E^*\left[\frac{\partial u}{\partial s} - \frac{w}{R} + \nu\frac{\partial v}{\partial y}\right], \qquad N_{sy} = \tfrac{1}{2}(1-\nu)E^*\left[\frac{\partial v}{\partial s} + \frac{\partial u}{\partial y}\right] \qquad (2.31)$$

We can also form the resultant moments per unit length

$$M_{ss} \equiv -\int \sigma_{ss} z\, dz\,, \qquad M_{sy} \equiv -\int \sigma_{sy} z\, dz$$

Again, substituting for the stresses and strains in terms of our approximations leads to

$$M_{ss} = D\left[\frac{\partial^2 w}{\partial s^2} + \nu\frac{\partial^2 w}{\partial y^2} + \frac{1}{R}\frac{\partial u}{\partial s}\right], \qquad M_{sy} = \tfrac{1}{2}D(1-\nu)\frac{\partial}{\partial y}\left[\frac{u}{R} + 2\frac{\partial w}{\partial s}\right]$$

Comparing these expressions to those for the boundary conditions, we see that the natural boundary conditions are equivalent to specifying

$$\begin{aligned}
Q_u &= N_{ss} + \frac{1}{R}M_{ss}\\
Q_v &= N_{sy}\\
Q_w &= -\frac{\partial M_{ss}}{\partial s} - 2\frac{\partial M_{sy}}{\partial y} = V_{sz}\\
Q_m &= M_{ss}
\end{aligned} \qquad (2.32)$$

The first of these resembles the resultant load expression used for curved beams [23], while the third resembles the Kirchhoff shear stress relation.

Example 2.11:   Use the Lamé solution for pressurized cylinders to assess some membrane aspects of the shell theory.

This is one of the very few shell problems for which there is an exact solution. Consider a hollow cylinder subjected to uniform internal pressure p_i and assume there are no variations along the length. This is an axisymmetric problem and therefore there is no bending. However, there is considerable membrane action, and this is what we can investigate.

Because the problem is plane and axisymmetric, we take the stress function as

$$\phi = A\log_n r + Cr^2$$

which gives the stresses

$$\sigma_{rr} = \frac{A}{r^2} + 2C\,, \qquad \sigma_{\theta\theta} = -\frac{A}{r^2} + 2C\,, \qquad \sigma_{r\theta} = 0$$

The boundary conditions are zero tractions on the outside, only a normal traction on the inside. The zero shear traction boundary conditions are automatically satisfied by the solution, leaving us with the normal traction conditions

$$r = a: \quad t_r = -\sigma_{rr} = p_i = -\frac{A}{a^2} - 2C, \qquad r = b: \quad t_r = \sigma_{rr} = 0 = \frac{A}{b^2} + 2C$$

This gives two equations for two unknowns. After solving for the coefficients, we can write the stresses as

$$\sigma_{rr} = \frac{p_i}{(1 - a^2/b^2)} \left[\frac{a^2}{b^2} - \frac{a^2}{r^2} \right]$$

$$\sigma_{\theta\theta} = \frac{p_i}{(1 - a^2/b^2)} \left[\frac{a^2}{b^2} + \frac{a^2}{r^2} \right]$$

This is the Lamé solution.

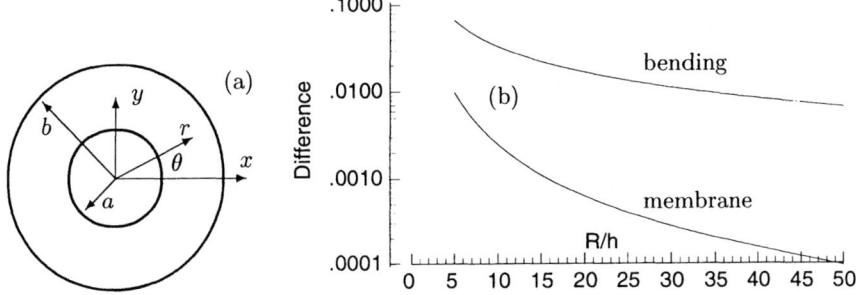

Figure 2.12: Testing the thin-wall approximation. (a) Geometry for pressurized cylinder. (b) Difference between the exact and thin-walled approximation.

Let the shell be thin so that $a = R - h/2$, $b = R + h/2$, $r = R + \xi$, where R is the average radius of the cylinder and h is the thickness. Substitute these in the geometry terms and simplify assuming $h/R << 1$ to get

$$\frac{a^2}{b^2} = \frac{(R - h/2)^2}{(R + h/2)^2} \approx 1 - \frac{2h}{R}, \qquad \frac{a^2}{r^2} = \frac{(R - h/2)^2}{(R + \xi)^2} \approx 1 - \frac{h}{R} + \frac{2\xi}{R}$$

Substituting into the Lamé solution gives the approximation for the stresses

$$\sigma_{rr} \approx \frac{p_i R}{h}[0 + O(h/R)] \approx 0, \qquad \sigma_{\theta\theta} \approx \frac{p_i R}{h}[1 + O(h/R)] \approx \frac{p_i R}{h}$$

The hoop stress is the dominant stress and is almost uniform on the cross-section. The radial stress is approximately zero (in comparison to the hoop stress) even though it is closely associated with the applied pressure.

The difference between the exact hoop stress and the thin-wall approximation, $\Delta = 1 - \sigma_t/\sigma_c$, is shown in Figure 2.12 as a function of R/h. When $R/h = 5$, the difference is about 1%, but when $R/h > 20$, the difference has dropped well below 0.1%. We therefore take that the transition to thin-wall behavior is around $R/h \approx 10$.

Example 2.12:   Use Golovin's solution for curved beams to assess some bending aspects of the shell theory.

Consider a curved beam subjected to end moments M_o as shown in the Figure 2.13(a). From the moment balance condition, it is evident that the moment on any radial cross-section along the beam is constant. In addition, the surface tractions are independent of θ. Hence, this is an axisymmetric problem in stress (although not necessarily in displacements).

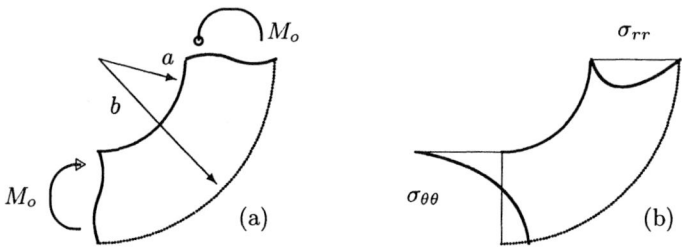

Figure 2.13: Curved beam with resultant moments. (a) Cylindrical coordinates. (b) Stress distributions in curved beam.

The Airy stress function for the problem is

$$\phi(r,\theta) = \phi(r) = A\log_n r + Br^2\log_n r + Cr^2$$

This gives the stresses

$$\sigma_{rr} = \frac{1}{r}\frac{\partial\phi}{\partial r} = \frac{A}{r^2} + B(1 + 2\log_n r) + 2C$$

$$\sigma_{\theta\theta} = \frac{\partial\phi^2}{\partial r^2} = -\frac{A}{r^2} + B(3 + 2\log_n r) + 2C$$

with $\sigma_{r\theta} = 0$ everywhere. There are three coefficients to be solved for. The boundary conditions on the lateral faces are

$$\sigma_{rr}|_{r=a} = 0 = \frac{A}{a^2} + B(1 + 2\log_n a) + 2C$$

$$\sigma_{rr}|_{r=b} = 0 = \frac{A}{b^2} + B(1 + 2\log_n b) + 2C$$

One more equation is needed. We cannot impose tractions as the boundary conditions on the ends simply because we do not know them. So we impose conditions on the resultants instead. That is,

$$F_\theta = \int_a^b \sigma_{\theta\theta}\,dr = 0, \qquad M = \int_a^b \sigma_{\theta\theta}\,r\,dr = M_o$$

On substituting for the stresses, this leads to only one nontrivial equation

$$F_\theta = \int \frac{\partial^2\phi}{\partial r^2}\,dr = \left.\frac{\partial\phi}{\partial r}\right|_b - \left.\frac{\partial\phi}{\partial r}\right|_a = b\,\sigma_{rr}(b) - a\,\sigma_{rr}(a) = 0$$

$$M_o = \int \frac{\partial^2\phi}{\partial r^2}r\,dr = \left.r\frac{\partial\phi}{\partial r}\right|_a^b - \int\frac{\partial\phi}{\partial r}\,dr = b^2\sigma_{rr}(b) - a^2\sigma_{rr}(a) - \phi(b) + \phi(a)$$

$$= -A\log_n(\frac{b}{a}) - B[b^2\log_n b - a^2\log_n a] - C[b^2 - a^2]$$

Solving these equations for the coefficients in terms of M_o gives

$$\sigma_{rr} = \frac{4M_o b^2}{N} \left[\frac{a^2}{r^2} \log_n\left(\frac{b}{a}\right) + \log_n\left(\frac{r}{b}\right) + \frac{a^2}{b^2} \log_n\left(\frac{a}{r}\right) \right]$$

$$\sigma_{\theta\theta} = \frac{4M_o b^2}{N} \left[-\frac{a^2}{r^2} \log_n\left(\frac{b}{a}\right) + \log_n\left(\frac{r}{b}\right) + \frac{a^2}{b^2} \log_n\left(\frac{a}{r}\right) + 1 - \frac{a^2}{b^2} \right]$$

$$N = (b^2 - a^2)^2 - 4b^2 a^2 \left(\log_n \frac{b}{a}\right)^2$$

This is Golovin's solution for curved beams. The stress distribution is shown in Figure 2.13(b). Note the very large increase in hoop stress at the inner radius.

A significant aspect of this solution is that the hoop stress $\sigma_{\theta\theta}$ is not linearly distributed on the cross-section. However, if we use the thin-wall assumption as done in the previous example, we wind up with the approximations

$$\sigma_{rr} \approx 0, \qquad \sigma_{\theta\theta} \approx \frac{12M_o\xi}{h^3}$$

The hoop stress, again, is the dominant stress and is almost linearly distributed on the cross-section.

The difference between the exact hoop stress and the thin-wall approximation, $\Delta = 1 - \sigma_t/\sigma_c$, is also shown in Figure 2.12 as a function of R/h. When $R/h = 5$, the difference is about 6%, but we have to have $R/h > 40$, for the difference to drop below 1.0%. This rate of change is significantly different than for the membrane stress. It is worth keeping in mind, however, that while the stress is overestimated on one side of the plate, it is underestimated on the other side, so that the linear approximation represents a very good average. Indeed, a comparison of the (absolute) averages does not register in Figure 2.12.

We therefore take that the transition to thin-wall behavior is around $R/h \approx 20$.

Example 2.13: Specialize the curved plate equations to those for a curved beam.

With reference to the coordinates of Figure 2.11(b), a beam has no dependence on the y coordinate and has no displacement v. This results in the two equations

$$\bar{E}\left[\frac{\partial^2 u}{\partial s^2} - \frac{1}{R}\frac{\partial w}{\partial s}\right] + \frac{D}{R^2}\left[\frac{\partial^2 u}{\partial s^2} + R\frac{\partial^3 w}{\partial s^3}\right] = \rho h \frac{\partial^2 u}{\partial t^2}$$

$$-\bar{E}\left[\frac{1}{R}\frac{\partial u}{\partial s} - \frac{w}{R^2}\right] + D\left[\frac{\partial^4 w}{\partial s^4}\right] + \frac{D}{R}\left[\frac{\partial^3 u}{\partial s^3}\right] = -\rho h \frac{\partial^2 w}{\partial t^2}$$

To have these equations resemble those for a beam bending in the $x - y$ plane, we make the association $w \to v$. Furthermore, multiply across by the beam depth b and replace

$$hb \equiv A, \qquad \bar{E}b = \frac{Ehb}{(1 - \nu^2)} \equiv EA, \qquad Db = \frac{Eh^3 b}{12(1 - \nu^2)} \equiv EI$$

Then the equations become

$$EA\frac{\partial^2 u}{\partial s^2} + \frac{1}{R^2}\left[EI\frac{\partial^2 u}{\partial s^2} - EAR\frac{\partial v}{\partial s} + EIR\frac{\partial^3 v}{\partial s^3}\right] = \rho A \frac{\partial^2 u}{\partial t^2}$$

$$EI\frac{\partial^4 v}{\partial s^4} + \frac{1}{R^2}\left[EAv - EAR\frac{\partial u}{\partial s} + EIR\frac{\partial^3 u}{\partial s^3}\right] = -\rho A \frac{\partial^2 v}{\partial t^2} \qquad (2.33)$$

If the beam depth b is small, then Poisson's ratio effect can be neglected in the definitions of EA and EI.

The resultant axial force F, shear force V, and bending moment M, are related to the deformation by

$$F = EA\left[\frac{\partial u}{\partial s} - \frac{v}{R}\right], \qquad V = -EI\left[\frac{\partial^3 w}{\partial s^3} + \frac{1}{R}\frac{\partial^2 u}{\partial s^2}\right], \qquad M = EI\frac{\partial}{\partial s}\left[\frac{\partial v}{\partial s} + \frac{u}{R}\right]$$

These are used in specifying the boundary conditions.

Discussion

A thick-walled curved plate has two features of significance. First, the hoop strain is not linearly distributed but is parabolic with the larger value being on the inside radius. Second, there is coupling between the membrane and bending effects. As we approach a thin-wall formulation, the previous developments show that we obtain a linear distribution of strain but that there are still strong coupling effects. Anticipating our future developments for nonlinear large deflection problems, we would prefer not to have to deal with strongly coupled equations such as Equation (2.30) or Equation (2.33).

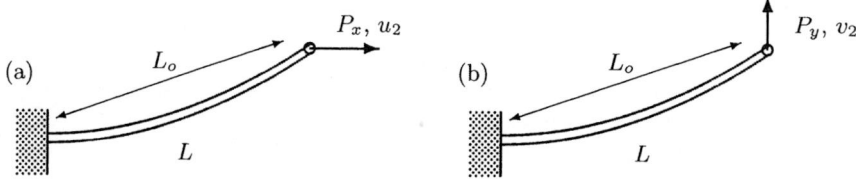

Figure 2.14: Slightly curved beam problems. (a) Horizontal load. (b) Vertical load.

Consider the slightly curved beam problems shown in Figure 2.14. For a given span L_o, it is clear that as the radius is decreased (so that there is a deeper arch) the effective stiffness of P_x against u_2 and P_y against v_2 decreases. The issue we are interested in is, to what extent this curved beam can be replaced with a straight segment and still give a reasonable approximation to the stiffness.

The strain energy of the segment is

$$U = \frac{1}{2}\int_L EA\left[\frac{\partial u}{\partial s} - \frac{v}{R}\right]ds + \frac{1}{2}\int_L EI\left[\frac{\partial^2 v}{\partial s^2} + \frac{1}{R}\frac{\partial u}{\partial s}\right]ds$$

Let us assume displacement shapes

$$u(s) = f_2(s)u_2, \qquad v(s) = g_3(s)v_2 + g_4(s)\phi_2$$

where $f_i(s)$ and $g_i(s)$ are, respectively, the rod and beam shape functions taken from Section 1.8. This discretization of strain energy will lead to a $[3 \times 3]$ stiffness relation. The first entry is associated with u_2 and leads to

$$k_{cuu} = \frac{\partial^2 U}{\partial u_2^2} = \frac{EA}{L} + 12\frac{EI}{L^3}\frac{L^2}{12R^2}$$

The angle subtended by the segment is $\theta = L/R$, the straight beam has an orientation half of this, and from the stiffness of a generally oriented member [22] (also see Chapter 3), we have

$$k_{suu} = \frac{EA}{L_o}\cos^2(\theta/2) + 12\frac{EI}{L_o^3}\sin^2(\theta/2) \approx \frac{EA}{L_o} + 12\frac{EI}{L_o^3}\frac{L_o^2}{4R^2}$$

There is also the geometric approximation that $L \approx L_o[1+L_o^2/(8R^2)]$. Thus, if L_o is smaller than $R/10$, then the difference in length is on the order of 0.1%, which is negligible. To put this into perspective, if a complete circular ring is replaced with 64 piecewise straight segments (or approximately 5° segments), then the criterion is met. This does not seem an unreasonable density of elements.

Taking the ratio of the two stiffnesses and assuming $L_o \approx L$, we get

$$\frac{k_{suu}}{k_{cuu}} \approx 1 - \frac{h^2}{6R^2}$$

where h is the beam thickness. If h is smaller than $R/20$ (which is a typical assumption for the thin-wall theory to hold), then the effect on the stiffness is of the order of 0.1%, which is negligible.

Doing a similar development for the vertical force, we get

$$k_{cvv} = \frac{\partial^2 U}{\partial v_2^2} = 12\frac{EI}{L^3} + \frac{EA}{L}\frac{L^2}{3R^2} \qquad \longleftrightarrow \qquad k_{svv} = 12\frac{EI}{L_o^3} + \frac{EA}{L_o}\frac{L_o^2}{4R^2}$$

The two stiffness expressions are very similar, the difference occurring only in the axial stiffness contribution. However, the axial stiffness is usually significantly larger than the flexural stiffness and therefore the seemingly small difference could actually be quite large. The ratio of the stiffnesses in this case becomes

$$\frac{k_{svv}}{k_{cvv}} \approx 1 - \frac{1}{6}\frac{L^2}{h^2}\frac{L^2}{R^2} = 1 - \frac{1}{6}\frac{R^2}{h^2}\frac{L^4}{R^4}$$

This time h appears in the denominator. For the thin-wall approximation with $h < R/20$, an element length of $L < R/10$ then gives a stiffness difference of about 0.007. This is quite small.

Based on these considerations, the plan in the remainder of the book is to develop only a straight frame and flat plate element formulation and treat all curved structures as assemblages of these. As it turns out, other considerations arising out of the nonlinear dynamics description also require a small element and therefore our treatment of general structures is not unduly inefficient.

2.5   Discretization Using Triangular Regions

We need to discretize the plates and shells as part of our finite element formulation of problems. There are a variety of discretization schemes available, but we will consider only at triangular regions. As will be shown, this is more than adequate for our needs.

Area and Natural Coordinates

Since we are dealing with triangular elements, it is convenient to work in area coordinates. Consider a triangle divided into three areas where the common apex is at (x, y) as shown in Figure 2.15(a). Define

$$h_1 = A_1/A, \qquad h_2 = A_2/A, \qquad h_3 = A_3/A, \qquad h_i = h_i(x, y)$$

with the obvious constraint that $h_1 + h_2 + h_3 = 1$. The areas of these triangles define uniquely the position of the common apex.

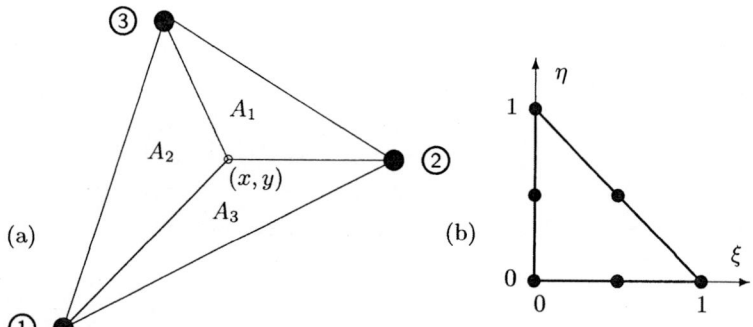

Figure 2.15: Area and natural coordinates.

The position of a point (x, y) in the triangle can be written as

$$\begin{Bmatrix} 1 \\ x \\ y \end{Bmatrix} = \begin{bmatrix} 1 & 1 & 1 \\ x_1 & x_2 & x_3 \\ y_1 & y_2 & y_3 \end{bmatrix} \begin{Bmatrix} h_1 \\ h_2 \\ h_3 \end{Bmatrix} \tag{2.34}$$

where the subscripts $1, 2, 3$ refer to the counterclockwise nodes of the triangle. We can invert this to get the expressions for the area coordinates

$$\begin{Bmatrix} h_1 \\ h_2 \\ h_3 \end{Bmatrix} = \frac{1}{2A} \begin{bmatrix} x_2 y_3 - x_3 y_2 & y_{23} & x_{32} \\ x_3 y_1 - x_1 y_3 & y_{31} & x_{13} \\ x_1 y_2 - x_2 y_1 & y_{12} & x_{21} \end{bmatrix} \begin{Bmatrix} 1 \\ x \\ y \end{Bmatrix} = \frac{1}{2A} \begin{bmatrix} a_1 & b_1 & c_1 \\ a_2 & b_2 & c_2 \\ a_3 & b_3 & c_3 \end{bmatrix} \begin{Bmatrix} 1 \\ x \\ y \end{Bmatrix}$$

with $2A \equiv x_{21} y_{31} - x_{31} y_{21}$, $x_{ij} \equiv x_i - x_j$, and so on. From this, it is apparent that functions of (x, y) can equally well be written as functions of (h_1, h_2, h_3).

That is, any function of interest can be written as

$$u(x, y) = \sum_i^3 h_i(x, y) u_i$$

where u_i are the nodal values. Note that special cases of this are for the coordinates themselves

$$x = \sum_i^3 h_i(x, y) x_i, \qquad y = \sum_i^3 h_i(x, y) y_i$$

which is obvious from Equation (2.34).

When performing differentiation, we use the chain rule as

$$\frac{\partial u}{\partial x} = \sum_i \frac{\partial u}{\partial h_i} \frac{\partial h_i}{\partial x}$$

A very useful formula when performing the integrations of the functions is the relation

$$\int_A h_1^i h_2^j h_3^k \, dA = 2A \frac{i! j! k!}{(2 + i + j + k)!}$$

where ! means factorial. If h_3, for example, does not appear in the integration, then we simply ignore the k exponent. Another useful integration formula is when the function is written in centroidal coordinates. We have [18]

$$\int_A x^i y^j \, dA = C_{i+j} A [x_1^i y_1^j + x_2^i y_2^j + x_3^i y_3^j],$$

$i + j : 1$	2	3	4	5
$C_{i+j} : 0$	$\frac{1}{12}$	$\frac{1}{30}$	$\frac{1}{30}$	$\frac{2}{105}$

In order for the coordinate description h_1, h_2, h_3 to describe the two coordinates x, y, it must be supplemented by the constraint $h_1 + h_2 + h_3 = 1$. We can invoke this constraint explicitly by introducing *natural coordinates* given as

$$h_1 = 1 - \xi - \eta, \qquad h_2 = \xi, \qquad h_3 = \eta \qquad (2.35)$$

These are shown in Figure 2.15(b). We have for a typical function

$$u(x, y) = \sum_i^3 h_i(\xi, \eta) u_i$$

The element strains are obtained in terms of derivatives of element displacements. Using the natural coordinate system, we get, for example,

$$\frac{\partial}{\partial x} = \frac{\partial}{\partial \xi} \frac{\partial \xi}{\partial x} + \frac{\partial}{\partial \eta} \frac{\partial \eta}{\partial x}$$

But to evaluate the derivatives of ξ, η with respect to x, y we need to have the explicit relation between the two sets of variables. This is obtained as

$$\left\{ \begin{array}{c} \frac{\partial}{\partial \xi} \\ \frac{\partial}{\partial \eta} \end{array} \right\} = \left[\begin{array}{cc} \frac{\partial x}{\partial \xi} & \frac{\partial y}{\partial \xi} \\ \frac{\partial x}{\partial \eta} & \frac{\partial y}{\partial \eta} \end{array} \right] \left\{ \begin{array}{c} \frac{\partial}{\partial x} \\ \frac{\partial}{\partial y} \end{array} \right\} \qquad \text{or} \qquad \{\frac{\partial}{\partial \xi}\} = [\ J\ ]\{\frac{\partial}{\partial x}\}$$

where $[\ J\ ]$ is called the *Jacobian operator* relating the natural coordinates to the local coordinates. This is essentially the same Jacobian introduced in Chapter 1. The relation for the derivatives requires

$$\{\frac{\partial}{\partial x}\} = [J^{-1}]\{\frac{\partial}{\partial \xi}\}$$

which requires that $[J^{-1}]$ exists. In most cases, the existence is clear; however, in cases where the element is much distorted or folds back on itself the Jacobian transformation can become singular.

Example 2.14: Evaluate the integral $\int_A h_1 h_2 \, dA$.
We use the general formula and since $0! = 1$, we get

$$\int_A h_1 h_2 \, dA = \int_A h_1^1 h_2^1 h_3^0 dA = 2A \frac{1! 1! 0!}{(2+1+1+0)!} = 2A \frac{1}{4 \times 3 \times 2 \times 1} = A \frac{1}{12}$$

Example 2.15: Determine the Jacobian operator for the linear triangle.
Noting that
$$h_1 = 1 - \xi - \eta, \qquad h_2 = \xi, \qquad h_3 = \eta$$
and from Equation 2.34, we get some of the derivatives as

$$\frac{\partial x}{\partial \xi} = x_1[-1] + x_2[1] + x_3[0] = x_{21}, \qquad \frac{\partial y}{\partial \eta} = y_1[-1] + y_2[0] + y_3[1] = y_{31}$$

The complete operator and its inverse is

$$[\ J\ ] = [\frac{\partial x}{\partial \xi}] = \begin{bmatrix} x_{21} & y_{21} \\ x_{31} & y_{31} \end{bmatrix}, \qquad [J^{-1}] = \frac{1}{2A} \begin{bmatrix} y_{31} & -y_{21} \\ -x_{31} & x_{21} \end{bmatrix}$$

Note that $\det[\ J\ ] = x_{21}y_{31} - x_{31}y_{21} = 2A$, thus the Jacobian becoming singular is equivalent to the area becoming zero and then negative. Therefore, in situations where there are large displacements, it is of value to check the sign of the area. This is all the more necessary in cases involving iterative solution strategies where some of the iterates (not yet being converged) can lead to physically unrealistic intermediate configurations.

Example 2.16: Evaluate the derivative of a function $u(x,y)$ whose values are given at the three nodal points: $(0,0)$, $(4,1)$, $(1,3)$.
We have from Equation (2.34) that

$$x = 4h_2 + h_3 = 4\xi + \eta, \qquad y = 1h_2 + 3h_3 = \xi + 3\eta$$

The Jacobian operator and its inverse are then

$$[\,J\,] = [\frac{\partial x}{\partial \xi}] = \begin{bmatrix} 4 & 1 \\ 1 & 3 \end{bmatrix}, \qquad [J^{-1}] = \frac{1}{11} \begin{bmatrix} 3 & -1 \\ -1 & 4 \end{bmatrix}$$

Let the function be given by the linear interpolation

$$u(x, y) = [h_1, h_2, h_3]\{u_1, u_2, u_3\}^T$$

but we will treat $h_i = h_i(\xi, \eta)$. The derivatives are

$$\left\{ \begin{array}{c} \frac{\partial u}{\partial x} \\ \frac{\partial u}{\partial y} \end{array} \right\} = [J^{-1}] \left[\left\{ \begin{array}{c} h_{1,\xi} \\ h_{1,\eta} \end{array} \right\}, \left\{ \begin{array}{c} h_{2,\xi} \\ h_{2,\eta} \end{array} \right\}, \left\{ \begin{array}{c} h_{3,\xi} \\ h_{3,\eta} \end{array} \right\} \right] \{u\}$$

$$= \frac{1}{11} \begin{bmatrix} 3 & -1 \\ -1 & 4 \end{bmatrix} \left[\left\{ \begin{array}{c} -1 \\ -1 \end{array} \right\}, \left\{ \begin{array}{c} 1 \\ 0 \end{array} \right\}, \left\{ \begin{array}{c} 0 \\ 1 \end{array} \right\} \right] \{u\}$$

where $\{u\} = \{u_1, u_2, u_3\}^T$ and the subscript comma indicates partial differentiation. The derivatives are constant.

Higher-Order Interpolations

The previous ideas can be generalized by considering higher-order interpolation functions. A possible sequence of higher-order functions is arranged in the form of *Pascal's triangle* as

$$
\begin{array}{ccccccc}
& & & 1 & & & \\
& & x & & y & & \\
& x^2 & & xy & & y^2 & \\
x^3 & & x^2 y & & xy^2 & & y^3
\end{array}
$$

where adding each complete level forms a complete representation. To utilize these, we must use more nodes, and a possible sequence is shown in Figure 2.16. Note that the cubic requires 10 nodes so as to have a complete Pascal triangle.

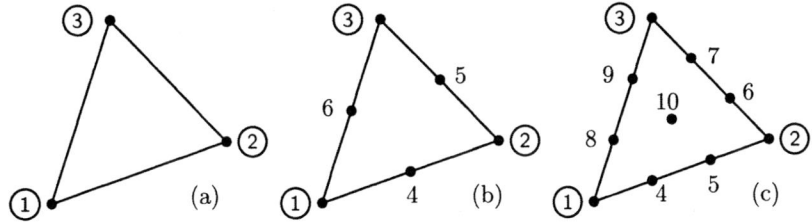

Figure 2.16: Some higher-order elements. (a) Linear. (b) Quadratic. (c) Cubic.

For the linear interpolation functions, we need three nodes for the complete representation. The representation is then

$$u(x, y) = \sum_{i}^{3} N_i u_i$$

where the shape functions are given by

$$
\begin{aligned}
N_1 &= h_1 \\
N_2 &= h_2 \\
N_3 &= h_3
\end{aligned}
\tag{2.36}
$$

For the quadratic interpolation functions, we need six nodes for the complete representation. The representation is then

$$
u(x, y) = \sum_i^6 N_i u_i
$$

where the shape functions correspond to a six-noded triangle (three apex nodes and three mid-side nodes) and are given by

$$
\begin{aligned}
N_1 &= (2h_1 - 1)h_1, & N_4 &= 4h_1 h_2 \\
N_2 &= (2h_2 - 1)h_2, & N_5 &= 4h_2 h_3 \\
N_3 &= (2h_3 - 1)h_3, & N_6 &= 4h_3 h_1
\end{aligned}
\tag{2.37}
$$

For the cubic interpolation functions, we need ten nodes for the complete representation. The representation is then

$$
u(x, y) = \sum_i^{10} N_i u_i
$$

where the shape functions are given by

$$
\begin{aligned}
N_1 &= \tfrac{1}{2}(3h_1 - 1)(3h_1 - 2)h_1, & N_4 &= \tfrac{9}{2}(3h_1 - 1)h_1 h_2, & N_7 &= \tfrac{9}{2}(3h_3 - 1)h_3 h_2 \\
N_2 &= \tfrac{1}{2}(3h_2 - 1)(3h_2 - 2)h_2, & N_5 &= \tfrac{9}{2}(3h_2 - 1)h_2 h_1, & N_8 &= \tfrac{9}{2}(3h_3 - 1)h_3 h_1 \\
N_3 &= \tfrac{1}{2}(3h_3 - 1)(3h_3 - 2)h_3, & N_6 &= \tfrac{9}{2}(3h_2 - 1)h_2 h_3, & N_9 &= \tfrac{9}{2}(3h_1 - 1)h_1 h_3
\end{aligned}
$$

and $N_{10} = 27 h_1 h_2 h_3$.

It is also possible to write each of these interpolation functions in terms of natural coordinates.

Example 2.17: Compare the quadratic and linear interpolations.

Figure 2.17 shows contours of a function interpolated using the linear and quadratic interpolations. The function varies as $\sin(x)\cos(y)$ but does not have an interior maximum or minimum. Note that if the function were linear, then both interpolations would give the same result. The original triangular region is modeled with six nodes in each case, which means using four linear interpolation regions.

It is clear that, for a given level of discretization (i.e., with the same number of degrees of freedom), the higher-order interpolations will out-perform the

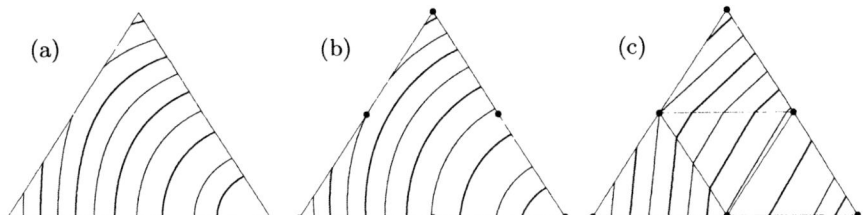

Figure 2.17: Comparison of the different interpolations. (a) Exact. (b) Quadratic. (c) Linear.

lower-order ones. The figure shows how the curved contours are approximated by piecewise linear segments.

Two additional factors, however, must be considered when choosing interpolation functions for elements. The first is that the computational cost increases for the higher-order functions, and it may well be that the simpler functions can afford to use more elements. The second is that, as we develop elements for complicated mechanics problems such as nonlinear deflections or elastic/plastic material behavior, it becomes increasingly more difficult to develop a higher-order consistent theory. We will generally opt, therefore, to choose the simpler functions and pay the price of having to use more elements.

2.6 Membrane Triangle Elements

Perhaps the simplest continuum element to formulate is that of the constant strain triangle (CST) element. We therefore begin with a discussion of this element. We also discuss a three-noded triangular element, which correctly implements the drilling DoF (ϕ_z) and therefore makes it suitable for 3-D applications.

Constant Strain Triangle Element

Consider a triangle with three nodes. The basic assumption in the formulation is that the displacements have the same description as the coordinates. That is,

$$x(x,y) = \sum_i^3 h_i(x,y)x_i , \qquad u(x,y) = \sum_i^3 h_i(x,y)u_i$$

$$y(x,y) = \sum_i^3 h_i(x,y)y_i , \qquad v(x,y) = \sum_i^3 h_i(x,y)v_i$$

where $h_i(x,y)$ are the linear triangle interpolation functions. These interpolation functions will be applied at all times during the deformation.

The displacement gradients are given by

$$\frac{\partial u}{\partial x} = \sum \frac{\partial h_i}{\partial x} u_i = \frac{1}{2A} \sum_i b_i u_i, \qquad \frac{\partial v}{\partial x} = \sum \frac{\partial h_i}{\partial x} v_i = \frac{1}{2A} \sum_i b_i v_i$$

$$\frac{\partial u}{\partial y} = \sum \frac{\partial h_i}{\partial y} u_i = \frac{1}{2A} \sum_i c_i u_i, \qquad \frac{\partial v}{\partial y} = \sum \frac{\partial h_i}{\partial y} v_i = \frac{1}{2A} \sum_i c_i v_i$$

where the coefficients b_i and c_i are understood to be evaluated with respect to the original configuration.

The strains are

$$\epsilon_{xx} = \frac{\partial u}{\partial x}, \qquad \epsilon_{yy} = \frac{\partial v}{\partial y}, \qquad 2\epsilon_{xy} = \gamma_{xy} \frac{\partial u}{\partial y} + \frac{\partial v}{\partial x}$$

which are expressed in matrix form as

$$\left\{ \begin{array}{c} \epsilon_{xx} \\ \epsilon_{yy} \\ 2\epsilon_{xy} \end{array} \right\} = \frac{1}{2A} \left[\begin{array}{cccccc} b_1 & 0 & b_2 & 0 & b_3 & 0 \\ 0 & c_1 & 0 & c_2 & 0 & c_3 \\ c_1 & b_1 & c_2 & b_2 & c_3 & b_3 \end{array} \right] \left\{ \begin{array}{c} u_1 \\ v_1 \\ u_2 \\ v_2 \\ u_3 \\ v_3 \end{array} \right\} \qquad \text{or} \quad \{\epsilon\} = [B_L]\{u\}$$

The stresses are related to the strains by the plane stress Hooke's law

$$\left\{ \begin{array}{c} \sigma_{xx} \\ \sigma_{yy} \\ \sigma_{xy} \end{array} \right\} = \frac{E}{1-\nu^2} \left[\begin{array}{ccc} 1 & \nu & 0 \\ \nu & 1 & 0 \\ 0 & 0 & (1-\nu)/2 \end{array} \right] \left\{ \begin{array}{c} \epsilon_{xx} \\ \epsilon_{yy} \\ 2\epsilon_{xy} \end{array} \right\} \qquad \text{or} \quad \{\sigma\} = [D]\{\epsilon\}$$

The virtual work for a plate in plane stress is

$$\delta W = \int_V [\sigma_{xx}\delta\epsilon_{xx} + \sigma_{yy}\delta\epsilon_{yy} + \sigma_{xy}\delta\gamma_{xy}] \, dV \qquad \text{or} \qquad \delta W = \int_V \{\sigma\}^T \delta\{\epsilon\} \, dV$$

Substituting for the stresses and strains in terms of the degrees of freedom gives

$$\delta W = \int_V \{u\}^T [B_L]^T [D][B_L]\delta\{u\} \, dV$$

Noting that none of the quantities inside the integral depends on the position coordinates, we then have

$$\delta W = \{u\}^T [B_L]^T [D][B_L]\delta\{u\} V$$

The virtual work of the nodal loads is

$$\delta W = \{F\}^T \delta\{u\}$$

From the equivalence of the two, we conclude that the nodal forces are related to the degrees of freedom through

$$\{F\} = [\ k\ ]\{u\}, \qquad [\ k\ ] \equiv [B_L]^T[\ D\ ][B_L]\,V \qquad (2.38)$$

The $[6 \times 6]$ square matrix $[\ k\ ]$ is called the stiffness matrix for the *Constant Strain Triangle* (CST) element.

For plane problems, the coordinate system used to formulate the element (that is, the local coordinate system) is also the global coordinate system, hence we do not need to do any rotation of the element stiffness before assemblage. The structural stiffness matrix is simply

$$[\ K\ ] = \sum_m [\ k\ ]_m$$

where the element stiffnesses are suitably augmented to conform with the global system. The coding associated with this procedure in given in Reference [22].

Example 2.18: Using the CST element, determine the nodal forces for the two-element assemblage shown in Figure 2.18.

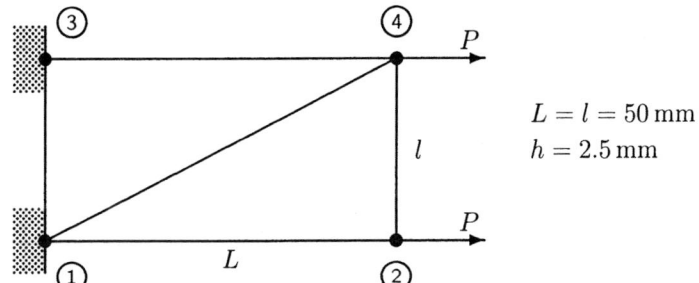

Figure 2.18: A two-element system with fixed end.

The two stiffness matrices are given by

$$[k_{124}] = Eh \begin{bmatrix} .267 & .000 & -.267 & .133 & .000 & -.133 \\ .000 & .100 & .200 & -.100 & -.200 & .000 \\ -.267 & .200 & .667 & -.333 & -.400 & .133 \\ .133 & -.100 & -.333 & 1.167 & .200 & -1.067 \\ .000 & -.200 & -.400 & .200 & .400 & .000 \\ -.133 & .000 & .133 & -1.067 & .000 & 1.067 \end{bmatrix}$$

$$[k_{143}] = Eh \begin{bmatrix} .400 & .000 & .000 & -.200 & -.400 & .200 \\ .000 & 1.067 & -.133 & .000 & .133 & -1.067 \\ .000 & -.133 & .267 & .000 & -.267 & .133 \\ -.200 & .000 & .000 & .100 & .200 & -.100 \\ -.400 & .133 & -.267 & .200 & .667 & -.333 \\ .200 & -1.067 & .133 & -.100 & -.333 & 1.167 \end{bmatrix}$$

Both matrices are clearly symmetric. The assembled stiffness matrix is for the degrees of freedom with the reductions

$$\{u\} = \{u_1, v_1;\ u_2, v_2;\ u_3, v_3;\ u_4, v_4\}$$
$$= \{0, 0;\ u_2, v_2;\ 0, 0;\ u_4, v_4\}$$

We first augment each element matrix to size $[8 \times 8]$ as

$$[k_{124}] \implies \{u_1, v_1;\ u_2, v_2;\ 0, 0;\ u_4, v_4\}$$
$$[k_{143}] \implies \{u_1, u_1;\ 0, 0;\ u_3, u_3;\ u_4, v_4\}$$

and simply add. This give

$$[K] = Eh \begin{bmatrix} .667 & .000 & -.267 & .133 & -.400 & .200 & .000 & -.333 \\ .000 & 1.167 & .200 & -.100 & .133 & -1.067 & -.333 & .000 \\ -.267 & .200 & .667 & -.333 & .000 & .000 & -.400 & .133 \\ .133 & -.100 & -.333 & 1.167 & .000 & .000 & .200 & -1.067 \\ -.400 & .133 & .000 & .000 & .667 & -.333 & -.267 & .200 \\ .200 & -1.067 & .000 & .000 & -.333 & 1.167 & .133 & -.100 \\ .000 & -.333 & -.400 & .200 & -.267 & .133 & .667 & .000 \\ -.333 & .000 & .133 & -1.067 & .200 & -.100 & .000 & 1.167 \end{bmatrix}$$

The reduced structural stiffness matrix is obtained by "scratching" the rows and columns corresponding to the zero degrees of freedom. The consequent system of equations is

$$[K^*]\{u\} = Eh \begin{bmatrix} .667 & -.333 & -.400 & .133 \\ -.333 & 1.167 & .200 & -1.067 \\ -.400 & .200 & .667 & .000 \\ .133 & -1.067 & .000 & 1.167 \end{bmatrix} \begin{Bmatrix} u_2 \\ v_2 \\ u_4 \\ v_4 \end{Bmatrix} = \begin{Bmatrix} P \\ 0 \\ P \\ 0 \end{Bmatrix}$$

Solving this system gives

$$\{u\}^T = \{.00200, .00025, .00187, .00000\}\frac{P}{Eh}$$

Note that although the problem is symmetric, the displacements are not because the element mesh arrangement is nonsymmetric.

The nodal forces are given by

$$\{F\} = [\ k\ ]\{u\}$$

Augmenting the displacement vectors with the zero degrees of freedom and then multiplying out gives

$$\{F\}_{124}^T = \{-1.00, .00, 1.00, .00, .00, .00\}P$$
$$\{F\}_{143}^T = \{.00, -.50, 1.00, .00, -1.00, .50\}P$$

These are shown in Figure 2.19. Note that each element is in equilibrium. The element nodes are connected to the large nodes to which the applied loads are attached. It is clear that the element nodal forces add up to the applied load. At the fixed end the element nodal forces add up to give the boundary reactions.

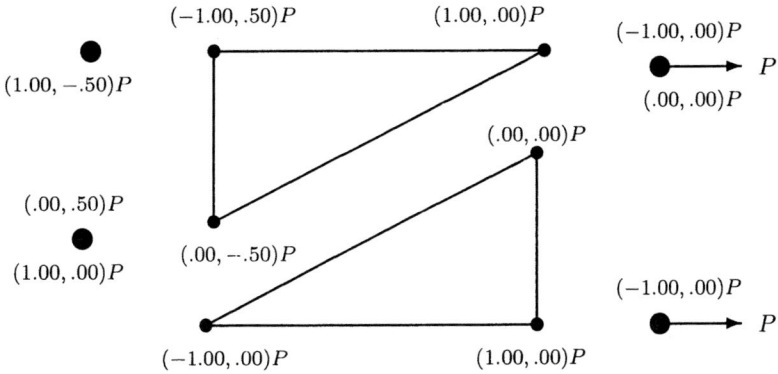

Figure 2.19: Nodal forces.

Membrane Element with Rotational DoF

Because we are interested in analyzing thin-walled 3-D structures comprising a mixture of frame and plate substructures, it simplifies the implementation when both structural types are modeled in a compatible way. The in-plane behavior of the plate is analogous to that of a plane 2-D frame. Thus, at each node we want the DoF to be

$$\{u\} = \{u,\ v,\ \phi_z\}$$

The constant strain triangle (CST) element has only the two displacements in its formulation. We now discuss a three-noded triangular element, which is shown to have superior in-plane performance over the CST. More importantly from our perspective, however, is the fact that it correctly implements the drilling DoF (ϕ_z) and therefore makes it suitable for 3-D applications. The "rotation" implemented is actually that taken from continuum mechanics

$$\phi_z = \tfrac{1}{2}\left(\frac{\partial v}{\partial x} - \frac{\partial u}{\partial y}\right)$$

This element was developed in References [13, 14], a similar element was developed in Reference [2], and a nice comparison of the performance of these is given in Reference [55].

I: Lumping Matrix

Consider a small triangular element removed from the deformed plate. There are tractions along each of the edges. Furthermore, let the deformation of an edge (as indicated in Figure 2.20) be represented in local coordinates as

$$u(s) = \left[1 - \left(\frac{s}{L}\right)\right]u_1 + \left[2\left(\frac{s}{L}\right)\right]Lu_2$$
$$\equiv f_1(s)u_1 + f_2(s)u_2$$

$$
\begin{aligned}
v(s) &= \left[1 - 3(\tfrac{s}{L})^2 + 2(\tfrac{s}{L})^3\right]v_1 + (\tfrac{s}{L})\left[1 - 2(\tfrac{s}{L}) + (\tfrac{s}{L})^2\right]L\phi_1 \\
&\quad + (\tfrac{s}{L})^2\left[3 - 2(\tfrac{s}{L})\right]v_2 + (\tfrac{s}{L})^2\left[-1 + (\tfrac{s}{L})\right]L\phi_2 \\
&\equiv g_1(s)v_1 + g_2(s)\phi_1 + g_3(s)v_2 + g_4(s)\phi_2 \qquad\qquad (2.39)
\end{aligned}
$$

The functions $f_n(s)$ and $g_n(s)$ are, respectively, the rod and beam shape functions from Section 1.8. In comparison to the CST element, we are allowing the normal displacements of the edge to be higher order.

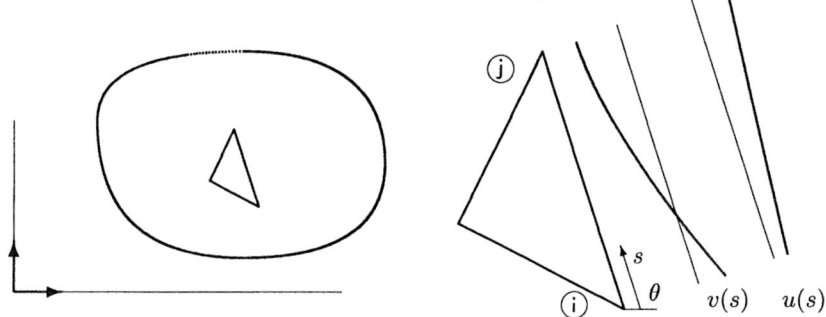

Figure 2.20: Assumed displacements on an edge.

We would like to obtain the strain energy due to this system. It is apparent that in the limit of small element size, we would want the stress state to approach that of a constant strain triangle. Let us assume that the stress is uniform, then the tractions on an edge at an orientation of θ is related to the stresses as

$$
\left\{\begin{array}{c} t_n \\ t_t \end{array}\right\} = \left[\begin{array}{ccc} \cos^2\theta & \sin^2\theta & 2\cos\theta\sin\theta \\ -\sin\theta\cos\theta & \sin\theta\cos\theta & \cos^2\theta - \sin^2\theta \end{array}\right] \left\{\begin{array}{c} \sigma_{xx} \\ \sigma_{yy} \\ \sigma_{xy} \end{array}\right\}
$$

which shows that the tractions are constant. The virtual work of the tractions is given by

$$
\begin{aligned}
\delta W &= \int [\bar{t}_n \delta\bar{v}(s) + \bar{t}_t \delta\bar{u}(s)]ds \\
&= \bar{t}_n \int [g_1(s)\delta\bar{v}_1 + g_2(s)L\delta\bar{\phi}_1 + g_3(s)\delta\bar{v}_2 + g_4(s)L\delta\bar{\phi}_2]ds \\
&\quad + \bar{t}_t \int [f_1(s)\delta\bar{u}_1 + f_2(s)\delta\bar{u}_1]ds
\end{aligned}
$$

where the integration is over the three edges. These evaluate for a typical side to

$$
\delta W_{12} = \tfrac{1}{2}t_n L_{12}[\delta v_1 + \tfrac{1}{6}\alpha L\delta\phi_1 + \delta v_2 + \tfrac{1}{6}\alpha L\delta\phi_2] + \tfrac{1}{2}t_t[\delta u_1 + \delta u_2]
$$

where α was introduced as an adjustable parameter on the rotation. We get similar expressions for the other sides. The total virtual work is the sum for the three edges, and this must be equal to the virtual work of the nodal loads acting on the degrees of freedom.

The virtual work of the nodal loads is

$$\delta W = \sum_i [F_{xi}\delta u_i + F_{yi}\delta v_i + M_{zi}\delta \phi_i]$$

Introducing a lumped matrix defined as

$$[\, L\,] = \Big[[\, L_1\,], [\, L_2\,], [\, L_3\,]\Big]^T$$

where

$$[\, L_j\,] = \frac{1}{2}\begin{bmatrix} y_{ki} & 0 & x_{ik} \\ 0 & x_{ik} & y_{ki} \\ \frac{1}{6}\alpha(y_{ji}^2 - y_{kj}^2) & \frac{1}{6}\alpha(x_{ij}^2 - x_{jk}^2) & \frac{1}{3}\alpha(x_{ij}y_{ji} - x_{jk}y_{kj}) \end{bmatrix}$$

using cyclic permutation on i, j, k. The first three contributions can be written as

$$\{F\} = [\, k_b\,]\{u\}, \qquad [\, k_b\,] = [\, L\,]^T [\, D\,][\, L\,]$$

The upper-left $[6 \times 6]$ portion of this array is identical to that of the constant strain triangle. The rest is the bending-like contribution. This stiffness-like matrix, however, cannot be used on its own because it is rank deficient.

II: Higher-Order Displacement Modes

The displacements will be conceived as made up of three components: rigid body modes, constant strain modes, and higher-order modes corresponding to bending. The first two would lead to the constant strain triangle of the last section. The total displacements are

$$\begin{Bmatrix} u \\ v \end{Bmatrix} = \begin{bmatrix} 1 & 0 & -\eta \\ 0 & 1 & \xi \end{bmatrix} \begin{Bmatrix} q_1 \\ q_2 \\ q_3 \end{Bmatrix} + \begin{bmatrix} \xi & 0 & \eta \\ 0 & \eta & \xi \end{bmatrix} \begin{Bmatrix} q_4 \\ q_5 \\ q_6 \end{Bmatrix}$$
$$+ \sum_{i=1}^{3} \begin{bmatrix} a_{1i}\xi^2 & a_{2i}\xi\eta & a_{3i}\eta^2 \\ b_{1i}\xi^2 & b_{2i}\xi\eta & b_{3i}\eta^2 \end{bmatrix} q_{r+i}$$

where $\{q\}$ are generalized degrees of freedom and q_7, q_8, q_9 are associated with the higher modes. The coordinates are defined as

$$\xi = \lambda(x - x_c), \qquad \eta = \lambda(y - y_c), \qquad \lambda = \sqrt{A}$$

and the higher modes have the coefficients

$$a_{1i} = -\tfrac{1}{2}S_i C_i^2, \qquad\qquad a_{2i} = \tfrac{1}{2}C_i^3, \qquad\qquad a_{3i} = \tfrac{1}{2}S_i^3 + S_i C_i^2$$
$$b_{1i} = -S_i^2 C_i - \tfrac{1}{2}C_i^3, \qquad b_{2i} = -\tfrac{1}{2}S_i^3, \qquad b_{3i} = \tfrac{1}{2}C_i^3 - C_i S_i^2$$

where the angles are given by

$$S_i = (y_m - y_i)/L_{mi}, \qquad x_m = \tfrac{1}{2}(x_j + x_k)$$
$$C_i = (x_m - x_i)/L_{mi}, \qquad y_m = \tfrac{1}{2}(y_j + y_k), \qquad L_{mi} = \sqrt{(x_m - x_i)^2 - (y_m - y_i)^2}$$

These higher modes can be thought of as bending about axes emanating from each apex. Note that these displacements are not consistent with the assumed displacements along the edge of the element.

The nodal degrees of freedom are related to the generalized DoF as

$$
\begin{Bmatrix} u_j \\ v_j \\ \phi_j \end{Bmatrix} =
\begin{bmatrix} 1 & 0 & -\eta_j \\ 0 & 1 & \xi_j \\ 0 & 0 & \lambda \end{bmatrix}
\begin{Bmatrix} q_1 \\ q_2 \\ q_3 \end{Bmatrix} +
\begin{bmatrix} \xi_j & 0 & \eta_j \\ 0 & \eta_j & \xi_j \\ 0 & 0 & 0 \end{bmatrix}
\begin{Bmatrix} q_4 \\ q_5 \\ q_6 \end{Bmatrix}
$$
$$
+ \sum_{i=1}^{3}
\begin{bmatrix} a_{1i}\xi_j^2 + a_{2i}\xi_j\eta_j + a_{3i}\eta_j^2 \\ b_{1i}\xi_j^2 + b_{2i}\xi_j\eta_j + b_{3i}\eta_j^2 \\ -\lambda C_i \xi_j - \lambda S_i \eta_j \end{bmatrix} q_{r+i}
$$

or

$$\{u\} = [\,G\,]\{q\}$$

This can be inverted (numerically) to give

$$\{q\} = [G^{-1}]\{u\} = [\,H\,]\{u\}$$

The following developments will be in terms of the generalized DoF but the above can be used at any stage to express them in terms of the global DoF.

The strains are

$$
\begin{Bmatrix} \epsilon_{xx} \\ \epsilon_{yy} \\ \gamma_{xy} \end{Bmatrix} = \lambda
\begin{bmatrix} 1 & 0 & 0 \\ 0 & 1 & 0 \\ 0 & 0 & 2 \end{bmatrix}
\begin{Bmatrix} q_4 \\ q_5 \\ q_6 \end{Bmatrix} + \lambda \sum_{i=1}^{3}
\begin{bmatrix} 2a_{1i}\xi_j + a_{2i}\eta_j \\ b_{2i}\xi_j + 2b_{3i}\eta_j \\ -4b_{3i}\xi - 4a_{1i}\eta \end{bmatrix} q_{6+i}
$$

or

$$\{\epsilon\} = [B_{rc}]\{q_{rc}\} + [B_h]\{q_h\} = [\,B\,]\{q\}$$

The virtual work of the internal stresses is

$$\delta W = \int_V \sigma_{ij}\delta\epsilon_{ij}\, dV = \tfrac{1}{2}\int_V \{\delta q\}^T [\,B\,]^T [\,D\,][\,B\,]\{q\}\, dV$$

Substituting and multiplying out gives

$$
\delta W = \int_V \Big[\{\delta q_{rc}\}^T [B_{rc}]^T [\,D\,][B_{rc}]\{q_{rc}\} + \{\delta q_h\}^T [B_h]^T [\,D\,][B_{rc}]\{q_{rc}\}
$$
$$
+ \{\delta q_{rc}\}^T [B_{rc}]^T [\,D\,][B_h]\{q_h\} + \{\delta q_h\}^T [B_h]^T [\,D\,][B_h]\{q_h\} \Big] dV
$$

The two middle terms evaluate to zero leaving

$$\delta W = \{\delta q_{rc}\}^T [\,k_{rc}\,]\{q_{rc}\} + \{\delta q_h\}^T [\,k_h\,]\{q_h\}$$

The first term is precisely that of the constant strain triangle and the second term is the higher-order contribution. Let us introduce as our potential for the element a combination of the lumped representation and the higher-order term. That is

$$U = \tfrac{1}{2}\{u\}^{T}[\,L\,]^{T}[\,D\,][\,L\,]\{u\} + \tfrac{1}{2}\beta\{u\}^{T}[\,H\,]^{T}[\,k_h\,][\,H\,]\{u\}$$

where β is introduced as another adjustable parameter. Minimizing this with respect to the DoF gives the total stiffness as

$$[\,k\,] = [\,L\,]^{T}[\,D\,][\,L\,] + \beta[\,H\,]^{T}[\,k_h\,][\,H\,]$$

The higher-order contribution is given by

$$[k_h(i,j)] \;=\; J_{\xi\xi}[B_{\xi i}]^{T}[\,D\,][B_{\xi i}] + J_{\xi\eta}([B_{\xi i}]^{T}[\,D\,][B_{\eta i}][B_{\eta i}]^{T}[\,D\,][B_{\xi i}])$$
$$+\, J_{\eta\eta}[B_{\eta i}]^{T}[\,D\,][B_{\eta i}]$$

where

$$J_{\xi\xi} \;=\; \int_{A} \xi^{2}\, dA = -\tfrac{1}{6}A(\xi_1\xi_2 + \xi_2\xi_3 + \xi_3\xi_1)$$

$$J_{\xi\eta} \;=\; \int_{A} \xi\eta\, dA = -\tfrac{1}{12}A(\xi_1\eta_1 + \xi_2\eta_2 + \xi_3\eta_3)$$

$$J_{\eta\eta} \;=\; \int_{A} \eta^{2}\, dA = -\tfrac{1}{6}A(\eta_1\eta_2 + \eta_2\eta_3 + \eta_3\eta_1)$$

The stresses are obtained from the strains, which are obtained from differentiation of the displacements. Coding for this element can be found in References [13, 14]. We will refer to this membrane element with rotations as the MRT element.

As for the CST element, the coordinate system used to formulate the element is also the global coordinate system, hence the structural stiffness matrix is simply the summation of the element stiffnesses suitably augmented to conform with the global system.

Example 2.19: Show that the MRT element passes the patch test.

Consider a rectangular block loaded as shown in Figure 2.21. The expected displacements are based on simple uniaxial stress with $\Delta L = \epsilon L = \sigma L / EA$ and the transverse behavior related to the Poisson's ratio contraction to give

$$u_4 = 0.0004, \quad v_4 = -0.0002, \quad \phi_4 = 0.0; \quad \sigma_{xx} = 2P/A$$

This is an interesting problem as regards specifying the appropriate boundary conditions in an FEM context. The intuitive boundary conditions are to specify

$$u_1 = 0,\ v_1 = 0,\ \phi_1 = 0; \quad u_2 = 0,\ v_2 = \text{free},\ \phi_2 = \text{free}$$

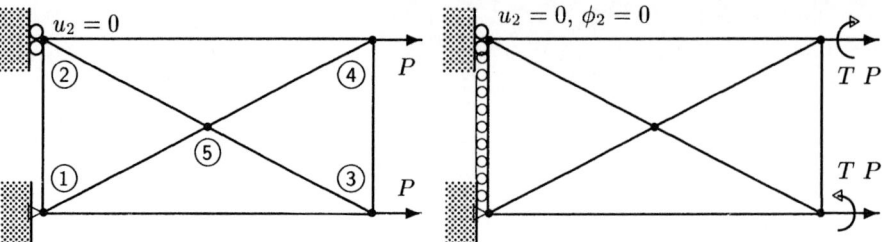

Figure 2.21: Two sets of boundary conditions for the patch test.

and the load condition as

$$P_3 = 1000, \ V_3 = 0, \ M_3 = 0; \qquad P_4 = 1000, \ V_4 = 0, \ M_4 = 0$$

The results for these loads are

$$u_4 = 0.00056, \quad v_4 = -0.00026, \quad \phi_4 = 0.003$$

The contours are shown in Figure 2.22(a). It is clear that the concentrated loads are causing a good deal of localized rotations. Furthermore, the element is too flexible.

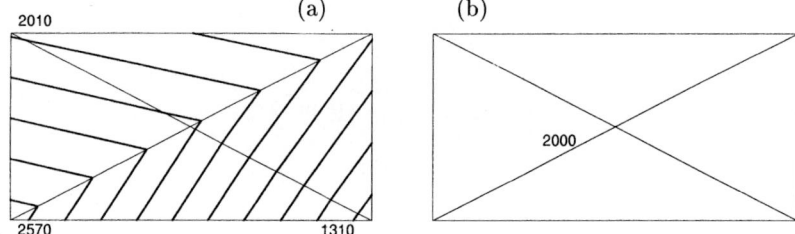

Figure 2.22: Contours of σ_{xx} stress for the patch test problem. (a) Incorrect boundary conditions. (b) Correct boundary conditions.

The appropriate boundary conditions to specify are

$$u_1 = 0, \ v_1 = 0, \ \phi_1 = 0; \qquad u_2 = 0, \ v_2 = \text{free}, \ \phi_2 = 0$$

Note that $\phi_2 = 0$ and so this would correspond to a line of symmetry. The load conditions are

$$P_3 = 1000, \ V_3 = 0, \ T_3 = \frac{\alpha}{12}\sigma L^2; \qquad P_4 = 1000, \ V_4 = 0, \ T_4 = -\frac{\alpha}{12}\sigma L^2$$

The edge of this element is like a beam and the uniform stress is like a distributed load on the beam. Hence, the edge also has moments. It is necessary to include the α in the effective moment because it was used in obtaining the relation between the nodal loads and the internal reactions. In general, the nodal loads are related to the tractions by

$$P_1 = \int_L g_1(x)\sigma(x)\,dxh = \frac{1}{2}\sigma_o Lh\,, \quad T_1 = \alpha\int_L g_4(x)\sigma(x)\,dxh = \frac{\alpha}{12}\sigma_o L^2 h$$

$$P_2 = \int_L g_2(x)\sigma(x)\,dxh = \frac{1}{2}\sigma_o Lh\,, \quad T_2 = -\alpha\int_L g_4(x)\sigma(x)\,dxh = -\frac{\alpha}{12}\sigma_o L^2 h$$

The lack of contours in Figure 2.22 clearly indicates the uniform state of deformation.

The patch test is necessary for the convergence properties of an element. In this problem, if the middle node is moved, the exact same results are obtained.

Example 2.20: The cantilever beam shown in Figure 2.23 has a parabolically distributed load on the end. Do a convergence test to demonstrate the performance of the MRT element.

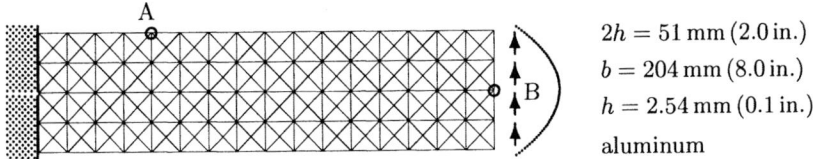

A

$2h = 51\,\text{mm}\ (2.0\,\text{in.})$

$b = 204\,\text{mm}\ (8.0\,\text{in.})$

$h = 2.54\,\text{mm}\ (0.1\,\text{in.})$

aluminum

Figure 2.23: Mesh geometry $[4 \times 16]$.

This is a problem that has a significant amount of bending (and hence rotation) and therefore is a good test case [18] for examining the performance of the rotational contribution of the MRT element. This is a problem solved earlier in this chapter and we will use that solution for comparison.

One of the most important characteristics of an element is its *convergence* properties, that is, it should converge to the exact result in the limit of small element size. When looking at convergence, it is necessary to change the mesh in a systematic way. The mesh shown is made of $[4 \times 16]$ modules where each module is made of four elements. The other meshes are similar except that the number of modules per side was changed.

The results for the tip deflection at B and stresses at A are shown in Figure 2.24 where there are compared to the exact solution. The normalizations are with respect to the exact solution for v_B and σ_A. The performance of the simpler constant strain triangle is also shown in the plot.

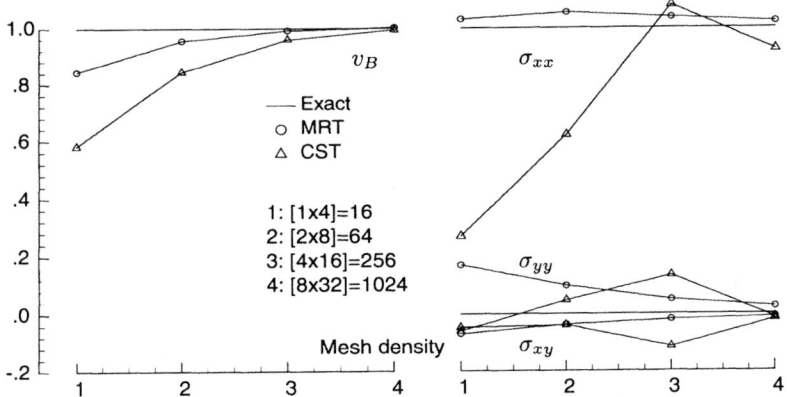

Figure 2.24: Convergence of displacements and stresses.

Both elements exhibit convergence to the exact solution. What is apparent, however, is that the MRT element gives good results even for relatively few modules through the depth. The CST element, on the other hand, is overly stiff.

The difference in performance of the stress is even more dramatic. These reported stresses are nodal averages. It is interesting to observe the other stresses on the boundary; these should be zero but it is only in the limit of very small element do they go to zero.

Example 2.21: Investigate the stress concentration around a hole in a plate.

Whenever there are cut-outs or changes in the geometry, stress concentrations occur that necessitate a finer mesh to accurately model the stress gradients. An ideal mesh is one that is uniform and very fine everywhere. This, of course, is not practical, so it is usual to increase the mesh density only in locations near the stress concentrations.

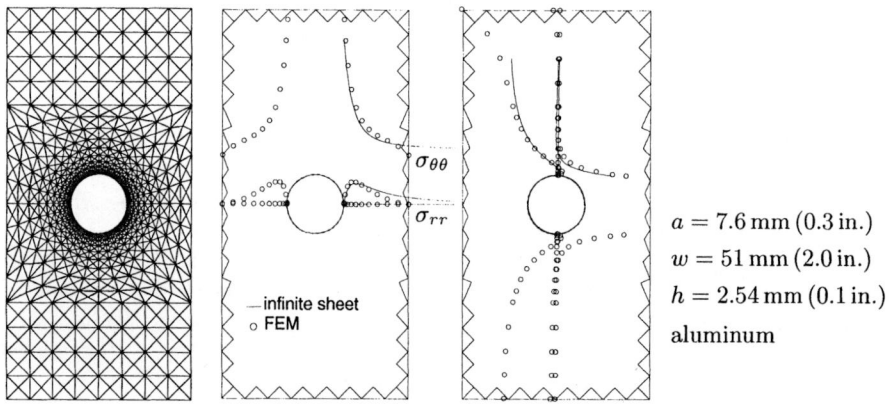

$a = 7.6\,\text{mm}\,(0.3\,\text{in.})$

$w = 51\,\text{mm}\,(2.0\,\text{in.})$

$h = 2.54\,\text{mm}\,(0.1\,\text{in.})$

aluminum

Figure 2.25: Stress distribution around a hole.

An example of a non-uniform mesh is shown in Figure 2.25. There is a gradient of element sizes that varies approximately as $\sqrt{r}$ where r is the distance from the center of the hole. A general purpose mesh generator would allow a choice of gradients.

The stresses computed using this mesh are also shown in Figure 2.25. The very high gradient of stress justifies the need for the very fine mesh. Note that the reported stresses are nodal averages.

Also shown in the figure is the Kirsch solution for an infinite plate. The finite width plate shows a higher stress concentration factor.

2.7   Triangular Plate Bending Elements

In this section we illustrate some of the pitfalls that can occur with an improperly formulated element. While the element we develop has very poor performance characteristics and should never be used, its derivation is instructive. We then derive a properly formulated element.

A Poor Performing Plate Bending Element

We want an element that has a node at each corner of the triangle and has the degrees of freedom

$$\{u\}^T = \{w_1, \phi_{x1}, \phi_{y1}; w_2, \phi_{x2}, \phi_{y2}; w_3, \phi_{x3}, \phi_{y3}\}$$

as shown in Figure 2.26. The rotations are related to the deflections by

$$\phi_x = \frac{\partial w}{\partial y}, \qquad \phi_y = -\frac{\partial w}{\partial x}$$

In local coordinates, the three-noded triangle has a total of nine degrees of freedom. The basic problem with triangular elements for flexure of plates is that if we represent the deflection in terms of polynomials, we have

$$w(x, y) = q_1 + q_2 x + q_3 y + q_4 x^2 + q_5 xy + q_6 y^2 + q_7 x^3 + q_8 x^2 y + q_9 xy^2 + q_{10} y^3$$

where q_i are the generalized degrees of freedom. This has 10 terms but we only have nine degrees of freedom. Simply deleting one of the terms will cause anisotropy or convergence problems. We illustrate this.

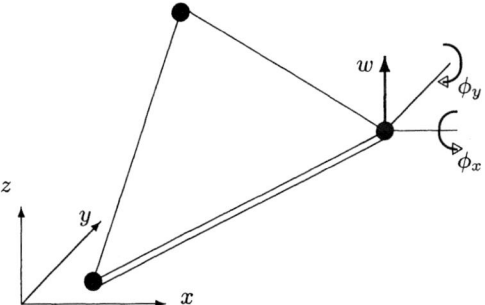

Figure 2.26: Element with nodal degrees of freedom.

One simple possibility is to take the polynomial in the form

$$w(x, y) = q_1 + q_2 x + q_3 y + q_4 x^2 + q_5 xy + q_6 y^2 + q_7 x^3 + q_8 (x^2 y + xy^2) + q_9 y^3$$

where the cross terms $x^2 y$ and xy^2 share a common coefficient. At Node i, we can establish the relationships

$$\begin{Bmatrix} w_j \\ \dfrac{\partial w_j}{\partial y} \\ -\dfrac{\partial w_j}{\partial x} \end{Bmatrix} = \begin{bmatrix} 1 & x_i & y_i & x_i^2 & x_i y_i & y_i^2 & x_i^3 & (x_i^2 y_i + x_i y^2) & y_i^3 \\ 0 & 0 & 1 & 0 & x_i & 2y_i & 0 & (x_i^2 + 2x_i y) & 3y_i^2 \\ 0 & -1 & 0 & -2x_i & -y_i & 0 & -3x_i^2 & (-2x_i y_i - y^2) & 0 \end{bmatrix} \{q\}$$

After evaluating this at the three nodes we can assemble the $[9 \times 9]$ system

$$\{u\} = [\, G\,]\{q\}$$

Inverting this numerically, we get

$$\{q\} = [\, G\,]^{-1}\{u\}$$

To get the $[\, B\,]$ matrix that relates the strains to the nodal DoF, we first relate the strains to the generalized DoF according to

$$\{\epsilon\} = \left\{ \begin{array}{c} -\dfrac{\partial^2 w}{\partial x^2} \\[6pt] -\dfrac{\partial^2 w}{\partial y^2} \\[6pt] -2\dfrac{\partial^2 w}{\partial x \partial y} \end{array} \right\} = [\, C\,]\{q\}$$

where

$$[\, C\,] = \begin{bmatrix} 0 & 0 & 0 & -2 & 0 & 0 & -6x & -2y & 0 \\ 0 & 0 & 0 & 0 & 0 & -2 & 0 & -2x & -6y \\ 0 & 0 & 0 & 0 & -2 & 0 & 0 & -4(x+y) & 0 \end{bmatrix}$$

Thus

$$\{\epsilon\} = [\, C\,][\, G\,]^{-1}\{u\} = [\, B\,]\{u\}$$

The stiffness is obtained in a manner similar to that of the CST element and is given by

$$[\, k\,] = \int_V [\, B\,]^T [\, D\,][\, B\,]\, dV$$

It remains now to perform the integrations.

Decompose the $[\, B\,]$ matrix as

$$[\, B\,] = [\, B_c\,] + [\, B_x\,]x + [\, B_y\,]y$$

The stiffness integral becomes

$$[\, k\,] = [B_c^T D B_c] \int dV + [B_c^T D B_x + B_x^T D B_c] \int x\, dV + [B_c^T D B_y + B_y^T D B_c] \int y\, dV$$

$$+ [B_x^T D B_x] \int x^2\, dV + [B_y^T D B_y] \int y^2\, dV + [B_x^T D B_y + B_y^T D B_x] \int xy\, dV$$

If we shift the coordinates to a centroidal system, then the integrals evaluate to

$$\int x^i y^j\, dV = A\frac{1}{i!\,j!}$$

where A is the area. There are a good many terms in the stiffness expression and therefore we will not write them out explicitly. This element is assembled analogously to that of the CST element.

We leave a discussion of the performance of this element until later.

Discrete Kirchhoff Triangular Element

The basic idea of the discrete Kirchhoff triangular element (DKT) is to treat the plate element in flexure as composed of a series of plane stress triangular laminas stacked on top of each other. From the previous section, we know that we can describe each lamina adequately using the CST or higher element, hence it remains then only to impose the constraints that the laminas form a structure in flexure.

I: Shape Functions

In deriving the DKT element, the complete polynomial is used in the form

$$w(x, y) = h_1^2[h_1 + 3h_2 + 3h_3)w_1 - h_1^2[x_{21}h_2 - x_{13}h_3]\phi_{y1}$$
$$-h_1^2[y_{31}h_3 - y_{12}h_2]\phi_{x1} + ... + \alpha h_1 h_2 h_3$$

where α is a generalized coefficient, and the additional six terms for $w(x, y)$ are obtained by cyclic permutation. We obtain α by imposing the Kirchhoff conditions discretely.

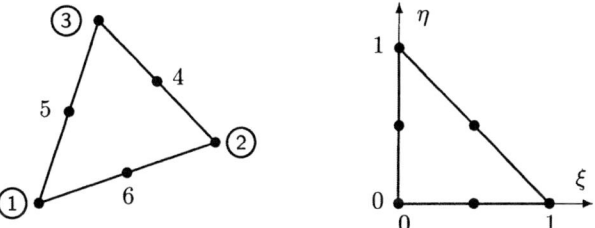

Figure 2.27: Element coordinates.

Let the rotations have the complete representations

$$\phi_x = \sum_i^6 N_i \phi_{xi}, \qquad \phi_y = \sum_i^6 N_i \phi_{yi}$$

where the shape functions correspond to a six-noded triangle (three apex nodes and three mid-side nodes) and are given by

$$N_1 = (2h_1 - 1)h_1 = (1 - \xi - \eta)(1 - 2\xi - 2\eta), \quad N_2 = (2h_2 - 1)h_2 = (2\xi - 1)\xi$$
$$N_3 = (2h_3 - 1)h_3 = (2\eta - 1)\eta), \quad N_4 = 4h_2h_3 = 4\xi\eta$$
$$N_5 = 4h_3h_1 = (1 - \xi - \eta)4\eta, \quad N_6 = 4h_1h_2 = (1 - \xi - \eta)4\xi \qquad (2.40)$$

(Note that we are using the convention of Reference [8] which puts node 4 opposite node 1 as shown in Figure 2.27.) Consider a laminate of thickness dz located a distance z from the midsurface of the plate. The displacements of this laminate are represented by

$$\bar{u}(x,y) = -z\phi_y \quad = \quad +z\Big[h_1(2h_1-1)\phi_{y1} + h_2(2h_2-1)\phi_{y2}$$

$$+h_3(2h_2-1)\phi_{y3} + 4h_2h_3\phi_{y4} + 4h_3h_1\phi_{y5} + 4h_1h_2\phi_{y6}\Big]$$

$$\bar{v}(x,y) = -z\phi_x \quad = \quad +z\Big[h_1(2h_1-1)\phi_{x1} + h_2(2h_2-1)\phi_{x2}$$

$$+h_3(2h_2-1)\phi_{x3} + 4h_2h_3\phi_{x4} + 4h_3h_1\phi_{x5} + 4h_1h_2\phi_{x6}\Big]$$

In this, there are a total of 12 degrees of freedom. Thus, initially there are more degrees of freedom than will appear in the final form of the element. The extra degrees of freedom are associated with the mid nodes and will be eliminated.

A nine-degree-of-freedom element is obtained by requiring the transverse shear strains to be zero at the corners

$$\frac{\partial w_i}{\partial x} = -\phi_{yi}, \qquad \frac{\partial w_i}{\partial y} = +\phi_{xi}, \qquad i=1,2,3$$

and along the sides of the element

$$\frac{\partial w_i}{\partial s} = \phi_{si}, \qquad i=4,5,6$$

where s is an edge tangent coordinate. We also impose that the slope normal to the element at the middle of the side is one half the sum of the values at the corner nodes

$$\phi_{n4} = \tfrac{1}{2}\big[\frac{\partial w_1}{\partial n} + \frac{\partial w_2}{\partial n}\big]$$

with others obtained by cyclic permutation. The following geometrical relations are needed on each side:

$$\left\{\begin{array}{c}\phi_x \\ \phi_y\end{array}\right\} = \begin{bmatrix} C & -S \\ S & C \end{bmatrix} \left\{\begin{array}{c}\phi_n \\ \phi_s\end{array}\right\}, \qquad \left\{\begin{array}{c}\dfrac{\partial w}{\partial s} \\ \dfrac{\partial w}{\partial n}\end{array}\right\} = \begin{bmatrix} C & S \\ S & -C \end{bmatrix} \left\{\begin{array}{c}\phi_x \\ \phi_y\end{array}\right\}$$

where $C \equiv \cos(\hat{x}, \hat{n}_{ij})$ and $S \equiv \sin(\hat{x}, \hat{n}_{ij})$ and interelement compatibility in terms of displacement and slope are still satisfied after these equations are applied. Note that since w varies cubically along the sides, $\partial w/\partial s$ varies quadratically and so does ϕ_s. Since $\partial w/\partial s$ matches ϕ_s at three points along the side, the Kirchhoff hypothesis (no shear strain) is satisfied along the entire boundary. Also, if this formulation is applied to a 1-D beam, the stiffness of the Bernoulli-Euler beam is recovered; this means that the DKT element is suitable for use with Bernoulli-Euler beam elements in complex structures.

It is this sequence of constraints that leads to the coupling of the rotations with the deflection. After the mid-side nodes are reduced, the resulting shape functions are given by

$$\phi_x = \{H_x(\xi,\eta)\}^T\{u\}, \qquad \phi_y = \{H_y(\xi,\eta)\}^T\{u\}$$

where the nine components are

$$
\{H_x\} = \left\{
\begin{array}{c}
-\frac{3}{2}(a_5 N_5 - a_6 N_6) \\
b_5 N_5 + b_6 N_6 \\
N_1 - c_5 N_5 - c_6 N_6 \\
-\frac{3}{2}(a_6 N_6 - a_4 N_4) \\
b_6 N_6 + b_4 N_4 \\
N_2 - c_6 N_6 - c_4 N_4 \\
-\frac{3}{2}(a_4 N_4 - a_5 N_5) \\
b_4 N_4 + b_5 N_5 \\
N_3 - c_4 N_4 - c_5 N_5
\end{array}
\right\}, \quad
\{H_y\} = \left\{
\begin{array}{c}
-\frac{3}{2}(d_5 N_5 - d_6 N_6) \\
-N_1 + e_5 N_5 + e_6 N_6 \\
-b_5 N_5 - b_6 N_6 \\
-\frac{3}{2}(d_6 N_6 - d_4 N_4) \\
-N_2 + e_6 N_6 + e_4 N_4 \\
-b_6 N_6 - b_4 N_4 \\
-\frac{3}{2}(d_4 N_4 - d_5 N_5) \\
-N_3 + e_4 N_4 + e_5 N_5 \\
-b_4 N_4 - b_5 N_5
\end{array}
\right\}
$$

and

$$
\begin{aligned}
a_k &= -x_{ij}/L_{ij}^2, & b_k &= \tfrac{3}{4}x_{ij}y_{ij}/L_{ij}^2, & c_k &= (\tfrac{1}{4}x_{ij}^2 - \tfrac{1}{2}y_{ij}^2)/L_{ij}^2 \\
d_k &= -y_{ij}/L_{ij}^2, & e_k &= (\tfrac{1}{4}y_{ij}^2 - \tfrac{1}{2}x_{ij}^2)/L_{ij}^2
\end{aligned}
$$

The strains are obtained by differentiating the functions $\{H_x\}$ and $\{H_y\}$.

II: Stiffness Relation

We are now in a position to obtain the stiffness relation. The curvatures are given by, for example,

$$
\kappa_{xx} = \frac{\partial \phi_x}{\partial x} \qquad \text{or} \qquad \{\kappa\} = [\,B\,]\{u\}
$$

where

$$
[B(\xi, \eta)] = \frac{1}{2A}
\begin{bmatrix}
+y_{31}\{H_{x,\xi}\}^T + y_{12}\{H_{x,\eta}\}^T \\
-x_{31}\{H_{y,\xi}\}^T - x_{12}\{H_{y,\eta}\}^T \\
-x_{31}\{H_{x,\xi}\}^T - x_{12}\{H_{x,\eta}\}^T + y_{31}\{H_{y,\xi}\}^T + y_{12}\{H_{y,\eta}\}^T
\end{bmatrix}
$$

The vectors appearing in this strain-displacement matrix are obtained from the derivatives as

$$
\{H_{x,\xi}\} = \left\{
\begin{array}{c}
p_6(1 - 2\xi) + (p_5 - p_6)\eta \\
q_6(1 - 2\xi) - (q_5 + q_6)\eta \\
w_1 + r_6(1 - 2\xi) - (r_5 + r_6)\eta \\
-p_6(1 - 2\xi) + (p_4 + p_6)\eta \\
q_6(1 - 2\xi) - (q_6 - q_4)\eta \\
w_2 + r_6(1 - 2\xi) + (r_4 - r_6)\eta \\
-(p_5 + p_4)\eta \\
(q_4 - q_5)\eta \\
-(r_5 - r_4)\eta
\end{array}
\right\}, \quad
\{H_{y,\xi}\} = \left\{
\begin{array}{c}
t_6(1 - 2\xi) + (t_5 - t_6)\eta \\
1 + r_6(1 - 2\xi) - (r_5 + r_6)\eta \\
-q_6(1 - 2\xi) + \eta(q_5 + q_6) \\
-t_6(1 - 2\xi) + (t_4 + t_6)\eta \\
-1 + r_6(1 - 2\xi) + (r_4 - r_6)\eta \\
-q_6(1 - 2\xi) - \eta(q_4 - q_6) \\
-(t_4 + t_5)\eta \\
(r_4 - r_5)\eta \\
-(q_4 - q_5)\eta
\end{array}
\right\}
$$

$$\{H_{x,\eta}\} = \left\{\begin{array}{c} -p_5(1-2\eta) - (p_6 - p_5)\xi \\ q_5(1-2\eta) - (q_5 + q_6)\xi \\ w_1 + r_5(1-2\eta) - (r_5 + r_6)\xi \\ (p_4 + p_6)\xi \\ (q_4 - q_6)\xi \\ -(r_6 - r_4)\xi \\ p_5(1-2\eta) - (p_4 + p_5)\xi \\ q_5(1-2\eta) + (q_4 - q_5)\xi \\ w_2 + r_5(1-2\eta) + (r_4 - r_5)\xi \end{array}\right\}, \quad \{H_{y,\eta}\} = \left\{\begin{array}{c} -t_5(1-2\eta) - (t_6 - t_5)\xi \\ 1 + r_5(1-2\eta) - (r_5 + r_6)\xi \\ -q_5(1-2\eta) + (q_5 + q_6)\xi \\ (t_4 + t_6)\xi \\ (r_4 - r_6)\xi \\ -(q_4 - q_6)\xi \\ t_5(1-2\eta) - (t_4 + t_5)\xi \\ -1 + r_5(1-2\eta) + (r_4 - r_5)\xi \\ -q_5(1-2\eta) - \xi(q_4 - q_5) \end{array}\right\}$$

In these, we have that

$$p_k = -6x_{ij}/L_{ij}^2 = 6a_k, \qquad r_k = 3y_{ij}^2/L_{ij}^2, \qquad w_1 = -4 + 6(\xi + \eta)$$
$$q_k = 3x_{ij}y_{ij}/L_{ij}^2 = 4b_k, \qquad t_k = -6y_{ij}/L_{ij}^2, \qquad w_2 = -2 + 6\xi$$

where $k = 4, 5, 6$ for $ij = 23, 31, 12$, respectively, and $L_{ij} = (x_{ij}^2 + y_{ij}^2)^{1/2}$.
The stiffness matrix is then given by

$$[\,k\,] = 2A \int_0^1 \int_0^{1-\eta} [\,B\,]^T [\,D\,] [\,B\,] \, d\xi \, d\eta$$

This can be integrated exactly using three numerical integration points. However, an explicit formulation is given in Reference [9] and coding is given in References [18, 36].

This element, now called the Discrete Kirchhoff Triangular (DKT) element was first introduced by Stricklin, Haisler, Tisdale, and Gunderson in 1968 [71]. It has been widely researched and documented as being one of the more efficient flexural elements. Batoz, Bathe, and Ho [8] performed extensive testing on three different elements including the hybrid stress model (HSM), DKT element, and a selective reduced integration (SRI) element. Comparisons between the different element types were made based on the results from different mesh orientations and different boundary and loading conditions. The authors concluded that the DKT and HSM elements are the most effective elements available for bending analysis of thin plates. Of these two elements, the DKT element was deemed superior to the HSM element based on the comparison between the experimental and theoretical results [8].

Once the nodal displacements have been determined, the bending moments at any point in the element are then obtained from

$$\{M\}(x,y) = [\,D\,][\,B\,](x,y)\{u\}$$

where

$$x = x_1 + \xi x_{21} + \eta x_{31}, \qquad y = y_1 + \xi y_{21} + \eta y_{31}$$

Note that the moment is not unique along the boundary shared by two elements. Consequently, nodal averaging will lead to improved results. Finally, assemblage of the structural stiffness matrix is simply the summation of the element stiffnesses suitably augmented to conform with the global system.

Example 2.22: Investigate the sensitivity of the simple and DKT elements to size aspect ratio.

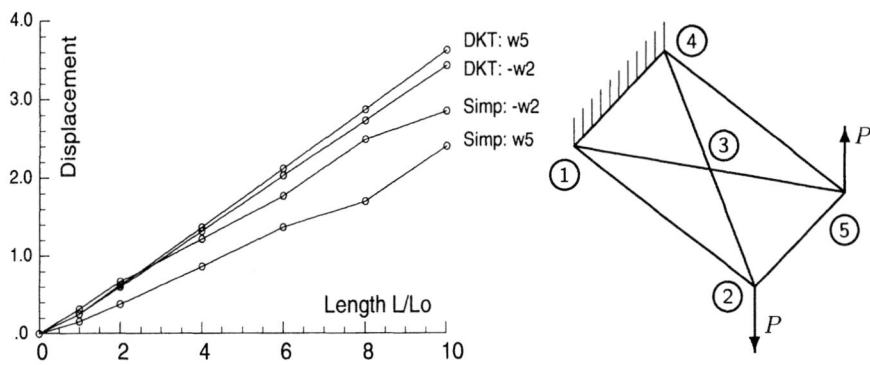

Figure 2.28: Twisting of an element.

The problem we will look at is that of a cantilevered plate with equal but opposite forces at its tips. This causes the plate to twist out of the plane.

Beginning with a length to width ratio of 1 : 1, we increase the length but keep the same number of elements. The two tips should have equal but opposite deflections.

As seen from Figure 2.28 the simple element has very poor aspect ratio performance. The DKT element on the other hand shows very good performance.

Example 2.23: Show that the DKT element passes the patch test and assess its aspect ratio performance.

The patch test is a simple numerical scheme for testing the convergence properties of an element formulation. A number of elements are assembled into a small patch with at least one node within the patch and is shared by two or more elements. The mesh of Figure 2.28 is a suitable arrangement. The boundary nodes are loaded consistently to a state of constant stress. If the computed stresses, throughout the element, agree exactly with the exact solution irrespective of the placement of the middle node, then the patch test is passed.

When a patch test is passed, there is some assurance that when the element is used to model complex structures that mesh refinement will produce a sequence of approximate solutions that converges to the exact solution.

For a plate element in bending, the test subjects the patch to constant bending moments. For the patch of Figure 2.28, a suitable set of boundary conditions and loads are:

$$w = 0, \qquad \phi_x = \text{free}, \qquad \phi_x = \text{free}$$

$$
\begin{aligned}
T_x &= T_o \quad \text{at Nodes 1, 2}, \qquad & T_x &= -T_o \quad \text{at Nodes 4, 5} \\
T_y &= T_o \quad \text{at Nodes 1, 4}, \qquad & T_y &= -T_o \quad \text{at Nodes 2, 5}
\end{aligned}
$$

The computed moments everywhere are

$$M_{xx} = \tfrac{1}{4}T_o, \qquad M_{yy} = \tfrac{1}{4}T_o, \qquad M_{xy} = 0$$

Figure 2.28 shows the performance of the DKT element on the twist test. Clearly it does not degrade very much as the aspect ratio is changed.

Example 2.24:   Assess the convergence properties of the DKT element.

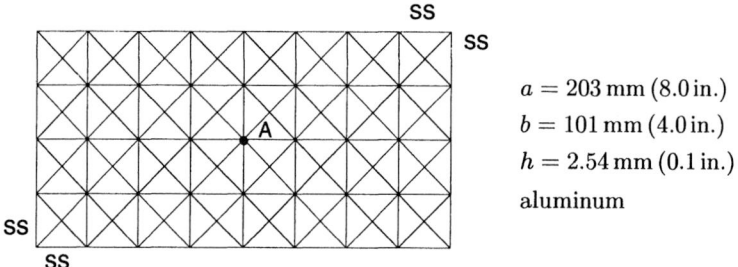

$a = 203\,\text{mm}\ (8.0\,\text{in.})$

$b = 101\,\text{mm}\ (4.0\,\text{in.})$

$h = 2.54\,\text{mm}\ (0.1\,\text{in.})$

aluminum

Figure 2.29: Generic $[4 \times 8]$ mesh used in convergence study.

The problem we will consider is that of a simply supported plate (with mesh as shown in Figure 2.29) and the transverse load applied uniformly. When looking at convergence, it is necessary to change the mesh in a systematic way. The mesh shown is made of $[4 \times 8]$ modules, where each module is made of four elements. The other meshes are variations of this by uniformly changing the number of modules.

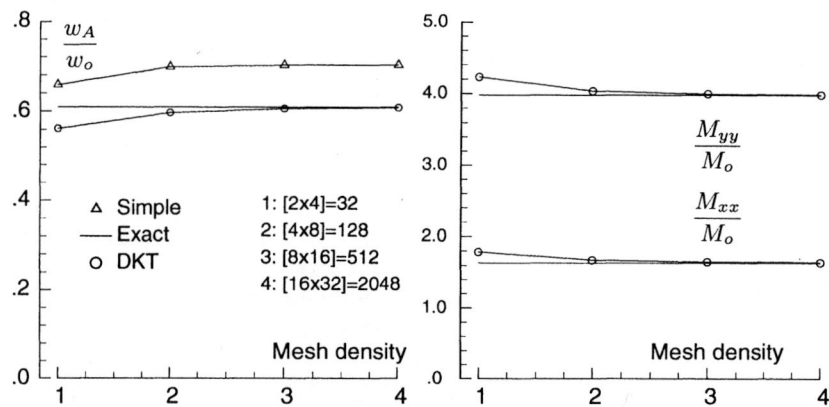

Figure 2.30: Convergence study for displacements and moments.

The results are shown in Figure 2.30 where it is compared to the exact solution and the performance of the simpler plate element. The normalizations for the displacements and moments are

$$w_o = \frac{16 q_o b^4}{D \pi^6}, \qquad M_o = \frac{q_o b^2}{4 \pi^2}$$

The DKT element exhibits excellent convergence, whereas the simple element converges to a value that is off by about 15%. In fact, the coarsest mesh gives the best results; such an element should never be used. It is pleasing to see that the DKT element gives good results even for relatively few modules.

The performance of the DKT element for the moments is also very good. These reported moments are nodal averages. Because a lumped load was used in this example, we conclude that a lumped representation is adequate when a suitable refined mesh is used.

Example 2.25: Analyze the plate with a hole shown Figure 2.31, under the action of edge moments.

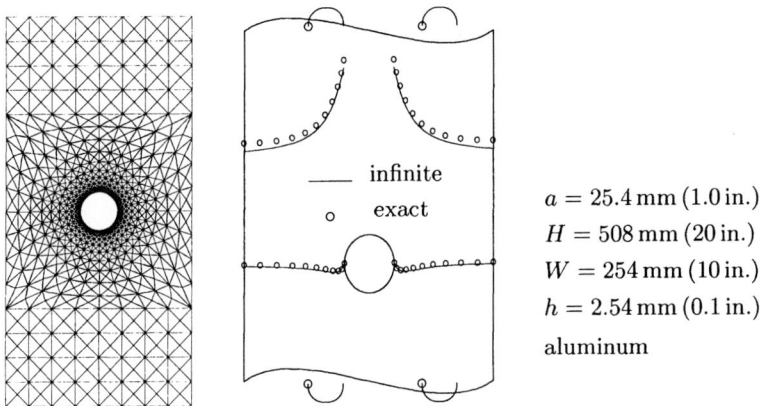

infinite
o   exact

$a = 25.4\,\text{mm}\,(1.0\,\text{in.})$
$H = 508\,\text{mm}\,(20\,\text{in.})$
$W = 254\,\text{mm}\,(10\,\text{in.})$
$h = 2.54\,\text{mm}\,(0.1\,\text{in.})$
aluminum

Figure 2.31: Moment distribution around a hole.

Just as for the membrane loading case, whenever there are cut-outs or changes in the geometry, stress concentrations occur. As pointed out before, for general loadings an ideal mesh is one that is uniform and very fine everywhere. This, of course, is not practical, so it is usual to increase the mesh density only in locations near the stress concentrations.

An example of a non-uniform mesh is shown in Figure 2.31 which is the same as in Figure 2.25 and therefore the same comments apply.

The stresses computed using this mesh are also shown in Figure 2.31. The very high gradient of stress justifies the need for the very fine mesh. Of particular significance is that the M_{rr} moment does not achieve a significant value. Note that the reported stresses are nodal averages.

Also shown in the figure is the infinite plate solution. As in the membrane case, the finite width plate shows a higher stress concentration factor.

Applied Distributed Loads

The applied loads fall into two categories. The first are point forces and moments; these do not need any special treatment. The others arise from distributed loads such as pressures, and these are the ones of interest here.

To get the equivalent nodal loads, we will equate the virtual work of the nodal loads to that of the distributed load. Consider a single element with transverse distributed load $q(x, y)$, the virtual work is

$$\{P\}^T\{\delta w\} = \int_A q\delta w\, dA = \int_A q\{N\}^T\{\delta w\}\, dA$$

where A is the area of the element. This leads to

$$\{P\} = \int \{N\}^T q\, dA \qquad (2.41)$$

In implementing this, shape functions different than what are used for calculating the stiffness matrix can be used. We will illustrate this with examples.

Using area coordinates for the triangle, we start by assuming a displacement shape in the form

$$
\begin{aligned}
w(x, y) = \quad & c_1 h_1 + c_2 h_2 + c_3 h_3 \\
+ \quad & c_4(h_1 h_2^2 + h) + c_5(h_2 h_3^2 + h) + c_6(h_3 h_1^2 + h) \\
+ \quad & c_7(h_1^2 h_2 + h) + c_8(h_2^2 h_3 + h) + c_9(h_3^2 h_1 + h) \qquad (2.42)
\end{aligned}
$$

where $2h \equiv h_1 h_2 h_3$. In this way the additional term is distributed among the other nine. If we now determine the nine coefficients in terms of the nine nodal degrees of freedom then we get the associated shape functions as

$$
\begin{aligned}
N_1 &= h_1 + (h_1^2 h_2 - h_1 h_2^2) + (h_1^2 h_3 - h_1 h_3^2) \\
N_2 &= -y_{12}(h_1^2 h_2 + h) + y_{31}(h_1^2 h_3 + h) \\
N_3 &= -x_{21}(h_1^2 h_2 + h) + x_{13}(h_1^2 h_3 + h) \qquad (2.43)
\end{aligned}
$$

The other six are obtained by permutation. Substituting these into Equation (2.41) then gives the nodal loads. These loads are a combination of forces and moments.

An alternative loading scheme is to simply lump the loads at the nodes. That is, let

$$P_i = \tfrac{1}{3}\int_A q\, dA, \qquad M_{xi} = 0, \qquad M_{yi} = 0$$

As shown in the example for the rectangular plate, this is usually adequate when reasonable mesh refinements are used.

The difference in the performance is only noticeable for coarse meshes, we will generally opt to use the simpler lumped approach. There is also another reason for preferring the lumped approach: as shown in the next chapter, during nonlinear deformations, the applied moments are nonconservative loads and therefore require special treatment.

Example 2.26: Analyze the circular plate shown Figure 2.32, under the action of a uniform pressure.

Figure 2.9 shows a comparison with the analytical solution obtained earlier. There is excellent agreement in the displacement distribution. The moment distribution is also very good except at the boundary. This is not surprising since we are replacing a circular boundary with a piece-wise linear boundary.

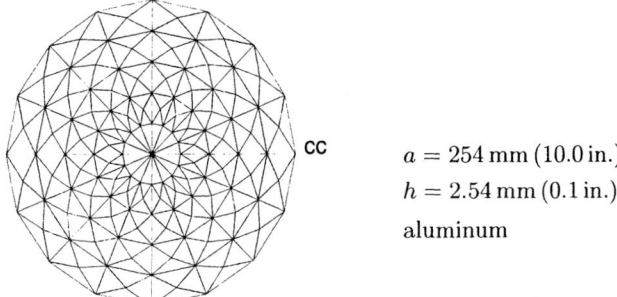

$a = 254 \, \text{mm} \, (10.0 \, \text{in.})$

$h = 2.54 \, \text{mm} \, (0.1 \, \text{in.})$

aluminum

Figure 2.32: Mesh for circular plate.

It is worth pointing out that the nature of the mesh configuration forced elements with a variety of aspect ratios. The good distribution results indicate that the element is reasonably robust as regards aspect ratio.

2.8   Shell and Frame Structures

The purpose of this section is to show how the stiffness of a general structure is constructed from the element stiffnesses in local coordinates. Once this assemblage procedure is established, then quite complicated structures can be formed simply by adding more of the basic elements.

Element Stiffness Matrix in Local Axes

In our formulation, the displacement of each node of a space frame or shell is described by three translational and three rotational components of displacement, giving six degrees of freedom at each unrestrained node. Corresponding to these degrees of freedom are six nodal loads. The notations we will use for the displacement and force vectors at each node are, respectively,

$$\{u\}_i = \begin{Bmatrix} u \\ v \\ w \\ \phi_x \\ \phi_y \\ \phi_z \end{Bmatrix}_i \quad , \quad \{F\}_i = \begin{Bmatrix} F_x \\ F_y \\ F_z \\ M_x \\ M_y \\ M_z \end{Bmatrix}_i$$

where i ranges $1, 2$ for frames and $1, 2, 3$ for shells. For each element in local coordinates, the forces are related to the displacements by

$$\{\bar{F}\} = \{\bar{u}\} = [\ \bar{k}\ ]\{\bar{u}\}$$

where the overhead "bar" designates local quantities and $[\ \bar{k}\ ]$ is of size $[12 \times 12]$ for frames and $[18 \times 18]$ for shells. The space frame is a combination of axial, two bending, and torsion effects; the upper left quadrant is

$$\frac{1}{L} \begin{bmatrix} EA & 0 & 0 & 0 & 0 & 0 \\ 0 & 12EI_z/L^2 & 0 & 0 & 0 & 6EI_z/L \\ 0 & 0 & 12EI_y/L^2 & 0 & -6EI_y/L & 0 \\ 0 & 0 & 0 & GI_x & 0 & 0 \\ 0 & 0 & -6EI_y/L & 0 & 4EI_y & 0 \\ 0 & 6EI_z/L & 0 & 0 & 0 & 4EI_z \end{bmatrix}$$

For regularity of notation, the torsional stiffness is written as $GI_x = GJ$. That the separate stiffnesses can be added like this follows from the assumed linearity of the formulation. Likewise, the shell element is a combination of the $[9 \times 9]$ MRT and $[9 \times 9]$ DKT elements.

Assemblage in Global Axes

The transformation of the components of a vector $\{v\}$ from the local to the global axes is given by

$$\{\bar{v}\} = [\, R \,]\{v\}$$

We will discuss the specific form of $[\, R \,]$ later. The same matrix will transform the vectors of nodal forces and displacements. To see this, note that the element nodal displacement vector is composed of four separate vectors, namely,

$$\{u\} = \left\{ \{u_1, v_1, w_1\}; \{\phi_{x1}, \phi_{y1}, \phi_{z1}\}; \{u_2, v_2, w_2\}; \{\phi_{x2}, \phi_{y2}, \phi_{z2}\}; \cdots \right\}$$

Each of these is separately transformed by the $[3 \times 3]$ rotation matrix $[\, R \,]$. Hence the complete transformation is

$$\{\bar{F}\} = [\, T \,]\{F\}, \qquad \{\bar{u}\} = [\, T \,]\{u\}$$

where

$$[\, T \,] \equiv \begin{bmatrix} R & 0 & 0 & \\ 0 & R & 0 & \\ 0 & 0 & R & \cdots \\ & & \vdots & \ddots \end{bmatrix}$$

is an $[18 \times 18]$ matrix for the shell. Substituting for the barred vectors into the element stiffness relation leads to the global stiffness as

$$[\, k \,] = [\, T \,]^T [\, \bar{k} \,][\, T \,]$$

We take advantage of the special nature of $[\, T \,]$ to reduce this further to

$$[k_{11}] = [\, R \,]^T [\bar{k}_{11}][\, R \,], \qquad [k_{12}] = [\, R \,]^T [\bar{k}_{12}][\, R \,], \qquad \cdots$$

This is a transform of the $[3 \times 3]$ partial stiffnesses.

The formal assembly process is that each element stiffness is rotated to the global coordinate system and then augmented to the system size. The structural stiffness matrix is then

$$[K] = \sum_m [T]_m^T [\bar{k}]_m [T]_m$$

where the summation is over each element. The rows and columns associated with the zero degrees-of-freedom are then "scratched" leaving the reduced structural stiffness matrix. This is the system of equations that is then solved. A point to note is that since the frame and shell elements share similar degrees of freedom, then there are no additional complications arising when structures are assembled from combinations of frame and plated elements.

In practice, there is no need to augment the member stiffness since we assemble the reduced global stiffness directly. The coding for doing this is given in Reference [22].

Determining the Rotation Matrix

The rotation matrix required to transform one cartesian coordinate system to another sharing a common origin is

$$[R] = \begin{bmatrix} l_x & m_x & n_x \\ l_y & m_y & n_y \\ l_z & m_z & n_z \end{bmatrix}$$

where l_x, m_x, and n_x are the direction cosines — the cosines of the angles that the $\bar{x}, \bar{y}, \bar{z}$ axes make with the global x, y, z axes, respectively, as shown in Figure 2.33. We have slightly different formulations for frames as for plates.

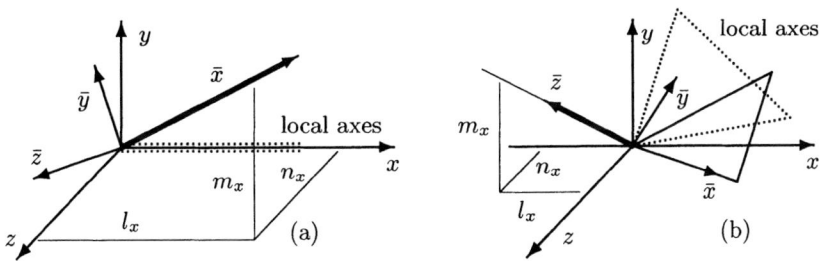

Figure 2.33: Direction cosines for 3-D elements. (a) Frame. (b) Shell.

Let the member axis of the frame coincide with $\bar{x}$, then the direction cosines of the first row can be determined as

$$l_x = (x_j - x_i)/L_{ij}, \qquad m_x = (y_j - y_i)/L_{ij}, \qquad n_x = (z_j - z_i)/L_{ij} \quad (2.44)$$

where (x_i, y_i, z_i) and (x_j, y_j, z_j) are the coordinates of the first and second nodes, respectively, and L_{ij} is the length of the member. The problem here is to find the remaining elements of $[R]$.

Assume that the element arrived at its current position by successive rotations of the element $\bar{x}$ axis assumed to be initially oriented along the x axis. The first rotation is through an angle α about the z axis. The second is a rotation through an angle β about the $\bar{y}$ axis. (This sequence leaves the element $\bar{y}$-axis always oriented so as to lie in the global $x - y$ plane.) The resulting rotation matrix is therefore

$$[R] = [R_\beta][R_\alpha] = \begin{bmatrix} \cos\beta & 0 & \sin\beta \\ 0 & 1 & 0 \\ -\sin\beta & 0 & \cos\beta \end{bmatrix} \begin{bmatrix} \cos\alpha & \sin\alpha & 0 \\ -\sin\alpha & \cos\alpha & 0 \\ 0 & 0 & 1 \end{bmatrix}$$

Multiplying these matrices together, leads to

$$[R] = \begin{bmatrix} \cos\beta\cos\alpha & \cos\beta\sin\alpha & \sin\beta \\ -\sin\alpha & \cos\alpha & 0 \\ -\sin\beta\cos\alpha & -\sin\beta\sin\alpha & \cos\beta \end{bmatrix}$$

Equating the first row to the direction cosines of the member gives

$$l_x = \cos\beta\cos\alpha, \qquad m_x = \cos\beta\sin\alpha, \qquad n_x = \sin\beta$$

Therefore, the functions $\cos\alpha$, $\sin\alpha$, $\cos\beta$, and $\sin\beta$ may be expressed in terms of the direction cosines of the member by

$$\cos\alpha = \frac{l_x}{D}, \qquad \sin\alpha = \frac{m_x}{D}, \qquad \cos\beta = D, \qquad \sin\beta = n_x, \qquad D \equiv \sqrt{l_x^2 + m_x^2}$$

Substitution of these expressions into the above then gives

$$[R] = \begin{bmatrix} l_x & m_x & n_x \\ -m_x/D & l_x/D & 0 \\ -l_x n_x/D & -m_x n_x/D & D \end{bmatrix}$$

This rotation matrix is expressed in terms of the direction cosines of the element axis (which are readily computed from the coordinates of the joints, Equation (2.44)).

When the element axes are specified in the manner just described, there is no ambiguity about their orientations except in the special case of an element $\bar{x}$-axis oriented along the global z-axis. There is no unique rotation to get to that orientation, e.g., $\alpha = 0°, \beta = 90°$ or $\alpha = 90°, \beta = 90°$. To overcome this difficulty, the additional specification will be made that the element $\bar{y}$-axis is always taken to be along the global y-axis for these cases. That is, $\alpha = 0°, \beta = 90°$. The complete set of direction cosines is therefore

$$[R_z] = \begin{bmatrix} 0 & 0 & n_x \\ 0 & 1 & 0 \\ -n_x & 0 & 0 \end{bmatrix}$$

All that is necessary is to substitute for the direction cosine n_x its appropriate value, which is either 1 or -1.

We handle the plate element in a similar manner except that the orientation of the element is characterized by the local $\bar{z}$-axis, which is normal to the element as shown in Figure 2.33(b). This vector is easily established from the vector cross-products of two vectors representing the 1-2 and 1-3 sides of the element — this is covered in more detail in Section 3.2.

Again, assume that the element arrived at its current position by successive rotations but this time of the element $\bar{z}$-axis assumed to be initially oriented along the z-axis. The first rotation is about the x-axis, and the second is a rotation about the $\bar{y}$-axis. (This sequence leaves the element $\bar{y}$-axis always in the global $y - z$ plane.) Working as before, the resulting rotation matrix is then given as

$$[\,R\,] = \begin{bmatrix} l_x n_x/D & m_x n_x/D & -D \\ -m_x/D & l_x/D & 0 \\ l_x & m_x & n_x \end{bmatrix}$$

Note that the 1-2 edge of the element in local coordinates does not coincide with the $\bar{x}$-axis; this is of no consequence because the formulations for both the MRT and DKT elements were for arbitrarily oriented elements in the local coordinate system.

Example 2.27: A truss structure has joints at the following coordinates:

node	x	y	z
1:	0	0	0
2:	100	0	0
3:	100	−200	0
4:	100	−200	100

Determine the rotation matrices for elements with connectivities $1 - 4$ and $3 - 4$.
    The length of element 1-4 is

$$L = \sqrt{(100 - 0)^2 + (-200 - 0)^2 + (-100 - 0)^2} = 100\sqrt{6}$$

The direction cosines are

$$\begin{aligned} l_x &= (100 - 0)/(100\sqrt{6}) = 1/\sqrt{6} \\ m_x &= (-200 - 0)/(100\sqrt{6}) = -2/\sqrt{6} \\ n_x &= (-100 - 0)/(100\sqrt{6}) = -1/\sqrt{6} \end{aligned}$$

This gives $D = \sqrt{5/6}$ and the rotation matrix is

$$[\,R\,] = \begin{bmatrix} .4082 & -.8165 & -.4082 \\ .8944 & .4472 & .0000 \\ .1826 & -.3652 & .3727 \end{bmatrix}$$

The length of element 3-4 is

$$L = \sqrt{(100 - 100)^2 + (-200 + 200)^2 + (-100 - 0)^2} = 100$$

The direction cosines are

$$l_x = (100-100)/100 = 0\,, \quad m_x = (-200+200)/100 = 0\,, \quad n_x = (-100-0)/100 = -1$$

This is the special case with $n_x = -1$, hence the rotation matrix is

$$[\,R\,] = \begin{bmatrix} 0 & 0 & -1 \\ 0 & 1 & 0 \\ 1 & 0 & 0 \end{bmatrix}$$

Boundary Conditions and Constraints

Having developed the analysis for the general case of a space frame and shell, we now mention some special considerations that are of value when dealing with practical problems.

There are six degrees of freedom at each node. Many problems, however, do not need this many; for example, the plane frame only requires three, while the plane truss uses two. Obviously to analyze a 2-D structure as a 3-D frame is a waste of computer resources.

The key to understanding the reduction of the general case is the idea of imposing constraints. In the case of fixed boundary conditions, we specify the degree of freedom as zero, and consequently "scratch" the associated rows and columns in the stiffness relation. In essence we do the same here; we specify constraints on the degrees of freedom as if they were boundary conditions. For example, consider the reductions for a grid structure. The grid is essentially a plane frame but with the loads applied laterally to its plane — the frame equivalent of a plate. Consequently, elements must also support axial twisting as well as bending. To recover this behavior from the space frame, we assume the grid lies in the $x - y$ plane and impose the constraints

$$u = v = 0\,, \qquad \phi_z = 0$$

at each node. The non-zero degrees of freedom are the out-of-plane displacement w, and the two in-plane rotations ϕ_x, ϕ_y.

We have formulated the stiffness approach in terms of a global coordinate system. Therefore, the allowable constraints must also be in terms of the global coordinates. Consider the case of a frame with oblique supports, that is, the frame is attached to rollers on an inclined surface. The boundary condition is that the displacement normal to the surface is zero. This is a constraint condition written as

$$u_{\text{normal}} = 0 = -u\sin\theta + v\cos\theta$$

where θ is the slope of the incline. Thus neither of the global degrees of freedom is zero. It is possible, of course, to reformulate the stiffness relation to allow for the incorporation of such constraint conditions. A simpler approach, however, is to use a boundary element. That is, we replace the actual support by a relatively stiff member having its longitudinal axis in the direction normal to the inclined surface. If this member is pinned at both ends, then all the motion will be perpendicular to it, thus simulating the effect of an inclined roller.

Variations on this idea can be used to simulate other types of boundary conditions. Keep in mind, however, that this is an essentially linear, small deflections idea that will not carry over to most nonlinear problems.

In a practical problem, some of the displacements may be obvious by inspection. For example, there is no w displacement in a plane frame. In other cases, we can infer this information from the symmetry (or antisymmetry) of the geometry and loading conditions. We then implement these constraints by use of equivalent boundary conditions and thereby reduce the size of the problem.

The use of symmetry and antisymmetry does not involve any approximation and therefore when the opportunity arises, advantage should be taken of it. It is worth keeping in mind that this can be done as long as the structure is symmetric; the load need not be symmetric, because any unsymmetrical loading can be decomposed into the sum of a symmetric and antisymmetric load. A word of caution for vibration and stability problems: the actual loads in these cases are not just the applied loads and sometimes assuming symmetry (of the loading) can lead to erroneous results.

Thin-Walled Reinforced Structures

Aerospace structures, for example, must not only be capable of withstanding the applied loads; but in addition, they must be light-weight. Consequently, such fundamental structural elements as beams are redesigned to maximize the bending resistance and minimize the weight. This is done by distributing the material away from the neutral axis. Unfortunately, the beam is then weak to shear and torsional loads and very susceptible to buckling-type failures. The resistance to these loads is greatly improved by the incorporation of shear webs and by the use of stringers and frames to form panels.

Following is a simple analysis of such webbed structures. The analysis is not intended as a substitute for an FEM analysis; rather, its purpose is to give a global understanding of the deformations and stresses, and thereby enhance the interpretation of the FEM results.

I: Two-stringer Beam

Consider a cantilevered beam of rectangular cross-section $[h \times W]$, where W is the height, and let it be aligned along the z-axis as shown in Figure 2.34(a). The bending and shear stress distributions due to an end load V are

$$\sigma_{zz} = -\frac{My}{I}, \qquad I = \int_A y^2 \, dA = \tfrac{1}{12}hW^3$$

$$\sigma_{yz} = \frac{VQ}{Ih} = \frac{V}{I}[\tfrac{1}{4}W^2 - y^2], \qquad Q = \int_{\text{above } y} y^* \, dA^* = [\tfrac{1}{4}W^2 - y^2]h$$

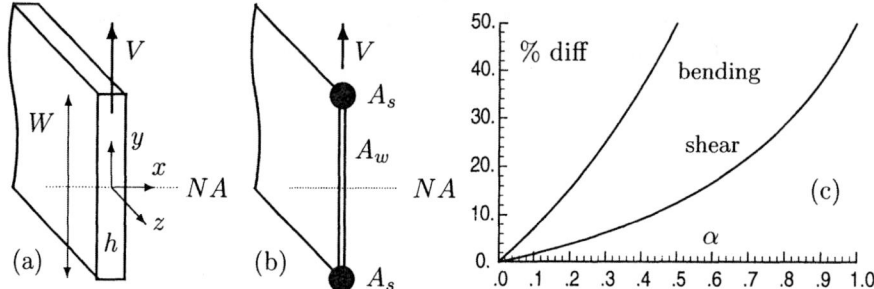

Figure 2.34: Forming a thin-wall beam. (a) Rectangular cross-section. (b) Single web beam. (c) Difference between full and simple theory.

The bending stress is linearly distributed while the shear stress distribution is parabolic with the maximum occurring at $y = 0$. We will now re-distribute the cross-sectional area so as to make the beam more efficient.

Consider the special beam shown in Figure 2.34(b) where most of the material is shifted (symmetrically) away from the neutral axis. The concentration of material is generally referred to as a *stringer*, whereas the material in between is called a *shear web*. The purpose of this arrangement is to maximize the bending resistance; the function of the shear web is to separate the stringers and to supply the resistance to the shear force.

In the simple analysis to follow, the area of the stringers is assumed concentrated at a point. Let the re-distribution of area be parameterized as

$$A_w = \alpha A = \alpha h W, \qquad A_s = \tfrac{1}{2}(1 - \alpha)A$$

where $0 < \alpha < 1$ is our parameter and we keep the total area A and beam height W constant. The moments of area are

$$
\begin{aligned}
I &= 2A_s y_s^2 + \tfrac{1}{12}\alpha h W^3 = \tfrac{1}{12}(3 - 2\alpha)AW^2 \\
Q &= A_s y_s + \tfrac{1}{2}\alpha h(y_s^2 - y^2) = \left[\tfrac{1}{4}(1 - \alpha) + \tfrac{1}{8}\alpha[1 - 4y^2/W^2]\right]AW
\end{aligned}
$$

and the stress distributions become

$$\sigma_{zz} = -\frac{4My}{(1 - \tfrac{2}{3}\alpha)AW^2}, \qquad \sigma_{yz} = \frac{V}{W\alpha h}\left[(\frac{1 - \alpha}{1 - \tfrac{2}{3}\alpha}) + (\frac{\tfrac{1}{2}\alpha}{1 - \tfrac{2}{3}\alpha})(1 - 4y^2/W^2)\right]$$

The bending stress remains linear, but the shear stress now has two parts: a constant part and a parabolic part. As the web thickness is made very thin ($\alpha h \to 0$), the parabolic part becomes relatively small and therefore we can neglect it. Introducing the concept of shear flow defined as $q \equiv \sigma_{yz}\alpha h$, then in the limit of very small α, we get

$$\sigma_{zz} = -\frac{4My}{AW^2}, \qquad q = \frac{V}{W}$$

The shear flow is constant and given simply as the applied load divided by the length of the web. The difference between the simple theory and the more full theory is shown in Figure 2.34(c). The simple theory seems quite reasonable for $\alpha < 0.05$.

In this simple theory, all the bending is carried by the stringers, while all the shear is carried by web. In the limit of a very thin web (when the theory is expected to be most appropriate), this implies that the shear stress goes to infinity (but the shear flow would remain constant). This would not occur in practice, since, at some stage as $\alpha \to 0$, the flexural shear resistance of the stringers would come into play and end up supporting all the load. We therefore conclude that the simple theory is not expected to be numerically accurate but provides a conceptual framework within which to understand actions happening in a thin-wall reinforced structure.

II: Three-Stringer Structure

The two-stringer beam can only withstand loads applied along the web and is very weak to laterally applied loads. To circumvent this, stringers are usually distributed in space and connected by a system of webs. We illustrate this with the three-stringer beam of Figure 2.35(a).

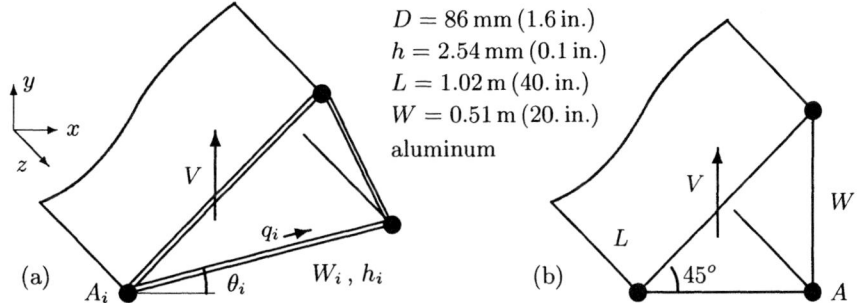

$D = 86\,\text{mm}\,(1.6\,\text{in.})$
$h = 2.54\,\text{mm}\,(0.1\,\text{in.})$
$L = 1.02\,\text{m}\,(40.\,\text{in.})$
$W = 0.51\,\text{m}\,(20.\,\text{in.})$
aluminum

Figure 2.35: Triangular thin-walled structure. (a) General case. (b) Test case.

Most of the bending is resisted by the stringers and most of the shear by the webs. Assume that the webs are essentially constant shear webs, then by equilibrium

$$\sum F_x = 0 \quad \Rightarrow \quad \sum_i q_i W_i \cos \theta_i = 0$$

$$\sum F_y = 0 \quad \Rightarrow \quad \sum_i q_i W_i \sin \theta_i - V = 0$$

There are three shear flows but only two equilibrium equations, hence another equation is required to solve for them. In general, the angle of twist due to shear loading of a closed section is given by

$$\frac{\Delta\phi}{\Delta z} = \frac{1}{2G\bar{A}} \oint \frac{qds}{h}$$

where the integral is taken around all the webs and $\bar{A}$ is the enclosed area. If the beam is restricted to being loaded through the shear center so that the angle of twist is zero, then the third equation is

$$0 = \sum_i \frac{q_i W_1}{h_i}$$

This gives three equations and three unknowns; consequently, all the shear flows can be solved for.

Because the body is loaded through the shear center, the resultant moment about any point is zero. That is,

$$\sum M = 0 \quad \Rightarrow \quad \sum_i q_i 2\bar{A}_i - Ve = 0 \quad \text{or} \quad e = \sum_i q_i 2\bar{A}_i / V$$

where $\bar{A}_i$ is the area enclosed by W_i and the pivot point. The shear center is now obtained as

$$e = \sum_i q_i 2\bar{A}_i / V$$

which is, of course, independent of the level of the load for linear problems. It is also independent of the stringer areas.

We obtain the bending stresses by assuming that all the bending inertia (second moment of area) is in the stringers. The centroid is

$$x_c = \frac{1}{A} \sum_i x_i A_i, \qquad y_c = \frac{1}{A} \sum_i y_i A_i, \qquad A = \sum_i A_i$$

The bending moment of inertias are

$$I_{xx} = \sum_i [y_i - y_c]^2 A_i, \qquad I_{yy} = \sum_i [x_i - x_c]^2 A_i, \qquad I_{xy} = \sum_i [x_i - x_c][y_i - y_c] A_i$$

Let there be moments about the x- and y-axes, then the stress distribution is

$$\sigma_{zz} = \frac{I_{xx} M_y - I_{xy} M_x}{I_{xx} I_{yy} - I_{xy}^2}[x - x_c] + \frac{I_{yy} M_x - I_{xy} M_y}{I_{xx} I_{yy} - I_{xy}^2}[y - y_c]$$

Even if there is only a moment about the x-axis, the neutral axis ($\sigma_{zz} = 0$) forms a line in space that does not coincide with the x-axis.

Example 2.28: Specialize the thin-web beam equations to a beam with a right-triangular cross-section. Also, let each skin thickness and stringer area be the same.

The angles and web lengths for this case are

$$\theta_1 = 0^\circ, \quad \theta_2 = 90^\circ, \quad \theta_3 = -135^\circ; \qquad W_1 = W, \quad W_2 = W, \quad W_3 = \sqrt{2}W$$

The system of equations becomes

$$0 = q_1 W_1 + 0 - q_3 W_3 \frac{1}{\sqrt{2}}, \quad 0 = 0 + q_2 W_2 - q_3 W_3 \frac{1}{\sqrt{2}} + V, \quad 0 = q_1 W_1 + q_2 W_2 + q_3 W_3$$

Solving these gives

$$q_1 = -\frac{V}{W_1}\frac{1}{(2+\sqrt{2})} = -\frac{V}{W}\frac{1}{(2+\sqrt{2})}$$

$$q_2 = -\frac{V}{W_2}\frac{(1+\sqrt{2})}{(2+\sqrt{2})} = \frac{V}{W\sqrt{2}}$$

$$q_3 = -\frac{V}{W_3}\frac{\sqrt{2}}{(2+\sqrt{2})} = -\frac{V}{W}\frac{1}{(2+\sqrt{2})}$$

Note that, in each case, the shear flow is given by a relation of the form $q = V/W^*$, but that the effective length is modified in each case. As expected, Web 2 (the vertical web) carries the greatest shear.

The shear center is obtained as

$$e = q_2 2\bar{A}_2/V = W/\sqrt{2} = 0.707\,W$$

This location is shown in Figure 2.36(a). The centroid is at

$$x_c = \frac{AW + AW}{3A} = \tfrac{2}{3}W, \qquad y_c = \frac{AW}{3A} = \tfrac{1}{3}W$$

In this case, the shear center and centroid are close, but, in general, they are quite distinct. The moments of inertia about the centroid are

$$I_{xx} = [-\tfrac{1}{3}W]^2 A + [-\tfrac{1}{3}W]^2 A + [\tfrac{2}{3}W]^2 A = \tfrac{2}{3}W^2 A, \quad I_{yy} = \tfrac{2}{3}W^2 A, \quad I_{xy} = \tfrac{1}{3}W^2 A$$

There is only a moment about the x-axis given by $M_x = VL$, then the bending stress distribution becomes

$$\sigma_{zz} = \left[-(x - x_c) + 2(y - y_c) \right] \frac{VL}{W^2}$$

The orientation of the neutral axis is shown in Figure 2.36(b).

Example 2.29: Investigate the simple thin-web beam analysis using FEM.

The reinforced structure is composed of a combination of flat plate and frame elements. All frame members have the same diameter of 20 mm (0.8 in.). The length is 1 m (40. in.) and both width and depth are 500 mm (20. in.), 20 mm (0.8 in.). The skin thickness is 2.54 mm (0.1 in.). All materials are aluminum.

A relatively stiff end-plate was attached to the free end so that the load could be applied without causing severe local effects. The load was applied to a variable length stiff frame attached to the end-plate.

The vertically applied load was placed at different positions and the resulting end rotations are shown in Figure 2.36(a). The shear center (the load position

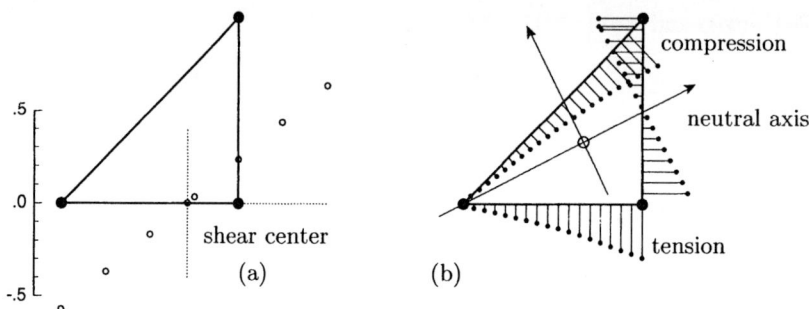

Figure 2.36: FEM results for a triangular thin-webbed beam. (a) End rotation for different positions of load. (b) Bending stress distribution.

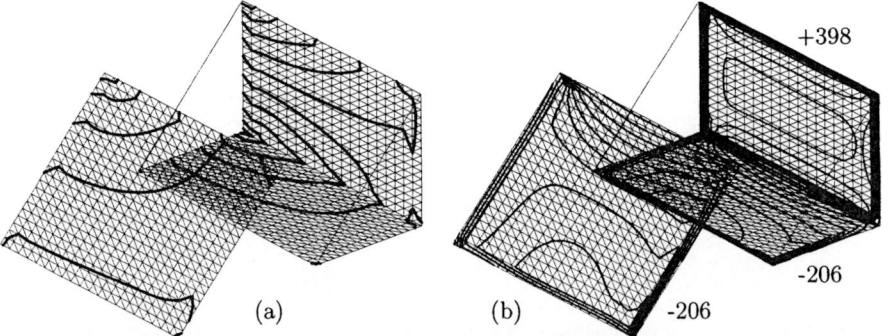

Figure 2.37: Contours of stresses when the load is applied through the shear center. (a) Bending stress distribution. (b) Shear stress distribution.

where the rotation is zero) coincides quite closely to that of the simple theory. As expected for a linear analysis, the rotation is linear with the distance from the shear center.

The axial stresses were sampled at a location $L/8$ from the fixed end and plotted on the cross-section in Figure 2.36(b). The bottom is in tension and the top in compression as expected from a vertically applied load. The orientation of the neutral axis follows that from the simple theory.

The contours of axial and shear stress are shown in Figure 2.37. The panels are approximately constant shear panels. The numbers correspond to the central value and it is seen that the ratio agrees reasonably well with the simple theory.

Problems

2.1 Consider the stress function

$$\phi(x,y) = \tfrac{1}{2}Py^2[1 - y^2/(6b^2)]$$

- Show that while the stress function gives stresses that are in equilibrium, the corresponding strains are not compatible.

2.2 Consider the Airy stress function

$$\phi(x,y) = A\log_n(\sqrt{x^2 + y^2})$$

- Sketch the stress distributions along a few coordinate lines.
- What are the tractions along the surface $x^2 + y^2 = a^2$?

2.3 Consider the Airy stress function

$$\phi(x, y) = Ax^2 + Bxy + Cy^2$$

- What class of problems is solved by this function?

2.4 Consider the following polynomial stress function

$$\phi(x, y) = Axy + Bx^3 + Cx^3y + Dxy^3 + Ex^3y^3 + Fxy^5$$

- Under what circumstance(s) is it bi-harmonic?
- Use it to solve the problem of pure bending of a prismatic bar.
- Show that it can be used to solve the rectangular dam problem. Note that because this polynomial is not symmetric in x and y, the orientation of the axes must be chosen appropriately.

2.5 Motivated by the desire to use Fourier series to represent the applied tractions, it is proposed to use the following stress function

$$\phi(x, y) = \cos(n\pi x/L) f(y)$$

where $n = 0, 1, \ldots$ and L is a constant.
- Determine the allowable form for $f(y)$ for this to be an acceptable Airy stress function.
- If the applied tractions are represented as $P(x) \approx \sum_n a_n \cos(n\pi x/L)$, determine the stress function in terms of a_n.
- Show that the stress σ_{xx} at a point on the surface of a half-plane is a compression equal to the applied pressure at that point.

2.6 Consider the stress functions

$$\phi(r, \theta) = [Ar^3 + Br^{-1} + Cr + Dr\log_n r]\{\sin\theta, \cos\theta\}$$

- Show that it can solve the problem of a curved cantilever beam with an end shear force.
- Compare the solution with the results of an FEM analysis.
- Show that it can solve the problem of a curved cantilever beam with an end normal force.
- Compare the solution with the results of an FEM analysis.

2.7 Consider the stress function

$$\phi(r, \theta) = Ar^2[\theta - \sin\theta\cos\theta]$$

- Show that it can solve the problem of a uniform load over half of the half-plane.

2.8 Flamant's problem is that of a point load on a half-plane.
- Show that it can be solved with the following stress function

$$\phi(r, \theta) = Ar\theta\sin\theta$$

- Compare the solution with the results of an FEM analysis.
- Investigate the need for mesh refinement.

2.9 A ring is split and the two ends moved radially apart an amount Δ.
- Show that the following stress function solves the problem.

$$\phi(r,\theta) = \frac{E\Delta}{4\pi} r \log_n r \sin \theta$$

- Compare the solution with the results of an FEM analysis.

2.10 A rigid disk has a resultant moment applied to it.
- What is the simplest distribution of traction on its edge that will keep it in equilibrium?

2.11 A rigid disk is solidly embedded in an infinite sheet.
- Determine the stress distribution in the sheet due to an applied moment acting on the disk.

2.12 A circular plate of radius b is rigidly supported on a radius a.
- Determine the deflections when a uniform pressure is applied.
- Compare the solution with the results of an FEM analysis.

2.13 A circular plate of outer radius b and inner radius a is simply supported on all edges.
- Determine the deflections when a uniform transverse pressure is applied.
- Compare the solution with the results of an FEM analysis.

2.14 The Gaussian curvature term in the strain energy for plates is

$$\frac{1}{2} \int \int D2(1-\nu)\left[\left(\frac{\partial^2 w}{\partial x^2}\right)\left(\frac{\partial^2 w}{\partial y^2}\right) - \left(\frac{\partial^2 w}{\partial x \partial y}\right)^2 \right] dxdy$$

- Show, by integration by parts, that this term is zero for a large class of boundary value problems for rectangular plates.

3
Nonlinear Static Analysis

In the general case of thin-walled structures, we can have both large displacements and large strains. This renders the governing equations highly nonlinear and therefore they can only be solved using computational methods. Furthermore, because of the complicated load history dependence, this suggests a time or load incremental solution. We combine these two requirements into an incremental/iterative solution algorithm. The total Lagrangian and corotational schemes are introduced as specific examples of solution schemes. These are combined with a Newton-Raphson iteration scheme for the actual solution of the nonlinear equations.

We are most interested in the case of large deflections and rotations but relatively small strains. An example of such a deformation is shown in Figure 3.1 for an elastica. The corotational scheme (where the reference axes rotate with the deforming body) seems quite appropriate for this type of problem and this is the main scheme we develop. References [19, 20, 55] were relied upon for much of the formulation.

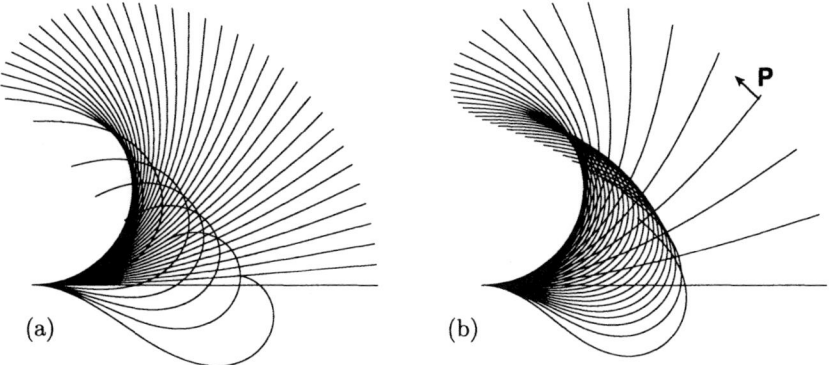

Figure 3.1: Deformed shapes for a cantilever beam with a transverse follower load. (a) Theory for equal increments of rotation. (b) FEM for equal load increments.

3.1 Truss and Elastica Problems

A truss is composed of slender members that support only axial load; consequently, these members must be triangulated for equilibrium under normal loads. An *elastica*, on the other hand, is a slender member that supports both axial and bending loads; however, it does not experience any axial stretching. We use the truss an introductory example to illustrate the effect of axial loads on the stiffness properties of a structure, and use the elastica as an example of large rotations with small strains. Both will also serve as test cases for our finite element formulations.

Trusses

The previous chapter introduced the elastic stiffness for plates in terms of geometry and material properties. When we deal with nonlinear problems, we must introduce the very important concept of the tangent stiffness. Unlike the elastic stiffness, this changes as the load changes giving rise to some surprising consequences. We use the truss to introduce some of the basic ideas.

I: Small Deflection Example

Consider the simple truss whose geometry is shown in Figure 3.2. The members are of original length L_o and the unloaded condition has the apex at a height of h. The two ends are on pinned rollers.

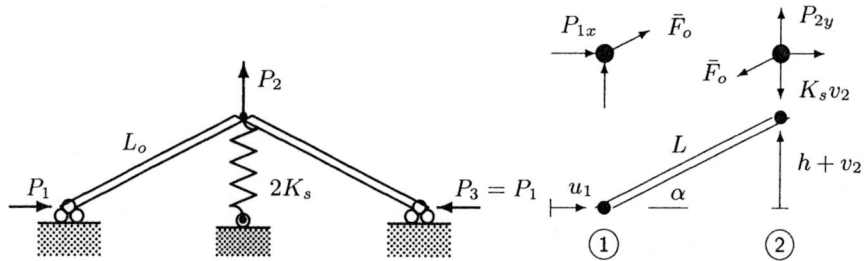

Figure 3.2: Simple pinned truss with a grounded spring.

Let the height h be small relative to the member length, and let the deflections be somewhat small; then we have the geometric approximations

$$L_x = L\cos\alpha \approx L_o - u_1, \qquad L_y = L\sin\alpha \approx h + v_2, \qquad u_1 << v_2$$

The deformed length of the member is

$$L = \sqrt{(L_o - u_1)^2 + (h + v_2)^2} \approx L_o - u_1 + \frac{h}{L_o}v_2 + \tfrac{1}{2}L_o\left(\frac{v_2}{L_o}\right)^2$$

The axial force is computed from the strain as

$$\bar{F}_o = EA_o\epsilon = EA_o\frac{L - L_o}{L_o} = EA_o\left[-\frac{u_1}{L_o} + \frac{h}{L_o}\frac{v_2}{L_o} + \tfrac{1}{2}\left(\frac{v_2}{L_o}\right)^2\right]$$

Note that we consider the parameters of the constitutive relation to be unchanged during the deformation.

Look at equilibrium in the deformed configuration; specifically, consider the resultant horizontal force at Node 1 and vertical force at Node 2 giving

$$0 = P_{1x} + \bar{F}_o\cos\alpha \approx P_{1x} + \bar{F}_o$$

$$0 = P_{2y} - \bar{F}_o\sin\alpha - K_s v_2 \approx P_{2y} - \beta\bar{F}_o - K_s v_2, \qquad \beta \equiv \frac{h + v_2}{L_o}$$

We rewrite these in vector form as

$$\begin{Bmatrix} 0 \\ 0 \end{Bmatrix} = \begin{Bmatrix} P_{1x} \\ P_{2y} \end{Bmatrix} + \begin{Bmatrix} \bar{F}_o \\ -\beta\bar{F}_o \end{Bmatrix} - \begin{Bmatrix} 0 \\ K_s v_2 \end{Bmatrix} \qquad \text{or} \qquad \{\mathcal{F}\} = \{P\} - \{F\} = \{0\}$$

We refer to the last form of the equation as the *loading equation*; $\{P\}$ is the vector of applied loads, $\{F\}$ is the vector of element nodal forces, and $\{\mathcal{F}\}$ is the vector of out-of-balance forces. For equilibrium, we must have that $\{\mathcal{F}\} = \{0\}$, but, as we will see, this is not necessarily (numerically) true during an incremental approximation of the solution.

Example 3.30: Determine the deflections when the loads are $P_{1x} = P$, $P_{2y} = 0$.

For this special case, we get $\bar{F}_o = -P$ and the two deflections are

$$u_1 = \left[\frac{P}{EA} + \left(\frac{h}{L_o}\right)^2\frac{P}{K_s L_o - P} + \tfrac{1}{2}\left(\frac{h}{L_o}\right)^2\left(\frac{P}{K_s L_o - P}\right)^2\right]L_o$$

$$v_2 = \left[\frac{h}{L_o}\frac{P}{K_s L_o - P}\right]L_o$$

The load deflection relations are nonlinear even though the deflections are assumed to be somewhat small. Furthermore, when the applied load is close to $K_s L_o$, we get very large deflections. (This is inconsistent with our above stipulation that the deflections are "somewhat small," let us ignore that issue for now and accept the results as indicated.) That is, at $P = P_{cr} = K_s L_o$, we get very large deflections meaning that the structure has become unstable. We say P has reached a critical value.

The full solutions are shown plotted in Figure 3.3 for different values of h. The effect of a decreasing h is to cause the transition to be more abrupt. Also shown are the behaviors for $P > K_s L_o$. These solutions could not be reached using monotonic loading, but they do in fact represent equilibrium states that can cause difficulties for a numerical scheme that seeks the equilibrium path approximately. That is, it is possible to accidentally converge on these spurious equilibrium states.

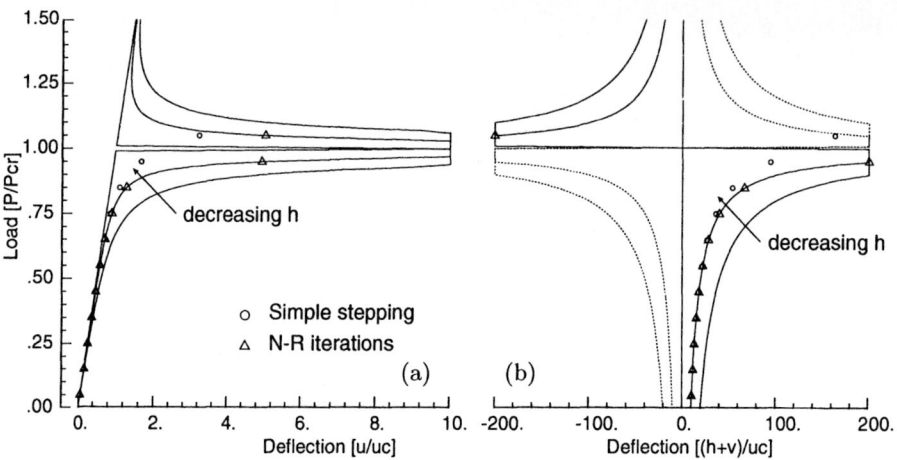

Figure 3.3: Load/deflection behavior for the simple truss. (a) Horizontal displacement u_1. (b) Vertical displacement v_2.

II: Large Deflection Example

The previous example predicted an infinite displacement at the critical load. Obviously this cannot occur in real structures, so we now look at an example where we take the large changes of geometry into account. We will consider the simple two-member truss shown in Figure 3.4; all joints are pinned.

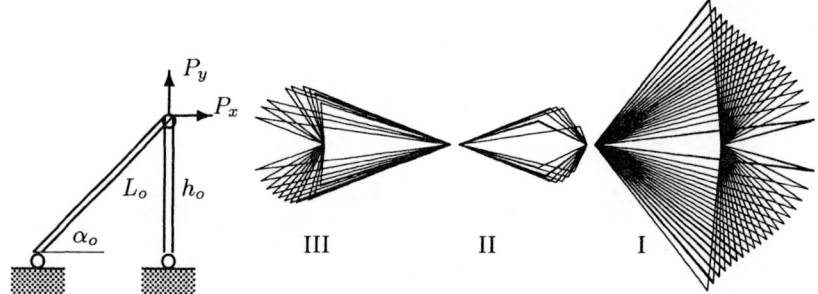

Figure 3.4: Pinned truss and various deformed shapes.

Let the loaded node have the two displacements u and v; then from geometrical considerations, we get that the new lengths are

$$L = \sqrt{(W_o + u)^2 + (h_o + v)^2}, \qquad h = \sqrt{(u)^2 + (h_o + v)^2}$$

where $W_o = L_o \cos \alpha_o$. The axial strain in each member is, respectively,

$$\epsilon_1 = \frac{L - L_o}{L_o}, \qquad \epsilon_2 = \frac{h - h_o}{h_o}$$

These strains are uniformly distributed, hence the total potential of the problem is

$$\Pi = \frac{EA}{2L_o}\left[\sqrt{(W_o + u)^2 + (h_o + v)^2}-L_o\right]^2 + \frac{EA}{2h_o}\left[\sqrt{(u)^2 + (h_o + v)^2}-h_o\right]^2 - P_x u - P_y v$$

Note that we consider the parameters of the constitutive relation to be unchanged during the large deformation.

There are two degrees of freedom, equilibrium is obtained by setting

$$\mathcal{F}_u = \frac{\partial \Pi}{\partial u} = \frac{EA}{L_o}\left[1 - \frac{L_o}{L}\right](w_o + u) + \frac{EA}{h_o}\left[1 - \frac{h_o}{h}\right](u) - P_x = 0$$

$$\mathcal{F}_v = \frac{\partial \Pi}{\partial v} = \frac{EA}{L_o}\left[1 - \frac{L_o}{L}\right](h_o + v) + \frac{EA}{h_o}\left[1 - \frac{h_o}{h}\right](h_o + v) - P_y = 0$$

This is a set of coupled nonlinear equations (because L and h depend on u and v). Some numerical scheme must be invoked in order to solve them.

Example 3.31: Plot all the equilibrium paths (load/deflection curves) for the simple truss structure shown in Figure 3.4 when $P_x = 0$ and $P_y = P$.

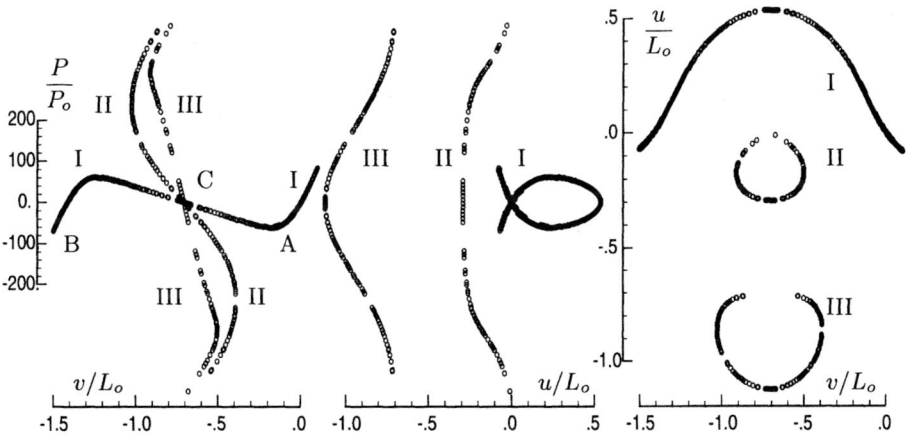

Figure 3.5: Equilibrium paths when $P_x = 0$.

Later in this chapter, we will develop an incremental/iterative solution scheme for nonlinear problems such as these. Here, however, we are interested in all possible equilibrium paths and therefore will choose a different tact. Consider this as a displacement driven problem: that is, determine the loads P_x and P_y as a function of all of the possible displacements u and v, then select only those solutions for which $P_x \approx 0$.

The results are shown in Figure 3.5. There are three equilibrium paths identified and correspond to when the apex tips to the right (I), goes through the center (II), and tips to the left (III), respectively, as shown in Figure 3.4. For path I, when v is positive, the structure stiffens, but when v is negative, a limit point (A) is reached beyond which the load cannot increase. At this stage, the structure is unstable. Under load control, if the load is increased, the next equilibrium point is at B which is a large displacement away. This phenomenon is called *snap-through*.

Under displacement control, the path including C can be traced, but the portion A-C is unstable.

Paths II and III can only be arrived at through a nonproportional loading scheme; for example, P_x can be increased negatively, then P_y increased negatively until the member lies horizontally. At this stage P_x can be decreased to zero. As pointed out in the previous example problem, although these solutions should not be reached using a path following method, they do in fact represent equilibrium states that can cause difficulties for a numerical scheme that seeks the equilibrium path approximately.

Basic Equations for the Elastica

Some of the results to follow can also be found in Reference [42]. With reference to Figure 3.6, let s be the distance along the elastica, hence we have

$$\frac{dx}{ds} = \cos\phi, \qquad \frac{dy}{ds} = \sin\phi$$

where ϕ is the slope. A point originally at position $x^o = s$, $y^o = 0$, moves to a location x, y a distance s along the elastica since the elastica is inextensible. Hence, the two displacements are given by

$$u = x - x^o = x - s, \qquad v = y - y^o = y - 0$$

We can put this in differential form as

$$\frac{du}{ds} = \frac{dx}{ds} - 1 = \cos\phi - 1, \qquad \frac{dv}{ds} = \frac{dy}{ds} = \sin\phi$$

Hence if we can determine $\phi(s)$, integration of these two equations will lead to the deflections.

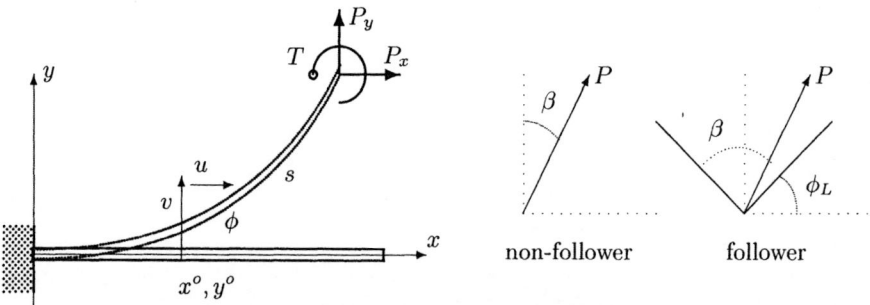

Figure 3.6: An elastica with tip loads.

Let the tip have an applied moment T and two components of force P_x and P_y. Moment equilibrium of the segment shown in Figure 3.6 leads to

$$M(s) = T - P_x[v_L - y] + P_y[L - x + u_L] = T - P_x[v_L - v(s)] + P_y[L - s + u_L - u(s)]$$

Additionally, we make the elementary beam theory assumption that the slope and moment are related through

$$EI\frac{d\phi}{ds} = M(s)$$

Differentiating this and using the equilibrium equation gives

$$EI\frac{d^2\phi}{ds^2} = -P_x[-\frac{dv}{ds}] + P_y[-1 - \frac{du}{ds}] = P_x \sin\phi - P_y \cos\phi$$

This is our governing equation and we will now consider some special cases.

I: Applied Tip Moment

We begin by considering when $P = 0$, the governing equation becomes

$$EI\frac{d^2\phi}{ds^2} = 0$$

Integrating twice and imposing the conditions

$$\text{at } s = 0: \quad \phi = 0, \qquad \text{at } s = L: \quad M = EI\frac{d\phi}{ds} = T$$

leads to

$$EI\phi(s) = Ts$$

Substitute this into the expressions for the displacements to get

$$u(s) = \frac{EI}{T}\sin(\frac{Ts}{EI}) - s, \qquad v(s) = \frac{EI}{T}[1 - \cos(\frac{Ts}{EI})] \qquad (3.1)$$

The tip value for all the variables are

$$u_L = \frac{EI}{T}\sin(\frac{TL}{EI}) - L, \qquad v_L = \frac{EI}{T}[1 - \cos(\frac{TL}{EI})], \qquad \phi_L = \frac{TL}{EI}$$

The deformed shape is that of a circle of radius EI/T and center located along the y axis.

Example 3.32: Determine the deformed shape of a cantilever beam with a tip moment.

We will present results for a beam that is of dimensions $[254 \times 25.4 \times 2.54\,\text{mm}^3]$ ($[10. \times 1.0 \times 0.1\,\text{in.}^3]$) made of aluminum. The results are shown in Figure 3.7. Note that at the final position the tip has rotated 360 deg. Also note that ϕ varies linearly with the load.

The comparison with the linear theory shows very good agreement up to a load of about 50. This is typical in these types of problems.

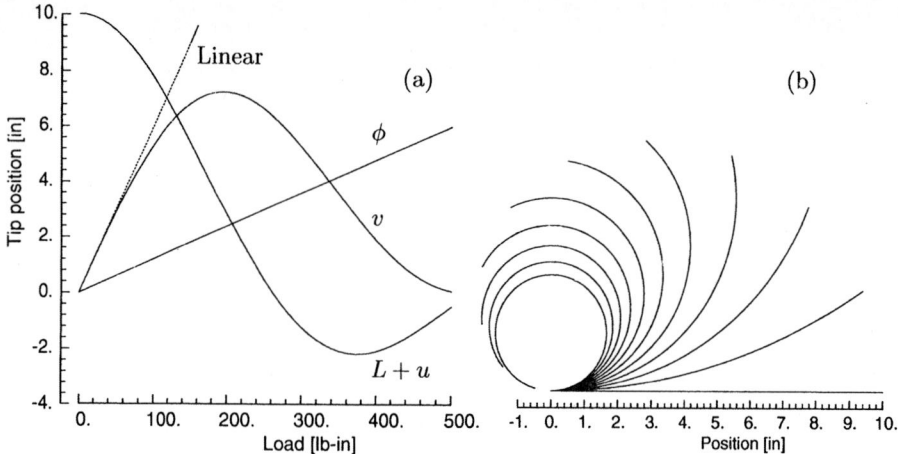

Figure 3.7: Response of an elastica to tip moment loading. (a) Tip positions. (b) Deformed shape.

II: Applied Tip Force

Let the applied force be P acting at an angle β to the vertical; this does not change its direction during the loading. The two components are then

$$P_x = P \sin \beta, \qquad P_y = P \cos \beta$$

The governing equation is

$$EI \frac{d^2 \phi}{ds^2} = P \sin \beta \sin \phi - P \cos \beta \cos \phi = -P \cos(\phi + \beta)$$

Rewrite this as

$$\frac{d^2 \phi}{ds^2} = -\frac{\alpha^2}{L^2} \cos(\phi + \beta), \qquad \alpha \equiv \sqrt{\frac{PL^2}{EI}}$$

and noting that

$$\frac{d}{ds} [(\frac{d\phi}{ds})^2] = 2 \frac{d\phi}{ds} \frac{d^2 \phi}{ds^2}$$

we can write the above as

$$\frac{d}{ds} [(\frac{d\phi}{ds})^2] = -2 \frac{\alpha^2}{L^2} \cos(\phi + \beta) \frac{d\phi}{ds} = -2 \frac{\alpha^2}{L^2} \frac{d}{ds} [\sin(\phi + \beta)]$$

This leads to a first integration

$$(\frac{d\phi}{ds})^2 = -2 \frac{\alpha^2}{L^2} \sin(\phi + \beta) + c_1$$

The constant of integration is obtained by imposing that there is no moment at $s = L$, hence $d\phi/ds = 0$ leading to

$$\frac{d\phi}{ds} = \sqrt{2} \frac{\alpha}{L} \sqrt{\sin(\phi_L + \beta) - \sin(\phi + \beta)}$$

We can re-arrange this as

$$\sqrt{2}\frac{\alpha}{L}\int_0^L ds = \int_0^{\phi_L} \frac{d\phi}{\sqrt{\sin(\phi_L + \beta) - \sin(\phi + \beta)}} = J_1(\phi_L)$$

The right-hand-side integral must be evaluated numerically for different values of final rotation. The left-hand-side integral is simply the length L, hence substituting for α, we get

$$\sqrt{2}\alpha = J_1(\phi_L) \qquad \text{or} \qquad P = \frac{EI}{2L^2}J_1^2(\phi_L)$$

This is an implicit relation between the applied load and the tip rotation.

Once we know the rotation, we can determine the tip deflections. For example,

$$\sin\phi = \frac{dv}{ds} = \frac{dv}{d\phi}\frac{d\phi}{ds} = \frac{dv}{d\phi}\sqrt{2}\frac{\alpha}{L}\sqrt{\sin(\phi_L + \beta) - \sin(\phi + \beta)}$$

Integrating then gives

$$\sqrt{2}\frac{\alpha}{L}v_L = \int_0^{\phi_L} \frac{\sin\phi\, d\phi}{\sqrt{\sin(\phi_L + \beta) - \sin(\phi + \beta)}} = J_2(\phi_L)$$

Again, J_2 is evaluated numerically. Similarly for the other deflection, we have

$$\cos\phi - 1 = \frac{du}{d\phi}\frac{d\phi}{ds} = \frac{du}{d\phi}\sqrt{2}\frac{\alpha}{L}\sqrt{\sin(\phi_L + \beta) - \sin(\phi + \beta)}$$

Integrating then gives

$$\sqrt{2}\frac{\alpha}{L}u_L = \int_0^{\phi_L} \frac{(\cos\phi - 1)\, d\phi}{\sqrt{\sin(\phi_L + \beta) - \sin(\phi + \beta)}} = J_3(\phi_L)$$

During numerical evaluation of the integrals, it is useful to test the quality of the results by observing how well equilibrium is satisfied. Returning to the equilibrium equation, we have that

$$\begin{aligned}
EI\frac{d\phi}{ds} &= -P\sin\beta[v_L - v(s)] + P\cos\beta[L - s + u_L - u(s)] \\
&= EI\sqrt{2}\frac{\alpha}{L}\sqrt{\sin(\phi_L + \beta) - \sin(\phi + \beta)}
\end{aligned}$$

The slope and displacements are zero at $s = 0$, hence

$$\frac{\alpha}{L}[-v_L\sin\beta + [L + u_L]\cos\beta] - \sqrt{2}\sqrt{\sin(\phi_L + \beta) - \sin(\beta)} = 0$$

Typically, this evaluates to 1.0E-6 during the computations.

Example 3.33: Analyze the response of the beam when the loads are either predominantly vertical or predominantly horizontal.

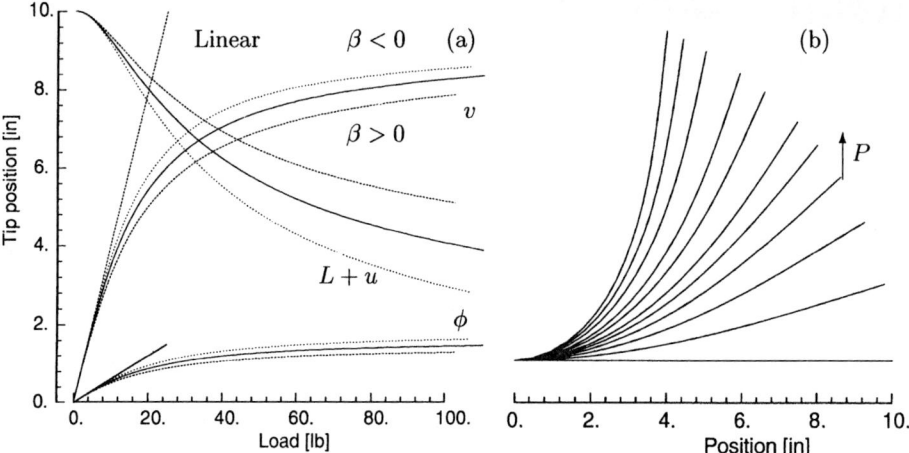

Figure 3.8: Response of an elastica when loads are predominantly vertical. (a) Tip deflections. (b) Deformed shapes.

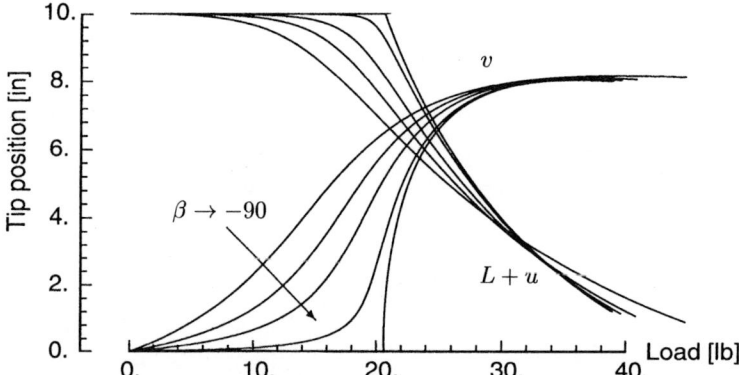

Figure 3.9: Tip deflections when the loads are predominantly horizontal.

The results are shown in Figure 3.8. When the applied force is predominantly vertical, we see essentially a stiffening of the beam. The beam is slightly more flexible as the load is rotated counterclockwise.

Very interesting behavior is observed when the load is almost horizontal but acting toward the beam ($\beta = -90\,\mathrm{deg}$). The results are shown in Figure 3.9. We see a definite limiting effect — there is a load value beyond which the initial horizontal configuration cannot be maintained and a new equilibrium configuration involving a large out-of-plane displacement (v) is found. This is an example of a *static instability*, where the structure changes rapidly (with respect to load) from one configuration to another. We will consider these situations in greater detail in Chapters 6 and 7 dealing with stability, but for now it is worth observing that while every point plotted is a valid equilibrium configuration, the rate of change with respect to the applied load (which is a measure of stiffness) is different for each load misalignment (β) and it is this difference that makes the situations more critical. The limiting case, for example, shows that a very small increase in load will cause a relatively large out-of-plane displacement. Obviously, such very rapid changes would, in actuality, be accompanied by significant dynamic effects. We will leave that discussion to later chapters.

III: Follower Loads

In the previous problem, the orientation of the load was considered to remain unchanged during the deformation. There are many cases where this is not true; pressure loading on a deforming surface is a familiar example. As a first study of this, let the applied force P be acting at a fixed angle β relative to the end of the beam. The solution procedure has much in common with the fixed orientation case, so we will omit some of the steps.

The two components of force are

$$P_x = P\sin(\beta - \phi_L), \qquad P_y = P\cos(\beta - \phi_L)$$

The governing equation is

$$EI\frac{d^2\phi}{ds^2} = P\sin(\beta - \phi_L)\sin\phi - P\cos(\beta - \phi_L)\cos\phi = -P\cos(\phi + \beta - \phi_L)$$

Rewrite this as

$$\frac{d^2\phi}{ds^2} = -\frac{\alpha^2}{L^2}\cos(\phi + \beta - \phi_L), \qquad \alpha \equiv \sqrt{\frac{PL^2}{EI}}$$

As was done before, we can get a first integration as

$$(\frac{d\phi}{ds})^2 = -2\frac{\alpha^2}{L^2}\sin(\phi + \beta - \phi_L) + c_1$$

The constant of integration is obtained by imposing that there is no moment at $s = L$, hence $d\phi/ds = 0$, leading to

$$\frac{d\phi}{ds} = \sqrt{2}\frac{\alpha}{L}\sqrt{\sin(\phi_L - \phi - \beta) + \sin(\beta)}$$

We can re-arrange this as

$$\sqrt{2}\frac{\alpha}{L}\int_0^L ds = \int_0^{\phi_L} \frac{d\phi}{\sqrt{\sin(\phi_L - \phi - \beta) + \sin(\beta)}} = J_4(\phi_L)$$

The right-hand-side integral must be evaluated numerically for different values of final rotation. The left-hand-side integral is simply the length L, hence substituting for α we get

$$\sqrt{2}\alpha = J_4(\phi_L) \qquad \text{or} \qquad P = \frac{EI}{2L^2}J_4^2(\phi_L)$$

As before, this is an implicit relation between the applied load and the tip rotation.

Knowing the rotation allows determining the tip deflections. Proceeding as before, leads to

$$\sqrt{2}\frac{\alpha}{L}v_L = \int_0^{\phi_L} \frac{\sin\phi\,d\phi}{\sqrt{\sin(\phi_L - \phi - \beta) + \sin(\beta)}} = J_5(\phi_L)$$

$$\sqrt{2}\frac{\alpha}{L}u_L = \int_0^{\phi_L} \frac{(\cos\phi - 1)\,d\phi}{\sqrt{\sin(\phi_L - \phi - \beta) + \sin(\beta)}} = J_6(\phi_L)$$

Mathematically, these results for the deflections and rotation resemble those found for the fixed orientation case; indeed they are related through the association $-\beta \leftrightarrow \beta + \phi$. But the responses are quite different.

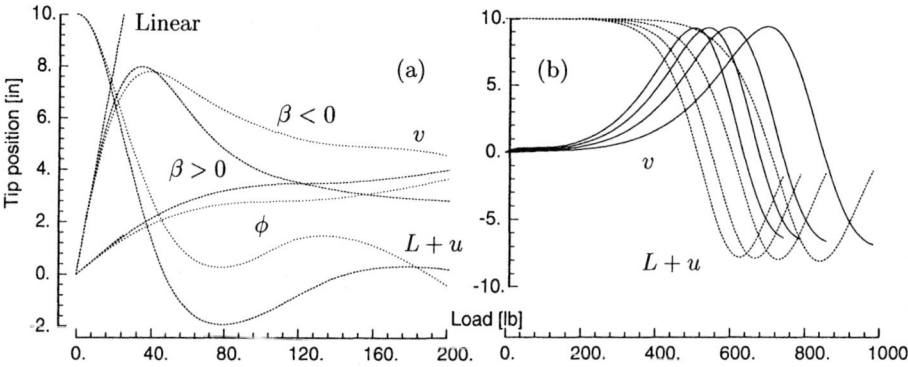

Figure 3.10: Comparison of tip deflections for follower loads. (a) Loads are predominantly transverse. (b) Loads are predominantly axial.

The equilibrium equation leads to

$$\frac{\alpha}{L}[-v_L\sin(\beta - \phi_L) + [L + u_L]\cos(\beta - \phi_L)] - \sqrt{2}\sqrt{\sin(\phi_L - \beta) + \sin(\beta)} = 0$$

This has some interesting special cases. For example, when the load is transverse ($\beta = 0$) and the tip rotation is an integer multiple of π, we conclude that $u_L = -L$.

Example 3.34: Analyze the response of the beam when the follower loads are either predominantly transverse or predominantly axial.

The results are shown in Figure 3.10. When the applied force is predominantly vertical, we see an initial stiffening of the beam but shortly thereafter the rotations dominate. What is surprising is that a slight clockwise rotation of the force leads to larger rotations.

Unlike the previous case, we do not observe any instability effects when the load is almost axial and acting toward the beam ($\beta = -90\,\text{deg}$). All that we see

is a definite increase in the limiting load way beyond that observed previously —
note that any transverse component will eventually be large enough to cause a
noticeable transverse deflection. We conclude that this situation does not have a
static instability. This is a surprising conclusion given that the loading for $\beta \approx$
90 deg resembles that of the fixed orientation case. We will reconsider this in
greater detail in Chapter 7 when we view the problem dynamically.

3.2   Finite Rotations

The three-dimensional thin-walled structures of interest undergo relatively large
deflections and rotations in the style of the elastica we just analyzed. As will be
shown, large rotations are not vector quantities. What this means is that in an
incremental scheme, the total rotation is not simply the sum of all the rotation
increments. We need to develop a proper method to update the orientation of the
elements. A comprehensive discussion of finite rotations is given in Reference [5].

Geometric Description of 3-D Structures

Later, when we formulate our incremental approach to nonlinear deformation
problems, we will need to be able to keep track of the deflections of nodes and
their rotation. This section summarizes some of the considerations.

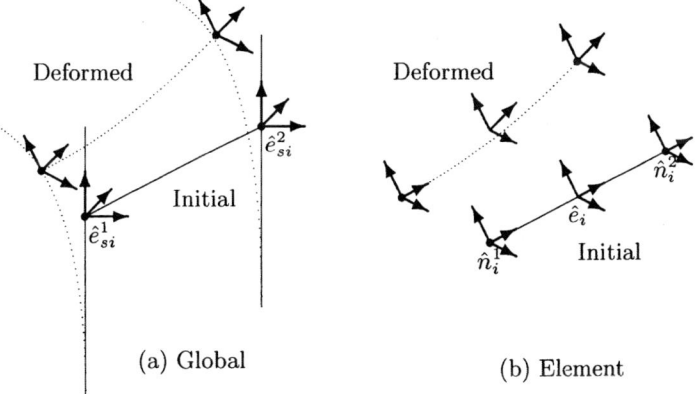

Figure 3.11: Triads used to describe the orientation of structural members. (a) Global
nodal triads. (b) Element nodal and element triads.

We will describe the orientation of a structure by use of an orthogonal triad
that can be associated with the three edges of a cube. Each triad comprises three
vectors with three components each, giving a total of nine numbers. Only six
numbers are independent (since any vector is simply the cross-product of the
other two) but we find it convenient to carry all nine components.

At the global level, the geometry of the deformation is described in terms of the nodal displacements and rotations. The rotations are described more specifically in terms of triads associated with the nodes. As shown in Figure 3.11(a), each node has a triad that is originally oriented with respect to the global coordinate system. As the deformation proceeds, these nodal triads are updated according to

$$[\ \hat{e}_s\ ] = [R(\Delta\phi)][\ \hat{e}_s\ ]$$

(We will define $[\ R\ ]$ presently.) Note that it is the orientation of the triad, and not the accumulated angle, that is stored.

At the element level, as shown in Figure 3.11(b), there is a triad for each node and an additional one that describes the general orientation of the element. These are used to keep track of the local deformation of the element.

We will define a local reference for each element, and the element nodal triads and element triad will initially have this orientation. Obviously, the element nodal triads are attached rigidly to the global nodal triad, but, as shown in Figure 3.11(a), different elements may share the same node, hence the algorithmic bookkeeping is simplified by disassociating the global and element nodal triads. In the case of frame members, the element nodal triads coincide with the principal values of the second moment of area. These triads are updated the same as the global triad. The element triad is recomputed based on the current nodal locations.

Rotations about Fixed Axes

When considering rotations, we can either view them as occurring about fixed axes in space or about a set of axes rigidly attached to the body. In either case, finite rotations are noncommutative. That is, different results occur depending on the sequence of the rotations.

As a simple illustration, consider the rotation about fixed axes of the cube shown in Figure 3.12. The first is a rotation about the z-axis followed by a rotation about the x-axis leaving the marked side facing z. The second is a rotation about the x-axis followed by a rotation about the z-axis leaving the marked side facing y. It is clear that the final position is different for the two cases. If the two rotations were applied simultaneously, the cube would end up in yet a third orientation.

The consequence of this is that if we define a vector rotation such that

$$\hat{\phi} \equiv \phi_x\hat{\imath} + \phi_y\hat{\jmath} + \phi_z\hat{k}$$

then a compound rotation consisting of two rotations is not given by the vector addition

$$\hat{\phi} \neq \hat{\phi}_1 + \hat{\phi}_2$$

For this reason, such a vector is called a *pseudo-vector*.

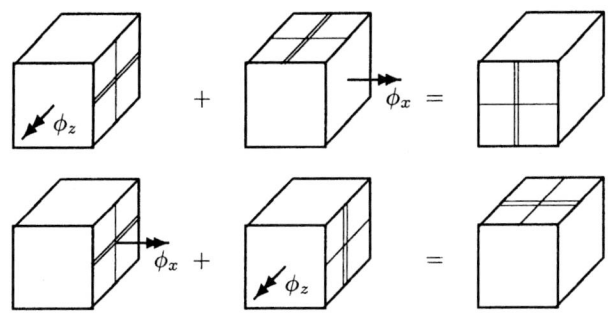

Figure 3.12: Noncommutivity of finite rotations.

Rotation Matrix

Consider the rotation of the vector $\hat{v}$ about an axis represented by the unit vector $\hat{e}$ as shown in Figure 3.13. The tip of the original vector, P, moves to Q; both of which are at a radius $r = |\hat{e} \times \hat{v}|$ from the axle.

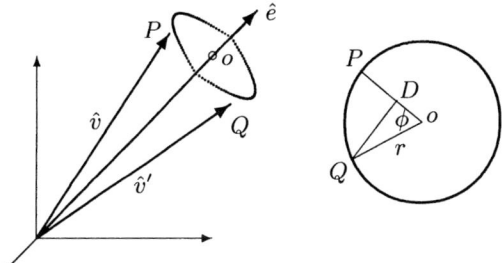

Figure 3.13: Rotation of a vector.

The new vector is given by

$$\hat{v}' = \hat{v} + \hat{PQ}$$

We will break the vector $\hat{PQ}$ into segments $\hat{PD}$ and $\hat{DQ}$, where $\hat{DQ}$ is perpendicular to the plane formed by $\hat{v}$ and $\hat{e}$, while $\hat{PD}$ lies in the plane and is perpendicular to $\hat{e}$. The unit vectors of these segments are

$$\hat{e}_{DQ} = \frac{\hat{e} \times \hat{v}}{|\hat{e} \times \hat{v}|} = \frac{\hat{e} \times \hat{v}}{r}, \qquad \hat{e}_{PD} = \hat{e} \times \hat{e}_{DQ} = \frac{\hat{e} \times (\hat{e} \times \hat{v})}{r}$$

The lengths of the segments are

$$|PD| = r - r\cos\phi, \qquad |DQ| = r\sin\phi$$

The new vector is now given by

$$
\begin{aligned}
\hat{v}' &= \hat{v} + |DQ|\hat{e}_{DQ} + |PD|\hat{e}_{PD} \\
&= \hat{v} + \sin\phi(\hat{e} \times \hat{v}) + (1 - \cos\phi)(\hat{e} \times (\hat{e} \times \hat{v}))
\end{aligned}
\tag{3.2}
$$

Introduce the pseudo-vector,

$$\phi \hat{e} \equiv \phi_x \hat{\imath} + \phi_y \hat{\jmath} + \phi_z \hat{k}, \qquad \phi = \sqrt{\phi_x^2 + \phi_y^2 + \phi_z^2}$$

so that the vector cross-products can be written as

$$\phi \hat{e} \times \hat{v} = (0 v_x - \phi_z v_y + \phi_y v_z) \hat{\imath} + (\phi_z v_x + 0 v_y - \phi_x v_z) \hat{\jmath} + (-\phi_y v_x + \phi_x v_y + 0 v_z) \hat{k}$$

We will often have a need to represent such vector cross-products in matrix form. To that end, we adopt the notation

$$\hat{a} \times \hat{b} \Longrightarrow \begin{bmatrix} 0 & -a_z & a_y \\ a_z & 0 & -a_x \\ -a_y & a_x & 0 \end{bmatrix} \begin{Bmatrix} b_x \\ b_y \\ b_z \end{Bmatrix} = [S(\hat{a})]\{b\} = -[S(\hat{b})]\{a\} \quad (3.3)$$

Consequently, our rotation has the matrix representation

$$\phi \hat{e} \times \hat{v} \Longrightarrow \begin{bmatrix} 0 & -\phi_z & \phi_y \\ \phi_z & 0 & -\phi_x \\ -\phi_y & \phi_x & 0 \end{bmatrix} \begin{Bmatrix} v_x \\ v_y \\ v_z \end{Bmatrix} = [S(\hat{\phi})]\{v\}$$

The total transformation is given by

$$\begin{Bmatrix} v_x' \\ v_y' \\ v_z' \end{Bmatrix} = \left[\begin{bmatrix} 1 & 0 & 0 \\ 0 & 1 & 0 \\ 0 & 0 & 1 \end{bmatrix} + \frac{\sin \phi}{\phi} \begin{bmatrix} 0 & -\phi_z & \phi_y \\ \phi_z & 0 & -\phi_x \\ -\phi_y & \phi_x & 0 \end{bmatrix} \right.$$
$$\left. + \frac{1 - \cos \phi}{\phi^2} \begin{bmatrix} -\phi_z^2 - \phi_y^2 & \phi_x \phi_y & \phi_x \phi_z \\ \phi_x \phi_y & -\phi_z^2 - \phi_x^2 & \phi_z \phi_y \\ \phi_x \phi_z & \phi_y \phi_z & -\phi_x^2 - \phi_y^2 \end{bmatrix} \right] \begin{Bmatrix} v_x \\ v_y \\ v_z \end{Bmatrix}$$

We will write this as

$$\{v'\} = \left[[\ I\ ] + \frac{\sin \phi}{\phi} [S(\phi)] + \frac{1 - \cos \phi}{\phi^2} [S(\phi)]^2 \right] \{v\} \equiv [R(\phi)]\{v\} \quad (3.4)$$

This elegant relation is known as *Rodrique's formula*.

Consider the compound rotation given by

$$\{v\}_1 = [R(\phi_1)]\{v\}_0, \qquad \{v\}_2 = [R(\phi_2)]\{v\}_1$$

then the result is

$$\{v\}_2 = [R(\phi_2)][R(\phi_1)]\{v\}_0 = [R(\phi)]\{v\}_0$$

Note, however, that $\phi \neq \phi_1 + \phi_2$. In fact, it is not even true if ϕ_2 is a small increment $\Delta \phi$.

Later, we will use this form as part of an incremental scheme where each rotation increment is not especially large. Under this circumstance, we could make the approximations

$$\frac{\sin\phi}{\phi} \approx \frac{\phi - \cdots}{\phi} = 1, \qquad \frac{1 - \cos\phi}{\phi^2} \approx \frac{1 - (1 - \frac{1}{2}\phi^2 + \cdots)}{\phi^2} = \frac{1}{2}$$

The rotation matrix is then

$$[R(\phi)] \approx \left[[\ I\ ] + [S(\phi)] + \tfrac{1}{2}[S(\phi)]^2 \right] \approx \left[[\ I\ ] + [S(\phi)] \right]$$

The computational savings, however, are minimal and the exact form might as well be used. Again later, we will have a need to obtain the derivative, or variation, of a vector under a small rotation. We can use the above approximations to obtain

$$\delta\hat{v} = \hat{v}' - \hat{v} = [S(\delta\phi)]\{v\} = -[S(\hat{v})]\{\delta\phi\} = -\hat{v} \times \delta\hat{\phi} \qquad (3.5)$$

The increment points in a direction perpendicular to the plane formed by the axis of rotation and the vector.

Example 3.35: Consider the two rotation histories about fixed axes

$$\phi_x = -\tfrac{1}{2}\pi(t + t^2 - t^3), \qquad \phi_y = -\pi(3t - t^2 - t^3)$$

Show their effect on a vector initially lying along the x-axis.

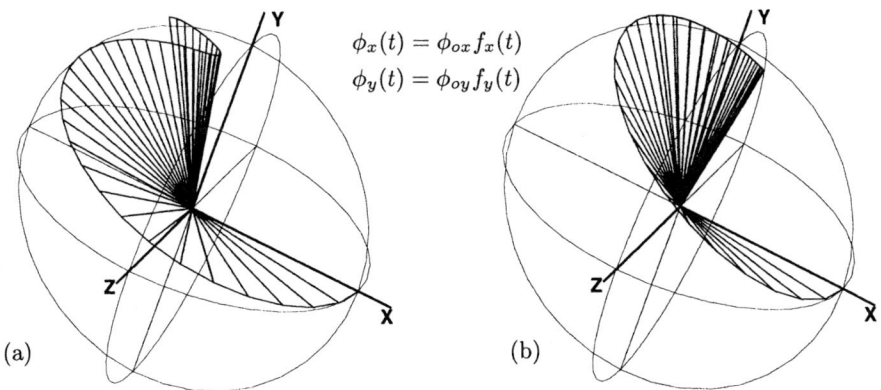

$$\phi_x(t) = \phi_{ox}f_x(t)$$
$$\phi_y(t) = \phi_{oy}f_y(t)$$

(a)   (b)

Figure 3.14: Rotations of a vector. (a) $f_x(t) \neq f_y(t)$. (b) $f_x(t) = f_y(t)$.

After one second, the rotation ϕ_y on its own places the vector in the negative x-direction, while the rotation ϕ_x on its own leaves the vector undisturbed (since the vector is initially along the x-axis). The resulting motion of the two components is shown in Figure 3.14(a).

Also shown in the figure is the resulting motion if both components have the same time history. That is, both components are derived from a single rotation vector whose orientation does not change; this situation is commutative.

Angle Between Two Triads

Consider an orthogonal triad $[\,P\,]$ arbitrarily oriented in space. It is comprised of three vectors given by

$$[\,P\,] \equiv [\{p\}_1\{p\}_2\{p\}_3]$$

We can think of these vectors as having being obtained by a rigid body rotation of a set of local vectors

$$\{\bar p\}_1^T = \{1,\,0,\,0\}, \qquad \{\bar p\}_2^T = \{0,\,1,\,0\}, \qquad \{\bar p\}_3^T = \{0,\,0,\,1\}$$

In other words, we can view the triad itself as a rotation transformation of the components of a vector in local coordinates to components in global coordinates

$$\{v\} = [\,P\,]\{\bar v\}, \qquad \{\bar v\} = [\,P\,]^T\{v\}$$

This will be very useful when we deal with local components.

Now consider the important problem of determining the angle between two arbitrarily oriented triads. That is, consider the rotation of a triad $\hat p_i$ into the triad $\hat q_i$ and we wish to know the relative angles.

Begin by defining the matrices formed from the triads

$$[\,P\,] \equiv [\{p\}_1\{p\}_2\{p\}_3], \qquad [\,Q\,] \equiv [\{q\}_1\{q\}_2\{q\}_3]$$

Note that both of these are orthogonal. The rotation matrix transforms one vector into another, hence we can write

$$[\{q\}_1\{q\}_2\{q\}_3] = [\,R\,][\{p\}_1\{p\}_2\{p\}_3] \qquad \text{or} \qquad [\,Q\,] = [\,R\,][\,P\,]$$

Knowing the two triads, we therefore obtain the rotation matrix as

$$[\,R\,] = [\,Q\,][\,P\,]^T \qquad \text{or} \qquad R_{ij} = \sum_k Q_{ik}P_{jk}$$

This rotation matrix has implicitly the information we require.

The trace (sum of diagonal terms) of the rotation matrix gives

$$\text{Tr} = R_{11} + R_{22} + R_{33} = 3 + \frac{1 - \cos\phi}{\phi^2}[-2\phi^2] = 1 + 2\cos\phi$$

giving the cosine, sine, and the angle as, respectively,

$$\cos\phi = (Tr - 1)/2, \quad \sin\phi = \sqrt{(1 - \cos^2\phi)}, \quad \phi = \pm\cos^{-1}[(Tr - 1)/2] \quad (3.6)$$

This gives valid magnitudes up to $\phi = \pm\pi$.

The antisymmetric part of the rotation matrix is related to the matrix $[S(\phi)]$. That is,

$$\frac{\sin\phi}{\phi}[S(\phi)] = \frac{1}{2}[[\,R\,] - [\,R\,]^T]$$

This leads to the rotation components

$$\left\{ \begin{array}{c} \phi_x \\ \phi_y \\ \phi_z \end{array} \right\} = \frac{\phi}{2\sin\phi} \left\{ \begin{array}{c} R_{32} - R_{23} \\ R_{13} - R_{31} \\ R_{21} - R_{12} \end{array} \right\} \tag{3.7}$$

Note that the sine of the angle is also obtained as

$$\sin\phi = \frac{1}{2}\sqrt{(R_{32} - R_{23})^2 + (R_{13} - R_{31})^2 + (R_{21} - R_{12})^2} \tag{3.8}$$

which can act as a check on the earlier calculation.

The components of the rotation as obtained above are referred to global coordinates; an interesting result is obtained if we refer the components to the local coordinates of $[\ P\ ]$, say. The rotation matrix $[\ R\ ]$ transforms one vector into another both of which are referred to global coordinates

$$\{\,v\,\}_2 = [\ R\ ]\{\,v\,\}_1 = [QP^T]\{\,v\,\}_1$$

If we now refer both vectors locally to $[\ P\ ]$, we get

$$[\ P\ ]\{\,\bar{v}\,\}_2 = [QP^T][\ P\ ]\{\,\bar{v}\,\}_1 \qquad \text{or} \qquad \{\,\bar{v}\,\}_2 = [P^TQ]\{\,\bar{v}\,\}_1$$

The rotation matrix is then

$$[\ \bar{R}\ ] = [\ P\ ]^T[\ Q\ ] \qquad \text{or} \qquad \bar{R}_{ij} = \{p\}_i^T\{q\}_j = \hat{p}_i \cdot \hat{q}_j$$

That is, the matrix entries are the vector dot products of the triad vectors. The angle between the two triads, referred to the triad $[\ P\ ]$, are

$$\left\{ \begin{array}{c} \bar{\phi}_x \\ \bar{\phi}_y \\ \bar{\phi}_z \end{array} \right\} = \frac{\bar{\phi}}{2\sin\bar{\phi}} \left\{ \begin{array}{c} \hat{p}_3 \cdot \hat{q}_2 - \hat{p}_2 \cdot \hat{q}_3 \\ \hat{p}_1 \cdot \hat{q}_3 - \hat{p}_3 \cdot \hat{q}_1 \\ \hat{p}_2 \cdot \hat{q}_1 - \hat{p}_1 \cdot \hat{q}_2 \end{array} \right\} \approx \frac{1}{2} \left\{ \begin{array}{c} \hat{p}_3 \cdot \hat{q}_2 - \hat{p}_2 \cdot \hat{q}_3 \\ \hat{p}_1 \cdot \hat{q}_3 - \hat{p}_3 \cdot \hat{q}_1 \\ \hat{p}_2 \cdot \hat{q}_1 - \hat{p}_1 \cdot \hat{q}_2 \end{array} \right\}$$

The approximation is useful for when the relative rotations are small even though the absolute values may be large. The last of these relations can easily be confirmed for a rotation only about the z-axis. That is, let

$$\hat{p}_1 = 1\hat{\imath}, \qquad \hat{p}_2 = 1\hat{\jmath}, \qquad \hat{q}_1 = \cos\bar{\phi}_z\hat{\imath} + \sin\bar{\phi}_z\hat{\jmath}, \qquad \hat{q}_2 = -\sin\bar{\phi}_z\hat{\imath} + \cos\bar{\phi}_z\hat{\jmath}$$

which gives the local rotation about the z-axis as

$$\bar{\phi}_z = \frac{\bar{\phi}_z}{2\sin\bar{\phi}_z}\left[\sin\bar{\phi}_z - (-\sin\bar{\phi}_z)\right] = \bar{\phi}_z$$

as expected. We will need to know the local relative rotations when we look at deforming elements in the corotational scheme.

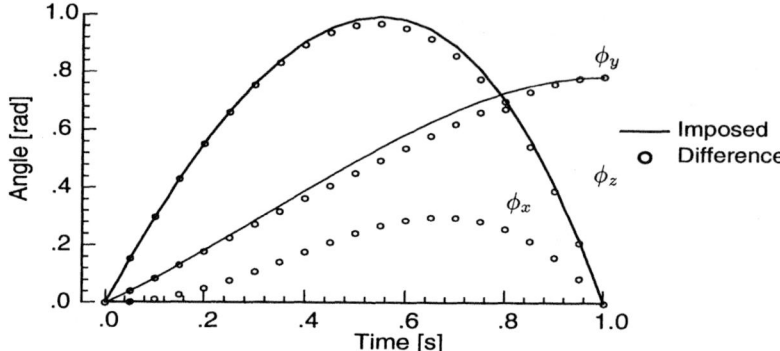

Figure 3.15: Angle between two triads rotating about fixed axes y and z, respectively.

Example 3.36: Consider rotations of two triads varying in time as

$$\phi_1 \hat{e}_1 = \tfrac{1}{4}\pi(t + t^2 - t^3)\{0,\, \hat{\jmath},\, 0\}^T, \qquad \phi_2 \hat{e}_2 = \tfrac{1}{2}\pi(2t - t^2 - t^3)\{0,\, 0,\, \hat{k}\}^T$$

Determine the angle between these triads.

Note that the triads are rotating about perpendicular axes and neither of them changes their axle of rotation.

Figure 3.15 shows an idea of the variation of angle. The rotations about the y and z axes are close (but not equal to) the imposed rotations; what is interesting is the significant component about the x-axis even though neither have components about that axis. It is easy to think of aligning the two triads by rotating first about the y-axis and then rotating about the z-axis and thus there would be no rotation about the x-axis. Remember, however, that what is actually done is a single rotation about an axle in space, the orientation of the axle has an x component.

Example 3.37: Rotate a known triad $\hat{p}_i$ such that its $\hat{p}_1$ vector coincides with the specified vector $\hat{q}_1$ of a second triad. That is, determine the remaining components of $\hat{q}_i$ by doing this in such a way that the rotation forms a minimum angle.

With reference to Figure 3.13, the two vectors are perpendicular to the axle of rotation. By the nature of the special vectors, we have

$$|\hat{p}_1 \times \hat{q}_1| = \sin\phi, \qquad \hat{p}_1 \cdot \hat{q}_1 = \cos\phi, \qquad \hat{e} = \frac{\hat{p}_1 \times \hat{q}_1}{|\hat{p}_1 \times \hat{q}_1|} = \frac{\hat{p}_1 \times \hat{q}_1}{\sin\phi}$$

We use the rotation matrix to transform the other components of the triads. For example,

$$\hat{p}_2 \longrightarrow \hat{q}_2 = \hat{p}_2 + \sin\phi(\hat{e} \times \hat{p}_2) + (1 - \cos\phi)(\hat{e} \times (\hat{e} \times \hat{p}_2))$$

Noting the vector relation

$$(\hat{a} \times \hat{b}) \times \hat{c} = (\hat{a} \cdot \hat{c})\hat{b} - (\hat{b} \cdot \hat{c})\hat{a}$$

we replace the vector cross products according to

$$\hat{e} \times \hat{p}_2 = \frac{1}{\sin\phi}((\hat{p}_1 \times \hat{q}_1) \times \hat{p}_2) = \frac{-1}{\sin\phi}(\hat{p}_2 \cdot \hat{q}_1)\hat{p}_1$$

and

$$\hat{e} \times (\hat{e} \times \hat{p}_2) = \frac{-1}{\sin\phi^2}(\hat{p}_2 \cdot \hat{q}_1)[\hat{q}_1 - \hat{p}_1 \cos\phi]$$

Hence substituting for these and simplifying a little leads to

$$\hat{q}_2 = \hat{p}_2 - \frac{(1 - \cos\phi)}{\sin\phi^2}(\hat{p}_2 \cdot \hat{q}_1)[\hat{p}_1 + \hat{q}_1]$$

Similarly for the third vector, we get

$$\hat{p}_3 \longrightarrow \hat{q}_3 = \hat{p}_3 - \frac{(1 - \cos\phi)}{\sin\phi^2}(\hat{p}_3 \cdot \hat{q}_1)[\hat{p}_1 + \hat{q}_1]$$

When these formulas are used in a circumstance where the angles are not very large, we can use the approximations $(1 - \cos\phi)/\sin\phi^2 \approx 1/2$ leading to

$$
\begin{aligned}
\hat{q}_2 &= \hat{p}_2 - \tfrac{1}{2}(\hat{p}_2 \cdot \hat{q}_1)[\hat{p}_1 + \hat{q}_1] \\
\hat{q}_3 &= \hat{p}_3 - \tfrac{1}{2}(\hat{p}_3 \cdot \hat{q}_1)[\hat{p}_1 + \hat{q}_1]
\end{aligned}
$$

The approximation is very good up angles of about 30 deg.

Element Triads

The final part of the description of the rotation is to establish the element triads. The triad for flat triangular shell elements is relatively straightforward to establish, but defining the orientation of a frame member is a nonunique process and will pose a challenge.

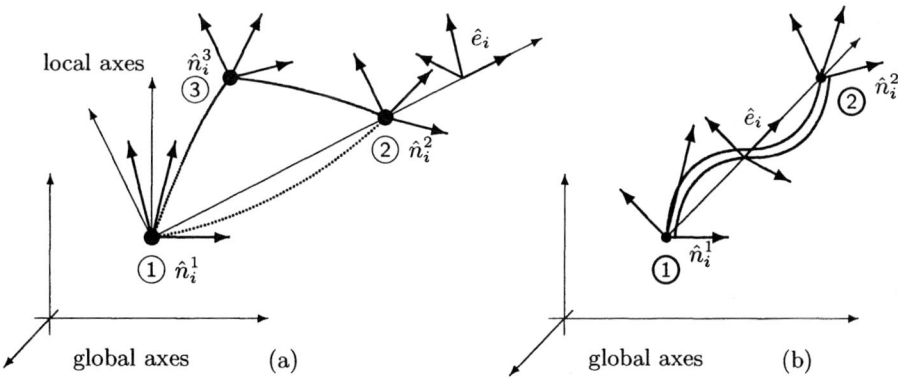

Figure 3.16: Element triads. (a) Flat triangular shell element. (b) Straight frame element.

With reference to Figure 3.16(a), we define the 1-axis as the orientation of the 1-2 side

$$\hat{e}_1 = [(x_2 - x_1)\hat{i} + (y_2 - y_1)\hat{j} + (z_2 - z_1)\hat{k}]/L_{21} = \frac{\hat{x}_{21} + \hat{u}_{21}}{L_{21}}$$

The 3-axis is perpendicular to the plane of the element. To determine this, form the two vectors associated with the 1-2 and 1-3 sides,

$$\hat{v}_1 = (x_2 - x_1)\hat{\imath} + (y_2 - y_1)\hat{\jmath} + (z_2 - z_1)\hat{k}, \qquad \hat{v}_2 = (x_3 - x_1)\hat{\imath} + (y_3 - y_1)\hat{\jmath} + (z_3 - z_1)\hat{k}$$

The vector area is given as

$$\hat{A} = \tfrac{1}{2}[\hat{v}_1 \times \hat{v}_2] = A_x\hat{\imath} + A_y\hat{\jmath} + A_z\hat{k}$$

where

$$
\begin{aligned}
A_x &= (x_1 - x_2)(z_3 - z_2) + (z_2 - z_1)(x_3 - x_2) \\
A_y &= (y_1 - y_2)(x_3 - x_2) + (x_2 - x_1)(y_3 - y_2) \\
A_z &= (z_1 - z_2)(y_3 - y_2) + (y_2 - y_1)(z_3 - z_2)
\end{aligned}
$$

We then have

$$\hat{e}_3 = (A_x/A)\hat{\imath} + (A_y/A)\hat{\jmath} + (A_z/A)\hat{k}$$

The 2-axis is simply given as the vector cross-product of the 3 and 1 vectors

$$\hat{e}_2 \equiv \hat{e}_3 \times \hat{e}_1 = (e_{3y}e_{1z} - e_{3z}e_{1y})\hat{\imath} + (e_{3z}e_{1x} - e_{3x}e_{1z})\hat{\jmath} + (e_{3x}e_{1y} - e_{3y}e_{1x})\hat{k}$$

This completes the specification of the shell triad.

With reference to Figure 3.16(b), we can define the 1-axis of the frame element as coinciding with the member axis

$$\hat{e}_1 = [(x_2 - x_1)\hat{\imath} + (y_2 - y_1)\hat{\jmath} + (z_2 - z_1)\hat{k}]/L = \frac{\hat{x}_{21} + \hat{u}_{21}}{L}$$

The 2- and 3-axes, however, need only be perpendicular to this and otherwise can be arbitrary. Consider the member to be bending and twisting, we could imagine therefore that the orientation of the nodal triads (while still coinciding with the principal directions) no longer coincide with each other. A reasonable assumption is to say that the member orientation lies somewhere as an average orientation of the ends. We will initially use this assumption.

First compute the rotation matrix and the angles between the ends

$$[\{e\}_1\{e\}_2\{e\}_3]_2 = [\,R\,][\{e\}_1\{e\}_2\{e\}_3]_1 \qquad \text{or} \qquad [\,e_2\,] = [\,R\,][\,e_1\,]$$

Knowing the two triads, we therefore obtain the rotation matrix as

$$[\,R\,] = [\,e_2\,][\,e_1\,]^T \qquad \text{or} \qquad R_{ij} = \sum_k e_{2ik}e_{1jk}$$

from which, according to Equations (3.6) and (3.7), we obtain the angle of rotation.

We now say that the average orientation is half of these component angles. Such a triad, however, will not have its 1-axis coinciding with the member axis, but as shown in the last subsection, we can do a smallest angle rotation so that the 1-axis does indeed coincide.

Let $\hat{a}$ be the average triad, and let the relative rotation between element nodes be small, then the element orientation is given as

$$
\begin{aligned}
\hat{a}_1 &\longrightarrow & \hat{e}_1 &= \hat{e}_1 \\
\hat{a}_2 &\longrightarrow & \hat{e}_2 &= \hat{a}_2 - \tfrac{1}{2}(\hat{a}_2 \cdot \hat{e}_1)[\hat{a}_1 + \hat{e}_1] \\
\hat{a}_3 &\longrightarrow & \hat{e}_3 &= \hat{a}_3 - \tfrac{1}{2}(\hat{a}_3 \cdot \hat{e}_1)[\hat{a}_1 + \hat{e}_1]
\end{aligned}
$$

While these triads are accurate, they lead to rather complicated expressions. Therefore, we opt instead to choose an alternative approximate set.

There are multiple choices for the triad $\hat{a}_i$; we will choose the triad at the first node, $\hat{n}_i^1$. Then, since $\hat{n}_1^1$ and $\hat{e}_1$ are nearly collinear, we have the approximation

$$
\begin{aligned}
\hat{e}_2 &= \hat{n}_2^1 - (\hat{n}_2^1 \cdot \hat{e}_1)\hat{e}_1 \\
\hat{e}_3 &= \hat{n}_3^1 - (\hat{n}_3^1 \cdot \hat{e}_1)\hat{e}_1
\end{aligned}
$$

These vectors are not of unit size, since, for example,

$$
\hat{e}_2 \cdot \hat{e}_2 = \hat{n}_2^1 \cdot \hat{n}_2^1 - 2(\hat{n}_2^1 \cdot \hat{e}_1)(\hat{n}_2^1 \cdot \hat{e}_1) + (\hat{n}_2^1 \cdot \hat{e}_1)^2(\hat{e}_1 \cdot \hat{e}_1) = 1 - (\hat{n}_2^1 \cdot \hat{e}_1)^2
$$

However, they are not very different from unity for small elements. To show this, consider the 2-D case where

$$
\{e_1\} = \{\cos\theta,\ \sin\theta,\ 0\}, \qquad \{n_2\} = \{-\sin\phi,\ \cos\phi,\ 0\}
$$

then

$$
\hat{e}_2 \cdot \hat{e}_2 = 1 - \sin^2(\phi_1 - \theta) = 1 - \sin^2(\bar{\phi}_1) \approx 1 - \bar{\phi}_1^2 \approx 1
$$

The approximation is based on the element being small, because then the relative twisting, $\bar{\phi}$, is also small. This is an approximation we will make multiple times in the following developments. The element triad has the following properties:

$$
\hat{e}_2 \cdot \hat{e}_1 = 0, \qquad \hat{e}_3 \cdot \hat{e}_1 = 0, \qquad \hat{e}_2 \cdot \hat{e}_3 = -(\hat{n}_2^1 \cdot \hat{e}_1)(\hat{n}_3^1 \cdot \hat{e}_1) \approx 0
$$

and are thus approximately orthogonal.

To describe the local deformation of the element (shell or frame), we need the relative twists between the node and element orientations. These local angles of twist are given in local coordinates (referred to the element) as

$$
\begin{aligned}
2\bar{\phi}_{jx} &= \hat{e}_3 \cdot \hat{n}_2^j - \hat{e}_2 \cdot \hat{n}_3^j \\
2\bar{\phi}_{jy} &= \hat{e}_1 \cdot \hat{n}_3^j - \hat{e}_3 \cdot \hat{n}_1^j \\
2\bar{\phi}_{jz} &= \hat{e}_2 \cdot \hat{n}_1^j - \hat{e}_1 \cdot \hat{n}_2^j
\end{aligned} \tag{3.9}
$$

for the j node — note that there is a relative twist at each node. This used the approximation that $\bar{\phi}/\sin\bar{\phi} \approx 1$, which is reasonable for the small strains of interest.

Example 3.38: Determine the element triad for a flat triangular shell element with the following coordinates:

$$
\begin{array}{cccc}
\text{node} & x & y & z \\
1: & 1 & 1 & 1 \\
2: & 2 & 1 & 2 \\
3: & 2 & 2 & 0
\end{array}
$$

Referring to Figure 3.16(a), the vectors of the two sides are

$$
\begin{aligned}
\hat{v}_{12} &= (2-1)\hat{\imath} + (1-1)\hat{\jmath} + (2-1)\hat{k} = 1\hat{\imath} + 0\hat{\jmath} + 1\hat{k} \\
\hat{v}_{13} &= (2-1)\hat{\imath} + (2-1)\hat{\jmath} + (0-1)\hat{k} = 1\hat{\imath} + 1\hat{\jmath} - 1\hat{k}
\end{aligned}
$$

The cross-product of these gives

$$
\hat{A} = -1\hat{\imath} + 2\hat{\jmath} + 1\hat{k}, \qquad A = \sqrt{6}
$$

Two vectors of the triad are given as

$$
\hat{e}_1 = \frac{\hat{v}_{12}}{|\hat{v}_{12}|} = \frac{1}{\sqrt{2}}\hat{\imath} + 0\hat{\jmath} + \frac{1}{\sqrt{2}}\hat{k}, \qquad \hat{e}_3 = \frac{\hat{A}}{|\hat{A}|} = \frac{-1}{\sqrt{6}}\hat{\imath} + \frac{2}{\sqrt{6}}\hat{\jmath} + \frac{1}{\sqrt{6}}\hat{k}
$$

The remaining vector is given as the cross-product

$$
\hat{e}_2 = \hat{e}_3 \times \hat{e}_1 = \frac{2}{\sqrt{6}}\hat{\imath} + \frac{2}{\sqrt{6}}\hat{\jmath} - \frac{2}{\sqrt{3}}\hat{k}
$$

The triad is now

$$
[\,e\,] = \frac{1}{\sqrt{6}}
\begin{bmatrix}
\sqrt{3} & \sqrt{2} & -1 \\
0 & \sqrt{2} & 2 \\
\sqrt{3} & -\sqrt{2} & 1
\end{bmatrix}
$$

These results can be easily verified by drawing the vectors and the element.

3.3 Solving Nonlinear Equations

All nonlinear problems are solved in an incremental/iterative manner with some sort of linearization done at each time or load step. In this section, using a simple truss as an example, we develop the basics of the method.

This also gives us an opportunity to introduce some notations that we will utilize in the later sections.

Incremental Solution Scheme

We formulate the solution in an incremental fashion. That is, we view the deformation as occurring in a sequence of steps associated with time increments Δt, and at each step it is the increment of displacements that are considered to be the unknowns.

To help fix ideas, look again at the truss in Figure 3.2. We have already shown that the equilibrium equation is

$$\begin{Bmatrix} 0 \\ 0 \end{Bmatrix} = \begin{Bmatrix} P_{1x} \\ P_{2y} \end{Bmatrix} + \begin{Bmatrix} \bar{F}_o \\ -\beta \bar{F}_o \end{Bmatrix} - \begin{Bmatrix} 0 \\ K_s v_2 \end{Bmatrix} \qquad \text{or} \qquad \{\mathcal{F}\} = \{P\} - \{F\} = \{0\}$$

$$(3.10)$$

and the axial force is the nonlinear function of the deformation

$$\bar{F}_o = EA \left[-\frac{u_1}{L_o} + \frac{h}{L_o}\frac{v_2}{L_o} + \tfrac{1}{2}\left(\frac{v_2}{L_o}\right)^2 \right]$$

Consider the equilibrium equation at time step t_{n+1}

$$\{\mathcal{F}\}_{n+1} = \{P\}_{n+1} - \{F(u)\}_{n+1} = \{0\}$$

We do not know the displacements $\{u\}_{n+1}$, hence we cannot compute the axial force $\bar{F}_o$ nor the nodal forces $\{F\}_{n+1}$. As is usual in such nonlinear problems, we linearize about a known state. That is, assume we know everything at time step t_n, then write the Taylor series approximation for the element nodal forces

$$\{F(u)\}_{n+1} \approx \{F(u)\}_n + [\frac{\partial F}{\partial u}]_n\{\Delta u\} + \cdots = \{F(u)\}_n + [K_T]_n\{\Delta u\} + \cdots \quad (3.11)$$

The square matrix $[K_T]$ is called the *tangent stiffness matrix*. The explicit form it takes for our truss problem is

$$[K_T]_n = [\frac{\partial F}{\partial u}]_n = \begin{bmatrix} \dfrac{\partial F_{1x}}{\partial u_1} & \dfrac{\partial F_{1x}}{\partial v_2} \\ \dfrac{\partial F_{2x}}{\partial u_1} & \dfrac{\partial F_{2x}}{\partial v_2} \end{bmatrix}_n = \begin{bmatrix} \dfrac{\partial \bar{F}_o}{\partial u_1} & -\dfrac{\partial \bar{F}_o}{\partial v_2} \\ \beta\dfrac{\partial \bar{F}_o}{\partial u_1} & \dfrac{1}{L_o}\bar{F}_o + \beta\dfrac{\partial \bar{F}_o}{\partial u_2} + K_s \end{bmatrix}_n$$

Performing the differentiations

$$\frac{\partial \bar{F}_o}{\partial u_1} = EA\left[-\frac{1}{L_o} \right], \qquad \frac{\partial \bar{F}_o}{\partial v_2} = EA\left[\frac{h}{L_o^2} + \frac{v_2}{L_o^2} \right]$$

then leads to the stiffness

$$[K_T] = [\frac{\partial F}{\partial u}] = \frac{EA}{L_o}\begin{bmatrix} 1 & -\beta \\ -\beta & \beta^2 + \dfrac{K_s L_o}{EA} \end{bmatrix} + \frac{\bar{F}_o}{L_o}\begin{bmatrix} 0 & 0 \\ 0 & 1 \end{bmatrix} = [K_E] + [K_G]$$

Note that both matrices are symmetric. The first matrix is the elastic stiffness — the elastic stiffness of a truss member oriented slightly off the horizontal by the angle $\beta = (h + v_2)/L_o$. The second matrix is called the *initial stress matrix* because it depends on the axial load $\bar{F}_o$. It is also called the *geometric stiffness matrix* because it arises due to the rotation of the member — it is this latter designation that we will adopt.

P/P_{cr}	u_1/u_{cr}	$(h+v_2)/u_{cr}$	P	$\mathcal{F}_{1x}$	$\mathcal{F}_{2y}$
.0500	.051000	10.5000	100.00	.050003	.989497E-02
.1500	.153401	11.6315	300.00	.256073	.446776E-01
.2500	.257022	13.1329	500.00	.450836	.589465E-01
.3500	.362372	15.0838	700.00	.761169	.759158E-01
.4500	.470656	17.7037	900.00	1.37274	.100333
.5500	.584416	21.3961	1100.0	2.72668	.137041
.6500	.709589	26.9600	1300.0	6.19165	.192208
.7500	.862467	36.1926	1500.0	17.0481	.257334
.8500	1.09898	53.8742	1700.0	62.5275	.938249E-01
.9500	1.66455	94.1817	1900.0	324.939	-4.00430
1.050	3.23104	163.411	2100.0	958.535	-24.0588

Table 3.1: Incremental results using simple stepping.

We are now in a position to solve for the increments of displacement; re-arrange the approximate equilibrium equation into a loading equation as

$$\{P\}_{n+1} - \{F\}_n - [K_T]\{\Delta u\} \approx 0 \qquad \Longrightarrow \qquad [K_T]\{\Delta u\} = \{P\}_{n+1} - \{F\}_n$$

Again, consider the special case when $P_{1x} = P$, $P_{2y} = 0$; then the system of equations to be solved is

$$\left[\frac{EA}{L_o} \begin{bmatrix} 1 & -\beta \\ -\beta & \beta^2 + \gamma \end{bmatrix} + \frac{\bar{F}_o}{L_o} \begin{bmatrix} 0 & 0 \\ 0 & 1 \end{bmatrix} \right]_n \left\{ \begin{matrix} \Delta u_1 \\ \Delta v_2 \end{matrix} \right\} = \left\{ \begin{matrix} P \\ 0 \end{matrix} \right\}_{n+1} - \left\{ \begin{matrix} -\bar{F}_o \\ -\beta \bar{F}_o + K_s v_2 \end{matrix} \right\}_n$$

with $\gamma = K_s L_o / EA$. A simple solution scheme, therefore, involves computing the increments at each step and updating the displacements as

$$u_{1(n+1)} = u_{1(n)} + \Delta u_1 , \qquad v_{2(n+1)} = v_{2(n)} + \Delta v_2$$

The axial force and orientation β also need to be updated as

$$\bar{F}_o|_{n+1} = EA \left[-\frac{u_1}{L_o} + \frac{h}{L_o} \frac{v_2}{L_o} + \frac{1}{2} \left(\frac{v_2}{L_o} \right)^2 \right]_{n+1} , \qquad \beta_{n+1} = \frac{h+v_2}{L_o} \bigg|_{n+1}$$

Table 3.1 and Figure 3.3 show the results using this simple stepping scheme, where $P_{cr} = K_s L_o$ and $u_{cr} = P_{cr}/EA$.

Table 3.1 also shows the out-of-balance force

$$\{\mathcal{F}\}_{n+1} = \{P\}_{n+1} - \{F\}_{n+1}$$

computed at the end of each step. Clearly, nodal equilibrium is not being satisfied and it deteriorates as the load increases. In order for this simple scheme to give reasonable results, it is necessary that the increments be small. This can be computationally prohibitive for large systems, because, at each step, the tangent stiffness must be formed and decomposed. A more-refined incremental version that uses an iterative scheme to enforce nodal equilibrium will now be developed.

i	u_1	v_2	$u_1 - u_{1ex}$	$v_2 - v_{2ex}$	$\mathcal{F}_{1x}$	$\mathcal{F}_{2y}$
1	-3.02304	76.0009	-3.12184	72.2009	.3207E+7	-.244E+7
2	29.0500	75.9986	28.9512	72.1986	.21923	-72.3657
3	-25.8441	3.95793	-25.9429	.157933	.2594E+7	-107896.
4	.105242	3.95793	.64416E-2	.157926	.6835E-2	-.158213
5	.09867	3.80001	-.12436E-3	.8344E-5	12.4694	-.498780
6	.09880	3.80000	.15646E-6	.4053E-5	.000000	-.3948E-5
7	.09880	3.80000	.000000	.000000	-.4882E-3	.1878E-4
exact	.09880	3.80000				

Table 3.2: Newton-Raphson iterations for a load step $0.95\,P_{cr}$.

Newton-Raphson Iterations

The increments in displacement are obtained by solving

$$\{K_T\}_n\{\Delta u\} = \{P\}_{n+1} - \{F\}_n$$

from which an estimate of the displacements is obtained as

$$\{u\}_{n+1} \approx \{u\}_n + \{\Delta u\}$$

As was just pointed out, if these estimates for the new displacements are substituted into Equation (3.10), then this equation will not be satisfied, because the displacements were obtained using only an approximation of the nodal forces given by Equation (3.11). What we can do, however, is repeat the above process at the same applied load level until we get convergence. That is, we repeat

$$\textbf{solve:} \qquad \{K_T\}_{n+1}^{i-1}\{\Delta u\}^i = \{P\}_{n+1} - \{F\}_{n+1}^{i-1}$$

$$\textbf{update:} \qquad \{u\}_{n+1}^i = \{u\}_{n+1}^{i-1} + \{\Delta u\}^i$$

$$\textbf{update:} \qquad \{K_T\}_{n+1}^i, \qquad \{F\}_{n+1}^i$$

until $\{\Delta u\}^i$ becomes less than some tolerance value. In this, i is the iteration counter. The iteration process is started (at each increment) using the starter values

$$\{u\}_{n+1}^0 = \{u\}_n, \qquad \{K_T\}_{n+1}^0 = \{K_T\}_n, \qquad \{F\}_{n+1}^0 = \{F\}_n$$

This basic algorithm is known as the *full Newton-Raphson method*.

Combined incremental and iterative results are given in Figure 3.3. We see that it gives the exact solution. Iteration results for a load level equal $0.95\,P_{cr}$ are given in Table 3.2, where the initial guesses correspond to the linear elastic solution. We see that convergence is quite rapid and the out-of-balance forces go to zero.

It is worth pointing out the converged value above P_{cr} in Figure 3.3; this corresponds to a vertical deflection where the truss has "flipped" over to the negative side. Such a situation would not occur physically, but does occur here due to a combination of linearizing the problem (i.e., the small angle approximation) and the nature of the iteration process (i.e., no restriction is placed on the size of the iterative increments).

Nonlinear Algorithm

In the following, we concentrate on the basic algorithm for the full Newton-Raphson method because it best illustrates the essential ingredients. This algorithm, for monotonically increasing loads, can be stated as:

Step 1: Specify parameters of the algorithm such as tolerances, and maximum iterations.

Step 2: Read the initial geometry and material properties.

Step 3: Specify load increments, number of steps.

Step 4: *Begin loop over time (load) increments:*

Step T.1: Increment the load vector $\{P\}_{t+\Delta t}$.

Step T.2: Initialize for equilibrium iterations

$$\{u\}^0_{t+\Delta t} = \{u\}_t , \qquad [K_T]^0_{t+\Delta t} = [K_T]_t , \qquad \{F\}^0_{t+\Delta t} = \{F\}_t$$

Step T.3: *Begin loop over iterations:*

Step I.1: ITERATE:

Step I.2: Assemble nodal force vector $\{F\}^i$.

Step I.3: Form the effective load vector

$$\{\Delta P_{eff}\}^i_{t+\Delta t} \equiv \{P\}_{t+\Delta t} - \{F\}^i_{t+\Delta t}$$

Step I.4: Test norm of effective load vector

$$\text{if} \quad |\{\Delta P_{eff}\}^i|/|\{P\}| > 1000 \quad \text{unstable, goto END}$$

Step I.5: Assemble the elastic stiffness matrix $[K_E]$.

Step I.6: Assemble the geometric stiffness matrix $[K_G]$.

Step I.7: Form the tangent stiffness matrix as

$$[K_T] = [K_E] + \gamma[K_G]$$

Step I.8: Decompose the tangent stiffness to

$$[K_T] = [\ U\ ]^T \lceil\ D\ \rfloor [\ U\ ]$$

Step I.9: Solve for the new displacement increments from

$$[\ U\ ]^T \lceil\ D\ \rfloor [\ U\ ]\{\Delta u\}^i = \{\Delta P_{eff}\}^i_{t+\Delta t}$$

Step I.10: Update the displacements

$$\{u\}^i_{t+\Delta t} = \{u\}^{i-1}_{t+\Delta t} + \beta\{\Delta u\}^i$$

Step I.11: Test for convergence.

if: $|\{\Delta u\}^i|/|\{u\}^i| < tol$ converged, goto UPDATE

if: $|\{\Delta u\}^i|/|\{u\}^i| > tol$ not converged, goto ITERATE

Step T.4: *End loop over iterations.*

Step T.5: UPDATE:

$$u_{t+\Delta t} = u_{t+\Delta t}^i , \qquad xyz_{t+\Delta t} = xyz_{t+\Delta t}^i$$

Step T.6: Store results for this time step.

Step T.7: If maximum load not exceeded continue looping over loads.

Step 5: *End loop over time (load) increments.*

Step 6: END

It is possible to enhance this algorithm by including automatic step changes, automatic testing for appropriate time step size, and monitoring the spectral properties of the tangent stiffness. The parameters β and γ can also be adjusted automatically.

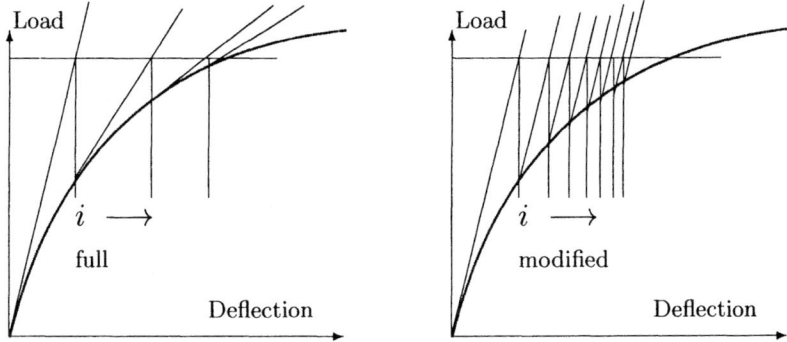

Figure 3.17: Full and modified Newton-Raphson methods.

The full Newton-Raphson method has the disadvantage that, during each iteration, the tangent stiffness matrix must be formed and decomposed. The cost of this can be quite prohibitive for large systems. Thus, effectively, the computational cost is like that of the incremental solution with many steps. It must be realized, however, that because of the quadratic convergence, six Newton-Raphson iterations, say, are more effective than six load increments.

The *modified Newton-Raphson method* is basically as above except that the tangent stiffness is not updated during the iterations but only after each load increment. This generally requires more iterations and sometimes is less stable but it is less computationally costly.

Both schemes are illustrated in Figure 3.17 where the starting point is from the zero load state. It is clear why the modified method will take more iterations. The plot for the modified method has the surprising implication that we

do not need to know the actual tangent stiffness in order to compute correct
results — this seems at odds with the previous chapter where care was taken
in order to derive good quality stiffness matrices. What must be realized in the
incremental/iterative scheme is that we are imposing equilibrium (iteratively)
in terms of the applied loads and resultant nodal forces; we need good quality
element stiffness matrices in order to get the good quality element nodal forces,
but the assembled tangent stiffness matrix is used only to suggest a direction for
the iterative increments. To get the correct converged results we need to have
good element stiffness relations, but not necessarily a good assembled tangent
stiffness matrix. Clearly, however, a good quality tangent stiffness will give more
rapid convergence as well as increase the radius of convergence. We look at this
again in some of the examples.

3.4   Total Lagrangian Incremental Formulation

In this section, we develop a total Lagrangian incremental formulation. There
are many other formulations, indeed for our 3-D thin-walled structures, we will
use a corotational scheme, but the present scheme is instructive in showing the
construction of the tangent stiffness matrix. It also gives us a comparison by
which to judge the corotational scheme.

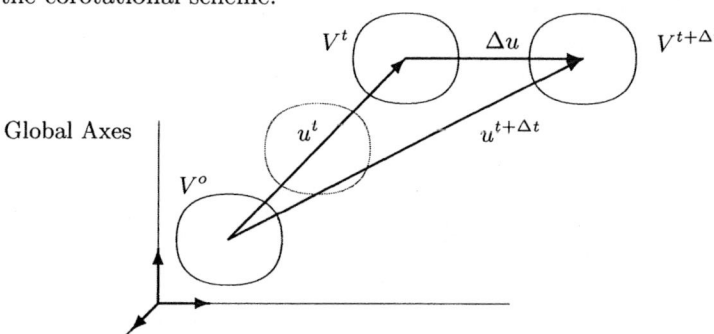

Figure 3.18: Decomposition of displacement.

Increments of Stress and Strain

With reference to Figure 3.18, consider the displacements u_i^t to be known at
time t. We decompose the deformation at the next time increment as

$$u_i^{t+\Delta t} = u_i^t + \Delta u_i$$

where Δu_i is the increment of displacement from the current value of u_i^t. The
increments Δu_i are the basic unknowns in the present formulation where it is assumed that everything at time t is known. Using this in the strain/displacement

relation allows the strain (at the next time) to be decomposed as

$$
\begin{aligned}
2E_{ij}^{t+\Delta t} &= \left(\frac{\partial u_i}{\partial x_j^o} + \frac{\partial u_j}{\partial x_i^o} + \sum_k \frac{\partial u_k}{\partial x_i^o} \frac{\partial u_k}{\partial x_j^o} \right)^{t+\Delta t} \\
&= \left(\frac{\partial u_i^t}{\partial x_j^o} + \frac{\partial \Delta u_i}{\partial x_j^o} \right) + \left(\frac{\partial u_j^t}{\partial x_i^o} + \frac{\partial \Delta u_j}{\partial x_i^o} \right) + \sum_k \left(\frac{\partial u_k^t}{\partial x_i} + \frac{\partial \Delta u_k}{\partial x_i} \right) \left(\frac{\partial u_k^t}{\partial x_j} + \frac{\partial \Delta u_k}{\partial x_j} \right) \\
&= \left(\frac{\partial u_i^t}{\partial x_j^o} + \frac{\partial u_j^t}{\partial x_i^o} + \sum_k \frac{\partial u_k^t}{\partial x_i^o} \frac{\partial u_k^t}{\partial x_j^o} \right) + \\
&\quad \left(\frac{\partial \Delta u_i}{\partial x_j^o} + \frac{\partial \Delta u_j}{\partial x_i^o} + \sum_k \frac{\partial u_k^t}{\partial x_i^o} \frac{\partial \Delta u_k}{\partial x_j^o} + \sum_k \frac{\partial \Delta u_k}{\partial x_i^o} \frac{\partial u_k^t}{\partial x_j^o} \right) + \sum_k \left(\frac{\partial \Delta u_k}{\partial x_i^o} \frac{\partial \Delta u_k}{\partial x_j^o} \right)
\end{aligned}
$$

The various collections of terms in parentheses are labeled as follows:

$$
E_{ij}^{t+\Delta t} = E_{ij}^t + \Delta E_{ij} + \eta_{ij}
$$

Note that ΔE_{ij} is an increment of strain from the current configuration but referenced to the zero configuration. It is linear in Δu_i but also contains components of the current displacement u_i^t. The wholly nonlinear term η_{ij} contains only the unknown displacement increment.

Let the constitutive relation at time t be written as

$$
\sigma_{ij}^{Kt} = \sum_{k,l} C_{ijkl}^t E_{kl}^t
$$

where C_{ijkl} is the (possibly) nonlinear anisotropic material tensor. The corresponding relation at time $t + \Delta t$ can be expanded as

$$
\begin{aligned}
\sigma_{ij}^{Kt+\Delta t} &= \sum_{k,l} C_{ijkl}^{t+\Delta t} E^{t+\Delta t} \\
&= \sum_{k,l} C_{ijkl}^{t+\Delta t} E_{kl}^t + \sum_{k,l} C_{ijkl}^{t+\Delta t} (\Delta E_{kl} + \eta_{kl}) \\
&\approx \sigma_{ij}^{Kt} + \Delta \sigma_{ij}^K
\end{aligned}
$$

We can thus view the next value of stress as the current stress value plus an increment.

Equation of Motion

We begin with the equations of motion, Equation (1.20), written at the next time

$$
\sum_{i,j} \int_{V^o} \sigma_{ij}^{t+\Delta t} \delta E_{ij}^{t+\Delta t} dV^o = \delta W_e^{t+\Delta t}
$$

In going from $t \to t + \Delta t$, we have that

$$
\delta E_{ij}^{t+\Delta t} = \delta E_{ij}^t + \delta[\Delta E_{ij} + \eta_{ik}] = \delta[\Delta E_{ij} + \eta_{ik}]
$$

since the strain at the current time, E_{ij}^t, is not varied. Hence, the equations of motion can be expanded as

$$\sum_{i,j,k,l} \int [\sigma_{ij}^{Kt} + C_{ijkl}^{t+\Delta t}(\Delta E_{kl} + \eta_{kl})]\delta[\Delta E_{ij} + \eta_{ij}]dV^o = \delta W_e^{t+\Delta t}$$

This can be further rearranged as

$$\sum_{i,j,k,l} \int C_{ijkl}^{t+\Delta t}(\Delta E_{kl} + \eta_{kl})\delta(\Delta E_{ij} + \eta_{ij})dV^o + \sum_{i,j} \int \sigma_{ij}^{Kt}\delta\eta_{ij}dV^o$$

$$= \delta W_e^{t+\Delta t} - \sum_{i,j} \int_{V^o} \sigma_{ij}^{Kt}\delta\Delta E_{ij}dV^o$$

These are the equations of motion in terms of the increment of displacement Δu_i where the right-hand side is essentially an increment of load.

The linearized version of these equations assumes that $\eta_{ij} \ll \Delta E_{ij}$ and leads to

$$\sum_{i,j,k,l} \int C_{ijkl}\Delta E_{kl}\delta\Delta E_{ij}dV^o + \sum_{i,j} \int \sigma_{ij}^{Kt}\delta\eta_{ij}dV^o = \delta W_e^{t+\Delta t} - \sum_{i,j} \int \sigma_{ij}^{Kt}\delta\Delta E_{ij}\,dV^o$$

Note that we still retain η_{ij} in the second term because (being quadratic in Δu_i) it leads essentially to $\Delta u_i \delta\Delta u_i$, which is linear in the unknown.

Stiffness Relations

The final step is to write the equations in the form of stiffness relations.

The basic idea is that, since the actual distribution of displacements is quite complicated, we will approximate it as a collection of piecewise simple regions as was done in the previous chapter. That is, let the displacement increments in a small region of volume V^o be represented by

$$\Delta u_i(x_1^o, x_2^o, x_3^o) = \sum_k h_k(x_1^o, x_2^o, x_3^o)\Delta U_{ik} = [\ h\ ]\{\Delta U_i\} \quad \text{or} \quad \{\Delta u\} = [H]\{\Delta U\}$$

where $h_k(x_1^o, x_2^o, x_3^o)$ are known shape functions and ΔU_{ik} are the unknown nodal values. All relevant quantities can now be written in terms of both of these. For example, the derivatives are given by

$$\frac{\partial\Delta u_i}{\partial x_j^o} = \sum_k \frac{\partial h_k}{\partial x_j^o}\Delta U_{ik} = [h']\{\Delta U\}$$

Hence the strain increments can be written symbolically as

$$\{\Delta E\} = [\ B\ ]\{\Delta U\}$$

where $[\,B\,]$ contain various spatial derivatives of h_k.

Substituting this representation for the displacement and strain increment into the linearized form of the equations of motion gives the following matrix relations. The linear or elastic contribution is

$$\sum_{i,j,r,s} \int_{V^o} C_{ijrs}\Delta E_{rs}\delta\Delta E_{ij}dV^o \quad\Rightarrow\quad [K_E^t]\{\Delta U\} = \Big[\int_{V^o}[B]^T[C][B]dV^o\Big]\{\Delta U\}$$

The nonlinear or geometric contribution is

$$\sum_{i,j} \int_{V^o} \sigma_{ij}^K \delta\eta_{ij}dV^o \quad\Rightarrow\quad [K_G^t]\{\Delta U\} = \Big[\int_{V^o}[B_G]^T\{\sigma^K\}[B_G]dV^o\Big]\{\Delta U\}$$

and the carry-over load term is

$$\sum_{i,j} \int_{V^o} \sigma_{ij}^K \delta\Delta E_{ij}dV^o \quad\Rightarrow\quad \{F\}^t = \Big\{\int_{V^o}[B]^T\{\sigma^K\}dV^o\Big\}$$

The virtual work of the applied loads leads to

$$\sum_i \int f_i\delta u_i\, dA^o \quad\Rightarrow\quad \{P\} = \Big\{\int_{A^o}[H]^T\{f\}dA^o\Big\}$$

Assemblage is done as in the linear case, and these give the stiffness relation for the increment of displacement as

$$[K_E^t + K_G^t]\{\Delta U\} = \{P\}^{t+\Delta t} - \{F\}^t$$

where $\{P\}$ are the externally applied loads. This equation is now solved in the usual fashion to obtain the nodal values of displacement increment. From this, all values of displacement, strain, and stress can be updated and then proceed to the next increment. However, this relation needs to be iterated until the internal stresses are in balance with the applied loads.

The explicit forms of the stiffness matrices depend on the particular forms chosen for the shape functions $h_k(x_1^o, x_2^o, x_3^o)$, which are also related to how the region is discretized into the smaller simpler regions.

Example 3.39: Specialize the total Lagrangian scheme for the two-dimensional linear displacement triangle.

The basic assumption in the formulation is that the displacements have the same description as the coordinates. That is,

$$x(x^o,y^o) = \sum_{i=1}^{3} h_i(x^o,y^o)x_i\,, \qquad y(x^o,y^o) = \sum_{i=1}^{3} h_i(x^o,y^o)y_i$$

$$u(x^o,y^o) = \sum_{i=1}^{3} h_i(x^o,y^o)u_i\,, \qquad v(x^o,y^o) = \sum_{i=1}^{3} h_i(x^o,y^o)v_i$$

$$\Delta u(x^o,y^o) = \sum_{i=1}^{3} h_i(x^o,y^o)\Delta u_i\,, \qquad \Delta v(x^o,y^o) = \sum_{i=1}^{3} h_i(x^o,y^o)\Delta v_i$$

where

$$\left\{ \begin{array}{c} h_1 \\ h_2 \\ h_3 \end{array} \right\} = \frac{1}{2A} \left[\begin{array}{ccc} x_2 y_3 - x_3 y_2 & y_{23} & x_{32} \\ x_3 y_1 - x_1 y_3 & y_{31} & x_{13} \\ x_1 y_2 - x_2 y_1 & y_{12} & x_{21} \end{array} \right] \left\{ \begin{array}{c} 1 \\ x \\ y \end{array} \right\} = \frac{1}{2A} \left[\begin{array}{ccc} a_1 & b_1 & c_1 \\ a_2 & b_2 & c_2 \\ a_3 & b_3 & c_3 \end{array} \right] \left\{ \begin{array}{c} 1 \\ x \\ y \end{array} \right\}$$

with $2A \equiv x_{21} y_{31} - x_{31} y_{21}$, $x_{ij} \equiv x_i - x_j$, and so on, exactly as used in Chapter 2. The displacement gradients can therefore be computed as

$$\frac{\partial u}{\partial x^o} = \sum_i \frac{\partial h_i}{\partial x^o} = \frac{1}{2A} \sum_i b_i u_i , \qquad \frac{\partial v}{\partial x^o} = \sum_i \frac{\partial h_i}{\partial x^o} = \frac{1}{2A} \sum_i c_i v_i$$

$$\frac{\partial u}{\partial y^o} = \sum_i \frac{\partial h_i}{\partial y^o} = \frac{1}{2A} \sum_i c_i u_i , \qquad \frac{\partial v}{\partial y^o} = \sum_i \frac{\partial h_i}{\partial y^o} = \frac{1}{2A} \sum_i c_i v_i$$

We express these in the matrix form

$$\left\{ \begin{array}{c} u_{,x} \\ u_{,y} \\ v_{,x} \\ v_{,y} \end{array} \right\} = \frac{1}{2A} \left[\begin{array}{cccccc} b_1 & 0 & b_2 & 0 & b_3 & 0 \\ c_1 & 0 & c_2 & 0 & c_3 & 0 \\ 0 & b_1 & 0 & b_2 & 0 & b_3 \\ 0 & c_1 & 0 & c_2 & 0 & c_3 \end{array} \right] \left\{ \begin{array}{c} u_1 \\ v_1 \\ u_2 \\ v_2 \\ u_3 \\ v_3 \end{array} \right\} \qquad \text{or} \quad \{u_{,x}\} = [B_D]\{u\}$$

where the comma indicates partial differentiation with respect to the subscripted variable.

With knowledge of the nodal displacements, we can get the displacement gradients and from these we can get the strains by

$$E_{xx} = \frac{\partial u}{\partial x^o} + \tfrac{1}{2} \left(\frac{\partial u}{\partial x^o}\right)^2 + \tfrac{1}{2} \left(\frac{\partial v}{\partial x^o}\right)^2$$

$$E_{yy} = \frac{\partial v}{\partial y^o} + \tfrac{1}{2} \left(\frac{\partial u}{\partial y^o}\right)^2 + \tfrac{1}{2} \left(\frac{\partial v}{\partial y^o}\right)^2$$

$$2E_{xy} = \frac{\partial u}{\partial y^o} + \frac{\partial v}{\partial x^o} + \frac{\partial u}{\partial x^o}\frac{\partial u}{\partial y^o} + \frac{\partial v}{\partial x^o}\frac{\partial v}{\partial y^o}$$

For definiteness, let the Kirchhoff stresses be related to the Lagrangian strains by

$$\left\{ \begin{array}{c} \sigma_{xx}^K \\ \sigma_{yy}^K \\ \sigma_{xy}^K \end{array} \right\} = \frac{E}{1 - \nu^2} \left[\begin{array}{ccc} 1 & \nu & 0 \\ \nu & 1 & 0 \\ 0 & 0 & (1-\nu)/2 \end{array} \right] \left\{ \begin{array}{c} E_{xx} \\ E_{yy} \\ 2E_{xy} \end{array} \right\} \qquad \text{or} \quad \{\sigma^K\} = [\,D\,]\{E\}$$

Note that, although this relation is linear, it takes the large rotations into account. The Cauchy stresses are obtained from the Kirchhoff stresses by

$$\sigma_{xx} = \frac{\rho^o}{\rho} \left[\left(1 + \frac{\partial u}{\partial x^o}\right)^2 \sigma_{xx}^K + 2\left(1 + \frac{\partial u}{\partial x^o}\right)\frac{\partial u}{\partial y^o}\sigma_{xy}^K + \left(\frac{\partial u}{\partial y^o}\right)^2 \sigma_{yy}^K \right]$$

$$\sigma_{yy} = \frac{\rho^o}{\rho} \left[\left(\frac{\partial v}{\partial x^o}\right)^2 \sigma_{xx}^K + 2\left(1 + \frac{\partial v}{\partial y^o}\right)\frac{\partial v}{\partial x^o}\sigma_{xy}^K + \left(1 + \frac{\partial v}{\partial y^o}\right)^2 \sigma_{yy}^K \right]$$

$$\sigma_{xy} = \frac{\rho^o}{\rho} \left[\left(1 + \frac{\partial u}{\partial x^o}\right)\frac{\partial v}{\partial x^o}\sigma_{xx}^K + \left(\frac{\partial u}{\partial y^o}\frac{\partial v}{\partial x^o} + \left(1 + \frac{\partial u}{\partial x^o}\right)\left(1 + \frac{\partial v}{\partial y^o}\right)\right)\sigma_{xy}^K \right.$$

$$\left. + \frac{\partial u}{\partial y^o}\left(1 + \frac{\partial v}{\partial x^o}\right)\sigma_{yy}^K \right]$$

Once we know the nodal displacements, we can determine all other quantities of interest.

The increment of strains can be written as

$$2\Delta E_{ij} = \left(\frac{\partial \Delta u_i}{\partial x_j^o} + \frac{\partial \Delta u_j}{\partial x_i^o} + \sum_k \frac{\partial u_k^t}{\partial x_i^o}\frac{\partial \Delta u_k}{\partial x_j^o} + \sum_k \frac{\partial \Delta u_k}{\partial x_i^o}\frac{\partial u_k^t}{\partial x_j^o}\right)$$

This expands out to

$$\Delta E_{xx} = \frac{\partial \Delta u}{\partial x^o} + \frac{\partial u^t}{\partial x^o}\frac{\partial \Delta u}{\partial x^o} + \frac{\partial v^t}{\partial x^o}\frac{\partial \Delta v}{\partial x^o}$$

$$\Delta E_{yy} = \frac{\partial \Delta v}{\partial y^o} + \frac{\partial u^t}{\partial y^o}\frac{\partial \Delta u}{\partial y^o} + \frac{\partial v^t}{\partial y^o}\frac{\partial \Delta v}{\partial y^o}$$

$$2\Delta E_{xy} = \frac{\partial \Delta u}{\partial y^o} + \frac{\partial \Delta v}{\partial x^o} + \frac{\partial u^t}{\partial x^o}\frac{\partial \Delta u}{\partial y^o} + \frac{\partial u^t}{\partial y^o}\frac{\partial \Delta u}{\partial x^o} + \frac{\partial v^t}{\partial y^o}\frac{\partial \Delta v}{\partial x^o} + \frac{\partial v^t}{\partial x^o}\frac{\partial \Delta v}{\partial y^o}$$

The strain increments contain initial displacement contributions such as

$$\frac{\partial u^t}{\partial x^o}, \quad \frac{\partial v^t}{\partial x^o}$$

in the ΔE_{xx} term. We can replace all functions using the interpolation functions and then express these in matrix form as

$$\left\{\begin{array}{c} \Delta E_{xx} \\ \Delta E_{yy} \\ 2\Delta E_{xy} \end{array}\right\} = \Big[[B_{L0}] + [B_{L1}]\Big] \left\{\begin{array}{c} \Delta u_1 \\ \Delta v_1 \\ \Delta u_2 \\ \Delta v_2 \\ \Delta u_3 \\ \Delta v_3 \end{array}\right\} \quad \text{or} \quad \{\Delta E\} = [\,B\,]\{\Delta u\}$$

where the two matrices are

$$[B_{L0}] \equiv \frac{1}{2A}\begin{bmatrix} b_1 & 0 & b_2 & 0 & b_3 & 0 \\ 0 & c_1 & 0 & c_2 & 0 & c_3 \\ c_1 & b_1 & c_2 & b_2 & c_3 & b_3 \end{bmatrix}$$

$$[B_{L1}] \equiv \frac{1}{2A}\begin{bmatrix} u_{,x}\,b_1 & v_{,x}\,b_1 & u_{,x}\,b_2 & v_{,x}\,b_2 & u_{,x}\,b_3 & v_{,x}\,b_3 \\ u_{,y}\,c_1 & v_{,y}\,c_1 & u_{,y}\,c_2 & v_{,y}\,c_2 & u_{,y}\,c_3 & v_{,y}\,c_3 \\ u_{,x}\,c_1+ & v_{,x}\,c_1+ & u_{,x}\,c_2+ & v_{,x}\,c_2+ & u_{,x}\,c_3+ & v_{,x}\,c_3+ \\ u_{,y}\,b_1 & v_{,y}\,b_1 & u_{,y}\,b_2 & v_{,y}\,b_2 & u_{,y}\,b_3 & v_{,y}\,b_3 \end{bmatrix}$$

In these expressions, the comma indicates partial differentiation with respect to the subscripted variable. These are associated with the deformation gradient, which is known at time t. The element stiffness matrix is now obtained as

$$[\,k_L\,] = \int_{V^o}[\,B_L\,]^T[\,D\,][\,B_L\,]\,dV^o = [\,B_L\,]^T[\,D\,][\,B_L\,]\,V^o$$

The integration is trivially performed because all contributions to $[\,B_L\,]$ are constant in space.

The nonlinear contribution to the stiffness comes from the virtual work term

$$\sum_{i,j}\sigma_{ij}^K\delta\eta_{ij} = \sigma_{xx}^K\delta\eta_{xx} + \sigma_{xy}^K\delta\eta_{xy} + \sigma_{yx}^K\delta\eta_{yx} + \sigma_{yy}^K\delta\eta_{yy}$$

Noting, for instance, that

$$\eta_{ij} = \sum_k \frac{\partial \Delta u_k}{\partial x_i^o} \frac{\partial \Delta u_k}{\partial x_j^o} , \qquad \delta\eta_{ij} = \sum_k \delta\frac{\partial \Delta u_k}{\partial x_i^o} \frac{\partial \Delta u_k}{\partial x_j^o} + \sum_k \frac{\partial \Delta u_k}{\partial x_i^o} \delta\frac{\partial \Delta u_k}{\partial x_j^o}$$

we can put the virtual work expression in the matrix form

$$\left\{ \begin{array}{c} \delta\Delta u_{,x} \\ \delta\Delta u_{,y} \\ \delta\Delta v_{,x} \\ \delta\Delta v_{,y} \end{array} \right\}^T \left[\begin{array}{cccc} \sigma_{xx}^K & \sigma_{xy}^K & 0 & 0 \\ \sigma_{xy}^K & \sigma_{yy}^K & 0 & 0 \\ 0 & 0 & \sigma_{xx}^K & \sigma_{xy}^K \\ 0 & 0 & \sigma_{xy}^K & \sigma_{yy}^K \end{array} \right] \left\{ \begin{array}{c} \Delta u_{,x} \\ \Delta u_{,y} \\ \Delta v_{,x} \\ \Delta v_{,y} \end{array} \right\} \quad \text{or} \quad \{\delta\Delta u_{,x}\}^T [\sigma^K]\{\Delta u_{,x}\}$$

where again the comma indicates partial differentiation with respect to the subscripted variable. The gradient increments can be expressed as

$$\left\{ \begin{array}{c} \Delta u_{,x} \\ \Delta u_{,y} \\ \Delta v_{,x} \\ \Delta v_{,y} \end{array} \right\} = \frac{1}{2A} \left[\begin{array}{cccccc} b_1 & 0 & b_2 & 0 & b_3 & 0 \\ c_1 & 0 & c_2 & 0 & c_3 & 0 \\ 0 & b_1 & 0 & b_2 & 0 & b_3 \\ 0 & c_1 & 0 & c_2 & 0 & c_3 \end{array} \right] \left\{ \begin{array}{c} \Delta u_1 \\ \Delta v_1 \\ \Delta u_2 \\ \Delta v_2 \\ \Delta u_3 \\ \Delta v_3 \end{array} \right\} \quad \text{or} \quad \{\Delta u_{,x}\} = [B_N]\{\Delta u\}$$

The nonlinear element stiffness matrix is now obtained as

$$[k_N] = \int_{V^o} [B_N]^T [\sigma^K][B_N]\, dV^o = [B_N]^T [\sigma^K][B_N]\, V^o$$

Again, the integration is trivially performed because all contributions to $[B_N]$ and $[\sigma^K]$ are constant in space.

The internal force vector is determined as

$$\sum_{i,j} \sigma_{ij}^K \delta\Delta E_{ij} = \sigma_{xx}^K \delta\Delta E_{xx} + \sigma_{xy}^K \delta\Delta E_{xy} + \sigma_{yx}^K \delta\Delta E_{yx} + \sigma_{yy}^K \delta\Delta E_{yy}$$

which can be put in the matrix form

$$\delta\{\Delta u\}^T \{B\}^T \{\sigma^K\}$$

This now leads to

$$\{F\} = \int_{V^o} [\, B\,]^T \{\sigma^K\}\, dV^o = [\, B\,]^T \{\sigma^K\}\, V^o$$

which completes the system of equations.

Example 3.40:   Treat the simple shear deformation of a block as a load control problem and show the relation between the Cauchy and Kirchhoff stresses.

A simple shear deformation parallel to the $x_1^o - x_2^o$ plane is shown in Figure 3.19 and given mathematically by

$$x_1 = x_1^o + k x_2^o , \qquad x_2 = x_2^o , \qquad x_3 = x_3^o$$

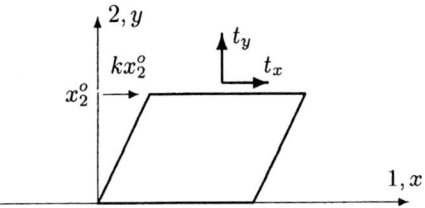

Figure 3.19: A block in simple shear.

The displacement components are readily obtained as

$$u_1 = k\,x_2^o, \qquad u_2 = 0, \qquad u_3 = 0$$

indicating that horizontal lines move horizontally only. The deformation gradients are

$$[\frac{\partial x_p}{\partial x_i^o}] = \begin{bmatrix} 1 & k & 0 \\ 0 & 1 & 0 \\ 0 & 0 & 1 \end{bmatrix} \quad \text{and} \quad [\frac{\partial x_p^o}{\partial x_i}] = \begin{bmatrix} 1 & -k & 0 \\ 0 & 1 & 0 \\ 0 & 0 & 1 \end{bmatrix}$$

Note that there is no volume change because $J = J_o = 1$. The Lagrangian strain tensor is

$$2E_{ij} = \sum_p \frac{\partial x_p}{\partial x_i^o}\frac{\partial x_p}{\partial x_j^o} - \delta_{ij} = \begin{bmatrix} 0 & k & 0 \\ k & k^2 & 0 \\ 0 & 0 & 0 \end{bmatrix}$$

Let the material have the following linear constitutive behavior:

$$\sigma_{ij}^K = 2\mu E_{ij} + \lambda\delta_{ij}\sum_k E_{kk}$$

where μ and λ are the Lamé constants. The Kirchhoff stress tensor, therefore, is

$$\sigma_{ij}^K = \begin{bmatrix} \frac{1}{2}\lambda k^2 & \mu k & 0 \\ \mu k & \mu k^2 + \frac{1}{2}\lambda k^2 & 0 \\ 0 & 0 & 0 \end{bmatrix} = \mu k \begin{bmatrix} \gamma k & 1 & 0 \\ 1 & (1+\gamma)k & 0 \\ 0 & 0 & 0 \end{bmatrix}$$

where $\gamma = \lambda/2\mu$. The tensile σ_{22}^K component arises from the fact that lines originally in the 2-direction are being stretched. The Cauchy stresses are obtained from

$$\sigma_{pq} = \sum_{i,j} \frac{\rho}{\rho^o}\sigma_{ij}^K \frac{\partial x_p}{\partial x_i^o}\frac{\partial x_q}{\partial x_j^o}$$

$$= \sigma_{11}^K \left[\frac{\partial x_p}{\partial x_1^o}\frac{\partial x_q}{\partial x_1^o}\right] + \sigma_{12}^K \left[\frac{\partial x_p}{\partial x_1^o}\frac{\partial x_q}{\partial x_2^o} + \frac{\partial x_p}{\partial x_2^o}\frac{\partial x_q}{\partial x_1^o}\right] + \sigma_{22}^K \left[\frac{\partial x_p}{\partial x_2^o}\frac{\partial x_q}{\partial x_2^o}\right]$$

Substituting for the deformation gradients leads to the complete stress tensor as

$$\sigma_{pq} = \mu k \begin{bmatrix} (2+\gamma)k + (1+\gamma)k^3 & 1 + (1+\gamma)k^2 & 0 \\ 1 + (1+\gamma)k^2 & (1+\gamma)k & 0 \\ 0 & 0 & 0 \end{bmatrix}$$

The Cauchy stress tensor, as expected, is symmetric.

The magnitude of the shear deformation is governed by the parameter k. It is worth noting that when it is small, both stress tensors approach the same values. Another point worth noting is that the simple constitutive relation $\sigma_{ij}^K = \mu E_{ij}$ ($\lambda = 0$) in the Lagrangian variables would not lead to an analogous simple relation between σ_{ij} and the Eulerian strain e_{ij}.

Imagine a free body cut parallel to the x-axis; this will expose two tractions related to the Cauchy stress by

$$t_x = \sigma_{xy}, \qquad t_y = \sigma_{yy}$$

The t_x traction, when multiplied by the area, gives a resultant horizontal force that we will consider to be the applied load. The resulting deformation is then related to the traction (and.hence load) as

$$\sigma_{xy} = \mu k[1 + (1+\gamma)k^2] = t_x = P/hL$$

where hL is the area over which the resulting force P acts. The deformation parameter k is a nonlinear function of the load, but we can easily solve for it using a Newton-Raphson iterative scheme as

$$k^{i+1} = k^i + \frac{P/hL - f_o}{\mu[1 + (1+\gamma)3(k^i)^2]}, \qquad f_o = \mu k^i[1 + (1+\gamma)(k^i)^2]$$

where i is the iteration counter. This converges very rapidly.

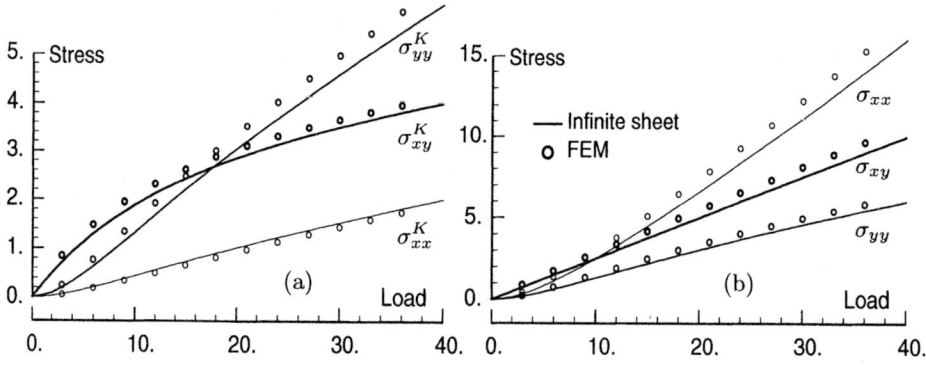

Figure 3.20: Stresses for a block under shear load control. (a) Kirchhoff stress. (b) Cauchy stress.

Once we know k we can then determine the Kirchhoff stresses. The results are shown plotted in Figure 3.20. It is interesting to note that the Cauchy σ_{xx} is the largest of the stresses.

Example 3.41: Obtain a numerical solution of the shear problem.

The analytical solution just developed was for a very large sheet under homogeneous deformation. This is impractical to achieve here so we will model the block as shown in Figure 3.21.

The top row of elements have a stiffness 1000 times that of the other elements, it is also constrained to move only horizontally. In the infinite sheet case the lateral

$L = 203\,\text{mm}\,(8.0\,\text{in.})$
$b = 101\,\text{mm}\,(4.0\,\text{in.})$
$h = 12.7\,\text{mm}\,(0.5\,\text{in.})$
aluminum

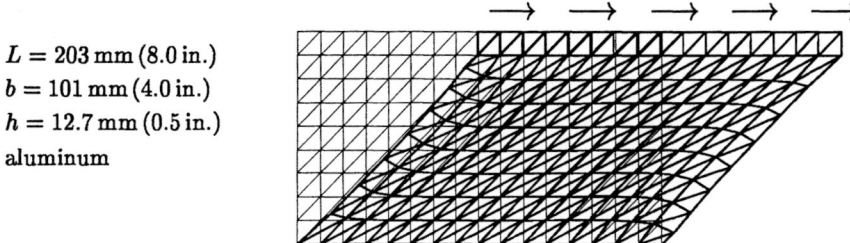

Figure 3.21: Undeformed and deformed shape of a block under shear.

sides have shear components but clearly in the present case the normal tractions are zero.

Figure 3.22 shows the contours of Cauchy stress at the maximum load drawn on the deformed block. What they all have in common is that they show a nearly uniform region of stress in the middle portion. We therefore expect to have a reasonable comparison with the infinite sheet solution in this region. Figure 3.20 shows a comparison of the stress histories with that for the infinite sheet — all the trends are in agreement.

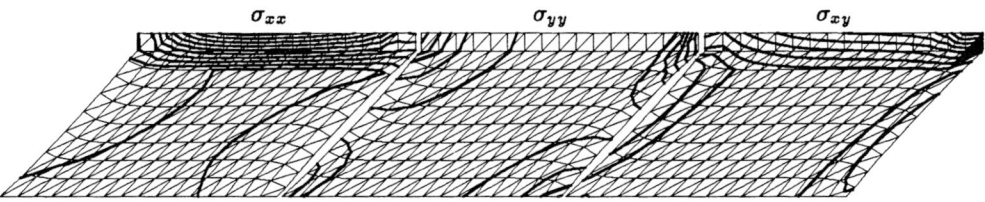

Figure 3.22: Contours of Cauchy stresses on the deformed block.

We do not expect the Cauchy σ_{xx} stress to go to zero at the boundaries because these boundaries are inclined in the deformed configuration

Discussion

Formulating the total Lagrangian scheme for two- and three-dimensional continuum problems is quite straightforward when the only degrees of freedom are the nodal displacements. The scheme gets complicated when applied to 3-D structures involving approximate structural theories (such as shells and frames) that use rotations as additional degrees of freedom. The application is considered in References [7, 20]. In the next section, we begin to formulate a different approach that seems particularly suited to the thin-walled structures of interest, which undergo relatively large deflections and rotations but rather small strains. As pointed out in Reference [24], the two formulations result in the same matrices but the corotational scheme seems easier to establish.

3.5   The Corotational Scheme

A particularly effective method for handling the analysis of structures is the corotational scheme. In this, a local coordinate system is envisioned as moving with each element, and, relative to this coordinate system, the element behaves linearly. Consequently, all of the nonlinearities of the problem are shifted into the description of the moving coordinates.

As a first step to analyzing 3-D structures, we begin by looking at 2-D trusses and frames and leave the more general case to the next section. This gives us an opportunity to illustrate the essential concepts of the corotational method without introducing the notational complications needed in the general cases. Early use of the corotational scheme is given in References [10, 11].

Illustration for 2-D Trusses

Consider the axial stretching and global rotation of the member shown in Figure 3.23. The figure indicates both stretching and twisting, but for now, we consider only the stretching.

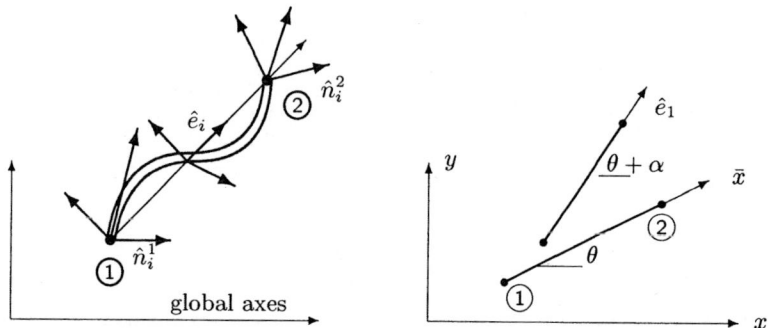

Figure 3.23: Arbitrarily oriented truss member.

The global degrees of freedom are

$$\{u\} = \{u_1\, v_1;\, u_2\, v_2\}^T$$

We establish a local coordinate system at the first node with the $\bar{x}$-axis directed along the member. Descriptions of quantities in the local coordinates will have an overhead bar. In this local description, the truss has the deformation system

$$\{\bar{u}\} = \{\bar{u}_1 = 0,\, \bar{v}_1 = 0;\, \bar{u}_2 = \bar{u},\, \bar{v}_2 = 0\}^T$$

That only a few of the local degrees of freedom are strain producing makes this approach appealing.

The main steps to be followed are: establishing stiffness relations in local coordinates, establishing the relation between the local variables and the global variables, and finally, establishing the stiffness relations in global variables.

I: Local Stiffness Relation

Although we derived the local stiffnesses for frames and trusses in Chapter 1, we will now re-derive it but in a slightly different manner.

In local coordinates, the axial strain is

$$\bar{\epsilon} = \frac{1}{L_o}(\bar{u}_2 - \bar{u}_1)$$

We express this in matrix form as

$$\bar{\epsilon} = \frac{1}{L_o}\{-1, 0; 1, 0\}\{\bar{u}_1, \bar{v}_1; \bar{u}_2, \bar{v}_2\}^T = \frac{1}{L_o}\{\bar{c}\}^T\{\bar{u}\} = \frac{1}{L_o}\{\bar{u}\}^T\{\bar{c}\}$$

The material behavior is assumed to be linear elastic, hence the axial stress is given by

$$\bar{\sigma} = E\bar{\epsilon} = \frac{E}{L_o}\{\bar{u}\}^T\{\bar{c}\}$$

The principle of virtual work can be used to determine a set of element nodal forces consistent with the internal stress as

$$\{\bar{F}\}^T\{\delta\bar{u}\} = \int \bar{\sigma}\,\delta\bar{\epsilon}\,dV_o = \int E\frac{1}{L_o}\{\bar{u}\}^T\{\bar{c}\}\frac{1}{L_o}\{\bar{c}\}^T\{\delta\bar{u}\}\,dV_o$$

The integration is performed trivially because all quantities are independent of original position. Since the virtual displacements are arbitrary, we get the nodal forces as

$$\{\bar{F}\}^T = \frac{EV_o}{L_o^2}\{\bar{u}\}^T\{\bar{c}\}\{\bar{c}\}^T \qquad \text{or} \qquad \{\bar{F}\} = \frac{EA_o}{L_o}\{\bar{c}\}\{\bar{c}\}^T\{\bar{u}\}$$

The local tangent stiffness is obtained from the variation of the nodal forces

$$\{\delta\bar{F}\} = [\frac{\partial\bar{F}}{\partial\bar{u}}]\{\delta\bar{u}\} = [\bar{k}]\{\delta\bar{u}\} \qquad \text{or} \qquad [\bar{k}] = [\frac{\partial\bar{F}}{\partial\bar{u}}] = \frac{EA_o}{L_o}\{\bar{c}\}\{\bar{c}\}^T$$

Multiplying the vectors gives

$$[\bar{k}] = \begin{bmatrix} \bar{k}_{11} & \bar{k}_{12} \\ \bar{k}_{21} & \bar{k}_{22} \end{bmatrix}, \qquad [\bar{k}_{11}] = [\bar{k}_{22}] = -[\bar{k}_{12}] = -[\bar{k}_{21}] = \frac{EA_o}{L_o}\begin{bmatrix} 1 & 0 \\ 0 & 0 \end{bmatrix}$$

This is the $[2 \times 2]$ submatrix of the $[4 \times 4]$ linear truss stiffness derived in Chapter 1. We thus write the local stiffness relation as

$$\{\bar{F}\} = [\bar{k}]\{\bar{u}\}$$

where the local force system is

$$\{\bar{F}\} = \{\bar{F}_1, 0; \bar{F}_2, 0\}^T = \{-1, 0; 1, 0\}^T\bar{\sigma}A_o = \{-1, 0; 1, 0\}^T\bar{F}_o = \{b\}^T\bar{F}_o$$

We now need to relate the local stiffness and force to the corresponding global versions.

II: Relation between Local and Global Variables

Locally, the only strain producing mode is the axial displacement. This displacement is related to the global variables by

$$\bar{u} = L - L_o = \sqrt{(\hat{x}_{21} + \hat{u}_{21}) \cdot (\hat{x}_{21} + \hat{u}_{21})} - \sqrt{(\hat{x}_{21}) \cdot (\hat{x}_{21})}$$

where the subscript notation means $\hat{x}_{21} = \hat{x}_2 - \hat{x}_1$, and so on. By taking the variation of this, we establish that

$$\delta\bar{u} = \frac{1}{L}(\hat{x}_{21} + \hat{u}_{21}) \cdot \delta\hat{u}_{21} = \hat{e}_1 \cdot \delta\hat{u}_{21} = -\hat{e}_1 \cdot \delta\hat{u}_1 + \hat{e}_1 \cdot \delta\hat{u}_2$$

where $\hat{e}_1$ is a unit vector directed along the member. This is expressed in matrix notation as

$$\delta\bar{u} = \{\{-e_1\}^T; \{e_1\}^T\}\{\delta u_1, \delta v_1; \delta u_2, \delta v_2\}^T = [\ B\ ]\{\delta u\}$$

Let the member be initially at an angle θ to the global x-axis and let it experience a rotation of α. Then,

$$[\ B\ ]^T = [-C, -S; C, S]$$

with the notations $S \equiv \sin(\theta + \alpha)$ and $C \equiv \cos(\theta + \alpha)$.

We also need to know the change of orientation of the member. This rotation is computed as

$$\sin\alpha\,\hat{k} = \frac{\hat{x}_{21}}{L_o} \times \frac{\hat{x}_{21} + \hat{u}_{21}}{L} = \frac{1}{L_o L}[x_{21}v_{21} - y_{21}u_{21}]\hat{k}$$

and

$$\cos\alpha = \frac{\hat{x}_{21}}{L_o} \cdot \frac{\hat{x}_{21} + \hat{u}_{21}}{L} = \frac{1}{L_o L}[L_o{}^2 + x_{21}u_{21} + y_{21}v_{21}]$$

Introducing the initial orientation, we can rewrite these as

$$\sin\alpha = \frac{1}{L}[-\sin\theta\,u_{21} + \cos\theta\,v_{21}], \qquad \cos\alpha = \frac{1}{L}[L_o + \cos\theta\,u_{21} + \sin\theta\,v_{21}]$$

The first variation of these gives

$$\cos\alpha\,\delta\alpha = \frac{1}{L}[-\sin\theta\,\delta u_{21} + \cos\theta\,\delta v_{21}], \qquad -\sin\alpha\,\delta\alpha = \frac{1}{L}[\cos\theta\,\delta u_{21} + \sin\theta\,\delta v_{21}]$$

Note that we do not take the variation of L, because it would lead to a higher-order effect (i.e., $\bar{u}\,\delta\bar{u}$). Multiply the first by $\cos\alpha$ and the second by $\sin\alpha$ and then subtract to get

$$\delta\alpha = \frac{1}{L}[-\sin(\theta + \alpha)\,\delta u_{21} + \cos(\theta + \alpha)\,\delta v_{21}] = \frac{1}{L}\hat{e}_2 \cdot \delta\hat{u}_{21}$$

where $\hat{e}_2$ is the unit vector perpendicular to $\hat{e}_1$. This relation could have been obtained directly by taking the component of $\delta\hat{u}_{21}$ resolved perpendicular to the member and then dividing by the length to get the tangent of the angle. Because $\delta u_{21} = \delta(u_2 - u_1)$ and $\delta v_{21} = \delta(v_2 - v_1)$, we can write the virtual rotation as

$$\delta\alpha = \frac{1}{L}\{S, -C; -S, C\}\{\delta u_1, \delta v_1; \delta u_2, \delta v_2\}^T \equiv \frac{1}{L}\{z\}^T\{\delta u\}$$

The only nonzero local force is the axial force $\bar{F}_o$. The virtual work in global variables must equal the virtual work in local variables, hence

$$\{F\}^T\{\delta u\} = \{\bar{F}\}^T\{\delta\bar{u}\} = \bar{F}\delta\bar{u} = \bar{F}_o[\,B\,]\{\delta u\}$$

From this we conclude that

$$\{F\} = [\,B\,]^T\bar{F}_o$$

III: Global Stiffness Relations

At the global level, the variation of the nodal forces leads to

$$\{\delta F\} = [\frac{\partial F}{\partial u}]\{\delta u\} = [\,k_T\,]\{\delta u\}$$

where $[\,k_T\,]$ is the element tangent stiffness matrix in global coordinates. Substitute for $\{F\}$ in terms of the local quantities to get

$$\{\delta F\} = [\,B\,]^T[\frac{\partial \bar{F}}{\partial \bar{u}}]\{\delta\bar{u}\} + \delta[\,B\,]^T\{\bar{F}\} = [\,B\,]^T[\,\bar{k}\,][\,B\,]\{\delta u\} + \bar{F}_o[\delta B]^T$$

Note that because the behavior is linear on the local level, the local tangent stiffness and elastic stiffness are the same. The first set of terms

$$[\,k_E\,] \equiv [\,B\,]^T[\,\bar{k}\,][\,B\,]$$

gives the elastic stiffness. On multiplying out, we get

$$[\,k_E\,] = \frac{EA_o}{L_o}\begin{bmatrix} C^2 & CS & -C^2 & -CS \\ CS & S^2 & -CS & -S^2 \\ -C^2 & -CS & C^2 & CS \\ -CS & -S^2 & CS & S^2 \end{bmatrix} \tag{3.12}$$

In local coordinates these reduce to

$$[\,\bar{k}_E\,] = \frac{EA_o}{L_o}\begin{bmatrix} 1 & 0 & -1 & 0 \\ 0 & 0 & 0 & 0 \\ -1 & 0 & 1 & 0 \\ 0 & 0 & 0 & 0 \end{bmatrix} \tag{3.13}$$

These are the standard element stiffnesses for the truss [22]. We therefore recognize a portion of the global tangent stiffness matrix as the components of the local stiffness matrix transformed to the current orientation of the member, that is,

$$[k_E] \equiv [\, T\,]^T [\, \bar{k}_E\,][\, T\,]$$

Returning now to the remaining term of the tangent stiffness,

$$\bar{F}_o[\delta B] = \bar{F}_o \delta[-C,\ -S;\ C,\ S] = \bar{F}_o[S,\ -C;\ -S,\ C]\delta\alpha = \frac{1}{L}\bar{F}_o\{z\}\{z\}^T\{\delta u\}$$

This gives the geometric contribution of the tangent stiffness as

$$[k_G] = \frac{1}{L}\bar{F}_o\{z\}\{z\}^T$$

On multiplying out, we get

$$[k_G] = \frac{\bar{F}_o}{L}\begin{bmatrix} S^2 & -CS & CS & -C^2 \\ -CS & C^2 & S^2 & -SC \\ -S^2 & CS & S^2 & -CS \\ CS & -C^2 & -CS & C^2 \end{bmatrix}$$

In local coordinates, these are

$$[\bar{k}_G] = \frac{\bar{F}_o}{L}\begin{bmatrix} 0 & 0 & 0 & 0 \\ 0 & 1 & 0 & -1 \\ 0 & 0 & 0 & 0 \\ 0 & -1 & 0 & 1 \end{bmatrix}$$

We recognize this as the geometric stiffness for a truss in local coordinates [22]. We therefore recognize the remainder of the global tangent stiffness matrix as the components of the local geometric stiffness matrix transformed to the current orientation of the member, that is,

$$[k_G] \equiv [\, T\,]^T [\, \bar{k}_G\,][\, T\,]$$

The assembled global tangent stiffness is given by

$$[K_T] = \sum_m \Big[[k_E]_m + [k_G]_m\Big] = \sum_m [\, T\,]_m^T \Big[[\bar{k}_E]_m + [\bar{k}_G]_m\Big] [\, T\,]_m$$

To summarize, at the local level, the nodal forces are obtained directly from the stresses or the elastic stiffness, their global components are simply their transformation to global coordinates. The global stiffness matrix contains the elastic stiffness plus a geometric contribution due to the rotation of the corotational axes.

Example 3.42: Determine the deflected shape of the simple two-member truss shown in Figure 3.24. Also show the effect of the geometric stiffness on the convergence rate of the Newton-Raphson method.

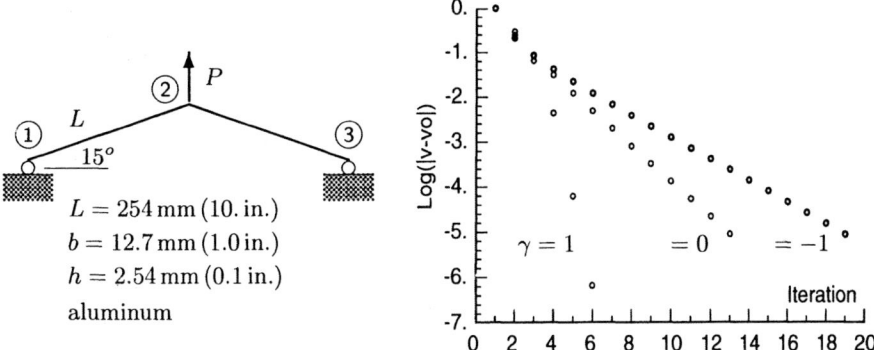

Figure 3.24: Convergence of the displacement error norm against number of iterations.

We first begin with the convergence properties. In evaluating the Newton-Raphson method, it must be realized that the tangent stiffness plays the role of the "slope of the function;" thus when equilibrium is achieved we have

$$[K_T]_{t+\Delta}^{i-1}\{\Delta u\}^i = \{P\}_{t+\Delta t} - \{F\}_{t+\Delta t}^{i-1} = 0$$

and we get $\{\Delta u\}^i = 0$ regardless of the tangent stiffness. That is, it is not necessary to have the exact stiffness in order to get convergence to the correct answer — this is the basis of many of the modified Newton-Raphson methods. Figure 3.24 shows the convergence rate when the tangent stiffness is formed as

$$[K_T] = [K_E] + \gamma[K_G]$$

for $\gamma = 1, 0, -1$. In each case, the algorithm converged to the correct value, but clearly the rate of convergence is affected. Implicit in this is that for more complicated problems, the modified methods are more likely to fail to converge than the full methods. There is, however, another important point to be learned from these results that will affect the developments of the next few sections. These results say that the more accurate the tangent stiffness, the better the convergence rate, but that it is not essential that the actual exact tangent stiffness be used. Consequently, if it is convenient to approximate the tangent stiffness, then the basic nonlinear formulation is not affected, only the convergence rate (and radius of convergence) of the algorithm is affected.

The results for many load increments are shown in Figure I.8. Note that the only reason many increments were used in this case is so that a better picture of the deformation history can be viewed.

The response against load can be divided into three stages. The first, at relatively low loads, shows agreement with the linear theory. The next stage shows the effect of changing geometry: the upward load causes a stiffening of the structure, while the downward load causes a softening of the structure. These are direct consequences of the $\bar{F}_o$ contribution to the geometric (and hence tangent) stiffness. The decrease in stiffness eventually leads to a very large displacement jump where the structure finds a new equilibrium position. In fact, as seen from the deformed shapes, the structure has "snapped-through" to the other side. This is an example of a limit point instability — a situation where the current structural configuration cannot sustain a further load increment without a significant change of shape — and we will consider it in more depth in Chapters 6 and 7 dealing with stability.

The comparison with the exact behavior confirms the correctness of the corotational formulation.

Illustration for 2-D Frames

We now analyze a 2-D frame to illustrate the effect moments have on the tangent stiffness. Much of what we will do follows from the truss example, but this time will also try to generalize the notations.

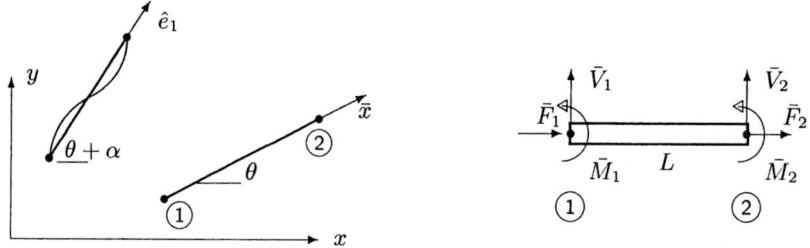

Figure 3.25: Arbitrarily oriented frame member.

Consider the axial stretching and rotation of the member shown Figure 3.25. The global degrees of freedom are

$$\{u\} = \{u_1, v_1, \phi_1; u_2, v_2, \phi_2\}^T$$

We establish a local coordinate system at the first node with the $\bar{x}$-axis directed along the member. Locally, the strain producing modes are the axial displacement and the relative twists. Hence, in this local description, the frame has the deformation system

$$\{\bar{u}\} = \{\bar{u}_1 = 0, \bar{v}_1 = 0, \bar{\phi}_1 = \bar{\phi}_1; \bar{u}_2 = \bar{u}, \bar{v}_2 = 0, \bar{\phi}_2 = \bar{\phi}_2\}^T$$

We take the local stiffness relation from Chapter 1 and concentrate on establishing the relation between the local variables and the global variables, and the stiffness relations in global variables.

I: Local Stiffness Relation

From Chapter 1, we have the stiffness relations

$$\bar{F}_2 = \frac{EA}{L}\hat{u}_2, \qquad \left\{\begin{matrix} \bar{M}_1 \\ \bar{M}_2 \end{matrix}\right\} = \frac{2EI}{L}\begin{bmatrix} 2 & 1 \\ 1 & 2 \end{bmatrix}\left\{\begin{matrix} \bar{\phi}_1 \\ \bar{\phi}_2 \end{matrix}\right\}$$

The moments and rotations are about the z-axis.

II: Relation between Local and Global Variables

We have already established most of the information we need to relate the local and global variables, in particular,

$$\delta\bar{u} = [\cos(\theta+\alpha)\delta u_{21}+\sin(\theta+\alpha)\delta v_{21}], \quad \delta\alpha = \frac{1}{L}[-\sin(\theta+\alpha)\delta u_{21}+\cos(\theta+\alpha)\delta v_{21}]$$

Introduce local and global displacement vectors defined, respectively, as

$$\{\delta\bar{u}\} \equiv \{\delta\bar{u}, \delta\bar{\phi}_1, \delta\bar{\phi}_2\}^T, \qquad \{\delta u\} \equiv \{\delta u_1, \delta v_1, \delta\phi_1; \delta u_2, \delta v_2, \delta\phi_2\}^T$$

then the above can be rewritten as

$$
\begin{aligned}
\delta\bar{u} &= [-C, -S, 0; C, S, 0]\{\delta u\} \equiv \{r\}^T\{\delta u\} \\
\delta\alpha &= \frac{1}{L}[S, -C, 0; -S, C, 0]\{\delta u\} \equiv \frac{1}{L}\{z\}^T\{\delta u\}
\end{aligned}
\qquad (3.14)
$$

with the notations $S \equiv \sin(\theta + \alpha)$ and $C \equiv \cos(\theta + \alpha)$. Noting that the local twists are related to the global angles as

$$\delta\bar{\phi}_1 = \delta\phi_1 - \delta\alpha, \qquad \delta\bar{\phi}_2 = \delta\phi_2 - \delta\alpha$$

then we can write the relation between the local and global variables as

$$\{\delta\bar{u}\} = [\,B\,]\{\delta u\} \qquad \text{with} \qquad [\,B\,] \equiv \frac{1}{L}\begin{bmatrix} -CL & -SL & 0 & CL & SL & 0 \\ -S & C & L & S & -C & 0 \\ -S & C & 0 & S & -C & L \end{bmatrix}$$

This fundamental relation will now be used to determine the relation between the global and local load terms.

The virtual work in global variables must equal the virtual work in local variables, hence

$$\{F\}^T\{\delta u\} = \{\bar{F}\}^T\{\delta\bar{u}\} = \{\bar{F}\}^T\{B\}\{\delta u\}$$

From this, we conclude that

$$\{F\} = \{B\}^T\{\bar{F}\}$$

In anticipation of generalization, note that, owing to equilibrium at the local level, we have

$$\bar{F}_1 = -\bar{F}_o, \qquad \bar{F}_2 = -\bar{F}_o, \qquad \bar{V}_1 = \frac{1}{L}(\bar{M}_1 + \bar{M}_2), \qquad \bar{V}_2 = -\frac{1}{L}(\bar{M}_1 + \bar{M}_2)$$

Augment the local force vector to $\{\bar{F}\} = \{\bar{F}_1, \bar{V}_1, \bar{M}_1; \bar{F}_2, \bar{V}_2, \bar{M}_2\}^T$, so that the first component of global force, for example, becomes

$$F_{1x} = \frac{1}{L}[-CL\bar{F}_o - S\bar{M}_1 - S\bar{M}_2] = [C\bar{F}_1 - S\bar{V}_1]$$

In this way, we can write the complete relation as

$$\{u\} = [\,T\,]\{\bar{u}\}, \quad \{F\} = [\,T\,]^T\{\bar{F}\}, \quad [\,T\,]^T = \begin{bmatrix} E & 0 \\ 0 & E \end{bmatrix}, \quad [\,E\,] = \begin{bmatrix} C & -S & 0 \\ S & C & 0 \\ 0 & 0 & 1 \end{bmatrix}$$

Thus $\{F\}$ is simply the components of $\{\bar{F}\}$ transformed to the global coordinates.

III: Global Stiffness Relations

The variation of the global nodal forces leads to

$$\{\delta F\} = [\frac{\partial F}{\partial u}]\{\delta u\} = [\,k_T\,]\{\delta u\}$$

where $[\,k_T\,]$ is the element tangent stiffness matrix in global coordinates. Substitute for $\{F\}$ in terms of local quantities to get

$$\begin{aligned}
\{\delta F\} &= [\,T\,]^T[\frac{\partial \bar{F}}{\partial \bar{u}}]\{\delta \bar{u}\} + \delta[\,T\,]^T\{\bar{F}\} \\
&= [\,T\,]^T[\,\bar{k}\,][\,T\,]\{\delta u\} + \bar{F}_o[\delta B]_1 + \bar{M}_1[\delta B]_2 + \bar{M}_2[\delta B]_3
\end{aligned}$$

where the subscript notation on $[\,B\,]$ indicates the column of $[\,B\,]^T$. The first set of terms

$$[\,k_E\,] \equiv [\,T\,]^T[\,\bar{k}\,][\,T\,]$$

gives the elastic stiffness. On multiplying out, we get

$$[\,k_E\,] = \frac{EA}{L}\begin{bmatrix}
C^2 & CS & 0 & -C^2 & -CS & 0 \\
CS & S^2 & 0 & -CS & -S^2 & 0 \\
0 & 0 & 0 & 0 & 0 & 0 \\
-C^2 & -CS & 0 & C^2 & CS & 0 \\
-CS & -S^2 & 0 & CS & S^2 & 0 \\
0 & 0 & 0 & 0 & 0 & 0
\end{bmatrix}$$
$$+ \frac{EI}{L^3}\begin{bmatrix}
12S^2 & -12CS & -6LS & -12S^2 & 12CS & -6LS \\
-12CS & 12C^2 & 6LC & 12CS & -12C^2 & 6LC \\
-6LS & 6LC & 4L^2 & 6LS & -6LC & 2L^2 \\
-12S^2 & 12CS & 6LS & 12S^2 & -12CS & 6LS \\
12CS & -12C^2 & -6LC & -12CS & 12C^2 & -6LC \\
-6LS & 6LC & 2L^2 & 6LS & -6LC & 4L^2
\end{bmatrix}$$

In local coordinates these reduce to

$$[\,\bar{k}_E\,] = \frac{EA}{L}\begin{bmatrix}
1 & 0 & 0 & 0 & 0 & 0 \\
0 & 0 & 0 & 0 & 0 & 0 \\
0 & 0 & 0 & 0 & 0 & 0 \\
-1 & 0 & 0 & 1 & 0 & 0 \\
0 & 0 & 0 & 0 & 0 & 0 \\
0 & 0 & 0 & 0 & 0 & 0
\end{bmatrix} + \frac{EI}{L^3}\begin{bmatrix}
0 & 0 & 0 & 0 & 0 & 0 \\
0 & 12 & 6L & 0 & -12 & 6L \\
0 & 6L & 4L^2 & 0 & 0 & 2L^2 \\
0 & 0 & 0 & 0 & 0 & 0 \\
0 & -12 & -6L & 0 & 12 & -6L \\
0 & 6L & 2L^2 & 0 & -6L & 4L^2
\end{bmatrix}$$

These are the standard element stiffnesses for the truss and beam [22], respectively. We therefore recognize part of the global stiffness matrix as the components of the local stiffness matrices transformed to the current orientation of the member, that is,

$$[\,k_E\,] \equiv [\,T\,]^T[\,\bar{k}\,][\,T\,]$$

Returning now to the remaining terms of the tangent stiffness, the first term gives

$$\bar{F}_o[\delta B]_1 = \bar{F}_o[S,\, -C,\, 0;\, -S,\, C,\, 0]^T\delta\alpha = \frac{1}{L}\bar{F}_o\{z\}\{z\}^T\{\delta u\}$$

which we already obtained for the truss. The second term gives

$$\begin{aligned}
\bar{M}_1[\delta B]_2 &= \bar{M}_1\delta\frac{1}{L}[-S,\, C,\, L;\, S,\, -C,\, 0]^T \\
&= \bar{M}_1\frac{1}{L}[-C,\, -S,\, 0;\, C,\, S,\, 0]^T\delta\alpha + \bar{M}_1\frac{1}{L^2}[S,\, -C,\, 0;\, -S,\, C,\, 0]^T\delta L \\
&= \frac{1}{L^2}\bar{M}_1(\{r\}\{z\}^T + \{z\}\{r\}^T)\{\delta u\}
\end{aligned}$$

where $\delta L = \delta\bar{u}$ was used. In like manner, we get $[\delta B]_3 = [\delta B]_2$. The collection of terms leads to the geometric stiffness

$$[\,k_G\,] = \frac{1}{L}\bar{F}\{z\}\{z\}^T + \frac{1}{L^2}(\bar{M}_1 + \bar{M}_2)\big[\{r\}\{z\}^T + \{z\}\{r\}^T\big]$$

However, for equilibrium of the element, we have

$$\frac{1}{L^2}(\bar{M}_1 + \bar{M}_2) = \frac{\bar{V}_1}{L} = -\frac{\bar{V}_2}{L} = -\frac{\bar{V}_o}{L}$$

because $V(\bar{x})$ is constant along the beam. On multiplying the vectors and replacing the moments, we get

$$[\,k_G\,] = \frac{\bar{F}_o}{L}\begin{bmatrix}
S^2 & -CS & 0 & CS & -C^2 & 0 \\
-CS & C^2 & 0 & S^2 & -SC & 0 \\
0 & 0 & 0 & 0 & 0 & 0 \\
-S^2 & CS & 0 & S^2 & -CS & 0 \\
CS & -C^2 & 0 & -CS & C^2 & 0 \\
0 & 0 & 0 & 0 & 0 & 0
\end{bmatrix}$$
$$-\frac{\bar{V}_o}{L}\begin{bmatrix}
-2CS & C^2-S^2 & 0 & 2CS & S^2-C^2 & 0 \\
C^2-S^2 & 2CS & 0 & S^2-C^2 & -2CS & 0 \\
0 & 0 & 0 & 0 & 0 & 0 \\
2CS & S^2-C^2 & 0 & -2CS & C^2-S^2 & 0 \\
S^2-C^2 & -2CS & 0 & C^2-S^2 & 2CS & 0 \\
0 & 0 & 0 & 0 & 0 & 0
\end{bmatrix}$$

In local coordinates, these are

$$[\,\bar{k}_G\,] = \frac{\bar{F}_o}{L}\begin{bmatrix}
0 & 0 & 0 & 0 & 0 & 0 \\
0 & 1 & 0 & 0 & -1 & 0 \\
0 & 0 & 0 & 0 & 0 & 0 \\
0 & 0 & 0 & 0 & 0 & 0 \\
0 & -1 & 0 & 0 & 1 & 0 \\
0 & 0 & 0 & 0 & 0 & 0
\end{bmatrix} - \frac{\bar{V}_o}{L}\begin{bmatrix}
0 & 1 & 0 & 0 & -1 & 0 \\
1 & 0 & 0 & -1 & 0 & 0 \\
0 & 0 & 0 & 0 & 0 & 0 \\
0 & -1 & 0 & 0 & 1 & 0 \\
-1 & 0 & 0 & 1 & 0 & 0 \\
0 & 0 & 0 & 0 & 0 & 0
\end{bmatrix}$$

We recognize the first term as the geometric stiffness for a truss [22] in local coordinates. The second term has nodal shears (and hence moments) and therefore we associate it with the bending action. For slender beams, $\bar{V}_o$ is generally not very large and, therefore, we can often neglect this contribution to the stiffness. What this points to is that the dominant contribution to the geometric component of the tangent stiffness comes from the axial load.

The assembled global tangent stiffness is given by

$$[K_T] = \sum_m \left[[k_E]_m + [k_G]_m\right] = \sum_m [T]_m^T \left[[\bar{k}_E]_m + [\bar{k}_G]_m\right] [T]_m$$

just as for the truss.

Example 3.43: Determine the deflected shape of a cantilever beam subjected, separately, to a tip moment and a tip transverse force.

The results for the applied moment are shown in Figure 3.26, and for the transverse load in Figure 3.27. The deformed shapes correspond to each load value and the comparisons are with the elastica solutions developed earlier in the chapter.

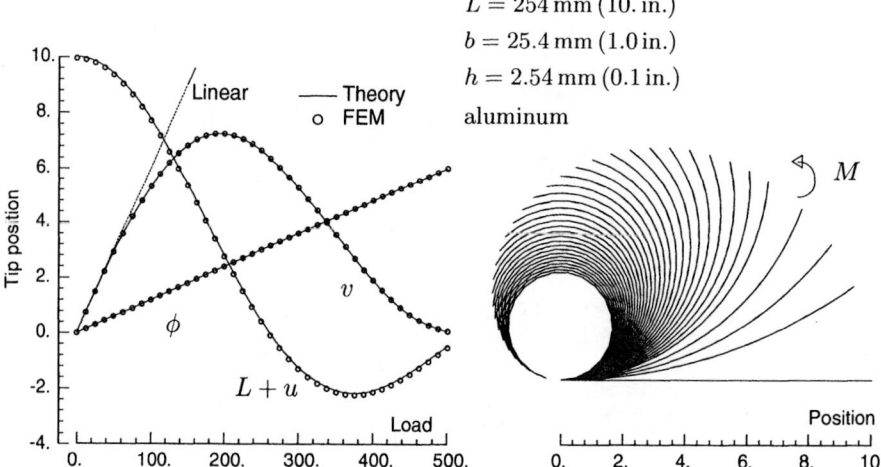

$L = 254\,\mathrm{mm}\,(10.\,\mathrm{in.})$

$b = 25.4\,\mathrm{mm}\,(1.0\,\mathrm{in.})$

$h = 2.54\,\mathrm{mm}\,(0.1\,\mathrm{in.})$

aluminum

Figure 3.26: Comparison of tip deflections and deformed shape at various stages of tip moment loading.

These problems are particularly challenging, because, initially, each load increment causes a large vertical only displacement (since the stiffnesses correspond to the linear case). Consequently, a very large axial force is generated that must be reduced through iteration. This problem can be alleviated in many ways: smaller load increments, under relaxation with $\beta < 1.0$, and using smaller elements. The last of these seems the most appropriate and the results in the figure are for the beam modeled with 20 elements. The trade-off is that the system size is larger but fewer iterations are used.

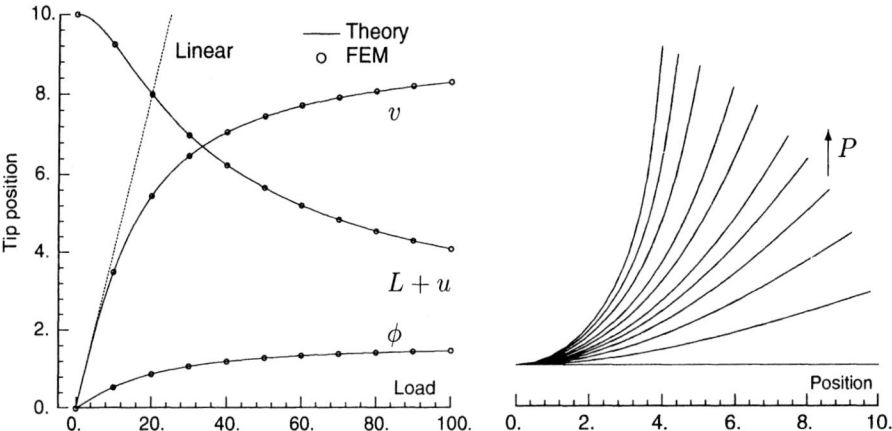

Figure 3.27: Comparison of tip deflections and deformed shape for vertically applied load.

3.6   Corotational Scheme for 3-D Structures

The 3-D structures of interest are thin-walled shells reinforced with slender member frames. We will model shells as a collection of many flat elements. These elements experience in-plane membrane behavior as well as bending action as covered in Chapter 2. In relation to the previous section, we need to pose the corotational scheme in a 3-D geometric description — the local behavior will be as described in Chapter 2. We will take advantage of the fact that the strains are small on the local level to simplify the formulation.

The approach follows the spin matrix formulation, which allows us to describe the effect of the rotating coordinates without making explicit reference to the element formulation.

Spin and Projector Matrix Formulation

We introduce a general formulation of nonlinear problems that captures the spirit of the corotational scheme in truly separating the local behaviors from the global behaviors. To help in the generalization, assume that each element has N nodes with three components of force at each node — we leave consideration of moments until later.

The main objectives are establishing the relation between the local variables and the global variables, from which we can establish the stiffness relations in global variables.

I: Relation between Local and Global Variables

There are $2N$ position variables we are interested in: the global position of each node before deformation $(\hat{x}_{oj})$, and after deformation $(\hat{x}_j)$, where the subscript

j enumerates the nodes. The local positions are given by

$$\hat{\bar{x}}_{oj} = [E_o^T]\{\hat{x}_{oj} - \hat{x}_{o1}\}, \qquad \hat{\bar{x}}_j = [E^T]\{\hat{x}_j - \hat{x}_{o1}\}$$

where $[E_o]$ and $[E]$ are the triads describing the orientation of the element before and after deformation. The local displacements are defined as

$$\hat{\bar{u}}_j = \hat{\bar{x}}_j - \hat{\bar{x}}_{oj}$$

Note that, unlike our formulations of the previous section, the local coordinates of the first node do not necessarily coincide; this will allow us to incorporate all degrees of freedom in our formulation and thus the formal presentation of the equations will appear simple.

The local virtual displacements are related to the global variables by

$$\delta\{\hat{\bar{u}}\}_j = \delta\{[E^T]\{\hat{x}_j - \hat{x}_{o1}\} - \hat{\bar{x}}_{oj}\} = [E^T]\{\delta\hat{u}_j\} + \delta[E^T]\{\hat{x}_j - \hat{x}_{o1}\}$$

Note that, for some vector $\hat{v}$,

$$\delta[E^T]\{v\} = [E^T][S(\hat{v})]\{\delta\beta\}$$

where $\{\beta\}$ is the small rotation spin and the notation $[S(\hat{v})]$ means the skew-symmetric matrix obtained with the components of the corresponding vector; that is, put the components of the vector $\hat{v}$ (i.e., v_x, v_y, v_z) into the rotation array defined in Equation (3.3). Using this, we get

$$\{\delta\hat{\bar{u}}\}_j = [E^T]\{\delta u\}_j + [E^T][S(\hat{x}_j - \hat{x}_{o1})]\{\delta\beta\}$$

This shows how the local displacements $\{\delta\bar{u}\}_j$, global spin $\{\delta\beta\}$, and global displacements $\{\delta u\}_j$ are interrelated. However, the spin is not independent of the displacements because we require that the local spin be zero (since it is rotating with the element). We now establish this constraint.

Imagine a container or wrapper surrounding the element; take the displacements of this container in local coordinates to be given by the interpolations

$$\bar{u}(\hat{r}) = \sum_j h_j(\hat{r})\bar{u}_j, \qquad \bar{v}(\hat{r}) = \sum_j h_j(\hat{r})\bar{v}_j, \qquad \bar{w}(\hat{r}) = \sum_j h_j(\hat{r})\bar{w}_j$$

where $\hat{r}$ means all three components of position. We will make the interpolations explicit in the examples to follow. The continuum rotations are

$$2\bar{\omega}_x = \frac{\partial\bar{w}}{\partial\bar{y}} - \frac{\partial\bar{v}}{\partial\bar{z}}$$

$$2\bar{\omega}_y = \frac{\partial\bar{u}}{\partial\bar{z}} - \frac{\partial\bar{w}}{\partial\bar{x}}$$

$$2\bar{\omega}_z = \frac{\partial\bar{v}}{\partial\bar{x}} - \frac{\partial\bar{u}}{\partial\bar{y}}$$

Introduce a rotation pseudo-vector defined by

$$\{\bar{\phi}\} \equiv [\ C\ ]\left\{\begin{matrix} \bar{\omega}_x \\ \bar{\omega}_y \\ \bar{\omega}_z \end{matrix}\right\} = \sum_j [\ G\ ]_j \{\bar{u}\}_j$$

where $[\ C\ ]$ is some constant matrix, and the $[3 \times 3]$ matrix $[\ G\ ]_j$ comes from evaluating the derivatives of the interpolation functions at each of the nodes. Strictly speaking, the orientation of the local axes should refer to the orientation of the zero spin axis at the centroid of the element; however, we will take it to correspond to the orientation of the 1-2 side of the element. This approximation is consistent with our interest of small strains but large deflections and rotations. For example, in local coordinates, we expect the deformation gradients not to exceed the largest strains, that is, we expect

$$\left|\frac{\partial \bar{u}}{\partial \bar{x}}\right|, \ \left|\frac{\partial \bar{v}}{\partial \bar{y}}\right|, \ \left|\frac{\partial \bar{w}}{\partial \bar{z}}\right| < \epsilon_{max} = \sigma_Y / E \approx 0.01$$

Thus the correction to the orientation, at most, is on the order of 0.01 radians.

Get the spin matrix by considering the variation

$$2\delta\{\bar{\omega}\} = \sum_j [\ G\ ]_j \{\delta\bar{u}\}_j = 0$$

Substitute the expression for $\{\delta\bar{u}\}_j$ into this

$$\sum_j [\ G\ ]_j [E^T]\{\delta u\}_j + \sum_j [\ G\ ]_j + [E^T][S(\hat{x}_j - \hat{x}_{o1})]\{\delta\beta\} = 0$$

from which we can solve for the spin as

$$\{\delta\beta\} = \left[- \sum_j [\ G\ ]_j [E^T][S(\hat{x}_j - \hat{x}_{o1})] \right]^{-1} \left[\sum_j [\ G\ ]_j [E^T]\{\delta\bar{u}\}_j \right] \equiv [\ V\ ]^T \{\delta u\}$$

where $\{\delta u\}$ is the vector of all nodal components and $[\ V\ ]^T$ is given by

$$[\ V\ ]^T = \left[- \sum_j [\ G\ ]_j [E^T][S(\hat{x}_j - \hat{x}_{o1})] \right]^{-1} \left[[\ G\ ]_1 [E^T], [\ G\ ]_2 [E^T], \cdots \right]$$

The matrix $[\ V\ ]^T$ is of size $[3 \times 3N]$ and is called the *spin matrix*.

We can now relate the local virtual displacements to only the global variables by

$$\{\delta\hat{u}\}_j = [E^T]\{\delta\hat{u}_j\} + [E^T][S(\hat{x}_j - \hat{x}_{o1})]([\ V\ ]^T\{\delta\hat{u}\})$$

or

$$\{\delta\bar{u}\} = [\ T\ ]\{\delta u\}, \quad [\ T\ ] = \left[\begin{matrix} [E^T] & & \\ & [E^T] & \\ & & \ddots \end{matrix} \right] + \left[\begin{matrix} [E^T][S_1] \\ [E^T][S_2] \\ \vdots \end{matrix} \right] [\ V\ ]^T$$

This fundamental relation will now be used to determine the relation between the global and local load terms.

In the 2-D developments of the corotational scheme, we saw the recurrence of the transformation associated with the transformation of coordinates. We will now make that aspect of the transformation explicit. First note we can establish by expansion that

$$[S(\hat{v}^*)] = [\ E\ ][S(\hat{\bar{v}})][E^T]\,, \qquad \{\hat{v}^*\} = [\ E\ ]\{\hat{\bar{v}}\}$$

The denominator in the expression for $[\ V\ ]$ becomes

$$[\ D\ ] = -\sum_j [\ G\ ]_j [E^T][S(\hat{x}_j - \hat{x}_{o1})] = -\sum_j [\ G\ ]_j [S(\hat{\bar{x}}_j)][\ E\ ]^T = [\ \bar{D}\ ][\ E\ ]^T$$

We will show through examples that $[\ \bar{D}\ ][\ E\ ]^T = [\ E\ ]^T$. The spin matrix simplifies to

$$[\ V\ ]^T = [\ E\ ]\left[[\ G\ ]_1[E^T],\ [\ G\ ]_2[E^T],\ \cdots\right]$$

The transformation matrix then simplifies to

$$[\ T\ ] = \left[\begin{bmatrix}[\ I_3\ ] & & \\ & [\ I_3\ ] & \\ & & \ddots\end{bmatrix} + \begin{bmatrix}[S(\hat{\bar{x}}_1)] \\ [S(\hat{\bar{x}}_2)] \\ \vdots\end{bmatrix}\Big[[\ G\ ]_1,\ [\ G\ ]_2,\ \cdots\Big]\right]\begin{bmatrix}[E^T] & & \\ & [E^T] & \\ & & \ddots\end{bmatrix}$$

We will write this expression as

$$[\ T\ ] = [\ P\ ][E^T]$$

and refer to $[\ P\ ]$ as a *projector matrix*. It depends only on the local coordinates. Discussions of the projector matrix can be found in References [52, 54, 55].

The virtual work in global variables must equal the virtual work in local variables, hence

$$\{F\}^T\{\delta u\} = \{\bar{F}\}^T\{\delta\bar{u}\} = \{\bar{F}\}^T[\ T\ ]\{\delta u\}$$

From this we conclude that

$$\{F\} = [\ T\ ]^T\{\bar{F}\} = [\ E\ ][\ P\ ]^T\{\bar{F}\}$$

The explicit form is

$$\left\{\begin{matrix}\hat{F}_1 \\ \hat{F}_2 \\ \vdots\end{matrix}\right\} = \left[\begin{bmatrix}[\ E\ ] & & \\ & [\ E\ ] & \\ & & \ddots\end{bmatrix} - [\ V\ ]\Big[[\ S_1\ ][\ E\ ],\ [\ S_2\ ][\ E\ ],\ \cdots\Big]\right]\left\{\begin{matrix}\hat{\bar{F}}_1 \\ \hat{\bar{F}}_2 \\ \vdots\end{matrix}\right\} \qquad (3.15)$$

Note that in the second term we have

$$\sum_j [\ S_j\ ][\ E\ ]\{\hat{\bar{F}}\}_j = \sum_j [S(\hat{x}_j - \hat{x}_{o1})]\{\hat{F}^*\}_j = \sum_j (\hat{x}_j - \hat{x}_{o1}) \times \hat{F}_j^* = \sum \text{moments}$$

with the last form coming from Equation (3.3). In this, $\hat{F}_j^*$ are the transformed components of the local force vector. As equilibrium is achieved, this term will tend to zero. Later, when we introduce approximate forms of our equations, the transformation relation will be approximated by *a priori* assuming equilibrium and dropping the term associated with [V]. We will not do that now because we intend to obtain the stiffness relations by essentially differentiating this relation.

What this discussion highlights, however, is the possibility (at least when equilibrium is achieved) that local quantities (including the stiffness matrix) have a simple coordinate rotation relation to their global counterparts. We will keep that in mind when we develop the stiffness relations.

Example 3.44: Establish an explicit form for the constituent arrays of the spin and projector matrices of a truss element.

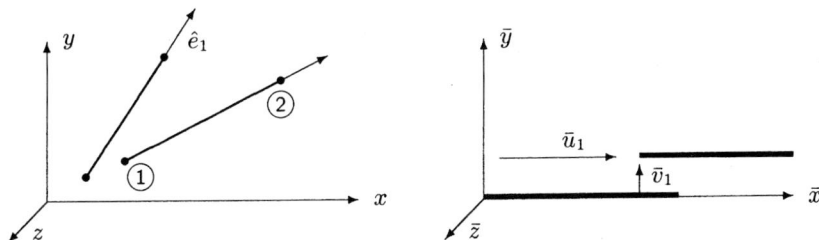

Figure 3.28: Global and local positions of a truss element.

Take the displacements in local coordinates of the element container as given by the simple interpolations

$$\bar{u}(\bar{x}) = \sum_j h_j(\bar{x})\bar{u}_j, \qquad \bar{v}(\bar{x}) = \sum_j h_j(\bar{x})\bar{v}_j, \qquad \bar{w}(\bar{x}) = \sum_j h_j(\bar{x})\bar{w}_j$$

where we have

$$h_1(\bar{x}) = [1 - \frac{\bar{x}}{L}], \qquad h_2(\bar{x}) = [\frac{\bar{x}}{L}]$$

Note that these describe the behavior of the container and not the element. The rotations are

$$2\bar{\omega}_x = \frac{\partial \bar{w}}{\partial \bar{y}} - \frac{\partial \bar{v}}{\partial \bar{z}} = 0$$

$$2\bar{\omega}_y = \frac{\partial \bar{u}}{\partial \bar{z}} - \frac{\partial \bar{w}}{\partial \bar{x}} = \frac{1}{L}[\bar{w}_1 - \bar{w}_2]$$

$$2\bar{\omega}_z = \frac{\partial \bar{v}}{\partial \bar{x}} - \frac{\partial \bar{u}}{\partial \bar{y}} = \frac{1}{L}[-\bar{v}_1 + \bar{v}_2]$$

The first equation is null, but we keep it in the formulation so that the structure of the equations remain 3-D. Introduce a rotation pseudo-vector defined by

$$\begin{Bmatrix} \bar{\phi}_x \\ \bar{\phi}_y \\ \bar{\phi}_z \end{Bmatrix} = 2 \begin{Bmatrix} \bar{\omega}_x \\ \bar{\omega}_y \\ \bar{\omega}_z \end{Bmatrix} = \frac{1}{L} \begin{bmatrix} 0 & 0 & 0 \\ 0 & 0 & 1 \\ 0 & -1 & 0 \end{bmatrix} \begin{Bmatrix} \bar{u} \\ \bar{v} \\ \bar{w} \end{Bmatrix}_1 + \frac{1}{L} \begin{bmatrix} 0 & 0 & 0 \\ 0 & 0 & -1 \\ 0 & 1 & 0 \end{bmatrix} \begin{Bmatrix} \bar{u} \\ \bar{v} \\ \bar{w} \end{Bmatrix}_2 \equiv \sum_j [\, G \,]_j \{ \bar{u} \}_j$$

The $[\, G \,]_j$ arrays are

$$[\, G \,]_1 = \frac{1}{L} \begin{bmatrix} 0 & 0 & 0 \\ 0 & 0 & 1 \\ 0 & -1 & 0 \end{bmatrix}, \qquad [\, G \,]_2 = \frac{1}{L} \begin{bmatrix} 0 & 0 & 0 \\ 0 & 0 & -1 \\ 0 & 1 & 0 \end{bmatrix}$$

and the $[\, S_j \,]$ arrays are

$$[S(\hat{\bar{x}}_1)] = [\, \mathbf{0}_3 \,], \qquad [S(\hat{\bar{x}}_2)] = \begin{bmatrix} 0 & 0 & 0 \\ 0 & 0 & -\bar{x}_2 \\ 0 & \bar{x}_2 & 0 \end{bmatrix}$$

and $\bar{x}_2 = L$. The denominator matrix becomes

$$[\, \bar{D} \,] = -\frac{1}{L} \begin{bmatrix} 0 & 0 & 0 \\ 0 & 0 & -1 \\ 0 & 1 & 0 \end{bmatrix} \begin{bmatrix} 0 & 0 & 0 \\ 0 & 0 & -L \\ 0 & L & 0 \end{bmatrix} = \begin{bmatrix} 0 & 0 & 0 \\ 0 & 1 & 0 \\ 0 & 0 & 1 \end{bmatrix}$$

The matrix $[\, V \,]^T$ is of size $[3 \times 6]$.

Example 3.45: Establish an explicit form for the constituent arrays of the spin and projector matrices for a membrane element.

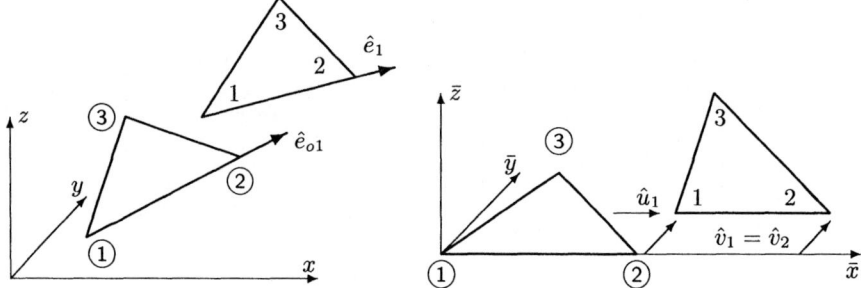

Figure 3.29: Global and local positions of an element.

We take the displacements in local coordinates of the element container to be given by

$$\bar{u}(\bar{x}, \bar{y}) = \sum_j h_j(\bar{x}, \bar{y}) \bar{u}_j, \qquad \bar{v}(\bar{x}, \bar{y}) = \sum_j h_j(\bar{x}, \bar{y}) \bar{v}_j, \qquad \bar{w}(\bar{x}, \bar{y}) = \sum_j h_j(\bar{x}, \bar{y}) \bar{w}_j$$

where the interpolations are those of the three noded triangle given in Chapter 2. Note that we are allowing local displacements in all three direction. The rotations are

$$2\bar{\omega}_x = \frac{\partial \bar{w}}{\partial \bar{y}} - \frac{\partial \bar{v}}{\partial \bar{z}} = \frac{1}{2A}\sum_j c_j w_j$$

$$2\bar{\omega}_y = \frac{\partial \bar{u}}{\partial \bar{z}} - \frac{\partial \bar{w}}{\partial \bar{x}} = \frac{1}{2A}\sum_j -b_j w_j$$

$$2\bar{\omega}_z = \frac{\partial \bar{v}}{\partial \bar{x}} - \frac{\partial \bar{u}}{\partial \bar{y}} = \frac{1}{2A}\sum_j b_j v_j - \frac{1}{2A}\sum_j c_j u_j$$

where the coefficients b_j and c_j are associated with the interpolation functions for the linear triangle and are evaluated with respect to the local deformed configuration. The spin $\bar{\omega}_z$ represents the average rotation of the element in the x-y plane as shown in Chapter 1. However, since the container is a thin lamina (and not a 3-D continuum) the spins $\bar{\omega}_x$ and $\bar{\omega}_x$ are only half of the rotations out of the x-y plane. Introduce a rotation pseudo-vector given by

$$\left\{\begin{matrix}\bar{\phi}_x\\\bar{\phi}_y\\\bar{\phi}_z\end{matrix}\right\} = \left\{\begin{matrix}2\bar{\omega}_x\\2\bar{\omega}_y\\\bar{\omega}_z\end{matrix}\right\} = \frac{1}{2A_o}\sum_j^3\begin{bmatrix}0 & 0 & c\\0 & 0 & -b\\-c\frac{1}{2} & b\frac{1}{2} & 0\end{bmatrix}_j\left\{\begin{matrix}\bar{u}\\\bar{v}\\\bar{w}\end{matrix}\right\}_j = \sum_j [\, G\,]_j\{\bar{u}\}_j$$

The nonzero local positions are $\bar{x}_2$, $\bar{x}_3$, $\bar{y}_3$, and the area is $2A = \bar{x}_2\bar{y}_3$. Using the definitions of b and c from Chapter 2, the $[\,G\,]_j$ arrays are

$$[\,G\,]_1 = \frac{1}{\bar{x}_2\bar{y}_3}\begin{bmatrix}0 & 0 & (\bar{x}_3 - \bar{x}_2)\\0 & 0 & -\bar{y}_3\\(\bar{x}_3 - \bar{x}_2)\frac{1}{2} & \bar{y}_3\frac{1}{2} & 0\end{bmatrix}$$

$$[\,G\,]_2 = \frac{1}{\bar{x}_2\bar{y}_3}\begin{bmatrix}0 & 0 & -\bar{x}_3\\0 & 0 & \bar{y}_3\\\bar{x}_3\frac{1}{2} & -\bar{y}_3\frac{1}{2} & 0\end{bmatrix}, \qquad [\,G\,]_3 = \frac{1}{\bar{x}_2\bar{y}_3}\begin{bmatrix}0 & 0 & -\bar{x}_2\\0 & 0 & 0\\\bar{x}_2\frac{1}{2} & 0 & 0\end{bmatrix}$$

and the $[\,S_j\,]$ arrays are

$$[S(\hat{\bar{x}}_1)] = \begin{bmatrix}0 & 0 & 0\\0 & 0 & 0\\0 & 0 & 0\end{bmatrix}, \quad [S(\hat{\bar{x}}_2)] = \begin{bmatrix}0 & 0 & 0\\0 & 0 & -\bar{x}_2\\0 & \bar{x}_2 & 0\end{bmatrix}, \quad [S(\hat{\bar{x}}_3)] = \begin{bmatrix}0 & 0 & \bar{y}_3\\0 & 0 & -\bar{x}_3\\-\bar{y}_3 & \bar{x}_3 & 0\end{bmatrix}$$

The denominator matrix becomes

$$[\,\bar{D}\,] = -\frac{1}{\bar{x}_2\bar{y}_3}\begin{bmatrix}0 & 0 & -\bar{x}_3\\0 & 0 & \bar{y}_3\\\bar{x}_3\frac{1}{2} & -\bar{y}_3\frac{1}{2} & 0\end{bmatrix}\begin{bmatrix}0 & 0 & 0\\0 & 0 & -\bar{x}_2\\0 & \bar{x}_2 & 0\end{bmatrix}$$

$$-\frac{1}{\bar{x}_2\bar{y}_3}\begin{bmatrix}0 & 0 & -\bar{x}_2\\0 & 0 & 0\\\bar{x}_2\frac{1}{2} & 0 & 0\end{bmatrix}\begin{bmatrix}0 & 0 & \bar{y}_3\\0 & 0 & -\bar{x}_3\\-\bar{y}_3 & \bar{x}_3 & 0\end{bmatrix} = \begin{bmatrix}1 & 0 & 0\\0 & 1 & 0\\0 & 0 & 1\end{bmatrix}$$

The matrix $[\,V\,]^T$ is of size $[3 \times 9]$.

II: Global Stiffness Relations

At the global level, the variation of the nodal forces leads to

$$\{\delta F\} = [\frac{\partial F}{\partial u}]\{\delta u\} = [\, k_T\,]\{\delta u\}$$

where $[\, k_T\,]$ is the element tangent stiffness matrix in global coordinates. Substituting for $\{F\}$ in terms of local variables, we get

$$
\begin{aligned}
\{\delta F\} \;=\;& [\, T\,]^T\{\delta \bar{F}\} + \delta[\, T\,]^T\{\bar{F}\} \\
=\;& [\, T\,]^T[\, \bar{k}\,]\{\delta \bar{u}\} + \left\{ \delta[\, E\,]\{\hat{\bar{F}}_1\}, \; \delta[\, E\,]\{\hat{\bar{F}}_2\}, \cdots \right\}^T \\
& -[\, V\,]\left[[S(\delta \hat{x}_1)][\, E\,], \; [S(\delta \hat{x}_2)][\, E\,], \cdots \right]\{\bar{F}\} \\
& -[\, V\,]\left[[\, S_1\,]\delta[\, E\,], \; [\, S_2\,]\delta[\, E\,], \cdots \right]\{\bar{F}\}
\end{aligned}
$$

We did not take the variation of $[\, V\,]$ because it multiplies a term that will go to zero at equilibrium and thus is associated with a negligible contribution. We see that the tangent stiffness relation is comprised of two parts: one is related to the elastic stiffness properties of the element, the other is related to the rotation of the element. Noting that

$$\delta[\, E\,]\{\hat{\bar{v}}\} = [S(\delta \beta)][\, E\,]\{\hat{\bar{v}}\} = [S(\delta \beta)]\{\hat{v}^*\} = -[S(\hat{v}^*)]\{\delta \beta\} = -[S(\hat{v}^*)][\, V\,]^T\{\delta u\}$$

gives

$$
\begin{aligned}
\{\delta F\} \;=\;& [\, T\,]^T[\, \bar{k}\,][\, T\,]\{\delta u\} \\
& - \left[[S(\hat{F}_1^*)], \; [S(\hat{F}_2^*)], \cdots \right]^T [\, V\,]^T\{\delta u\} + [\, V\,]\left[[S(\hat{F}_1^*)], \; [S(\hat{F}_2^*)], \cdots \right]\{\delta u\} \\
& + [\, V\,]\left[\sum_j [S(\hat{x}_j - x_{oj})][S(\hat{F}_j^*)] \right][\, V\,]^T\{\delta u\}
\end{aligned}
$$

Each term is post-multiplied by $\{\delta u\}$, and therefore we can associate each term with a stiffness matrix. The latter contribution is the geometric stiffness matrix. The first set of terms

$$[\, k_E\,] \equiv [\, T\,]^T[\, \bar{k}\,][\, T\,] = [\, E\,][\, P\,]^T[\, \bar{k}\,][\, P\,][\, E\,]^T$$

gives the elastic stiffness. We therefore recognize the global stiffness matrix as the components of the local stiffness matrices transformed to the current orientation of the element. However, contrary to the simpler cases of the previous section, it is not just the local stiffness but the local stiffness times the projector matrix.

The remaining set of terms gives the geometric contribution to the stiffness matrix. That is,

$$[k_G] = -\Big[[S(\hat{F}_1^*)],\ [S(\hat{F}_2^*)],\ \cdots\Big]^T [\ V\ ]^T + [\ V\ ]\Big[[S(\hat{F}_1^*)],\ [S(\hat{F}_2^*)],\ \cdots\Big]$$

$$+\ [\ V\ ]\Big[\sum_j [S(\hat{x}_j - \hat{x}_{oj})][S(\hat{F}_j^*)]\Big][\ V\ ]^T$$

The first two terms are the transpose of each other and therefore form a symmetric combination. The third term, in general, is nonsymmetric, but approaches symmetry as equilibrium is achieved. We demonstrate this in an example to follow.

To introduce the projector matrix description of the tangent stiffness, recall that for $\{\hat{v}^*\} = [\ E\ ]\{\hat{\bar{v}}\}$ we have the relations

$$[S(\hat{v}^*)] = [\ E\ ][S(\hat{\bar{v}})][E^T],\quad [\ V\ ]^T = [\ E\ ]\Big[[\ G\ ]_1[E^T],\ [\ G\ ]_2[E^T],\ \cdots\Big]$$

and $[\ D\ ] = [E^T]$. Substituting this into, for example, the second geometric stiffness term gives

$$[k_{G2}] = \Big[[\ G\ ]_1[E^T],\ [\ G\ ]_2[E^T],\ \cdots\Big]^T [E^T]\Big[[\ E\ ][S(\hat{\bar{F}}_1)][\ E\ ]^T,\ \cdots\Big]$$

$$= \begin{bmatrix} [\ E\ ] & & \\ & [\ E\ ] & \\ & & \ddots \end{bmatrix}\begin{bmatrix} [\ G\ ]_1^T \\ [\ G\ ]_2^T \\ \vdots \end{bmatrix}\Big[[S(\hat{\bar{F}}_1)],\ [S(\hat{\bar{F}}_2)],\ \cdots\Big]\begin{bmatrix} [\ E\ ]^T & & \\ & [\ E\ ]^T & \\ & & \ddots \end{bmatrix}$$

The core terms are referred only to the local coordinates. Similar substitutions for the other terms give a similar conclusion. The contribution to the geometric stiffness reduces to

$$[k_G] = [\ E\ ]\Big[[\bar{k}_{G1}] + [\bar{k}_{G2}] + [\bar{k}_{G3}]\Big][E^T]$$

Each matrix is of size $[3N \times 3N]$ and the sub $[3 \times 3]$ matrices are given by

$$[\bar{k}_{G1ij}] = -[S(\hat{\bar{F}}_i)][\ G_j\ ],\qquad [\bar{k}_{G2ij}] = [\ G_i\ ]^T[S(\hat{\bar{F}}_j)]$$

$$[\bar{k}_{G3ij}] = [\ G_i\ ]^T \sum_k [S(\hat{\bar{x}}_k)][S(\hat{\bar{F}}_k)][\ G_j\ ]$$

The first two terms are the transpose of each other and therefore combine to form a symmetric matrix. The third term is symmetric.

The second and third contributions may be combined to form a product with the projector matrix. Thus

$$[k_G] = -[\ E\ ]\Big[[S_9(\hat{\bar{F}})][\ G_9\ ] + [\ G_9\ ]^T[S_9(\hat{\bar{F}}_9)][\ P\ ]\Big][E^T]$$

where

$$[S_9(\hat{\bar{F}})] = \left[[S(\hat{\bar{F}}_1)], [S(\hat{\bar{F}}_2)], \cdots \right], \qquad [G_9] = \left[[\, G\,]_1^T, [\, G\,]_2^T, \cdots \right]$$

Example 3.46: Show that at equilibrium, the third contribution to the geometric stiffness becomes symmetric.

We first recall the useful formulas

$$[S(\hat{a} \times \hat{b})] = \hat{b}\hat{a}^T - \hat{a}\hat{b}^T, \qquad [S(\hat{a})][S(\hat{b})] = \hat{b}\hat{a}^T - (\hat{a} \cdot \hat{b})[\, I\,]$$

From this, we conclude that

$$\text{sym } [S(\hat{a})][S(\hat{b})] \;=\; \tfrac{1}{2}(\hat{b}\hat{a}^T + \hat{a}\hat{b}^T) - (\hat{a} \cdot \hat{b})[\, I\,]$$

$$\text{anti-sym } [S(\hat{a})][S(\hat{b})] \;=\; \tfrac{1}{2}(\hat{b}\hat{a}^T - \hat{a}\hat{b}^T) = \tfrac{1}{2}[S(\hat{a} \times \hat{b})]$$

The antisymmetric part of the stiffness inside the [V] brackets is then

$$\tfrac{1}{2} \sum_j [S((\hat{x}_j - \hat{x}_{o1}) \times (F_j^*))]$$

Remembering that $[S()]$ is a linear operator; we see that the antisymmetric part
is zero because the summation is a sum of moments.

Example 3.47: Determine the geometric stiffness matrix for a 3-D truss.

We take advantage of the fact that the strains are small to simplify some of the
above relations. Assume that all local relative displacements $\hat{\bar{u}}_j - \hat{\bar{u}}_1$ are small.
That is, assume $\hat{\bar{x}}_j - \hat{\bar{x}}_1 \approx \hat{\bar{x}}_{oj} - \hat{\bar{x}}_{o1}$ are small.

Introducing the notations

$$\bar{F}_{x1} = -\bar{F}_o, \quad \bar{F}_{x2} = +\bar{F}_o, \quad \bar{F}_{y1} = -\bar{V}_o, \quad \bar{F}_{y2} = +\bar{V}_o, \quad \bar{F}_{z1} = -\bar{W}_o, \quad \bar{F}_{z2} = +\bar{W}_o$$

which are based on the equilibrium conditions, we can obtain the $[S(\hat{\bar{F}}_i)]$ matrices
as

$$[S(\hat{\bar{F}}_1)] = \begin{bmatrix} 0 & \bar{W}_o & -\bar{V}_o \\ -\bar{W}_o & 0 & F_o \\ \bar{V}_o & -F_o & 0 \end{bmatrix}, \qquad [S(\hat{\bar{F}}_2)] = \begin{bmatrix} 0 & -\bar{W}_o & \bar{V}_o \\ \bar{W}_o & 0 & -F_o \\ -\bar{V}_o & F_o & 0 \end{bmatrix}$$

We already established the [G]$_j$ matrices. Performing the required multiplica-
tions leads to

$$[\bar{k}_{G1} + \bar{k}_{G2}]_{11} = \frac{1}{L}\begin{bmatrix} 0 & -\bar{V}_o & -\bar{W}_o \\ -\bar{V}_o & 2F_o & 0 \\ -\bar{W}_o & 0 & 2F_o \end{bmatrix}, \qquad [\bar{k}_{G3}]_{11} = \frac{1}{L}\begin{bmatrix} 0 & 0 & 0 \\ 0 & -F_o & 0 \\ 0 & 0 & -F_o \end{bmatrix}$$

The total geometric stiffness is then

$$[\, \bar{k}_G\,]_{11} = \frac{\bar{F}_o}{L}\begin{bmatrix} 0 & 0 & 0 \\ 0 & 1 & 0 \\ 0 & 0 & 1 \end{bmatrix} + \frac{\bar{V}_o}{L}\begin{bmatrix} 0 & -1 & 0 \\ -1 & 0 & 0 \\ 0 & 0 & 0 \end{bmatrix} + \frac{\bar{W}_o}{L}\begin{bmatrix} 0 & 0 & -1 \\ 0 & 0 & 0 \\ -1 & 0 & 0 \end{bmatrix}$$

The remainder of the $[6 \times 6]$ array is given by

$$[\, \bar{k}_G \,]_{12} = [\, \bar{k}_G \,]_{21} = -[\, \bar{k}_G \,]_{22} = -[\, \bar{k}_G \,]_{11}$$

It is worth noting that we recover the result for the 2-D frame (with bending) and not the 2-D truss (without bending). We will consider the effect of moments presently.

In frame problems with slender members, the transverse shear forces V_o and W_o are not very large, and their contribution to the geometric stiffness is often neglected.

Example 3.48: Illustrate some of the differences between a linear and a nonlinear analysis.

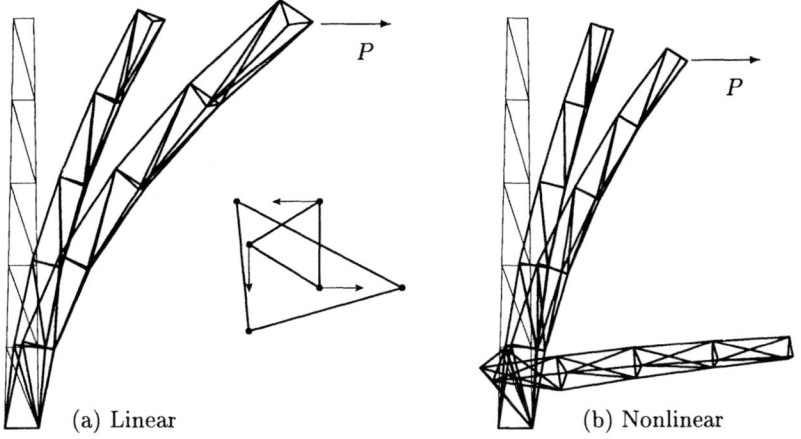

(a) Linear   (b) Nonlinear

Figure 3.30: Contrast between a linear and a nonlinear analysis for a 3-D frame structure.

Figure 3.30 shows a tower loaded by a horizontal force. This is a truss structure, hence the members are triangulated so as to avoid a mechanism. The cross-section is triangular.

A linear analysis is such that the displacements at each load level are proportional. A consequence of this is the exaggerated vertical motion. The nonlinear analysis, by contrast, shows a lowering of the tower. Even more erroneously, the tip members elongate. These results are especially evident when there are rotations. Consider the triangular cross-section experiencing a small rotation as shown in the inset: the small deflection analysis gives that two corners move horizontal but opposite, while the third corner moves vertically down. If this motion is not updated but extended, say, for the length of the triangle, then it is easy to see how the moved corners form a very large triangle. This example clearly shows the need to update the orientation of a structure during a deformation.

Also, quite significantly, a nonlinear analysis can predict failure. In this case, members on the right side are in compression, which means they experience a loss of stiffness with increasing load. Eventually, this leads to a buckling collapse of the structure. In Chapter 6, we discuss such nonlinear effects in greater detail.

Example 3.49: Determine an approximation to the geometric stiffness for membrane shells.

To simplify the relations, first note that the summations involving the interpolation coefficients b_j, c_j are unaffected by rigid body displacements, since

$$\sum_j b_j = 0, \qquad \sum_j c_j = 0$$

That is, we can add, respectively, $\hat{x}_{o1} - \hat{x}_1$ to each node without affecting the result. The elastic stiffness $[\bar{k}_{Eij}]$ has zero contributions associated with the w_i degree of freedom, hence the contribution to the tangent stiffness will be that of the geometric stiffness. Because $[\bar{k}_{G3ij}]$ has zeros on the diagonal and neglecting $\bar{F}_z$, then the only significant contribution is

$$\bar{k}_{G33} = \frac{1}{4A}[\bar{F}_x b + \bar{F}_y c]_1$$

From Chapter 2, we have that the nodal forces are related to the stresses by

$$\bar{F}_x = \frac{h}{2}[b_1 \sigma_{xx} + c_1 \sigma_{xy}], \qquad \bar{F}_y = \frac{h}{2}[c_1 \sigma_{yy} + b_1 \sigma_{xy}]$$

Then letting $\bar{N}_{xx} = \sigma_{xx} h$, and so on, we get

$$\bar{k}_{G33ij} = \frac{1}{4A}\bar{N}_{xx} b_i b_j + \frac{1}{4A}\bar{N}_{yy} c_i c_j + \frac{1}{4A}\bar{N}_{yy}(b_i c_j + c_i b_j)$$

All other components are zero. We discuss more about the geometric stiffness in Chapter 6.

Example 3.50: Determine the deflected shape of a deep cantilever beam subjected to a uniform tip transverse traction.

$L = 254\,\text{mm}\,(10.\,\text{in.})$
$b = 25.4\,\text{mm}\,(1.0\,\text{in.})$
$h = 2.54\,\text{mm}\,(0.1\,\text{in.})$
aluminum

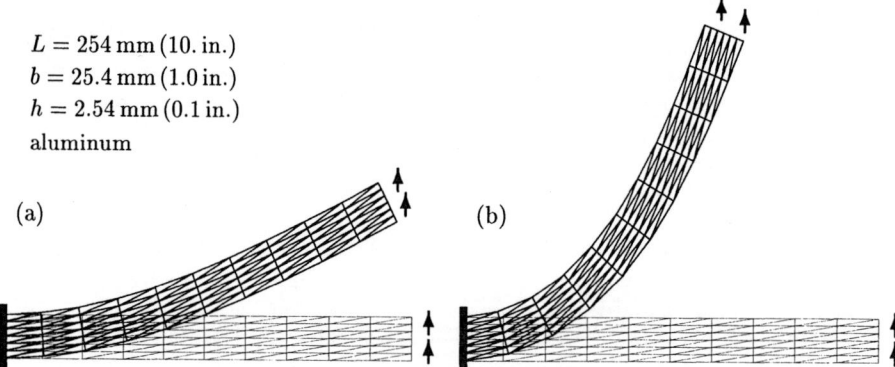

(a)   (b)

Figure 3.31: Large deflection of a deep cantilevered beam. (a) Linear displacement discretization. (b) Linear displacement discretization with rotational degrees of freedom.

The results for the uniformly applied traction on the end are shown in Figure 3.31. This problem has only the in-plane membrane action and therefore the difference is between the performance of the CST element and the MRT element. This problem is dominated by the in-plane rotation action and the comparison shows the advantage of the MRT element — many more CST elements would be required through the depth in order to get comparable results.

These results also show that the corotational scheme gives the benefit of utilizing the rotational degrees of freedom of the MRT element in contrast to to the total Lagrangian scheme.

Bending Behaviors

For our thin-walled structures, the geometry of the deformation is described in terms of the nodal displacements and rotations, and the nodal loads comprise both forces and moments. We now consider the effects of moments within the corotational formulation. Specifically, we consider the flat faceted element shown in Figure 3.32.

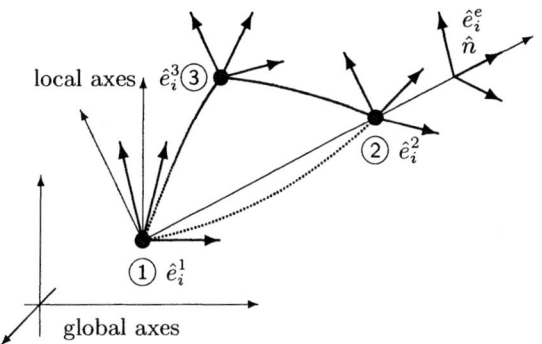

Figure 3.32: Triads associated with the triangular element.

At the local level, the strain producing bending degrees of freedom are

$$\{\bar{u}\}_j = \{\bar{u},\, \bar{v},\, \bar{w},\, \bar{\phi}_x,\, \bar{\phi}_y,\, \bar{\phi}_z\}_j^T$$

The corresponding nodal forces and moments are obtained from

$$\{\bar{F}\} = [\,\bar{k}\,]\{\bar{u}\}$$

where $[\,\bar{k}\,]$ is the $[9 \times 9]$ membrane element stiffness established in Chapter 2 and augmented to size $[18 \times 18]$, added to the $[9 \times 9]$ bending element stiffness also established in Chapter 3 and augmented to size $[18 \times 18]$. In the following discussion, the degrees-of-freedom for each element will be arranged as

$$\{\bar{u}\} = \{\{\bar{u}\}_1,\, \{\bar{\phi}\}_1;\, \{\bar{u}\}_2,\, \{\bar{\phi}\}_2;\, \{\bar{u}\}_3,\, \{\bar{\phi}\}_3\}^T$$

The local angles of twist in local coordinates are obtained as

$$2\bar{\phi}_{jx} \;=\; \hat{e}_3 \cdot \hat{n}_2^j - \hat{e}_2 \cdot \hat{n}_3^j$$

$$2\bar{\phi}_{jy} = \hat{e}_1 \cdot \hat{n}_3^j - \hat{e}_3 \cdot \hat{n}_1^j \quad \text{or} \quad 2 \begin{Bmatrix} \bar{\phi}_x \\ \bar{\phi}_y \\ \bar{\phi}_z \end{Bmatrix}_j = \begin{bmatrix} 0 & -\hat{n}_3 & \hat{n}_2 \\ \hat{n}_3 & 0 & -\hat{n}_1 \\ -\hat{n}_2 & -\hat{n}_1 & 0 \end{bmatrix}_j \begin{Bmatrix} \hat{e}_1 \\ \hat{e}_2 \\ \hat{e}_3 \end{Bmatrix}$$

$$2\bar{\phi}_{jz} = \hat{e}_2 \cdot \hat{n}_1^j - \hat{e}_1 \cdot \hat{n}_2^j$$

While the local twist $\bar{\phi}_{jz}$ does not contribute to the bending, we retain it in the formulation because it is the drilling degree of freedom that will contribute to the membrane action.

The key to relating the local variables to the global variables is to take the variation of the relation for the local angles of twist — we have already established the corresponding relations for the displacements. Noting relations such as

$$\delta\hat{e}_3 = -[S(\hat{e}_3)]\{\delta\beta\} = -[S(\hat{e}_3)][\,V\,]^T\{\delta u\}, \qquad \{\delta\hat{n}_3\} = -[S(\hat{n}_3)]\{\delta\hat{\phi}\}$$

we get

$$2\delta \begin{Bmatrix} \bar{\phi}_x \\ \bar{\phi}_y \\ \bar{\phi}_z \end{Bmatrix}_j = \begin{bmatrix} 0 & -\hat{n}_3 & \hat{n}_2 \\ \hat{n}_3 & 0 & -\hat{n}_1 \\ -\hat{n}_2 & -\hat{n}_1 & 0 \end{bmatrix}_j \begin{Bmatrix} [S(\hat{e}_1)] \\ [S(\hat{e}_2)] \\ [S(\hat{e}_3)] \end{Bmatrix} \{\beta\} + \begin{bmatrix} 0 & -\hat{e}_3 & \hat{e}_2 \\ \hat{e}_3 & 0 & -\hat{e}_1 \\ -\hat{e}_2 & -\hat{e}_1 & 0 \end{bmatrix} \begin{Bmatrix} [S(\hat{n}_1)] \\ [S(\hat{n}_2)] \\ [S(\hat{n}_3)] \end{Bmatrix}_j \{\delta\phi$$

which we write as

$$2\{\delta\bar{\phi}\}_j = [N^*][S^*(\hat{e})]\{\beta\} - [\,E^*\,][S^*(\hat{n})]\{\delta\phi\}$$

For small elements, where the orientation of the nodes are approximately the same as that of the element, we have the approximations

$$[N^*][S^*(\hat{e})] \approx 2[E^T], \qquad [\,E^*\,][S^*(\hat{n})] \approx 2[E^T]$$

from which we get that

$$\begin{aligned} \{\delta\bar{\phi}\}_j &= -[E^T]\{\delta\beta\} + [E^T]\{\delta\phi\}_j = -[E^T][\,V\,]^T\{\delta u\} + [E^T]\{\delta\phi\}_j \\ &= -\Big[[\,G\,]_1[\,E\,]^T, [\,G\,]_2[\,E\,]^T, \cdots\Big]\{\delta u\} + [E^T]\{\delta\phi\}_j \end{aligned}$$

which says that the local twist is the difference between the global twists and the rotation of the element; this is as expected. The total degrees of freedom are related through

$$\{\bar{u}\} = [\,P\,][E^T]\{u\}$$

but now the projector matrix is of size $[18 \times 18]$ and constructed as

$$[P_{ij}] = [\,I_6\,] - [\,S\,]_i[\,G\,]_j, \quad [\,S\,]_i = \begin{Bmatrix} -[S(\hat{\tilde{x}}_i)] \\ [\,I_3\,] \end{Bmatrix}, \quad [\,G\,]_i = \{[\,G\,]_i, [\,0_3\,]\}$$

The virtual work in global variables must equal the virtual work in local variables. At the global level, we have the $3N$ displacements $\{\delta u\}_j$ and the $3N$ rotations $\{\delta\phi\}_j$, and the corresponding forces and moments $\{F\}_j$, $\{M\}_j$. Equating the global and local virtual works we conclude as before

$$\{F\} = [\ T\ ]^T\{\bar{F}\} = [\ E\ ][\ P\ ]^T\{\bar{F}\}$$

Working in a manner similar to before, we get a result that is formally the same. That is,

$$[\ k\ ] = [\ E\ ]\Big[[\ P\ ]_T[\ k\ ][\ P\ ] - [S_{18}(\hat{\bar{F}})][G_{18}] - [G_{18}]^T[S_9(\hat{\bar{F}}_{18})][\ P\ ]\Big][E^T]$$

where

$$[S_{18}(\hat{\bar{F}})] = \Big[[S(\hat{\bar{F}}_1)],\ [S(\hat{\bar{M}}_1)],\ \cdots\Big], \qquad [S_{18}(\hat{\bar{F}})] = \Big[[S(\hat{\bar{F}}_1)],\ [S(\hat{\bar{0}}_1)],\ \cdots\Big]$$

and

$$[G_{18}] = \Big[[\ G\ ]_1^T,\ [\ G\ ]_2^T,\ \cdots\Big]$$

Now the matrices are of size $[18 \times 18]$.

Nonlinear Algorithm for 3-D Structures

We are now ready to put the pieces together to form an algorithm. The assembled global tangent stiffness is given by

$$[\ K\ ] = \sum_m \Big[[\ k_E\ ]_m + [\ k_G\ ]_m\Big] = \sum_m [\ T\ ]_m^T\Big[[\ \bar{k}_E\ ]_m + [\ \bar{k}_G\ ]_m\Big][\ T\ ]_m$$

As indicated before, we will formulate the solution in an incremental fashion. In the following, we concentrate on the basic algorithm for the full Newton-Raphson method because it best illustrates the essential ingredients. The algorithm can be stated as:

Step 1: Specify parameters of the algorithm such as tolerances.

Step 2: Read the initial geometry and material properties.

Step 3: Read the load vector $\{P\}_{t+\Delta t}$. It may be necessary to interpolate this from non-equispaced values.

Step 4: Initialize triads.

Step 5: *Begin loop over time (load) increments:*

    Step T.1: Increment the load vector $\{P\}_{t+\Delta t}$.

    Step T.2: Initialize for equilibrium iterations

$$\{u\}_{t+\Delta t}^0 = \{u\}_t, \qquad [K_T]_{t+\Delta}^0 = [K_T]_t, \qquad \{F\}_{t+\Delta t}^0 = \{F\}_t$$

Step T.3: *Begin loop over iterations:*

    Step I.1: ITERATE:

    Step I.2: Assemble nodal load vector $\{F\}^i$

    Step I.3: Form the effective load vector

$$\{P_{eff}\}_{t+\Delta t} \equiv \{P\}_{t+\Delta t} - \{F\}_{t+\Delta t}^{i-1}$$

    Step I.4: Test norm of effective load vector

$$\text{if} \quad |\{\Delta P_{eff}\}^i|/|\{P\}| > 1000 \quad \text{unstable, goto END}$$

    Step I.5: Assemble the tangent stiffness matrix as the transformed components of the local element stiffnesses

$$[K_T]^i = \sum_m [T^T]_m^i [[\,\bar{k}_E\,]_m^i][\,T\,]_m^i + \sum_m [T^T]_m^i [[\,\bar{k}_G\,]_m^i][\,T\,]_m^i$$

    Step I.6: Decompose the tangent stiffness to

$$[K_T]^i = [\,U\,]^T \lceil\,D\,\rfloor [\,U\,]$$

    Step I.7: Solve for the new displacement increments from

$$[\,U\,]^T \lceil\,D\,\rfloor [\,U\,]\{\Delta u\}^i = \{P_{eff}\}_{t+\Delta t}$$

    Step I.8: Update the displacements

$$\{u\}_{t+\Delta t}^i = \{u\}_{t+\Delta t}^{i-1} + \{\Delta u\}^i$$

    Step I.9: Increment geometry and triads

    Step I.10: Test for convergence.

$$\text{if:} \quad |\{du\}^i|/|\{u\}^i| < tol \quad \text{converged, goto UPDATE}$$
$$\text{if:} \quad |\{du\}^i|/|\{u\}^i| > tol \quad \text{not converged, goto ITERATE}$$

Step T.4: *End loop over iterations.*

Step T.5: UPDATE:

$$u_{t+\Delta t} = u_{t+\Delta t}^i$$
$$xyz_{t+\Delta t} = xyz_{t+\Delta t}^i$$

Step T.6: Compute orientation of global nodes.

Step T.7: Store results for this time step.

Step T.8: If maximum load not exceeded continue looping over loads.

Step 6: *End loop over time (load) increments.*

Step 7: END

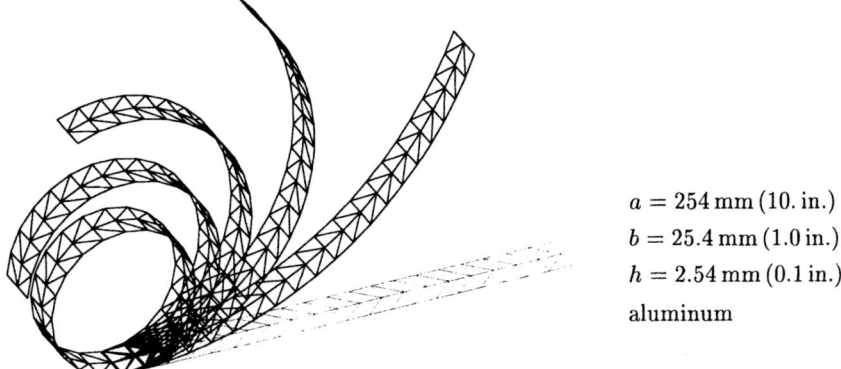

$a = 254 \, \text{mm} \, (10. \, \text{in.})$

$b = 25.4 \, \text{mm} \, (1.0 \, \text{in.})$

$h = 2.54 \, \text{mm} \, (0.1 \, \text{in.})$

aluminum

Figure 3.33: Deformed shapes of a plate with an end moment.

It is possible to enhance this algorithm by including automatic step changes, automatic testing for appropriate time step size, and monitoring the spectral properties of the tangent stiffness.

Example 3.51: Solve the elastica problem of a beam with an end moment as a plate problem.

The results are shown in Figure 3.33 where the shape corresponds to each load step. This problem has no membrane stresses and hence the geometric stiffness is zero. However, during the iterative stage, while there is only an approximation to the deformation, very large membrane stresses can be produced. This can severely restrict the radius of convergence for the iterative scheme.

For such problems, Reference [34] recommends using the previously converged values of membrane stress to estimate the current tangent stiffness. This is like a modified Newton-Raphson method and generally requires more load increments.

Example 3.52: Determine the deflection of a cylindrical shell, with free ends, and a concentrated transverse point load.

This is a problem which transitions from predominantly bending effects to predominantly in-plane membrane effects. It has been considered by many authors two of which are References [55, 57].

The results are shown in Figure 3.34 for a load up to about the stage where the membrane action begins to dominate. The dimensions and estimated displacement of the load point (indicated as circles in the figure) are taken from References [55]. The agreement is good.

3.7 Deformation-Dependent Loads

In the computational examples up to now, the applied load vector $\{P\}$ was taken as fixed in direction. There are many situations, however, where the load depends on the response. Some examples are wind, aerodynamic, and contact loadings. An introduction was given when we considered the elastica with follower loads.

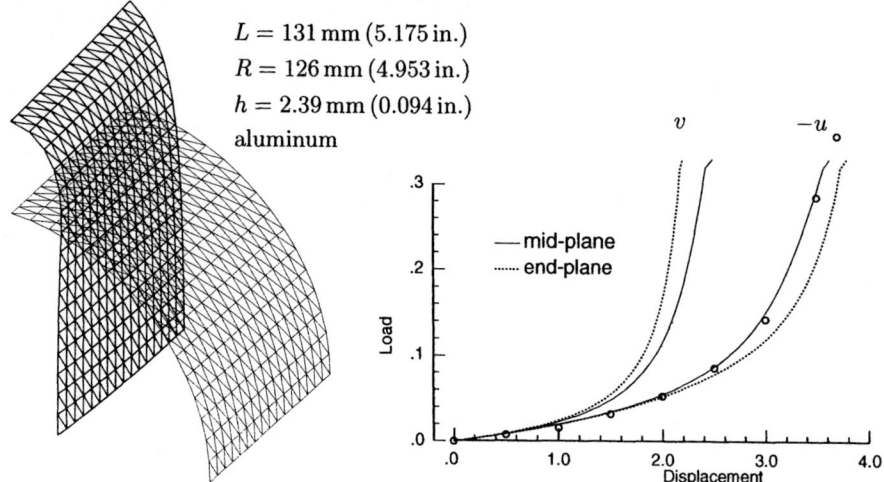

$L = 131 \, \text{mm} \, (5.175 \, \text{in.})$
$R = 126 \, \text{mm} \, (4.953 \, \text{in.})$
$h = 2.39 \, \text{mm} \, (0.094 \, \text{in.})$
aluminum

Figure 3.34: Circular cylinder with a central transverse load.

We now look at the computational difficulties that this type of loading can cause. Distributed load problems, such as pressure distributions, are typically modeled as a series of nodal loads, hence it is sufficient for us to only consider concentrated nodal loads.

Point Loads at a Nodes

Consider the simple situation of a cantilever beam loaded with a follower force as shown in Figure 3.35. At time t, the load vector is

$$\{P\} = \{-\sin\phi, \, \cos\phi, \, 0\}^T P$$

At the next time step, $t + \Delta t$, both P and its orientation, ϕ, will have changed. The new force is then

$$\{P'\} = \{-\sin(\phi + \Delta\phi), \, \cos(\phi + \Delta\phi), \, 0\}^T (P + \Delta P)$$

Expanding and regrouping gives

$$\{P'\} = \{-\sin\phi, \, \cos\phi, \, 0\}^T P + \{-\sin\phi, \, \cos\phi, \, 0\}^T \Delta P$$
$$+ \{-\cos\phi, \, -\sin\phi, \, 0\}^T P\Delta\phi + \cdots$$

Neglecting the higher-order terms, we see that the new load is comprised of three terms; the first is the load at the previous time, the second is the load increment but aligned with the previous orientation, and the third term includes the orientation increment. It is this last term that can create some difficulties.

To see why it creates difficulties, first write the new load in the form

$$\{P'\} = P \left\{ \begin{array}{c} -\sin\phi \\ \cos\phi \\ 0 \end{array} \right\} + \Delta P \left\{ \begin{array}{c} -\sin\phi \\ \cos\phi \\ 0 \end{array} \right\} + P \left[\begin{array}{ccc} 0 & 0 & -\cos\phi \\ 0 & 0 & -\sin\phi \\ 0 & 0 & 0 \end{array} \right] \left\{ \begin{array}{c} \Delta u \\ \Delta u \\ \Delta\phi \end{array} \right\}$$

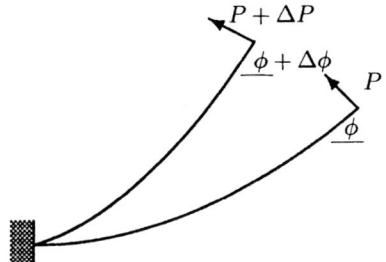

Figure 3.35: Incremental representation of a follower force.

The first two are familiar load vectors, but the third term acts as a contribution to the stiffness matrix because it depends on the deformation increment. Normally, this could be taken to the right-hand side of the equilibrium equation to result in an effective stiffness. However, the contribution results in a nonsymmetric stiffness matrix, which then changes the complete nature of the finite element programming, making them computationally expensive to solve.

One common approximate scheme is simply to use a symmetricized version of the effective stiffness. Another approximate scheme is to ignore the second and third terms but update the force orientation as part of the Newton-Raphson equilibrium iterations. That is, the applied force vector at time $t + \Delta t$ is

$$\{P'\} = \{-\sin \phi^i, \cos \phi^i, 0\}^T (P + \Delta P)$$

This has the advantage of simplicity, but it is at the expense of having to use a smaller load step size as well as more iterations. Furthermore, the iterations can "lock" in the sense that they seem to oscillate between two different states but with approximately the same convergence norm.

We can do a more general development by referring to Equation (3.4). That is, write the rotated load vector as

$$\hat{P}' = \hat{P} + \sin \phi (\hat{e} \times \hat{P}) + (1 - \cos \phi)(\hat{e} \times (\hat{e} \times \hat{P}))$$

where the rotation pseudo-vector is

$$\phi \hat{e} \equiv \phi_x \hat{\imath} + \phi_y \hat{\jmath} + \phi_z \hat{k}, \qquad \phi = \sqrt{\phi_x^2 + \phi_y^2 + \phi_z^2}$$

We are interested in small virtual variations, hence

$$\delta \hat{P} = \delta \phi \hat{e} \times \hat{P}$$

The vector cross-products can be expanded as

$$\delta \hat{P} = -\left[(0\delta\phi_x - P_z\delta\phi_y + P_y\delta\phi_z)\hat{\imath} + (P_z\delta\phi_x + 0\delta\phi_y - P_x\delta\phi_z)\hat{\jmath} + (-P_y\delta\phi_x + P_x\delta\phi_y - 0\delta\phi_z)\hat{k}\right]$$

This leads to the matrix representation for the virtual force change

$$\{\delta P\} = -\begin{bmatrix} 0 & -P_z & P_y \\ P_z & 0 & -P_x \\ -P_y & P_x & 0 \end{bmatrix} \begin{Bmatrix} \delta\phi_x \\ \delta\phi_y \\ \delta\phi_z \end{Bmatrix}$$

Suppose P is applied along one of the global coordinate directions; then, over time, it will have the components

$$\{P\} = \{e_x,\, e_y,\, e_z\}^T P$$

where the unit vector is obtained from the structural level nodal triad.

When the deformation-dependent load is brought to the left-hand side of the equilibrium equation, it has a stiffness contribution of

$$[\,k_G\,] = P \begin{bmatrix} 0 & -e_z & e_y \\ e_z & 0 & -e_x \\ -e_y & e_x & 0 \end{bmatrix}$$

This occupies the $[i, 3 + j]$, $i, j = 1, 3$ submatrix of the $[18 \times 18]$ element stiffness matrix with all other entries being zero. Clearly, the contribution is nonsymmetric and the loading is nonconservative. To avoid the computational cost associated with nonsymmetric matrices it is common practice to use a symmetricized version of the stiffness [18], but additional discussions are given in References [4, 20]. That is,

$$[\,k_G^*\,] = \tfrac{1}{2} \Big[[\,k_G\,] + [\,k_G\,]^T \Big]$$

Again, this is at the expense of increased iterations.

Example 3.53:   Determine the displacement histories and deformed shapes of a cantilever beam with a transverse follower force.

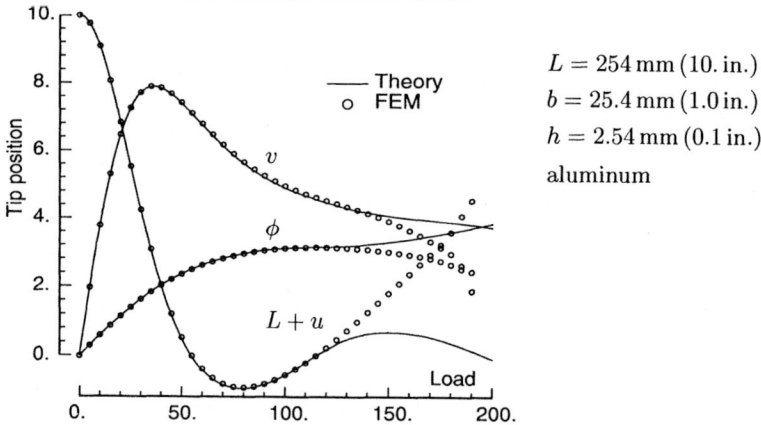

$L = 254\,\text{mm}\ (10.\,\text{in.})$

$b = 25.4\,\text{mm}\ (1.0\,\text{in.})$

$h = 2.54\,\text{mm}\ (0.1\,\text{in.})$

aluminum

Figure 3.36: Tip positions for a cantilever beam with transverse follower load.

This is one of the cases considered in Section 3.1. We use 10 elements. The results are shown in Figure 3.36. There is excellent agreement up to the point where the tip becomes horizontal. At that stage, the theory shows that the rotation continues to increase but the FEM solution shows a decreasing orientation. Also, the FEM solution had difficulty converging for these values of load.

The deformed shapes are shown in Figure 3.1. The FEM results are for equal load increments, while the analytical is for equal tip angle increments. The final load for both cases is about the same. Although only 10 elements were used, the curves look smooth because the beam shape functions were used to get intermediate values.

Nodal Moments

An applied moment about a fixed axis acting on a point on a structure which is free to rotate about an arbitrary axis is nonconservative. This is seen by considering the following situation. Let the body be rotated π about the z-axis. The same final position is obtained by the successive rotations of π first about the x-axis and then about the y-axis. Now suppose there is a moment vector in the z direction; the first scenario does work because there is rotation about z, the second scenario does no work since there is no rotation about z. Clearly, the loading is path dependent and hence nonconservative.

These cases are considered in Reference [5]; in the present case, we prefer to replace all moments with concentrated forces (forming couples) and therefore amenable to the developments just presented.

Problems

3.1 Reconsider the truss problem shown in Figure 3.4.
- Investigate the effect of an initial P_x force on the equilibrium paths.
- Use an FEM analysis to confirm the results.

3.2 Reconsider the beam/plate problem shown in Figure 3.33.
- Investigate the effect of an initial P_x force on the equilibrium paths.
- Use an FEM analysis to confirm the results.

3.3 For the truss shown:
- Establish the nonlinear equations necessary to determine the deflections at A.
- Use a numerical method to solve the equations.
- Confirm the results using a nonlinear FEM analysis.

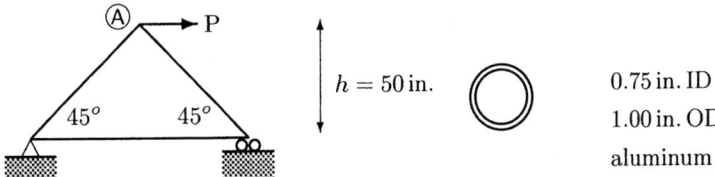

$h = 50 \, \text{in.}$ 0.75 in. ID
1.00 in. OD
aluminum

3.4 For the frame structure shown:
- Establish the nonlinear equations to determine the deflections at A,
- Use a numerical method to solve the equations.
- Confirm the results using a nonlinear FEM analysis.

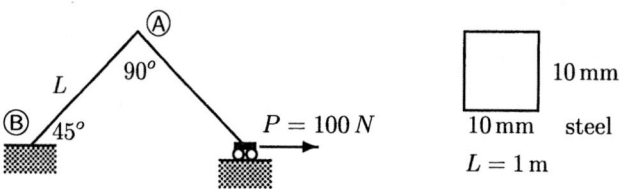

3.5 For the torque loaded symmetric frame structure shown:
- Establish the nonlinear equations to determine the deflections at A and the reactions at B.
- Use a numerical method to solve the equations.
- Confirm the results using a nonlinear FEM analysis.

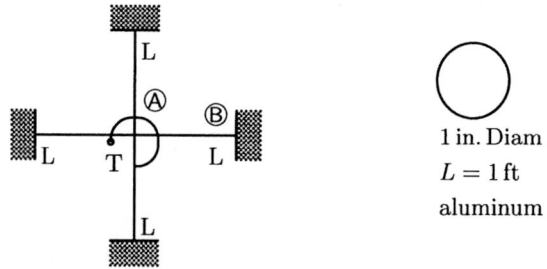

3.6 For the frame structure on rollers shown:
- Establish the nonlinear equations to determine the deflections at A.
- Use a numerical method to solve the equations.
- Confirm the results using a nonlinear FEM analysis.

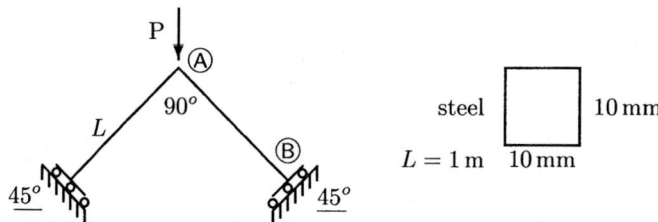

3.7 Reconsider the cantilever beam problem shown in Figure 3.31.
- Use an FEM analysis to investigate the effect of follower loads.

4
Vibrations of Structures

In this chapter, we look at the effect inertia has on the response of structures. For thin-walled structures and frames, the out-of-plane (or transverse) flexural vibration is more dominant than the in-plane, and we concentrate on analyzing this. Throughout, we alternate between the free and forced responses, although restricting ourselves to the linear behaviors.

The main goal of the chapter is consideration of the discretized form of the inertia necessary for our computational analysis. As a follow-on, we look at the modal analysis of structures; this will be valuable when we discuss the general properties of dynamical systems. Modal analysis allows us to order the dynamic effects; Figure 4.1 illustrates two of the vibrational mode shapes of a circular cylinder (the exploded view is intended to give a clearer picture of the three-dimensional mode shape). In interpreting the figure, the numbers indicate the particular mode. We conclude the chapter with a discussion of the relationship between a modal (vibrational) analysis and a transient (wave) analysis; this will help to put into perspective where the two types of analyses are applicable.

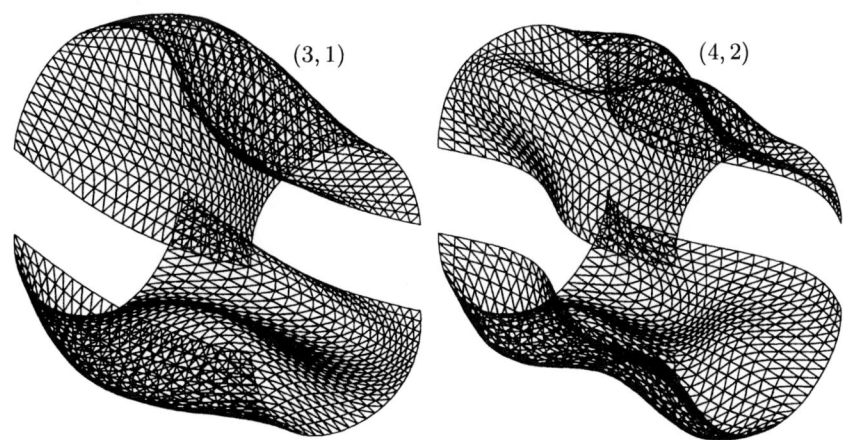

Figure 4.1: The (n, m) vibration mode shapes for a cylinder.

4.1   Free and Forced Vibrations

A *vibration* is a special form of dynamic behavior where the structure executes an
oscillatory motion about an equilibrium position. A vibration executed without
the presence of external forces is called a *free vibration*. A pendulum is a simple
example. Vibration that takes place under the excitation of periodic external
forces is called a *forced frequency* vibration. An example of forced vibration is
that due to unbalance in rotating machinery. This section is a brief review of the
elements of vibration analysis as well as some aspects of spectral analysis. More
detailed background can be found in References [23, 49, 76] and some interesting
historical issues can be found in Reference [56].

Harmonic Motion and Vibration

A vibration motion such as

$$u(t) = A \sin \omega t$$

is called *simple harmonic* motion with an amplitude A and an angular frequency
ω. A plot of this function is shown in Figure 4.2. The time for the response to
repeat itself is called the *period* and is given by $T = 2\pi/\omega$. The rate of repetition
is called the frequency $f = 1/T$. The relation between displacement, velocity,
and acceleration for the point undergoing harmonic motion is obtained simply
by differentiation, that is,

$$
\begin{aligned}
\text{displacement:} \quad u &= A\sin(\omega t) \\
\text{velocity:} \quad \dot{u} &= \omega A\cos(\omega t) &= \omega A\sin(\omega t - \pi/2) \\
\text{acceleration:} \quad \ddot{u} &= -\omega^2 A\sin(\omega t) &= \omega^2 A\sin(\omega t - \pi)
\end{aligned}
$$

We use the notation of a super dot to mean derivative with respect to time. The
behavior of all three responses is harmonic and is shown (scaled) in Figure 4.2.
It is obvious that they all have the same shape. What is different is their phase
— how much they need to be moved (in time) relative to each other so as to
overlap exactly. In the above case, for example, the velocity is 90 degrees ($\pi/2$
radians) out of phase with the displacement. Phase plays are very important
role in the analysis of vibrating systems. It is apparent from this that a general
expression for harmonic motion is $u(t) = A\sin(\omega t + \delta)$, where δ is a phase shift.

The description of the dynamic response of elastic systems will be motivated
by considering the simple case of a single spring/mass system. Consider the free
body diagram of the mass attached to the spring of stiffness K as shown in
Figure 4.3. We identify four forces acting on the displaced mass. The applied
force P is the agent causing the displacement, the elastic force Ku attempts to
return the mass to its original position, the inertia force $-M\ddot{u}$ acts so as to keep
the mass where it is, and finally the damping force F^d attempts to retard the
motion. The equation of motion for the mass is therefore

$$Ku + M\ddot{u} + F^d = P(t) \tag{4.1}$$

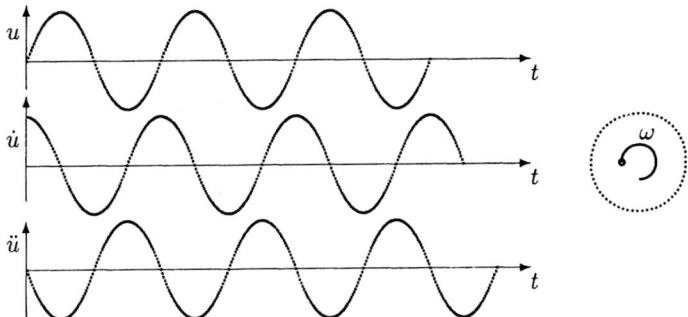

Figure 4.2: Simple harmonic motion.

where $P(t)$ is the externally applied load history.

All real structures experience some sort of dissipation of energy (or damping) when set in motion. This is due to such factors as friction with the surrounding air, and internal friction of the material itself. The scientific nature of friction is still not too well understood, therefore its treatment in vibration is approached from the point of view of convenience. We will consider the question in more detail later, but as a first attempt at modeling damping, we will look at *viscous* damping as represented mechanically by the dashpot. The dashpot exerts a retarding force which is proportional to the instantaneous velocity. Thus, we write $F^d = C\dot{u}$, where C is the damping constant. The equation of motion that we will mostly discuss is therefore

$$Ku + C\dot{u} + M\ddot{u} = P(t) \tag{4.2}$$

where we seek to find $u(t)$ when $P(t)$ is specified.

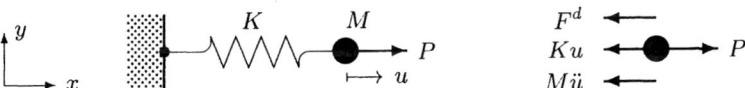

Figure 4.3: Simple spring/mass system.

Example 4.1: Determine the motion of a spring/mass system after it is displaced from its initial position and released. Assume no damping.

The differential equation of motion (after release) reduces to

$$Ku + M\ddot{u} = 0$$

This is a second order differential equation with constant coefficients. We expect solutions of the form

$$u(t) = A\cos\alpha t + B\sin\alpha t$$

where A and B are the constants of integration, and α is an as yet undetermined constant. Substitute the assumed solution into the differential equation to get

$$[K - M\alpha^2]A\cos\alpha t + [K - M\alpha^2]B\sin\alpha t = 0$$

Because the differential equation must be satisfied for any value of time, then we must have that

$$K - M\alpha^2 = 0$$

This specifies α to be

$$\alpha = \pm\sqrt{\frac{K}{M}} = \pm\omega_o$$

and gives the general solution as

$$u(t) = A\cos\omega_o t + B\sin\omega_o t$$

The arbitrary constants A and B are determined from the *initial conditions*. The problem as stated says that initially the mass is displaced and then released from rest. The initial conditions at $t = 0$ are therefore that

$$u(0) = u_o, \qquad \dot{u}(0) = 0$$

This gives $A = u_o$ and $B = 0$, and the solution

$$u(t) = u_o\cos\omega_o t, \qquad \dot{u}(t) = -\omega_o u_o\cos\omega_o t$$

This is shown plotted in Figure 4.4 for the case $K = 2$, $M = 1$, $u_o = 1/\sqrt{2}$. The system is exhibiting an harmonic motion of frequency $\omega = \omega_o = \sqrt{K/M}$. This value is called the *natural frequency*. A single-degree-of-freedom system, when set in free vibration motion, vibrates at only one frequency, and that frequency depends only on the material properties of the system. The phase-plane plot of displacement against velocity is an ellipse continuously repeated.

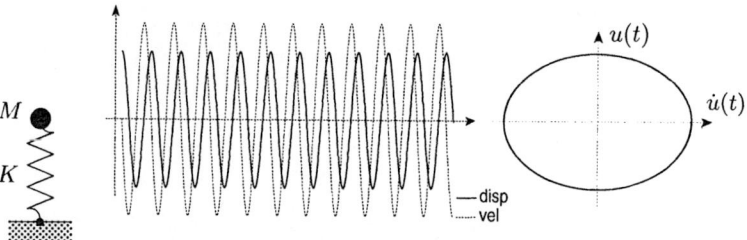

Figure 4.4: Free vibration response.

Example 4.2:   Let the mass of the previous example be already in motion at time $t = 0$. Describe the subsequent motion if there is no damping.

The complete dynamic state of a single-degree-of-freedom system is described in terms of its displacement and velocity. At the instant in time, $t = 0$ say, we let the initial conditions be $u(0) = u_o$ and $\dot{u}(0) = v_o$, then, irrespective of how the motion was originally initiated, a free vibration is described as the sum of a sine and cosine term in the form

$$u(t) = A\cos\omega_o t + B\sin\omega_o t$$

Using the initial conditions gives that $A = u_o$ and $B = v_o/\omega_o$ allowing the time history to be written as

$$u(t) = u_o\cos\omega_o t + \frac{v_o}{\omega_o}\sin\omega_o t = C\cos(\omega_o t - \delta)$$

where the amplitude and phase are given, respectively, by

$$C \equiv \sqrt{u_o^2 + v_o^2/\omega_o^2}, \qquad \delta \equiv \tan^{-1}(u_o\omega_o/v_o)$$

Such motion is also periodic and of frequency ω_o.

Example 4.3: Determine the response of the mass of Figure 4.3 to a sinusoidally varying load $P(t) = \hat{P}\sin\omega t$. Neglect damping.

Under this circumstance, the equation of motion becomes

$$Ku + M\ddot{u} = P(t) = \hat{P}\sin\omega t$$

This differential equation is *inhomogeneous* because of the nonzero on the right-hand side. Thus the solution will comprise two parts; the general solution obtained after setting $P = 0$, and the particular (or complementary) solution obtained so as to give $P(t)$.

We already know that the homogeneous solution is given by

$$u_h(t) = A\cos\omega_o t + B\sin\omega_o t$$

Look for particular solutions of similar form, that is, try

$$u_p(t) = C\cos\alpha t + D\sin\alpha t$$

where α is an as yet unspecified frequency and C, D are arbitrary constants. On substituting into the differential equation, get

$$[K - M\alpha^2]C\cos\alpha t + [K - M\alpha^2]D\sin\alpha t = \hat{P}\sin\omega t$$

This must be true at any value of time; hence separately equating the terms associated with the sines and cosines gives

$$C = 0, \qquad D = \frac{\hat{P}}{K - \omega^2 M}, \qquad \alpha = \omega$$

The total displacement response can therefore be written as

$$u(t) = A\cos\omega_o t + B\sin\omega_o t + \frac{\hat{P}}{K - \omega^2 M}\sin\omega t$$

Again, the coefficients are obtained from the initial conditions. Using the initial conditions of the last example gives the complete solution as

$$u(t) = u_o\cos\omega_o t + \frac{v_o}{\omega_o}\sin\omega_o t - \frac{\hat{P}\omega/\omega_o}{K - \omega^2 M}\sin\omega_o t + \frac{\hat{P}}{K - \omega^2 M}\sin\omega t$$

The first three terms carry the natural frequency ω_o, while the last term carries the forcing frequency ω. In any real system, where some slight damping always exists, the only motion that will persist is the motion described by the last term. Hence we call the last term the *steady-state* response, while the rest are called the *transients*.

An interesting feature of this solution is observed when the forcing frequency is varied; it is seen that the amplitude of the response changes. Indeed, when

$$\omega^2 = \frac{K}{M} = \omega_o^2$$

the response is infinite, even for very small values of excitation force. This situation is called *resonance*. Figure 4.5 shows how the steady-state amplitude ratio

$$\frac{\hat{u}}{\hat{P}/K} = \frac{1}{1 - \omega^2 M/K} = \frac{1}{1 - \omega^2/\omega_o^2}$$

varies as a function of frequency (the figure also shows the effect of damping ζ which is zero in the present case). As will be shown later, in practical situations there is always some damping and therefore an infinite response is never achieved as implied in the figure.

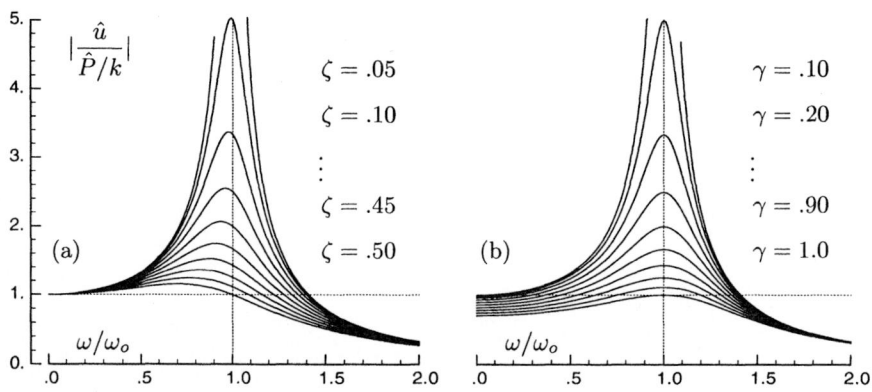

Figure 4.5: Forced frequency response of spring/mass system. (a) Viscous damping. (b) Hysteretic damping.

Complex Notation and Spectral Analysis

The use of complex algebra facilitates the mathematical analysis of vibration especially when we deal with phase shifts. We therefore find it propitious to introduce it at this stage.

A complex quantity is written as

$$z = a + ib, \qquad i \equiv \sqrt{-1}$$

This can be thought of as a vector with components a and b; a is the real part, b is the imaginary part. The magnitude and orientation is given by

$$|z| = \sqrt{a^2 + b^2} = A, \qquad \delta = \tan^{-1}(a/b)$$

Consequently, an alternative form for the complex number is

$$z = A(\cos\delta + i\sin\delta)$$

We can put this in a convenient form by noting the following relation for exponential functions. The Taylor series expansion of the exponential function e^{ix} is

$$
\begin{aligned}
e^{ix} &\approx 1 + (ix) + \tfrac{1}{2}(ix)^2 + \tfrac{1}{6}(ix)^3 + \tfrac{1}{24}(ix)^4 + \tfrac{1}{120}(ix)^5 + \tfrac{1}{720}(ix)^6 + \ldots \\
&= [1 - \tfrac{1}{2}x^2 + \tfrac{1}{24}x^4 - \tfrac{1}{720}x^6 + \ldots] + i[x - \tfrac{1}{6}x^3 + \tfrac{1}{120}x^5 + \ldots] \\
&= \cos x + i \sin x
\end{aligned}
$$

We can now write the complex number as

$$
z = A e^{i\delta}
$$

Some other relations of use are

$$
\cos \delta = \mathrm{Re}[\, e^{i\delta} \,] = [e^{i\delta} + e^{-i\delta}]/2 , \qquad \sin \delta = \mathrm{Im}[\, e^{i\delta} \,] = [e^{i\delta} - e^{-i\delta}]/2i
$$

where Re and Im stand for real part and imaginary part, respectively.

The addition, multiplication, and so on, of complex numbers follows the usual rules of vector algebra. For example, suppose we have two complex numbers

$$
z_1 = a + ib = A_1 e^{i\delta_1} , \qquad z_2 = c + id = A_2 e^{i\delta_2}
$$

Then addition is achieved by adding the components

$$
z_1 + z_2 = (a + c) + i(b + d) = A_1 e^{i\delta_1} + A_2 e^{i\delta_2}
$$

Multiplication is given by

$$
z_1 z_2 = (ac - bd) + i(ad + bc) = A_1 A_2 e^{i(\delta_1 + \delta_2)}
$$

The exponential form makes multiplication very simple.

To show how these ideas can help to simplify the description of harmonic motions, consider the equation of motion

$$
Kv + C\dot{v} + M\ddot{v} = P_o \cos(\omega t + \delta)
$$

where all terms are real. Now introduce the complementary equation of motion

$$
Kw + C\dot{w} + M\ddot{w} = P_o \sin(\omega t + \delta)
$$

Multiply the second equation by i and add it to the first. The result shows that the complex variable $u \equiv v + iw$ must satisfy the following differential equation:

$$
Ku + C\dot{u} + M\ddot{u} = P_o e^{i(\omega t + \delta)} = \hat{P} e^{i\omega t}
$$

In the last form for the load, we have incorporated the phase with the applied load so that $\hat{P}$, in general, is a complex quantity. If we solve this equation for u, then we can recover both v or w from

$$
v = \mathrm{Re}[\, u \,] , \qquad w = \mathrm{Im}[\, u \,]
$$

respectively. We emphasize that working with the complex variable u is equivalent to working with the real variable v; no information is gained or lost, it is just a matter of convenience.

The solution for harmonic motion is written simply as

$$u(t) = \hat{u}e^{i\omega t}$$

In the following, the super hat notation will designate the complex amplitude of each frequency component; these components are called the *amplitude spectrum*. It is understood that when the actual displacement is required, then the above is combined with its complex conjugate to give a real response.

Example 4.4: Determine the forced frequency response for the spring/mass system of Figure 4.3 taking damping into account.

The equation of motion for forced single frequency sinusoidal excitation may be written as

$$Ku + C\dot{u} + M\ddot{u} = \hat{P}e^{i\omega t}$$

where $\hat{P}$ is the excitation force and ω is the excitation frequency. Using a trial solution of the form

$$u(t) = \hat{u}\,e^{i\omega t}$$

gives the velocity and acceleration as

$$\begin{aligned}\dot{u}(t) &= i\omega\,\hat{u}e^{i\omega t} = i\omega\,u \\ \ddot{u}(t) &= (i\omega)^2\,\hat{u}e^{i\omega t} = -\omega^2\,\hat{u}e^{i\omega t} = -\omega^2\,u\end{aligned}$$

This shows that differentiation is accomplished by multiplying by $i\omega$. Therefore by substituting for $u(t)$ and canceling the common time factors, we get

$$[K + i\omega\,C - \omega^2\,M]\hat{u} = \hat{P}$$

This is solved to give

$$\hat{u} = \frac{\hat{P}}{[K - \omega^2\,M + i\omega\,C]} = \frac{\omega_o^2\,\hat{P}/K}{[\omega_o^2 - \omega^2 + i2\zeta\omega\omega_o]} = \frac{\hat{P}/K}{[1 - (\omega/\omega_o)^2 + i2\zeta\omega/\omega_o^2]}$$

where $\omega_o \equiv \sqrt{K/M}$ is the undamped natural frequency, $\zeta = C/2M\omega_0$ is the dimensionless damping ratio, and $\hat{P}/K$ is the static extension of the spring caused by the force. These are shown plotted in Figure 4.5(a).

The idea of representing the time variation of a function by a summation of harmonic functions is extended here to representing arbitrary functions of time and position resulting from the solution of our distributed systems. The approach is to remove the time variation by using the spectral representation of the solution. This leaves a new differential equation for the coefficients that, in many cases, can be integrated directly.

Consider the time variation of the solution at a particular point in space; it has the spectral representation

$$u(x_1, y_1, t) = f_1(t) = \sum C_{1n} e^{i\omega_n t}$$

At another point, the solution behaves as a second time function $f_2(t)$ and is represented by the Fourier coefficients C_{2n}. That is, the coefficients are different at each spatial point. Thus, the solution at an arbitrary position has the following spectral representation

$$u(x, y, t) = \sum \hat{u}_n(x, y, \omega_n) e^{i\omega_n t}$$

where $\hat{u}_n(x, y)$ are the spatially dependent Fourier coefficients. Note that these coefficients are functions of frequency ω_n, and thus there is no reduction in the total number of independent variables.

For shorthand, the summation and subscripts will often be understood and the function will be given the representation

$$u(x, y, t) \quad \Rightarrow \quad \hat{u}_n(x, y, \omega_n) \quad \text{or} \quad \hat{u}(x, y, \omega)$$

Sometimes, we will write the representation simply as $\hat{u}$.

The governing differential equations, in general, are in terms of both space and time derivatives. Because these equations are linear, it is then possible to apply the spectral representation to each term appearing. Thus, the spectral representation for the time derivative is

$$\frac{\partial u}{\partial t} = \frac{\partial}{\partial t} \sum \hat{u}_n e^{i\omega_n t} = \sum i\omega_n \hat{u}_n e^{i\omega_n t}$$

In shorthand this becomes

$$\frac{\partial u}{\partial t} \quad \Rightarrow \quad i\omega_n \hat{u}_n \quad \text{or} \quad i\omega \, \hat{u}$$

In fact, time derivatives of general order have the representation

$$\frac{\partial^m u}{\partial t^m} \quad \Rightarrow \quad (i\omega_n)^m \, \hat{u}_n \quad \text{or} \quad (i\omega)^m \, \hat{u}$$

Herein lies the advantage of the spectral approach to solving differential equations — time derivatives are replaced by algebraic expressions in the Fourier coefficients. That is, there is a reduction in the number of derivatives occurring.

Similarly, the spatial derivatives are represented by

$$\frac{\partial u}{\partial x} = \frac{\partial}{\partial x} \sum \hat{u}_n e^{i\omega_n t} = \sum \frac{\partial \hat{u}_n}{\partial x} e^{i\omega_n t}$$

and in shorthand notation

$$\frac{\partial u}{\partial x} \quad \Rightarrow \quad \frac{\partial \hat{u}_n}{\partial x} \quad \text{or} \quad \frac{\partial \hat{u}}{\partial x}$$

In this case there does not appear to be any reduction; as will be seen later, with the removal of time as an independent variable, these derivatives often become ordinary derivatives, and thus more amenable to integration.

Example 4.5:   Determine the general spectral solution for a vibrating rod. The equation of motion for a simple rod obtained in Chapter 1 is

$$EA\frac{\partial^2 u}{\partial x^2} - \rho A\frac{\partial^2 u}{\partial t^2} - \eta A\frac{\partial u}{\partial t} = 0$$

Assume a spectral representation of the solution in the form

$$u(x,t) = \sum \hat{u}_n(x,\omega_n)e^{i\omega_n t}$$

and substitute into the governing equation to get

$$\sum \left[EA\frac{d^2\hat{u}_n}{dx^2} + (\rho A\omega_n^2 - i\eta A\omega_n)\hat{u}_n \right] e^{i\omega_n t} = 0$$

Because the bases functions $e^{i\omega_n t}$ are independent, we conclude that this equation must be true for each ω_n to give

$$EA\frac{d^2\hat{u}_n}{dx^2} + (\rho A\omega_n^2 - i\eta A\omega_n)\hat{u}_n = 0$$

This is an ordinary differential equation with constant coefficients (note that ω_n just plays the role of a parameter) and therefore has solutions of the form

$$\hat{u}(x) = \mathbf{A}e^{-ikx}$$

where k (called the *wavenumber*) is as yet undetermined. This is determined by substituting into the above equation, which leads to

$$[-EAk^2 + (\rho A\omega_n^2 - i\eta A\omega_n)]\mathbf{A} = 0$$

This can only be true when

$$k^2 = (\rho A\omega_n^2 - i\eta A\omega_n)/EA$$

which gives two possible solutions

$$k_1 = +\sqrt{(\rho A\omega_n^2 - i\eta A\omega_n)/EA}, \qquad k_2 = -\sqrt{(\rho A\omega_n^2 - i\eta A\omega_n)/EA}$$

The general solution is then

$$\hat{u}(x) = \mathbf{A}e^{-ik_1 x} + \mathbf{B}e^{-ik_2 x} \qquad \text{and} \qquad u(x,t) = \sum [\mathbf{A}e^{-ik_1 x} + \mathbf{B}e^{-ik_2 x}]e^{i\omega t}$$

The coefficients $\mathbf{A}$ and $\mathbf{B}$ are complex and are determined from the boundary conditions. Specific examples will be considered later. A thorough application of this approach to wave propagation in structures is given in Reference [23].

Example 4.6:   Determine the free vibration response for a fixed-free rod. Neglect damping.

Since we neglect damping, we can write the solution in the real-only form

$$\hat{u}(x) = c_1 \cos(\beta x) + c_2 \sin(\beta x), \qquad \beta = \omega\sqrt{\rho A/EA}$$

The boundary condition at $x = 0$ is

$$\hat{u}(0) = 0 \quad \implies \quad 0 = c_1$$

The boundary condition at $x = L$ is

$$\hat{F}(L) = EA\frac{d\hat{u}}{dx} = 0 \quad \implies \quad 0 = EAc_2\beta\cos(\beta L)$$

The only nontrivial solution is when

$$\beta L = \tfrac{1}{2}\pi, \ \tfrac{5}{2}\pi, \ \cdots \qquad \text{or} \qquad \omega_n = (n + \tfrac{1}{2})\frac{\pi}{L}\sqrt{EA/\rho A}$$

for $n = 0, 1, \cdots$. The corresponding mode shape is

$$\hat{u}(x) = c_2 \sin[(n + \tfrac{1}{2})\pi x/L]$$

The coefficient c_2 is unknown.

It is worth noting that the differential equation dictated the form of the spectrum relation; but the boundary condition then determined those frequencies that are acceptable.

Damping

All real structures experience some sort of energy dissipation (or damping) when set in motion. This is due to such factors as friction with the surrounding air and internal friction of the material itself. This section considers some of the consequences of this on the motion.

There are two simple mathematical models for damping in a vibrating structure; the damping may be viscous or hysteretic. In the first, energy dissipation per cycle is proportional to the forcing frequency, while in hysteretic damping, it is independent of frequency. Mathematically, the two types are very similar; we shall therefore give a brief comparison of their effects, but concentrate on the viscous damping.

I: Critical Damping

Before we proceed with discussing the effects of damping, we would first like to get a measure of what is meant by small amounts of damping. To this end, consider the free vibration of the system with viscous damping. The equation of motion is

$$Ku + C\dot{u} + M\ddot{u} = 0$$

Look for particular solutions of this in the form $u(t) = Ae^{i\mu t}$. (A note on notation: when we expect the frequency of vibration to be real only, we will assume the harmonic response $Ae^{i\omega t}$, but if the frequency can be complex (as is usually the case with damped systems) we assume the response $Ae^{i\mu t}$.) Substitute into the differential equation and get the characteristic equation

$$A[K + iC\mu - M\mu^2] = 0$$

The value of μ that satisfies this is obtained by solving the quadratic equation and is

$$\mu = \frac{iC}{2M} \pm \frac{1}{2M}\sqrt{4MK - C^2}$$

The time response of the solution is affected by the sign of the radical term as

$$
\begin{aligned}
C^2 &> 4MK; \quad &\text{overdamped} \\
C^2 &= 4MK; \quad &\text{critical damping} \\
C^2 &< 4MK; \quad &\text{underdamped}
\end{aligned}
$$

Let the critical damping be defined by

$$C_c \equiv \sqrt{4MK} = 2M\omega_o, \qquad \omega_o \equiv \sqrt{K/M}$$

then the characteristic values of μ are given by

$$\mu = \omega_o[i\zeta \pm \sqrt{1 - \zeta^2}]$$

where $\zeta \equiv C/C_c$ is the ratio of the damping to critical damping. The free vibration solutions are

$$u(t) = e^{-\zeta\omega_o t}[Ae^{-i\omega_d t} + Be^{+i\omega_d t}]$$

where $\omega_d \equiv \omega_o\sqrt{1 - \zeta^2}$ is called the *damped natural frequency* and A and B are constants to be determined from the initial conditions. The critical point occurs when $\zeta = 1$, thus we say that the structure is lightly damped when $\zeta \ll 1$. This is the situation of most interest to us in structural analysis; measurement instruments (accelerometers, for example) are usually overdamped [48].

Example 4.7: Determine the motion of the mass of Figure 4.6 after it is displaced from its initial position and released. Assume the system is lightly damped.
  We use as initial conditions at $t = 0$ that

$$u(0) = u_o, \qquad \dot{u}(0) = 0$$

to determine the coefficients A and B. This gives the solution

$$u(t) = \tfrac{1}{2}u_o e^{-\omega_o\zeta t}[(1 + \frac{i\omega_o\zeta}{\omega_d})e^{-i\omega_d t} + (1 - \frac{i\omega_o\zeta}{\omega_d})e^{+i\omega_d t}]$$

which is shown plotted in Figure 4.6 for the case $K = 2$, $M = 1$, $C = 0.1$, and $u_o = 1/\sqrt{2}$. Note that the response eventually decreases to zero, but oscillates as it does so. The frequency of oscillation is $\omega_d = \omega_o\sqrt{1 - \zeta^2} \approx \omega_o(1 - \tfrac{1}{2}\zeta^2)$. Hence, for small amounts of damping this is essentially the undamped natural frequency. The rate of decay is dictated by the term $e^{-\omega_o\zeta t} = e^{-Ct/2M}$. The phase-plane plot of displacement against velocity is an elliptical spiral shrinking to zero.

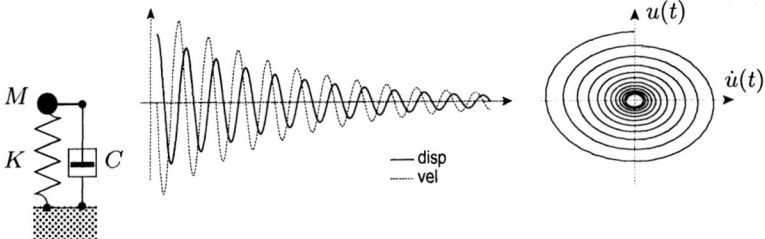

Figure 4.6: Damped response.

II: Viscous and Hysteretic Damping

We shall compare the forced frequency response of the system for both viscous and hysteretic (structural) damping; in both cases we assume that the system is only lightly damped.

The equation of motion for forced single frequency sinusoidal excitation of the system with viscous damping may be written as

$$Ku + C\dot{u} + M\ddot{u} = \hat{P}e^{i\omega t}$$

where $\hat{P}$ is the excitation force and ω is the excitation frequency. Using a trial solution of the form

$$u(t) = \hat{u}e^{i\omega t}$$

we can show by differentiation and substitution that

$$\hat{u} = \frac{\hat{P}}{K - \omega^2 M + i\omega C} = \frac{\omega_o^2 \hat{P}/K}{\omega_0^2 - \omega^2 + i2\zeta\omega_o\omega}$$

where $\omega_o = \sqrt{K/M}$ is the undamped natural frequency, $\zeta = C/2M\omega_o$ is the dimensionless damping ratio, and $\hat{P}/K$ is the extension in the spring caused by the force alone. Thus, the displacement history is

$$u(t) = \hat{u}e^{i\omega t} = \left[\frac{1}{1 - (\omega/\omega_o)^2 + i2\zeta\omega/\omega_o} \right] \frac{\hat{P}}{K} e^{i\omega t} = H(\omega)\hat{P}e^{i\omega t}$$

It can be seen that the displacement is proportional to the applied force, and the proportionality factor $H(\omega)$ is called the *frequency response function* (FRF) — it is complex and depends on frequency. The damping causes the response to lag behind the applied force. The phase difference is given by the angle

$$\delta = \tan^{-1} \frac{2\zeta\omega/\omega_o}{[1 - (\omega/\omega_o)^2]}$$

The solution can therefore, alternatively, be written in the form

$$u(t) = \left[\frac{1}{\sqrt{[1 - (\omega/\omega_o)^2]^2 + (2\zeta\omega/\omega_o)^2}} \right] \frac{\hat{P}}{K} e^{i(\omega t - \delta)} = |H(\omega)| \hat{P}e^{i(\omega t - \delta)}$$

which emphasizes the separate effects of amplitude and phase. The amplitude response is shown in Figure 4.5(a) for different values of damping.

Many materials, when subjected to cyclic strain, generate internal friction that dissipates energy per cycle which is relatively independent of the strain rate. In the present context, this means the damping force is taken as

$$F^d = h\frac{\dot{u}}{\omega}, \qquad \hat{F}^d = ih\,\hat{u}$$

It is important to realize that the hysteretic damping idealization is restricted to the forced frequency case because, otherwise, the frequency in its definition is undefined. If we take the forcing frequency as ω_o, the natural frequency, then this damping reduces to the viscous case.

The equation of motion for a single-degree-of-freedom system with structural damping is written in the time domain as

$$Ku + \frac{h}{\omega}\dot{u} + M\ddot{u} = P(t) = \hat{P}\sin\omega t$$

and in the spectral form as

$$\left[K(1+i\gamma) - \omega^2 M\right]\hat{u} = \hat{P} \qquad \text{or} \qquad \hat{u} = \frac{\hat{P}}{K}\left[\frac{1}{1 - (\omega/\omega_o)^2 + i\gamma}\right]$$

where $\gamma \equiv h/K$ is called the structural damping factor. The frequency response function is obtained from

$$u(t) = \hat{u}e^{i\omega t} = H(\omega)\hat{P}e^{i\omega t} = \left[\frac{1}{\sqrt{[1 - (\omega/\omega_o)^2]^2 + \gamma^2}}\right]\frac{\hat{P}}{K}e^{i(\omega t - \delta)}$$

where the displacement lags behind the force by the angle

$$\delta = \tan^{-1}\left[\frac{\gamma}{1 - (\omega/\omega_o)^2}\right]$$

For hysteretic damping, the maximum response occurs exactly at $\omega/\omega_o = 1$, independent of the damping. At very low frequencies, the response depends on the amount of damping, unlike the viscous case, as shown in Figure 4.5(b). When the system is vibrating at the natural frequency with $\omega/\omega_o = 1$, both the viscous and hysteretic models give the same results if we have $\gamma = 2\zeta$.

III: Effects of Damping

The frequency response function, $H(\omega)$, can be interpreted as a magnification factor between the input force and the output response. Figure 4.5 shows the absolute value of this as a function of the frequency ratio ω/ω_o for various values of the damping ratio ζ. We can see that increasing the damping diminishes the peak amplitudes. Furthermore, there is a shift of these peaks to the left of $w/w_o = 1$. In fact, the peaks occur at frequencies given by

$$\omega = \omega_o \sqrt{(1 - 2\zeta^2)}$$

and the peak value of $|H(\omega)|$ is given by

$$|H(\omega)| = \frac{1/K}{2\zeta\sqrt{1 - \zeta^2}} \approx \frac{1/K}{2\zeta}$$

This last relation is for light damping ($\zeta < 0.1$) and shows the sensitivity of the peak to damping. The points where the amplitude of $|H(\omega)|$ reduces to $1/\sqrt{2}$ of its peak value are called the *half power* points. The difference in the frequencies at the half power points for light damping can be shown to be

$$\Delta\omega = \omega_2 - \omega_1 = 2\zeta\omega_o$$

For this reason, the term 2ζ is sometimes called the *Loss Factor*.

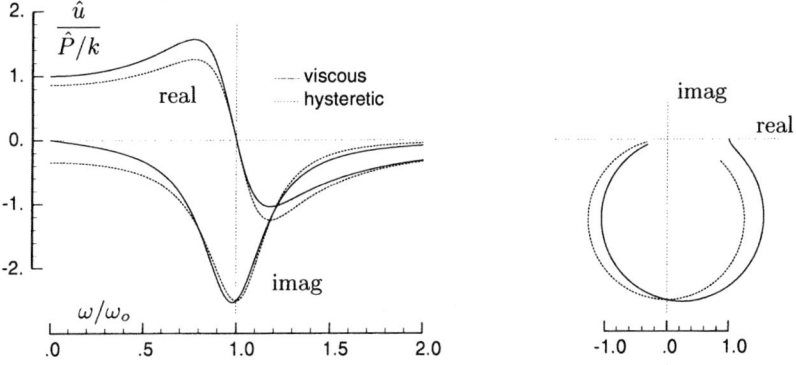

Figure 4.7: Forced frequency response of spring/mass system. (a) Real and imaginary components of the FRF. (b) Nyquist plot.

Because the frequency response function is a complex quantity, it can therefore be decomposed into its real and imaginary parts by multiplying the numerator and the denominator by its complex conjugate. Thus

$$H(\omega) = \left[\frac{1 - (\omega/\omega_o)^2}{[1 - (\omega/\omega_o)^2]^2 + (2\zeta\omega/\omega_o)^2} - \frac{i2\zeta\omega/\omega_o}{[1 - (\omega/\omega_o)^2]^2 + (2\zeta\omega/\omega_o)^2} \right] \frac{1}{K}$$
$$= H_R + iH_I$$

As shown in Figure 4.7, the real component of the FRF has a zero at $\omega/\omega_o = 1$, independent of damping and exhibits maxima at frequencies given by

$$\omega_1 = \omega_o\sqrt{1 - 2\zeta}, \qquad \omega_2 = \omega_o\sqrt{1 + 2\zeta}$$

These frequencies are often used to estimate the damping of the system from

$$2\zeta = \frac{1 - (\omega_1/\omega_2)^2}{1 + (\omega_1/\omega_2)^2}$$

The plot of the imaginary part of the FRF has a peak close to $\omega/\omega_o = 1$, which is sharper than that of the magnitude of $[H(\omega)]$.

A similar analysis can be done for the hysteretic damping. It must be kept in mind, however, that real structures exhibit neither viscous nor structural damping in its pure form. More likely, they exhibit a nonlinear combination of both, with the proportion of each probably depending on the frequency range. Additionally, much of the damping in structures comes from the joints and the interaction with attachments. As a consequence, the damping is not a material "constant" like the Young's modulus or density that can be determined by component testing. Because we deal with lightly damped structures, it is sufficient that we consider just the viscous model.

Example 4.8: The spring/mass system of Figure 4.6 is initially at rest. Find the damped response to the following step loading:

$$
\begin{aligned}
P(t) &= 0 & t < 0 \\
P(t) &= P_o & t > 0
\end{aligned}
$$

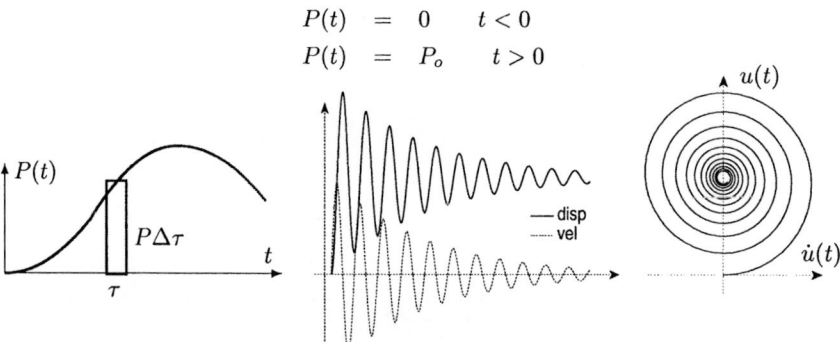

Figure 4.8: Response to impulsive loads.

An *impulsive* force is a force that acts over a short period of time. The time integral of the force is referred to as the *impulse* of the force. When the system is linear, we can obtain the response to an arbitrary force history $P(t)$ by considering it to be the sum of a sequence of impulses. Specifically, consider an arbitrary force history $P(t)$ as shown in Figure 4.8 with one of the impulses indicated. Each impulse is $P\Delta\tau$. The action of this impulse on the mass is to cause a change of momentum given by

$$M\Delta\dot{u} = P\Delta\tau \qquad \text{or} \qquad \Delta\dot{u} = \frac{P\Delta\tau}{M}$$

If the mass is initially at rest, then the change in velocity is the initial velocity for the motion. That is, we have

$$u(0) = u_o = 0, \qquad \dot{u}(0) = v_o = \frac{P\Delta\tau}{M}$$

The response to this impulse is

$$u(t) = \frac{P\Delta\tau}{M\omega_d}e^{-\zeta\omega_o(t-\tau)}\sin[\omega_d(t-\tau)]$$

The term $(t - \tau)$ takes into account the fact that the pulse occurs at time τ and not time zero. The actual force history is a series of these impulses at different times τ; hence, the cumulative effect is obtained by letting $\Delta\tau$ become very small and replacing the summation by an integral over the full time to give

$$u(t) = \frac{1}{\omega_d M}\int_0^t P(\tau)e^{-\zeta\omega_o(t-\tau)}\sin[\omega_d(t-\tau)]\,d\tau$$

This is called *Duhamel's integral* and represents a particular solution of the differential equation of motion subjected to an arbitrary forcing function. For simple forcing functions (for example, stepped loading) the integration may be performed exactly, but generally it must be done numerically.

The initial conditions for our problem are such that $u_o = 0$ and $v_o = 0$; if, however, the initial conditions are not zero, then the homogeneous solution must be added to complete the solution. The solution is obtained by substituting for the force into Duhamel's integral to get

$$u(t) = \frac{P_o}{\omega_d M}\int_0^t e^{-\zeta\omega_o t}\sin[\omega_d(t-\tau)]\,d\tau = \frac{P_o}{K}\left[1 - e^{-\zeta\omega_o t}(\cos\omega_d t + \frac{\omega_o\zeta}{\omega_d}\sin\omega_d t)\right]$$

This response is shown in Figure 4.8 for a value of damping of $\zeta = 0.04$. Note that the response oscillates about the static deflection position. The maximum deflection approaches two times the static value when the damping is very small.

4.2   Free Vibration of Plates and Beams

We use the thin-plate theory derived in Chapter 2. With reference to Figure 4.9, the first task is to extend the spectral analysis method so as to handle spatial variations in two dimensions. We will do this by introducing a Fourier series representation in the y and θ directions.

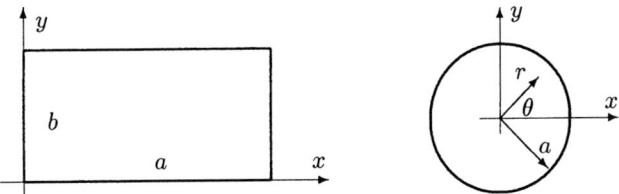

Figure 4.9: Rectangular and circular plates.

Free Vibration Analysis

It is easy for us to add some viscous damping to the governing equations, it is just a matter of modifying the inertia term to give

$$D
\nabla^2\nabla^2 w + Kw + \eta h\frac{\partial w}{\partial t} + \rho h\frac{\partial^2 w}{\partial t^2} = q\,, \qquad \nabla^2 \equiv \frac{\partial^2}{\partial x^2} + \frac{\partial^2}{\partial y^2} \qquad (4.3)$$

where η is the viscous damping per unit volume, K is the stiffness of an elastic foundation, and $D = Eh^3/(1 - \nu^2)$. For free vibration with $q = 0$, we have that a typical variable can be written as

$$w(x, y, t) = \hat{w}(x, y)e^{-i\omega t}$$

where ω is the angular frequency, $i = \sqrt{-1}$, and it is understood that $\hat{w}$ could be complex. The spectral form for the governing equation is then

$$[D\nabla^2\nabla^2 + K - \rho h\omega^2 + i\eta h\omega]\hat{w} = 0 \qquad (4.4)$$

The solution of this equation can be written as linear sums of solutions of the following two differential equations:

$$\nabla^2\hat{w}_1 + \beta^2\hat{w}_1 = 0\,, \quad \nabla^2\hat{w}_2 - \beta^2\hat{w}_2 = 0\,, \quad \beta^2 \equiv \sqrt{\frac{\rho h\omega^2 - i\eta h\omega - K}{D}} \qquad (4.5)$$

These form the basic equations for further analysis and emphasize that there are two fundamentally different modes.

A summary of the spectral form of Equations (2.17) is given by

$$
\begin{aligned}
\text{Displacement}: \quad & \hat{w} = \hat{w}(x, y, \omega) = \hat{w}_1(x, y) + \hat{w}_2(x, y)\\[2mm]
\text{Slope}: \quad & \hat{\psi}_x = \frac{\partial\hat{w}}{\partial x}\\[2mm]
\text{Moment}: \quad & \hat{M}_{xx} = +D\left[\frac{\partial^2\hat{w}}{\partial x^2} + \nu\frac{\partial^2\hat{w}}{\partial y^2}\right]\\[2mm]
\text{Shear}: \quad & \hat{V}_{xz} = -D\left[\frac{\partial^3\hat{w}}{\partial x^3} + (2 - \nu)\frac{\partial^3\hat{w}}{\partial x\partial y^2}\right]\\[2mm]
\text{Loading}: \quad & \hat{q} = D\nabla^2\nabla^2\hat{w} - (\rho h\omega^2 - i\eta h\omega - K)\hat{w} \qquad (4.6)
\end{aligned}
$$

A similar set for cylindrical coordinates can easily be constructed from Equations (2.22). Time domain responses are obtained by performing an inverse Fourier transform on $\hat{w}(x, y, \omega)$.

Example 4.9: Specialize the governing equations for the case of cylindrical bending.

For cylindrical bending about the y-axis, the displacement has the special form $w(x, y, t) = w(x, t)$. The differential equation for $\hat{w}$ then becomes

$$\frac{d^2\hat{w}}{dx^2} \pm \beta^2\hat{w} = 0$$

This has constant coefficients, hence e^{-ikx} is a kernel solution. The spectrum relations are then

$$k_1 = \pm\sqrt{+\beta^2}, \qquad k_2 = \pm\sqrt{-\beta^2}$$

For the undamped case and no elastic foundation, the relation between the wavenumber and frequency is given by

$$k_1 = \pm\sqrt{\omega}\left[\frac{\rho h}{D}\right]^{1/4}, \qquad k_2 = \pm i\sqrt{\omega}\left[\frac{\rho h}{D}\right]^{1/4}$$

This is the behavior of a beam [22] if the following associations are made:

$$\frac{EI}{\rho A} \Leftrightarrow \frac{D}{\rho h} \quad \text{or} \quad E \Leftrightarrow \frac{E}{(1-\nu^2)}$$

Thus the plate in cylindrical bending behaves as a beam in plane strain.

Example 4.10: Determine the effect of an elastic foundation on the vibration of a simply supported beam.

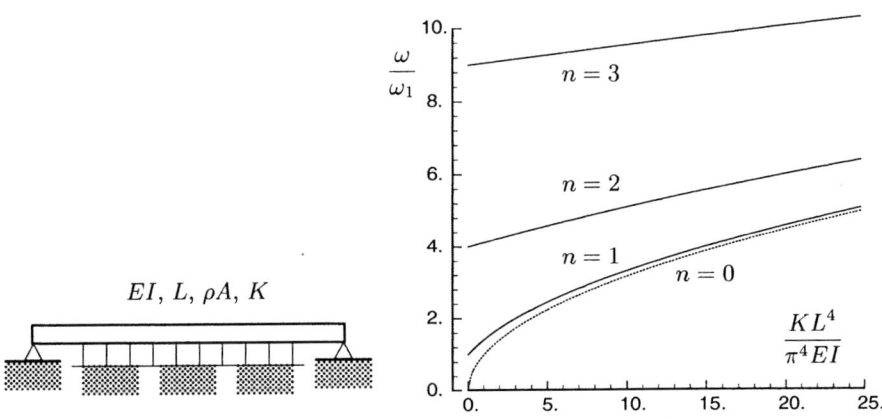

Figure 4.10: Pinned/pinned beam on an elastic foundation.

The governing equation is

$$EI\frac{\partial^4 v}{\partial x^4} + Kv + \rho A\frac{\partial^2 v}{\partial t^2} = 0$$

and the spectral form of the differential equation for $\hat{v}$ becomes

$$\frac{d^4\hat{v}}{dx^4} - \beta^4\hat{v} = 0, \qquad \beta^4 \equiv \frac{(\rho A\omega^2 - K)}{EI}$$

Because this has constant coefficients, we seek solutions of the form $\hat{v} = Ae^{ikx}$, which leads to the characteristic equation

$$k^4 - \beta^4 = 0$$

from which we obtain the four spectrum relations

$$k_{1,3}(\omega) = \pm\sqrt{+\tfrac{1}{2}\sqrt{4\beta^4}} \equiv \pm\alpha, \qquad k_{2,4}(\omega) = \pm\sqrt{-\tfrac{1}{2}\sqrt{4\beta^4}} \equiv \pm i\bar\alpha$$

Thus, the general solution is represented by

$$\hat v(x) = \mathbf{A}e^{-i\alpha x} + \mathbf{B}e^{-\bar\alpha x} + \mathbf{C}e^{+i\alpha x} + \mathbf{D}e^{+\bar\alpha x} \tag{4.7}$$

In analyzing problems without damping (where α and $\bar\alpha$ are likely to be real only), we will find it more convenient to use the solution form

$$\hat v(x) = c_1\cos(\alpha x) + c_2\sin(\alpha x) + c_3\cosh(\bar\alpha x) + c_4\sinh(\bar\alpha x)$$

Both forms, of course, will lead to the same answer.

The boundary conditions are that at $x = 0$, we have

$$v = 0 \quad\Rightarrow\quad \hat v = 0 = c_1 + c_3$$

$$M = 0 \quad\Rightarrow\quad \frac{d^2\hat v}{dx^2} = 0 = -\alpha^2 c_1 + \bar\alpha^2 c_3 = 0$$

This leads to $c_1 = 0$ and $c_3 = 0$. At $x = L$, we have

$$v = 0 \quad\Rightarrow\quad \hat v = 0 = c_1 S + c_3 S_h$$

$$M_{xx} = 0 \quad\Rightarrow\quad \frac{d^2\hat v}{dx^2} = 0 = -\alpha^2 c_1 S + \bar\alpha^2 c_3 S_h = 0$$

where $S \equiv \sin(\alpha L)$ and $S_h \equiv \sinh(\bar\alpha L)$. These two equations lead to the system

$$\begin{bmatrix} \sin\alpha L & \sinh\bar\alpha L \\ -\alpha^2\sin\alpha L & -\bar\alpha^2\sinh\bar\alpha L \end{bmatrix} \left\{ \begin{array}{c} c_2 \\ c_4 \end{array} \right\} = 0$$

The characteristic equation is obtained by setting the determinant equal to zero and gives

$$(\alpha^2 + \bar\alpha^2)\sin(\alpha L)\sinh(\bar\alpha L) = 0$$

This has the solutions

$$\alpha L = n\pi \qquad\text{or}\qquad \sqrt{\tfrac{1}{2}\sqrt{4\frac{(\rho A\omega^2 - K)}{EI}}}\,L = n\pi$$

Expanding and re-arranging gives

$$\omega_n = \sqrt{\frac{EI}{\rho A}}\sqrt{\left(\frac{n\pi}{L}\right)^4 + \frac{K}{EI}} \qquad\text{or}\qquad \frac{\omega_n}{\omega_1} = \sqrt{n^4 + \frac{KL^4}{\pi^4 EI}}, \qquad \omega_1 = \sqrt{\frac{EI}{\rho A}}\frac{\pi^2}{L^2}$$

The corresponding mode shapes are given by

$$\hat v_n(x) = c_2\sin(\alpha x) = c_2\sin\!\left(\frac{n\pi x}{L}\right)$$

These mode shapes do not depend on the elastic foundation.

The variation of resonance frequency with spring stiffness is shown in Figure 4.10. We see that the spring increases the frequency; this is typical structural behavior — added stiffness increases the vibrational frequencies. Furthermore, they all asymptote (from above) to the $n = 0$ line as the foundation stiffness is increased.

Rectangular Plate Solution

In this section, we extend the spectral analysis method so as to handle spatial variations of the deflected shape in two dimensions. This results in an extra summation over the space wavenumbers quite comparable to what was done in Chapter 2.

Since the deflected shapes, possibly, can have arbitrary shapes in space, consider a representation of the form

$$\hat{w}_1(x,y) = \frac{1}{b} \sum_m \tilde{w}_{1m} e^{-i\xi_m y}, \qquad \hat{w}_2 = \frac{1}{b} \sum_m \tilde{w}_{2m} e^{-i\xi_m y} \qquad (4.8)$$

with the space wavenumber given by $\xi_m = 2\pi m/b$. The differential equations governing the transformed displacements are

$$\frac{d^2\tilde{w}_1}{dx^2} + (\beta^2 - \xi_m^2)\tilde{w}_1 = 0, \qquad \frac{d^2\tilde{w}_2}{dx^2} - (\beta^2 + \xi_m^2)\tilde{w}_2 = 0, \qquad \beta^2 \equiv \sqrt{\frac{\rho h \omega^2 - i\eta h\omega - K}{D}}$$

where we allow β to be complex. The coefficients of the differential equations are constant, hence the solutions are exponentials of the form e^{-ikx}. The characteristic equations associated with these solutions are

$$-k_1^2 - \xi^2 + \beta^2 = 0, \qquad -k_2^2 - \xi^2 - \beta^2 = 0$$

giving the spectrum relations as

$$k_1(\omega,\xi) = \pm\sqrt{\beta_n^2 - \xi_m^2} \equiv \pm\alpha_m, \qquad k_2(\omega,\xi) = \pm i\sqrt{\beta_n^2 + \xi_m^2} \equiv \pm i\bar{\alpha}_m \qquad (4.9)$$

These spectrum relations are shown in Figure 4.11. It is noted that for a particular ξ the first mode shows a cut-off frequency with the lower-frequency components being purely complex.

The full solution for the plate becomes

$$w(x,y,t) = \sum_n \sum_m [\mathbf{A}e^{-ik_1 x} + \mathbf{B}e^{-ik_2 x} + \mathbf{C}e^{+ik_1 x} + \mathbf{D}e^{+ik_2 x}]e^{-i\xi_m y}e^{i\omega_n t} \qquad (4.10)$$

That is, the actual solution is obtained by summing kernel solutions of the above form for many values of ω_n and ξ_m. To gain an intuition for this solution, consider it as a plane wave in x (the bracketed term) modified in y. Then, for a particular ξ, the summation over ω is similar to that for a beam as shown in References [22, 23]. The corresponding spectrum relations, however, are modified by ξ as shown in Figure 4.11.

Sometimes it will be more convenient to represent the general solution by

$$\hat{w}(x,y) = \sum_m [c_1 \cos(\alpha_m x) + c_2 \sin(\alpha_m x) + c_3 \cosh(\bar{\alpha}_m x) + c_4 \sinh(\bar{\alpha}_m x)] f(\xi_m y)$$

where $f(\xi_m y)$ is either $\cos(\xi_m y)$ or $\sin(\xi_m y)$. For the case $m = 0$, we have beam-like behavior. For $m > 0$, we have beam-like behavior with a more complicated

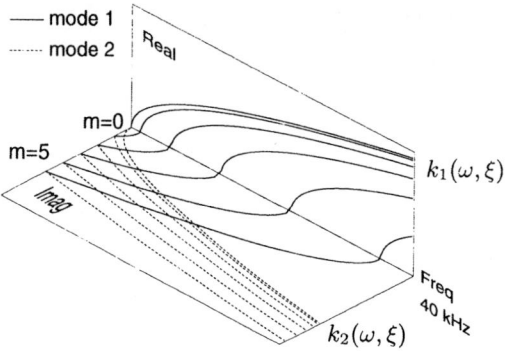

Figure 4.11: The spectrum relation for different values of the ξ_m wavenumber.

variation in y, and a more complicated spectrum relation. This association to beam theory will help in specifying the boundary conditions. Looking at the boundary conditions at $y = 0$ and $y = b$, we see that

$$\hat{w} = 0, \qquad \frac{\partial^2 \hat{w}}{\partial y^2} = 0$$

always. That is, this is true for each m term and implies that this particular solution can solve only those problems with simply supported lateral sides.

Example 4.11: Determine the resonance frequencies and mode shapes for a simply supported rectangular plate.

Consider a simply supported rectangular plate of size $[a \times b]$. Choose only the $\sin(\xi_m y)$ terms. We need only concentrate on the boundary conditions at $x = 0$ and $x = a$, since, as indicated above, the lateral boundary conditions are automatically satisfied.

At $x = 0$, we have

$$w = 0 \quad \Rightarrow \quad \tilde{w}_m = 0 = (c_1 + c_3) \sin(\xi_m y)$$

$$M_{xx} = 0 \quad \Rightarrow \quad \frac{\partial^2 \tilde{w}_m}{\partial x^2} + \nu \frac{\partial^2 \tilde{w}_m}{\partial y^2} = 0$$

$$\Rightarrow \quad [-\alpha_m^2 c_1 + \bar{\alpha}_m^2 c_3] \sin(\xi_m y) + \nu(c_1 + c_3)[-\xi_m^2 \sin(\xi_m y)] = 0$$

These two must be true for any value of y, hence together they give that $c_1 = 0$ and $c_3 = 0$. At $x = a$, we have

$$w = 0 \quad \Rightarrow \quad \tilde{w}_m = 0 = (c_2 S + c_4 S_h) \sin(\xi_m y)$$

$$M_{xx} = 0 \quad \Rightarrow \quad \frac{\partial^2 \tilde{w}_m}{\partial x^2} + \nu \frac{\partial^2 \tilde{w}_m}{\partial y^2} = 0$$

$$\Rightarrow \quad [-\alpha_m^2 c_2 S + \bar{\alpha}_m^2 c_4 S_h] \sin(\xi_m y) + \nu(c_2 S + c_4 S_h)[-\xi_m^2 \sin(\xi_m y)] = 0$$

where $S \equiv \sin(\alpha_m a)$ and $S_h \equiv \sinh(\bar{\alpha}_m a)$. These two equations lead to the system

$$\begin{bmatrix} \sin \alpha_m a & \sinh \bar{\alpha}_m a \\ -\alpha_m^2 \sin \alpha_m a & -\bar{\alpha}_m^2 \sinh \bar{\alpha}_m a \end{bmatrix} \begin{Bmatrix} c_2 \\ c_4 \end{Bmatrix} = 0$$

The characteristic equation is obtained by setting the determinant to zero and gives

$$(\alpha_m^2 + \bar{\alpha}_m^2) \sin(\alpha_m a) \sinh(\bar{\alpha}_m a) = 0$$

This has the solutions

$$\alpha_m a = n\pi \quad \Rightarrow \quad \left(\sqrt{\beta^2 - \xi_m^2}\right) a = n\pi \quad \Rightarrow \quad \left(\sqrt{\omega\sqrt{\frac{\rho h}{D}} - (\frac{m\pi}{b})^2}\right) a = n\pi$$

Expanding and re-arranging gives

$$\omega_{mn} = \sqrt{\frac{D}{\rho h}} \left[(\frac{n\pi}{a})^2 + (\frac{m\pi}{b})^2 \right]$$

This gives an ordered set of frequencies as m and n are varied.

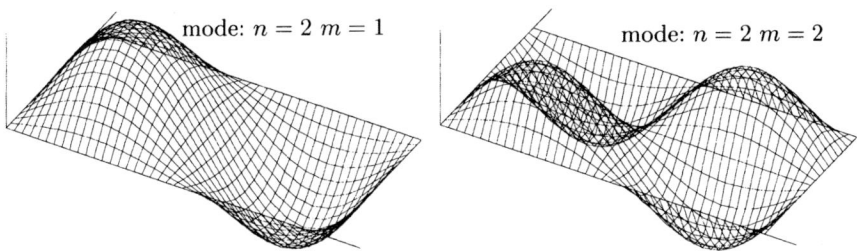

mode: $n = 2$ $m = 1$ mode: $n = 2$ $m = 2$

Figure 4.12: First two mode shapes for a rectangular plate.

The corresponding mode shapes are given by

$$\tilde{w}_{mn}(x,y) = c_2 \sin(\alpha_m x) \sin(\xi_m y) = c_2 \sin(\frac{n\pi x}{a}) \sin(\frac{n\pi y}{b})$$

These mode shapes form a regular 2-D pattern as shown in Figure 4.12. Thus, in comparison to a beam, we see not just different mode shapes at different frequencies but here we also see the pattern changing in both dimensions.

Circular Plate Solution

In cylindrical coordinates, the differential operator for the spatial variation is

$$\nabla^2 = \frac{\partial^2}{\partial r^2} + \frac{1}{r}\frac{\partial}{\partial r} + \frac{1}{r^2}\frac{\partial^2}{\partial \theta^2}$$

Again, we can consider the variation with respect to one of the coordinates to be represented in a Fourier series. For example, consider the form

$$\hat{w}_1(r,\theta) = \sum R_{1m}(r)e^{im\theta}$$

The differential equation for $\bar{w}_1$ becomes

$$\sum_m \left[\frac{d^2 R_{1m}}{dr^2} + \frac{1}{r}\frac{dR_{1m}}{dr} + (\frac{-m^2}{r^2} + \beta^2)R_{1m} \right] e^{im\theta} = 0$$

Setting this to be true for all components m leads to a differential equation for $R_{1m}(r)$ as

$$\frac{d^2 R_{1m}}{dr^2} + \frac{1}{r}\frac{dR_{1m}}{dr} + [\frac{-m^2}{r^2} + \beta^2]R_{1m} = 0$$

Make the substitution $z = \beta r$ and rearrange the differential equation as

$$\frac{d^2 R_{1m}}{dz^2} + \frac{1}{z}\frac{dR_{1m}}{dz} + \frac{1}{z^2}(z^2 - m^2)R_{1m} = 0$$

This is a *Bessel equation* and the solutions are

$$R_{1m}(r) = c_1 J_m(z) + c_2 Y_m(z), \qquad z = \beta r$$

The notation used for the Bessel functions is that of Reference [1].

A similar analysis of the second equation leads to the differential equation

$$\frac{d^2 R_{2m}}{dz^2} + \frac{1}{z}\frac{dR_{2m}}{dz} - \frac{1}{z^2}(z^2 + m^2)R_{2m} = 0$$

This is a *modified Bessel equation* and the solutions are

$$R_{2m}(r) = c_3 K_m(z) + c_4 I_m(z), \qquad z = \beta r$$

Combining the two solutions, we get a representation for the general solution as

$$\hat{w}(r,\theta) = \sum_m [c_1 J_m(\beta r) + c_2 Y_m(\beta r) + c_3 K_m(\beta r) + c_4 I_m(\beta r)] \left\{ \begin{array}{c} \cos(m\theta) \\ \sin(m\theta) \end{array} \right.$$

$$(4.11)$$

This solution can be used to solve a variety of plate problems including those with an inner circular hole. Note that when $m = 0$, we get the axisymmetric solutions, but just because the plate is geometrically axisymmetric does not mean that they are the only solutions. This will be seen in the next example.

Example 4.12: Determine the resonance frequencies and mode shapes for a clamped circular plate.

We require that the deflection and its various derivatives be finite at $r = 0$. Because of the singular nature of the functions Y_m and K_m at $r = 0$, this requires that $c_2 = 0$ and $c_4 = 0$. Also, it is sufficient to just use the $\cos(m\theta)$ terms. This gives the solution as

$$\hat{w}(r,\theta) = \sum_m [c_1 J_m(\beta r) + c_3 I_m(\beta r)] \cos(m\theta)$$

We will obtain the remaining coefficients from the boundary conditions at $r = a$.

The deflection and radial slope are zero at the outer edge, giving,

$$\tilde{w}_m(a,\theta) = 0 \quad \Rightarrow \quad [c_1 J_m(\beta a) + c_3 I_m(\beta a)]\cos(m\theta) = 0$$

$$\frac{\partial \tilde{w}_m(a,\theta)}{\partial r} = 0 \quad \Rightarrow \quad \left[c_1 J'_m(\beta a) + c_3 I'_m(\beta a) \right]\cos(m\theta) = 0$$

$$\Rightarrow \quad \left[c_1 \beta\{\frac{m}{\beta a}J_m - J_{m+1}\} + c_3 \beta\{\frac{m}{\beta a}I_m + I_{m+1}\} \right]\cos(m\theta) = 0$$

In this, the superscript prime indicates differentiation with respect to $z = \beta a$. We can put these equations in the form of a matrix; multiplying out the determinant and re-arranging gives the characteristic equation as

$$J_m(\beta a)I_{m+1}(\beta a) + I_m(\beta a)J_{m+1}(\beta a) = 0$$

This must be solved numerically. The results can be nondimensionalized using

$$\lambda_{nm} = (\beta a)^2_{nm} = \omega_{nm}\sqrt{\frac{\rho h}{D}} \quad \text{or} \quad \omega_{nm} = \sqrt{\frac{D}{\rho h}}\frac{(\beta a)^2_{nm}}{a^2} = \sqrt{\frac{D}{\rho h}}\frac{\lambda_{nm}}{a^2}$$

The values of λ_{nm} are given in Table 4.1. Additional values for λ_{nm} can be found in Reference [45].

n, m	0	1	2	3	4	5
1	10.21	21.26	34.88	51.04	69.66	90.73
2	39.77	60.82	84.58	111.0	140.1	171.8
3	89.10	120.0	153.8	190.3	229.5	271.4
4	158.1	199.0	242.7	289.1	338.4	390.4

Table 4.1: Values of $\lambda_{nm} = (\beta a)^2_{nm}$.

The mode shapes are given by

$$\tilde{w}_{nm}(r, \theta) = c_1\left[J_m(\beta_{nm}r) - \frac{J_m(\beta_{nm}a)}{I_m(\beta_{nm}a)}I_m(\beta_{nm}r)\right]\cos(m\theta) \tag{4.12}$$

with $\beta_{nm} \equiv \sqrt{\lambda_{nm}}/a$. These are shown in Figure 4.13 labeled as $[nm]$. Additional solutions for circular plates and for other plate geometries can be found in Reference [45].

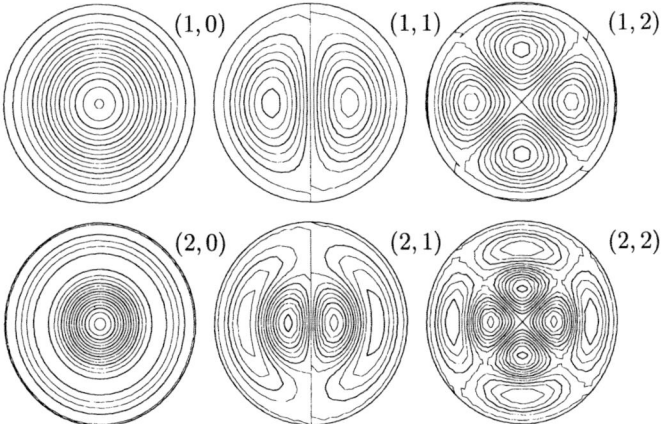

Figure 4.13: The (n, m) mode shapes for circular plate clamped on its outer edge.

Circular Cylinder Solution

As can be inferred from the flat-plate solutions, we are not in a position to solve general shell or curved plate problems. We therefore restrict ourselves to a complete cylindrical shell with the following support conditions at both ends:

$$u = 0, \qquad w = 0, \qquad N_{yy} = 0, \qquad M_{yy} = 0$$

These could be achieved physically by having a very rigid ring at both ends. Note that the axial motion is unconstrained.

The boundary conditions are satisfied by assuming solutions of the form

$$u = u_o \sin(ks)\sin(\xi y)e^{i\omega t}, \quad v = v_o \cos(ks)\cos(\xi y)e^{i\omega t}, \quad w = w_o \sin(ks)\sin(\xi y)e^{i\omega t}$$

where $k = n/R$ and $\xi = m\pi/L$. Substitution into the governing equations (2.30) gives the homogeneous system of equations

$$
\begin{bmatrix}
\alpha_1 - [k^2 + (1-\nu)\xi^2]\dfrac{D}{R^2} & \gamma & [\bar{E} + (k^2 + \xi^2)D]\dfrac{k}{R} \\[2ex]
\gamma & \alpha_2 & -\nu\bar{E}\dfrac{\xi}{R} \\[2ex]
-[\bar{E} + (k^2 + \xi^2)D]\dfrac{k}{R} & \nu\bar{E}\dfrac{\xi}{R} & \alpha_3 + \bar{E}\dfrac{1}{R^2}
\end{bmatrix}
\begin{Bmatrix} u_o \\ v_o \\ w_o \end{Bmatrix} = 0
$$

where

$$
\begin{aligned}
\alpha_1 &\equiv -\bar{E}[k^2 + \tfrac{1}{2}(1-\nu)\xi^2] + \rho h\omega^2, & \gamma &\equiv \tfrac{1}{2}(1+\nu)\bar{E}k\xi \\
\alpha_2 &\equiv -\bar{E}[\xi^2 + \tfrac{1}{2}(1-\nu)k^2] + \rho h\omega^2, & \alpha_3 &\equiv D[k^2 + \xi^2]^2 - \rho h\omega^2
\end{aligned}
$$

The α_1, α_2, and γ terms alone define the flat membrane problem, while α_3 alone defines the flat-plate flexural problem. All the other terms are couplings due to the curvature. The curved beam result is obtained by setting $\xi = 0$.

The characteristic equation, obtained from the determinant of this system, is cubic in ω^2. It is therefore simplest to solve for the resonance frequencies using some numerical scheme.

Example 4.13: Determine how the resonant frequencies are dependent on the aspect ratio of the cylinder.

One of the challenges in assimilating results for complicated systems such as a shell is to see the results as part of a pattern. The simple flat plates (both rectangular and circular) gave a nicely ordered system in terms of n and m. This is not the case for the cylinder. Figure 4.14 shows how the frequencies vary against length of cylinder.

A vertical line on this plot will give the ordered sequence obtained in an eigen-analysis. For example, the sequence of modes for two aspect ratios are:

$$
\begin{aligned}
L/R = 2.5: & \quad 41, 31, 51, 61, 21, 52, 62, 71, 42, 81, 32, 91, 22 \\
L/R = 4.0: & \quad 31, 41, 21, 51, 42, 52, 61, 32, 62, 71, 22, 81, 91
\end{aligned}
$$

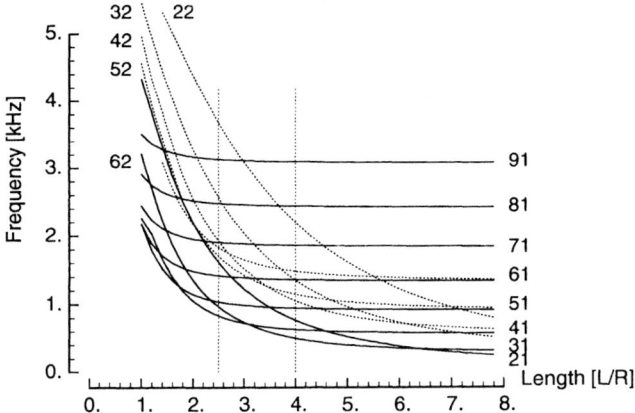

Figure 4.14: Variation of resonance frequency with length of cylinder. Solid lines are $m = 1$, dashed lines are $m = 2$.

Clearly, the simplest mode shape does not necessarily have the lowest frequency. Furthermore, in looking at the $L/R = 2.5$ line, in some cases, such as the (8,1) and (3,2) modes, the frequencies are the same, while in other cases, such as the (6,2), (7,1), and (4,2) modes, the frequencies form a cluster. Thus, in any practical analysis, slight variations in dimensions or material properties can have a significant effect on the observed sequence of mode shapes.

4.3 Matrix Representation of Inertia Forces

Inertia loads are a special case of body forces and therefore the matrix representation will follow directly as was done, for example, in Section 3.4. For rotational motion such as that of turbine blades or helicopter rotors, a corotational (or convected) frame of reference is often used. We also look at these cases to see how the reference frame affects the representation of inertia; the formulation for this is taken primarily from Reference [4].

Mass and Damping Matrix

By D'Alembert's principle, we can consider the external loads as comprising the applied loads and the inertia

$$\rho f_i \Rightarrow \rho f_i - \rho \ddot{u}_i - \eta \dot{u}_i$$

Thus, we can do a similar treatment as used for the body forces in Chapter 3. That is, using $u_i = \{h\}^T\{u\}_i$ or $\{\mathcal{U}\} = \{H\}\{u\}$ as the discretized representation of the displacements, the virtual work of the applied loads leads to

$$\sum_i \int f_i \delta u_i \, dA^o \quad \Rightarrow \quad \{P\} = [\int_{A^o} [H]^T\{f\}dA^o]$$

Applying this specifically to the inertia terms leads to

$$-\sum_i \int_{V^o} (\rho \ddot{u}_i + \eta \dot{u}_i) \delta u_i dV^o \quad \Rightarrow \quad -[\, m \,]\{\ddot{u}\} = -[\int_{V^o} \rho [H]^T [H] \, dV^o]\{\ddot{u}\}$$

$$\Rightarrow \quad -[\, c \,]\{\dot{u}\} = -[\int_{V^o} \eta [H]^T [H] \, dV^o]\{\dot{u}\}$$

where $[\, m \,]$ is called the element mass matrix and $[\, c \,]$ is called the element damping matrix. Note that the integrations are over the original geometry and that the resulting matrix forms for $[\, m \,]$ and $[\, c \,]$ are identical. That is,

$$[\, c \,] = \frac{\eta}{\rho A}[\, m \,]$$

This is an example of the damping matrix being proportional to the mass matrix on an element level.

When the shape functions $[\, H \,]$ are the same as used in the stiffness formulation, the mass and damping matrices are called *consistent*. Note that these masses do not necessarily have any simple interpretation of masses at nodes.

The assemblage process for the mass and damping matrices is done in exactly the same manner as for the linear elastic stiffness. As a result, the mass and damping matrices will exhibit all the symmetry and bandedness properties of the stiffness matrix. The result is that we get the equations of motion of the structure as a whole to be

$$[\, M \,]\{\ddot{u}\} + [\, C \,]\{\dot{u}\} = \{P\} - \{F\}$$

where $[\, M \,]$ is the structural mass matrix, $[\, C \,]$ is the assembled damping matrix, $\{F\} = [\, K \,]\{u\}$ for linear elastic problems, and $\{P\}$ are the externally applied loads not including the inertia contributions.

When the structural joints have concentrated masses, we need only amend the structural mass matrix as follows

$$[\, M \,] = \sum_i [m^{(i)}] + \lceil M_c \rfloor$$

where $\lceil M_c \rfloor$ is the collection of joint concentrated masses. This is a diagonal matrix. In the next chapter we consider the more complex nonlinear case when the joints have rotational inertia and undergo finite rotations.

For proportional damping at the structural level, we assume

$$[\, C \,] = \alpha [\, M \,] + \beta [\, K \,]$$

where α and β are constants chosen to best represent the physical situation. Note that this relation is not likely to hold for structures composed of different materials. However, for lightly damped structures it can be a useful approximation.

Example 4.14: Determine the consistent mass matrix for a frame element.

The displacements for the rod element can be written in terms of the nodal values as

$$u(x) = (1 - \frac{x}{L})u_1 + (\frac{x}{L})u_2 \equiv f_1(x)u_1 + f_2(x)u_2$$

Write the accelerations in matrix form as

$$\{ \ddot{u} \} = [\ f_1 \quad f_2 \ ] \left\{ \begin{array}{c} \ddot{u}_1 \\ \ddot{u}_2 \end{array} \right\} \qquad \text{or} \qquad \{\ddot{u}\} = [\ H \ ]\{\ddot{u}\}$$

with $[\ H \ ]$ being a $[1 \times 2]$ matrix. Then the mass matrix is

$$[\ m \ ] = \int_{V^o} \rho [\ H \ ]^T [\ H \ ] \, dV^o = \int_{V^o} \rho \begin{bmatrix} f_1 f_1 & f_1 f_2 \\ f_2 f_1 & f_2 f_2 \end{bmatrix} dV^o \ \Rightarrow \ [\ m \ ] = \frac{\rho A L}{6} \begin{bmatrix} 2 & 1 \\ 1 & 2 \end{bmatrix}$$

The element masses are also given by

$$m_{ij} = \int_0^L \rho A f_i(x) f_j(x) \, dx$$

It is clear that it is the symmetry of the terms $f_i(x)f_j(x)$ that ensures the symmetry of the mass matrix.

The procedure for determining the element mass matrix for beams proceeds as for the rod. Recall from Chapter 1 that the deflection can be represented in terms of the nodal values as

$$\begin{aligned} v(x) &= \left[1 - 3(\frac{x}{L})^2 + 2(\frac{x}{L})^3\right] v_1 + (\frac{x}{L})\left[1 - 2(\frac{x}{L}) + (\frac{x}{L})^2\right] L\phi_1 \\ &\quad + (\frac{x}{L})^2 \left[3 - 2(\frac{x}{L})\right] v_2 + (\frac{x}{L})^2 \left[-1 + (\frac{x}{L})\right] L\phi_2 \\ &= g_1(x)v_1 + g_2(x)\phi_1 + g_3(x)v_2 + g_4(x)\phi_2 \end{aligned}$$

We write this in the matrix form

$$\{ \ddot{v} \} = [\ g_1 \quad g_2 \quad g_3 \quad g_4 \ ] \left\{ \begin{array}{c} \ddot{v}_1 \\ \ddot{\phi}_1 \\ \ddot{v}_2 \\ \ddot{\phi}_2 \end{array} \right\} \qquad \text{or} \qquad \{\ddot{u}\} = [\ H \ ]\{\ddot{u}\}$$

with $[\ H \ ]$ now being a $[1 \times 4]$ matrix. Then the mass matrix is

$$[\ m \ ] = \int_{V^o} \rho [\ H \ ]^T [\ H \ ] \, dV^o \ \Rightarrow \ [\ m \ ] = \frac{\rho A L}{420} \begin{bmatrix} 156 & 22L & 54 & -13L \\ 22L & 4L^2 & 13L & -3L^2 \\ 54 & 13L & 156 & -22L \\ -13L & -3L^2 & -22L & 4L^2 \end{bmatrix}$$

We can also write the masses as

$$m_{ij} = \int_0^L \rho A g_i(x) g_j(x) \, dx$$

which again shows that it is the symmetry of the terms $g_i(x)g_j(x)$ that ensures the symmetry of the mass matrices.

The mass matrix of the frame is a composition of that of the rod and beam suitably augmented, for example, to $[6 \times 6]$ for a plane frame.

Example 4.15: Determine the mass matrix for a triangular plate element.

As was done in the earlier chapters dealing with plates and as just done with the frame element, we find it convenient to separate the behaviors into membrane and bending. For the membrane behavior we use the shape functions associated with the constant strain triangle. The accelerations are represented as

$$\ddot{u}(x,y) = \sum_{i=1}^{3} h_i(x,y)\ddot{u}_i , \qquad \ddot{v}(x,y) = \sum_{i=1}^{3} h_i(x,y)\ddot{v}_i$$

In matrix form

$$\left\{ \begin{array}{c} \ddot{u} \\ \ddot{v} \end{array} \right\} = \left[\begin{array}{cccccc} h_1 & 0 & h_2 & 0 & h_3 & 0 \\ 0 & h_1 & 0 & h_2 & 0 & h_3 \end{array} \right] \left\{ \begin{array}{c} \ddot{u}_1 \\ \ddot{v}_1 \\ \ddot{u}_2 \\ \ddot{v}_2 \\ \ddot{u}_3 \\ \ddot{v}_3 \end{array} \right\} \qquad \text{or} \qquad \{\ddot{u}\} = [\, H \,]\{\ddot{u}\}$$

Then the mass matrix is

$$[\, m \,] = \int_{V^o} \rho [\, H \,]^T [\, H \,] dV^o \quad \Longrightarrow \quad [\, m \,] = \frac{\rho A h}{12} \left[\begin{array}{cccccc} 2 & 0 & 1 & 0 & 1 & 0 \\ 0 & 2 & 0 & 1 & 0 & 1 \\ 1 & 0 & 2 & 0 & 1 & 0 \\ 0 & 1 & 0 & 2 & 0 & 1 \\ 1 & 0 & 1 & 0 & 2 & 0 \\ 0 & 1 & 0 & 1 & 0 & 2 \end{array} \right]$$

For the MRT element, we also have the three drilling degrees of freedom $\{\phi_1, \phi_2, \phi_3\}$. In the next subsection, we will treat this as diagonal and estimate it based on the lumped rotational inertia for a beam.

For the bending behavior, let the displacements be represented by

$$\{w(x,y)\} = [\, N \,]\{u\}, \qquad \{u\} = \{w_1, \phi_{x1}, \phi_{y1}; w_2, \phi_{x2}, \phi_{y2}; w_3, \phi_{x3}, \phi_{y3}\}$$

where the shape functions $N(x,y)$ are given from Equation (2.43). Again, we get

$$[\, m \,] = \int \rho h [\, N \,]^T [\, N \,] dA$$

The expressions are too lengthy to write here. We will generally find it more beneficial, anyway, to use the lumped mass matrix.

Example 4.16: Assemble the system of equations for the dynamic response of the fixed/fixed rod shown in Figure 4.15. Use two, then three, elements to represent the structure.

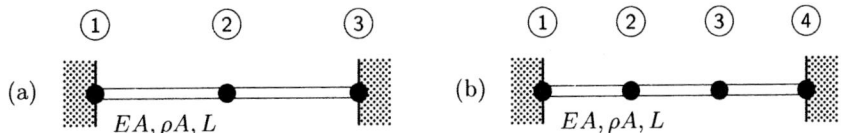

Figure 4.15: Fixed/fixed rod with two and three element models.

The element stiffness matrices are

$$[k^{(12)}] = \frac{EA}{L/2}\begin{bmatrix} 1 & -1 \\ -1 & 1 \end{bmatrix}, \qquad [k^{(23)}] = \frac{EA}{L/2}\begin{bmatrix} 1 & -1 \\ -1 & 1 \end{bmatrix}$$

giving the full assembled structural stiffness matrix as

$$[\,K\,] = \frac{2EA}{L}\begin{bmatrix} 1 & -1 & 0 \\ -1 & 2 & -1 \\ 0 & -1 & 1 \end{bmatrix}$$

Note that this is the same as if it were a static problem. The element mass matrices (using the consistent mass matrix) are

$$[m^{(12)}] = \frac{\rho AL/2}{6}\begin{bmatrix} 2 & 1 \\ 1 & 2 \end{bmatrix}, \qquad [m^{(23)}] = \frac{\rho AL/2}{6}\begin{bmatrix} 2 & 1 \\ 1 & 2 \end{bmatrix}$$

giving the full assembled structural mass matrix as

$$[\,M\,] = \frac{\rho AL}{12}\begin{bmatrix} 2 & 1 & 0 \\ 1 & 4 & 1 \\ 0 & 1 & 2 \end{bmatrix}$$

The equations of motion in full form for the dynamic response of the structure are

$$\frac{2EA}{L}\begin{bmatrix} 1 & -1 & 0 \\ -1 & 2 & -1 \\ 0 & -1 & 1 \end{bmatrix}\begin{Bmatrix} u_1 \\ u_2 \\ u_3 \end{Bmatrix} + \frac{\rho AL}{12}\begin{bmatrix} 2 & 1 & 0 \\ 1 & 4 & 1 \\ 0 & 1 & 2 \end{bmatrix}\begin{Bmatrix} \ddot{u}_1 \\ \ddot{u}_2 \\ \ddot{u}_3 \end{Bmatrix} = \begin{Bmatrix} P_1 \\ P_2 \\ P_3 \end{Bmatrix}$$

This is reduced in the usual manner by removing the fixed degrees of freedom. The boundary conditions are

$$u_1 = u_3 = 0, \quad \ddot{u}_1 = \ddot{u}_3 = 0$$

Consequently, the reduced structural matrices are

$$[\,K^*\,] = \frac{2EA}{L}[\,2\,], \qquad [\,M^*\,] = \frac{\rho AL}{12}[\,4\,]$$

The dynamic problem now simply becomes

$$\frac{2EA}{L}2u_2 + \frac{\rho AL}{12}4\ddot{u}_2 = P_2(t)$$

Consider the same problem but this time use three elements. Number the nodes as shown in Figure 4.15(b). Only consider the reduced matrices; the unknown displacements and known forces are, respectively,

$$\{u_u\} = \begin{Bmatrix} u_2 \\ u_3 \end{Bmatrix}, \qquad \{P_k\} = \begin{Bmatrix} P_2 \\ P_3 \end{Bmatrix}$$

The reduced element stiffnesses are

$$[k^{*(12)}] = \frac{3EA}{L}[\ 1\ ], \qquad [k^{*(23)}] = \frac{3EA}{L}\begin{bmatrix} 1 & -1 \\ -1 & 1 \end{bmatrix}, \qquad [k^{*(34)}] = \frac{3EA}{L}[\ 1\ ]$$

giving the reduced structural stiffness matrix as

$$[K^*] = \frac{3EA}{L}\begin{bmatrix} 2 & -1 \\ -1 & 2 \end{bmatrix}$$

The reduced element mass matrices are

$$[m^{*(12)}] = \frac{\rho AL}{18}[\ 2\ ], \qquad [m^{*(23)}] = \frac{\rho AL}{18}\begin{bmatrix} 2 & 1 \\ 1 & 2 \end{bmatrix}, \qquad [m^{*(34)}] = \frac{\rho AL}{18}[\ 2\ ]$$

giving the reduced structural mass matrix as

$$[M^*] = \frac{\rho AL}{18}\begin{bmatrix} 4 & 1 \\ 1 & 4 \end{bmatrix}$$

The dynamic system of equations becomes

$$\frac{3EA}{L}\begin{bmatrix} 2 & -1 \\ -1 & 2 \end{bmatrix}\begin{Bmatrix} u_2 \\ u_3 \end{Bmatrix} + \frac{\rho AL}{18}\begin{bmatrix} 4 & 1 \\ 1 & 4 \end{bmatrix}\begin{Bmatrix} \ddot{u}_2 \\ \ddot{u}_3 \end{Bmatrix} = \begin{Bmatrix} P_2 \\ P_3 \end{Bmatrix}$$

These equations can now be solved to obtain the dynamic response. We will do this in the next section.

Lumped Representations

It is useful to realize that because the mass matrix does not involve derivatives of the shape function, then we can be more lax about the choice of shape function than for the stiffness matrix. In fact, in many applications we will find it preferable to use a lumped mass (and damping) approximation where the only nonzero terms are on the diagonal. We show some examples here.

The simplest mass model is to consider only the translational inertias, which are obtained simply by dividing the total mass by the number of nodes and placing this value of mass at each node. Thus, the diagonal terms for the 3-D frame and plate are

$$\lceil m \rfloor = \frac{\rho AL}{2}\lceil 1, 1, 1, 0, 0, 0;\ 1, 1, 1, 0, 0, 0 \rfloor$$

$$\lceil m \rfloor = \frac{\rho Ah}{3}\lceil 1, 1, 1, 0, 0, 0;\ 1, 1, 1, 0, 0, 0;\ 1, 1, 1, 0, 0, 0 \rfloor$$

respectively. These neglect the rotational inertias of the flexural actions. Generally, these contributions are negligible and the above are quite accurate especially when the elements are small. There is, however, a very important circumstance when a zero diagonal mass is unacceptable and reasonable nonzero values are needed. In the next chapter, we develop an explicit numerical integration scheme where the time step depends on the highest resonant frequencies of the structure; these frequencies in turn are dictated by the rotational inertias.

First consider the frame. It is tempting to estimate the rotational inertia of a beam by taking the total rotational inertia, $\rho A L^3/12$, and placing half of it at each node. This would grossly overestimate the inertia because the lumped masses already contribute a significant rotary inertia. We instead will use the diagonal terms of the consistent matrix to form an estimate of the diagonal matrix. Note that the translation diagonal terms add up to only $\rho A L\, 312/420$. Hence, by scaling each diagonal term by $420/312$ we get

$$\lceil m \rfloor = \frac{\rho A L}{2}\lceil 1,\, 1,\, 1,\, \beta,\, \beta,\, \beta;\, 1,\, 1,\, 1,\, \beta,\, \beta,\, \beta \rfloor, \qquad \beta = \alpha L^2/40$$

where α is typically taken as unity. This scheme has the merit of correctly giving the translational inertias.

We treat the plate in an analogous manner as

$$\lceil m \rfloor = \frac{\rho A h}{3}\lceil 1,\, 1,\, 1,\, \beta,\, \beta,\, 20\beta;\, 1,\, 1,\, 1,\, \beta,\, \beta,\, 20\beta;\, 1,\, 1,\, 1,\, \beta,\, \beta,\, 20\beta \rfloor$$

with $\beta = \alpha L^2/40$. We estimate the effective length $L \approx \sqrt{A/\pi}$ as basically the radius of a disk of the same area as the triangle. Again, α is typically taken as unity.

Example 4.17: Assemble the system of equations for the dynamic response of the fixed/fixed rod shown in Figure 4.15. Use two, then three, lumped mass elements.

The procedure follows that of the previous example, hence we state just the mass results. The element mass matrices are

$$[m^{(12)}] = \frac{\rho A L/2}{2}\begin{bmatrix} 1 & 0 \\ 0 & 1 \end{bmatrix}, \qquad [m^{(23)}] = \frac{\rho A L/2}{2}\begin{bmatrix} 1 & 0 \\ 0 & 1 \end{bmatrix}$$

giving the full assembled structural mass matrix as

$$[\,M\,] = \frac{\rho A L}{4}\begin{bmatrix} 1 & 0 & 0 \\ 0 & 2 & 0 \\ 0 & 0 & 1 \end{bmatrix}$$

The equations of motion in full form for the dynamic response of the structure are

$$\frac{2EA}{L}\begin{bmatrix} 1 & -1 & 0 \\ -1 & 2 & -1 \\ 0 & -1 & 1 \end{bmatrix}\begin{Bmatrix} u_1 \\ u_2 \\ u_3 \end{Bmatrix} + \frac{\rho A L}{4}\begin{bmatrix} 1 & 0 & 0 \\ 0 & 2 & 0 \\ 0 & 0 & 1 \end{bmatrix}\begin{Bmatrix} \ddot{u}_1 \\ \ddot{u}_2 \\ \ddot{u}_3 \end{Bmatrix} = \begin{Bmatrix} P_1 \\ P_2 \\ P_3 \end{Bmatrix}$$

This is reduced as before to

$$\frac{2EA}{L}2u_2 + \frac{\rho AL}{4}2\ddot{u}_2 = P_2(t)$$

Consider the same problem but this time use three elements. Number the nodes as shown in Figure 4.15(b) and only consider the reduced matrices. The reduced element mass matrices are

$$[m^{*(12)}] = \frac{\rho AL}{6}[\ 1\ ], \qquad [m^{*(23)}] = \frac{\rho AL}{6}\begin{bmatrix} 1 & 0 \\ 0 & 1 \end{bmatrix}, \qquad [m^{*(34)}] = \frac{\rho AL}{6}[\ 1\ ]$$

giving the reduced structural mass matrix as

$$[M^*] = \frac{\rho AL}{6}\begin{bmatrix} 2 & 0 \\ 0 & 2 \end{bmatrix}$$

The dynamic system of equations becomes

$$\frac{3EA}{L}\begin{bmatrix} 2 & -1 \\ -1 & 2 \end{bmatrix}\begin{Bmatrix} u_2 \\ u_3 \end{Bmatrix} + \frac{\rho AL}{6}\begin{bmatrix} 2 & 0 \\ 0 & 2 \end{bmatrix}\begin{Bmatrix} \ddot{u}_2 \\ \ddot{u}_3 \end{Bmatrix} = \begin{Bmatrix} P_2 \\ P_3 \end{Bmatrix}$$

These equations can now be solved to obtain the dynamic response. This will be done in the next section where comparisons are made to the results for the consistent mass formulation.

Inertia in a Rotating Reference System

We are interested in describing the motion of a point in terms of variables relevant to a moving observer. The laws of mechanics must be written in terms of an inertial frame; what we need to do here is establish the relationship between the two sets of variables.

Consider two reference frames: the fixed absolute inertial frame $(_Ax,\ _Ay,\ _Az)$ is situated at A and has unit vectors $(\hat{i},\ \hat{j},\ \hat{k})$; the moving observer frame $(x,\ y,\ z)$ is situated at O with unit vectors $(_A\hat{e}_x,\ _A\hat{e}_y,\ _A\hat{e}_z)$ relative to A and is rotating with angular velocity $_A\omega$ relative to A. The components of a vector referred to unit vectors at A can be written in terms of components referred to unit vectors at O according to

$$_A\hat{v} = [\ T\ ]_O\hat{v}, \qquad _O\hat{v} = [\ T\ ]_A^T\hat{v}, \qquad [\ T\ ] \equiv {}_A[\hat{e}_x,\ \hat{e}_y,\ \hat{e}_z]$$

where the triad $[\ T\ ]$ has the components of the unit vectors at O referred to the unit vectors at A. The components of the unit vectors are related through

$$_A\hat{e}_x = \beta_{xx}\hat{i} + \beta_{xy}\hat{j} + \beta_{xz}\hat{k} = \sum_i \beta_{xi}\hat{e}_i$$

where β_{ij} are the direction cosines. The rate of change of this vector (and the triad) is

$$_A\dot{\hat{e}}_i = {}_A\hat{\omega} \times_A \hat{e}_i \qquad \Longrightarrow \qquad [\ \dot{T}\ ] = {}_A\hat{\omega} \times_A [\ T\ ]$$

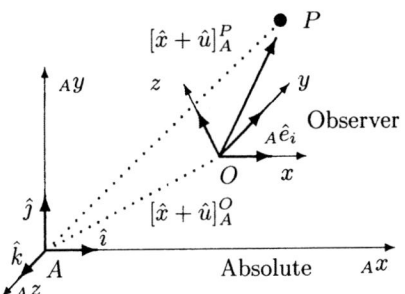

Figure 4.16: Rotating frame of reference.

as we saw in the Section 3.2 on finite rotations.

A point P initially at $\hat{x}_A^P$ goes in motion with the displacement $\hat{u}(t)_A^P$. The position of this point P relative to A and referred to unit vectors at A can be written in terms of position vectors relative to O and referred to unit vectors at O according to

$$_A[\hat{x} + \hat{u}]_A^P = {}_A[\hat{x} + \hat{u}]_A^O +_A [\hat{x} + \hat{u}]_O^P = {}_A[\hat{x} + \hat{u}]_A^O + [\ T\ ]_O[\hat{x} + \hat{u}]_O^P$$

where we are using the notation

$$_c[\quad]_b^a, \qquad a = \text{point}, \quad b = \text{relative to}, \quad c = \text{referred to}$$

The velocity of the point is given by

$$
\begin{aligned}
_A[\ \dot{u}\ ]_A^P &= {}_A[\ \dot{u}\ ]_A^O + [\ \dot{T}\ ]_O[\hat{x} + \hat{u}]_O^P + [\ T\ ]_O[\ \dot{u}\ ]_O^P \\
&= {}_A[\ \dot{u}\ ]_A^O +_A \hat{\omega} \times_A [\ T\ ]_O[\hat{x} + \hat{u}]_O^P + [\ T\ ]_O[\ \dot{u}\ ]_O^P
\end{aligned}
$$

We want to refer the components to the observer frame, hence multiply across by the transpose of the triad and re-arrange to get

$$_O[\ \dot{u}\ ]_A^P = {}_O[\ \dot{u}\ ]_O^P +_O [\ \dot{u}\ ]_A^O +_A \hat{\omega} \times_O [\hat{x} + \hat{u}]_O^P$$

The first term is called the relative velocity, while the next two are the *transport velocity* of O relative to $_AO$.

In a similar way, we can determine the acceleration of a point as

$$
\begin{aligned}
_O[\ \ddot{u}\ ]_A^P = {}_O[\ \ddot{u}\ ]_O^P +_O [\ \ddot{u}\ ]_A^O +_A \dot{\hat{\omega}} \times_O [\hat{x} + \hat{u}]_O^P +_A \hat{\omega} \times_A \hat{\omega} \times_O [\hat{x} + \hat{u}]_O^P \\
+ 2_A\hat{\omega} \times_O [\ \dot{u}\ ]_O^P
\end{aligned}
$$

The first term is called the relative acceleration, while the next three are the transport acceleration of O relative to $_AO$, and the final term is the Coriolis acceleration.

As before, using $u_i = \{h\}^T\{u\}_i$ or $\{\mathcal{U}\} = \{H\}\{u\}$ as the discretized representation of the displacements, the virtual work of the inertia force becomes

$$\sum_i \int_{V^o} \rho_A \ddot{u}_i \delta u_i dV^o \Rightarrow$$

$$\Rightarrow \quad [M]\{\ddot{u}\} = \left[\int_{V^o} \rho[H]^T[H] dV^o\right]\{\ddot{u}\}$$

$$\Rightarrow \quad [C_2]\{\dot{u}\} = \left[\int_{V^o} [H]^T 2_A[\omega][H] dV^o\right]\{\dot{u}\}$$

$$\Rightarrow \quad [K_2]\{u\} = \left[\int_{V^o} [H]^T_A[\dot{\omega}][H] dV^o\right]\{u\}$$

$$\Rightarrow \quad [K_3]\{u\} = \left[\int_{V^o} [H]^T_A[\omega][\omega][H] dV^o\right]\{u\}$$

$$\Rightarrow \quad \{P_R\} = \left\{\int_{V^o} [H]^T[\ddot{u}_{io} +_A[\dot{\omega}]\{x\} +_A[\omega][\omega]\{x\}] dV^o\right\}$$

where we have introduced the rotation matrix

$$[\omega] \equiv \begin{bmatrix} 0 & -\omega_z & \omega_y \\ \omega_z & 0 & -\omega_x \\ -\omega_y & \omega_x & 0 \end{bmatrix}, \quad \hat{\omega} \times \hat{v} \Rightarrow [\omega]\{v\}$$

to allow the vector crossproducts to be written in matrix form.

The system of equations can now be written as

$$[K + K_2 + K_3]\{u\} + [C + C_2]\{\dot{u}\} + [M]\{\ddot{u}\} = \{P\} - \{P_R\}$$

where it is clear that using a rotating reference frame has introduced both stiffness-like and damping-like terms. What is especially worth noting is that the matrices $[K_2]$ and $[C_2]$ are skew symmetric. This would add considerably to the computational cost of the solution. In the subsequent sections and chapters, we will always use an inertial reference frame because of the simpler nature of the matrices.

Example 4.18: Consider the dynamics of a rod loaded as shown in Figure 4.17. Determine the equations of motion in a fixed coordinate system when $\Omega = 0$.

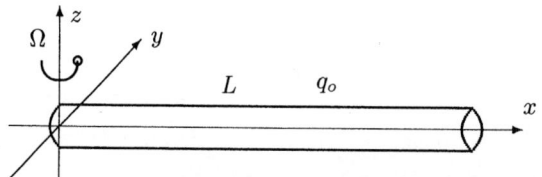

Figure 4.17: Rotating rod.

Based on the shape functions for a rod, we have

$$\bar{u}(x) = (1 - \frac{x}{L})u_1 + (\frac{x}{L})u_2 , \qquad \{h_x\} = \{(1 - \frac{x}{L}), (\frac{x}{L})\}$$

The derivatives are therefore

$$\partial = \frac{d}{dx} , \qquad \{\partial h_x\} = \{-\frac{1}{L}, \frac{1}{L}\}$$

This leads to the stiffness relation

$$[k] = \int \{-\frac{1}{L}, \frac{1}{L}\}^T [D]\{-\frac{1}{L}, \frac{1}{L}\} \, dV = \frac{EA}{L} \begin{bmatrix} 1 & -1 \\ -1 & 1 \end{bmatrix}$$

The mass matrix is determined as

$$[m] = \int \{(1 - \frac{x}{L}), (\frac{x}{L})\}^T \rho\{(1 - \frac{x}{L}), (\frac{x}{L})\} \, dV = \frac{\rho AL}{6} \begin{bmatrix} 2 & 1 \\ 1 & 2 \end{bmatrix}$$

If there is a distributed load $q(x) = q_o = $ constant, this evaluates to

$$\{P\} = \int \{(1 - \frac{x}{L}), (\frac{x}{L})\}^T q_o \, dx = \frac{q_o AL}{2} \begin{Bmatrix} 1 \\ 1 \end{Bmatrix}$$

The equations of motion then become

$$\frac{EA}{L} \begin{bmatrix} 1 & -1 \\ -1 & 1 \end{bmatrix} \begin{Bmatrix} u_1 \\ u_2 \end{Bmatrix} + \frac{\rho AL}{6} \begin{bmatrix} 2 & 1 \\ 1 & 2 \end{bmatrix} \begin{Bmatrix} \ddot{u}_1 \\ \ddot{u}_2 \end{Bmatrix} = \frac{q_o AL}{2} \begin{Bmatrix} 1 \\ 1 \end{Bmatrix}$$

Example 4.19: Now consider when the rod is rotating at an angular speed of Ω about the z-axis. Establish the equations of motion in a coordinate system rotating with the rod.

The mass and stiffness matrices remain the same, but we will get some additional matrices dependent on Ω.

At the instant when the rod is in the position shown where the absolute and observer frames coincide

$$\dot{\hat{u}}_o = 0, \qquad \hat{u}_o = u\hat{i}, \qquad \dot{\hat{\omega}}_o = 0, \qquad \omega_o = \Omega\hat{k}, \qquad \hat{g} = g_x$$

Evaluating each term, we get

$$[C_2] = 0$$
$$[K_2] = 0$$
$$[K_3] = \int_{V^o} \{g\}^T [\Omega\hat{k} \times \Omega\hat{k} \times g_x\hat{k}_i] \, dV^o = \int_{V^o} \rho\Omega^2\{g\}_x^T\{g\}_x \, dV^o = -\frac{\rho\Omega^2 AL}{6} \begin{bmatrix} 2 & 1 \\ 1 & 2 \end{bmatrix}$$
$$\{P_R\} = \int_{V^o} \rho\{g\}^T [0 + 0 + \Omega\hat{k} \times \Omega\hat{k} \times x\hat{k}_i] \, dV^o = \int_{V^o} \rho\Omega^2\{g\}_x \, dV^o = -\frac{\rho A\Omega^2 L^2}{6} \begin{Bmatrix} 1 \\ 2 \end{Bmatrix}$$

These lead to the equations of motion

$$\left[\frac{EA}{L} \begin{bmatrix} 1 & -1 \\ -1 & 1 \end{bmatrix} - \frac{\rho\Omega^2 AL}{6} \begin{bmatrix} 2 & 1 \\ 1 & 2 \end{bmatrix} \right] \begin{Bmatrix} u_1 \\ u_2 \end{Bmatrix} + \frac{\rho AL}{6} \begin{bmatrix} 2 & 1 \\ 1 & 2 \end{bmatrix} \begin{Bmatrix} \ddot{u}_1 \\ \ddot{u}_2 \end{Bmatrix} = -\frac{\rho A\Omega^2 L^2}{6} \begin{Bmatrix} 1 \\ 2 \end{Bmatrix}$$

Note that the applied force is larger at the second node than what occurred in the first example.

Consider the special case when the first node is restrained, then $u_1 = 0$. For free vibrations with $u_2 = \hat{u}_2 e^{i\mu t}$, we have that

$$[\frac{EA}{L} - \frac{\rho\Omega^2 AL2}{6} - \frac{\rho AL2}{6}\mu^2]\hat{u}_2 = 0$$

which gives the frequency

$$\mu = \sqrt{\frac{EA/L - \rho\Omega^2 AL/3}{\rho AL/3}}$$

An interesting situation to note is that it is possible to have a static instability effect when the angular speed reaches

$$\Omega = \frac{1}{L}\sqrt{\frac{3EA}{\rho A}}$$

To put the magnitude of this speed into perspective, the axial vibration of the rod is given by

$$\frac{EA}{L} - \frac{\rho AL2}{6}\omega^2 = 0 \qquad \text{or} \qquad \omega_c = \frac{1}{L}\sqrt{\frac{3EA}{\rho A}}$$

Thus the system must be rotating with an angular speed comparable to the resonance frequency in order to see the effect.

Matrix Form of Linear Dynamic Problems

The computer solution of the structural equations of motion is discussed in more detail in the next chapter, but it is of value now to consider some of the major problem types originating from our present matrix formulation.

The matrix form of the equations of motion for a linear system are

$$[K]\{u\} + [C]\{\dot{u}\} + [M]\{\ddot{u}\} = \{P\}$$

When the equations are written in an inertial frame, all the matrices are symmetric. This equation is to be interpreted as a system of differential equations in time for the unknown nodal displacements $\{u\}$, subject to the known forcing histories $\{P\}$, and a set of boundary and initial conditions. Generally, these require some numerical scheme for integration over time. Therefore, for *transient* dynamic problems, the matrix method approach becomes computationally intensive in two respects. First, a substantial increase in the number of elements must be used in order to model the mass distribution accurately. The other is that the complete system of equations must be solved at each time increment. These issues are dealt with in Chapter 5.

For the special case when the excitation force is harmonic, that is,

$$\{P\} = \{\hat{P}\}\, e^{i\omega t} \quad \text{or} \quad \left\{ \begin{array}{c} P_1 \\ P_2 \\ \vdots \\ P_n \end{array} \right\} = \left\{ \begin{array}{c} \hat{P}_1 \\ \hat{P}_2 \\ \vdots \\ \hat{P}_n \end{array} \right\} e^{i\omega t}$$

(note that many of the $\hat{P}_n$ could be zero) then the response is also harmonic and given by

$$\{u\} = \{\hat{u}\}\, e^{i\omega t} \quad \text{or} \quad \left\{ \begin{array}{c} u_1 \\ u_2 \\ \vdots \\ u_n \end{array} \right\} = \left\{ \begin{array}{c} \hat{u}_1 \\ \hat{u}_2 \\ \vdots \\ \hat{u}_n \end{array} \right\} e^{i\omega t}$$

This type of analysis is referred to as *forced frequency* analysis. Substituting these forms into the differential equations gives

$$[\,K\,]\{\hat{u}\}e^{i\omega t} + i\omega[\,C\,]\{\hat{u}\}e^{i\omega t} - \omega^2[\,M\,]\{\hat{u}\}e^{i\omega t} = \{\hat{P}\}e^{i\omega t}$$

or, after canceling through the common time factor,

$$\big[[\,K\,] + i\omega[\,C\,] - \omega^2[\,M\,]\big]\{\hat{u}\} = \{\hat{P}\} \quad \text{or} \quad [\,\hat{K}\,]\{\hat{u}\} = \{\hat{P}\}$$

The solution can be obtained analogous to the static problem; the difference is that the stiffness matrix is modified by the inertia term $\omega^2[\,M\,]$ and the complex damping term $i\omega[\,C\,]$. The matrix $[\,\hat{K}\,]$ is the discrete approximation of the dynamic structural stiffness; it is frequency dependent as well as being complex. This system of equations is now recognized as the spectral form of the equations of motion of the structure. One approach, then, to transient problems is to evaluate the above at each frequency and use the FFT [16] for time domain reconstructions. This is feasible, but a more full fledged spectral approach based on the exact dynamic stiffness is developed in Reference [23].

A case of very special interest is that of free vibrations. When the damping is zero, this case gives the mode shapes that are very important in a modal analysis. For free vibrations of the system, the applied loads $\{P\}$ are zero giving the equations of motion as

$$\big[[\,K\,] - \omega^2[\,M\,]\big]\{\hat{u}\} = 0$$

This is a system of homogeneous equations for the nodal displacements $\{\hat{u}\}$. For a nontrivial solution, the determinant of the matrix of coefficients must be zero. We thus conclude that this is an eigenvalue problem, ω^2 are the eigenvalues and the corresponding $\{\hat{u}\}$ the eigenvectors of the problem. Note that the larger the number of elements (for a given structure), the larger the system of equations; consequently, the more eigenvalues we can obtain.

Once the matrices are assembled, NonStaD uses the subspace iteration scheme [7, 22] to solve the eigenvalue problem. In this analysis, a reduced eigensystem is established by iteration on a set of Ritz vectors. The advantage in using subspace iteration (over vector iteration, say) is that the convergence of the subspace and not of individual iteration vectors is achieved. Consequently, it is less likely to miss any eigenvectors during the search.

Example 4.20: Consider the free vibration of the fixed/fixed rod shown in Figure 4.15. Neglecting damping, use two elements to find an approximate solution.

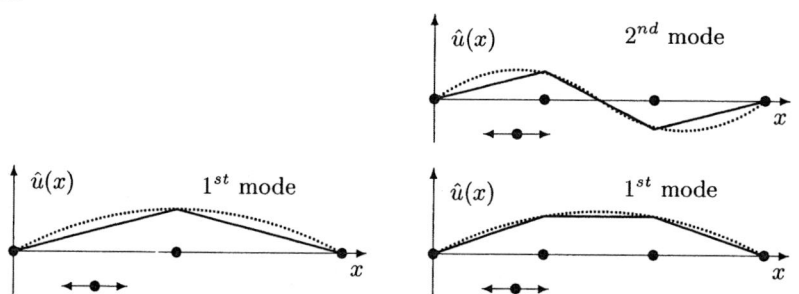

Figure 4.18: Mode shapes for a fixed/fixed rod modeled with two and three elements.

We already established the equations of motion in the previous section, we will now solve the free vibration problem. The equations of motion in full form for the free vibration of the structure are

$$\left[\frac{2EA}{L}\begin{bmatrix} 1 & -1 & 0 \\ -1 & 2 & -1 \\ 0 & -1 & 1 \end{bmatrix} - \omega^2\frac{\rho AL}{12}\begin{bmatrix} 2 & 1 & 0 \\ 1 & 4 & 1 \\ 0 & 1 & 2 \end{bmatrix}\right]\left\{\begin{array}{c} \hat{u}_1 \\ \hat{u}_2 \\ \hat{u}_3 \end{array}\right\} = 0$$

This is reduced by removing the fixed degrees of freedom. That is, the boundary conditions are used to determine the unknown degrees of freedom as

$$u_1 = u_3 = 0, \quad \ddot{u}_1 = \ddot{u}_3 = 0 \quad \Rightarrow \quad \{u_u\} = \{u_2\}, \quad \{\ddot{u}_u\} = \{\ddot{u}_2\}$$

Consequently, the reduced eigenvalue problem now simply becomes

$$\left[\frac{2EA}{L}2 - \omega^2\frac{\rho AL}{12}4\right]\hat{u}_2 = 0$$

allowing the resonant frequency to be obtained as

$$\omega = \frac{\sqrt{12}}{L}\sqrt{\frac{EA}{\rho A}} \simeq \frac{3.46}{L}\sqrt{\frac{EA}{\rho A}}, \qquad \omega_{\text{exact}} = \frac{\pi}{L}\sqrt{\frac{EA}{\rho A}}$$

There is a difference of about 10% in comparison with the exact solution. However, there is only one value computed — the two element formulation is incapable of giving any higher resonances.

The corresponding lumped mass result is

$$\omega = \frac{\sqrt{8}}{L}\sqrt{\frac{EA}{\rho A}} \simeq \frac{2.83}{L}\sqrt{\frac{EA}{\rho A}}$$

This value is an underestimate by about the same amount that the consistent mass is an overestimate. Thus it appears, from an accuracy point of view, there is no significant difference between the two approaches.

The mode shape for this solution is simply $\{0, 1, 0\}$. This corresponds to the first symmetric mode of the exact solution as shown in Figure 4.18.

Now consider the same problem but this time use three elements. Using the earlier results, the eigensystem of equations for the reduced system becomes

$$\left[\frac{3EA}{L}\begin{bmatrix} 2 & -1 \\ -1 & 2 \end{bmatrix} - \omega^2\frac{\rho AL}{18}\begin{bmatrix} 4 & 1 \\ 1 & 4 \end{bmatrix}\right]\begin{Bmatrix} \hat{u}_2 \\ \hat{u}_3 \end{Bmatrix} = 0$$

These equations can now be solved to obtain the eigenvalues. That is, the frequency equation is obtained by multiplying the determinant out, and rearranging to get

$$(\frac{54EA}{5L^2\rho A} - \omega^2)(\frac{54EA}{L^2\rho A} - \omega^2) = 0$$

The solutions of this are

First mode : $\qquad \omega = \dfrac{\sqrt{54/5}}{L}\sqrt{\dfrac{EA}{\rho A}} \simeq \dfrac{3.29}{L}\sqrt{\dfrac{EA}{\rho A}}$

Second mode: $\qquad \omega = \dfrac{\sqrt{54}}{L}\sqrt{\dfrac{EA}{\rho A}} \simeq \dfrac{7.35}{L}\sqrt{\dfrac{EA}{\rho A}}$

The accuracy of the first mode is improved, but also, an estimate of the second frequency is obtained. (Note that for this problem, the numerical factor for the theoretical solution varies as $n\pi$.) The lumped mass results are also improved giving

First mode : $\qquad \omega = \dfrac{\sqrt{9}}{L}\sqrt{\dfrac{EA}{\rho A}} \simeq \dfrac{3.0}{L}\sqrt{\dfrac{EA}{\rho A}}$

Second mode: $\qquad \omega = \dfrac{\sqrt{27}}{L}\sqrt{\dfrac{EA}{\rho A}} \simeq \dfrac{5.20}{L}\sqrt{\dfrac{EA}{\rho A}}$

Again, these lumped frequencies are on the lower side of theory by about the same amount that the consistent results are higher.

The corresponding mode shapes are (irrespective of the mass matrix)

$$\{\hat{u}\}_1 = \{0, 1, 1, 0\}, \qquad \{\hat{u}\}_2 = \{0, 1, -1, 0\}$$

It is apparent that these are estimates for the first symmetric and first anti-symmetric mode shapes, respectively, as shown in Figure 4.18.

Example 4.21: Use a convergence study to compare the performance of the lumped and consistent mass matrices for a plate in flexural vibration.

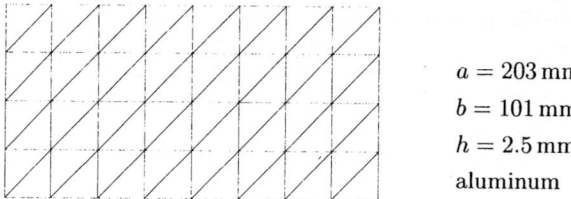

$a = 203$ mm

$b = 101$ mm

$h = 2.5$ mm

aluminum

Figure 4.19: Generic $[4 \times 8]$ mesh.

The generic mesh is shown in Figure 4.19. The other meshes are obtained by dividing this. We will use the simply supported plate as the test case.

The resonance frequencies are given in Figure 4.20. In each case, the mesh represented the complete plate and subspace iteration was used to determine the eigenvalues.

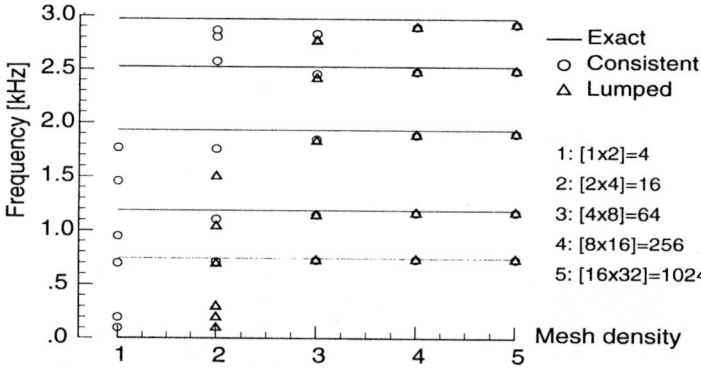

Figure 4.20: Convergence study for the resonance frequencies of a simply supported plate.

There is very little difference in the performance of the mass matrices except for the very coarse meshes. Hence, we can conclude that because of the attractive diagonal property of the lumped mass that, generally, this will be the mass matrix of choice.

Example 4.22: Test the performance of the flat platelet modeling of a circular cylinder.

The dimensions and mesh are shown in Figure 4.21. There are a total of 64 modules in the hoop direction and 24 in the length direction. This gives modules that are nearly square. The boundary conditions at each end are

$$\{u, v, w, \phi_x, \phi_y, \phi_z\} = \{0, 0, 1, 1, 1, 0\}$$

There is a rigid body mode in the z-direction. These boundary conditions correspond to the problem solved in Figure 4.14.

The inset table in Figure 4.21 shows the sequence of modes obtained and are compared with those obtained from Figure 4.14. The values are quite close thus

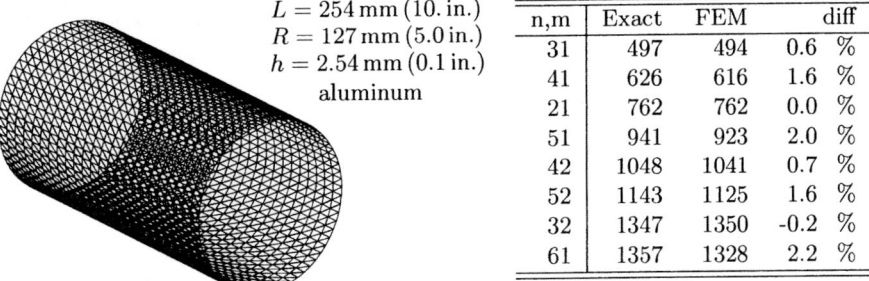

$L = 254\,\text{mm} \,(10.\,\text{in.})$
$R = 127\,\text{mm} \,(5.0\,\text{in.})$
$h = 2.54\,\text{mm} \,(0.1\,\text{in.})$
aluminum

n,m	Exact	FEM	diff	
31	497	494	0.6	%
41	626	616	1.6	%
21	762	762	0.0	%
51	941	923	2.0	%
42	1048	1041	0.7	%
52	1143	1125	1.6	%
32	1347	1350	-0.2	%
61	1357	1328	2.2	%

Figure 4.21: Dimensions of cylinder with a typical mesh.

validating the faceted element modeling. Since the complete cylinder was modeled, many of the computed modes are actually for double roots; the table only reports results for distinct mode shapes.

Two of the mode shapes are shown in Figure 4.1. The exploded view is intended to give a clearer picture of the three-dimensional mode. In interpreting the figure, recall that the boundary conditions imposed are that the ends do not change in diameter.

4.4 Modal Analysis

It is apparent that the analysis of complicated structures will involve systems that have very many degrees of freedom and therefore are described by a large number of equations. This is all the more true since the use of the approximate stiffness requires subdividing a given member into many small elements. This section develops some of the concepts that form the basis for the treatment and understanding of the dynamical behavior of large systems. Central to this development is the concept of the *modal matrix* because through it the system can be transformed into a set of uncoupled equations. We only consider the case when the system matrices are symmetric; nonsymmetric matrices can give rise to complex eigenvalues and we leave some of their discussion until Chapter 7.

Orthogonality of Free Vibration Mode Shapes

When an undamped system is excited, it will continue to vibrate long after the initial disturbance is over. Furthermore, it vibrates with a characteristic shape (called the *mode shape*) governed by the following system of equations:

$$[\,K\,]\{u\} + [\,M\,]\{\ddot{u}\} = 0$$

Since the motion is harmonic, then $\{u(t)\} = \{\hat{u}\}e^{i\omega t}$, and the characteristic shape satisfies the algebraic system of equations

$$[[\,K\,] - \lambda[\,M\,]]\,\{\hat{u}\} = 0 \tag{4.13}$$

where $\lambda = \omega^2$. These equations are homogeneous, hence the solutions, in general, are zero. The only time a nontrivial solution is obtained is when the determinant of the coefficients is zero. Thus Equation (4.13) is recognized as the familiar eigenvalue problem; λ are the eigenvalues and $\{\hat{u}\}$ are the eigenvectors. There are as many eigenvalues as the order of the system of equations. That is, the solution yields N eigenvalues λ_i and N corresponding eigenvectors $\{\hat{u}\}_i$.

It is apparent that if $\{\hat{u}\}$ is a solution, then $\alpha\{\hat{u}\}$ is also a solution, where α is a nonzero scalar constant. That is, the modal vector represents a shape rather than the absolute deflection of the structure; the ratio of the elements of the modal vector are fixed not their absolute value. If, however, one of the values is fixed, then the eigenvector becomes unique in an absolute sense also. The process of scaling the elements of the mode shape is called *normalization*; the resulting scaled modes are called *orthonormal* modes. There are several methods available for doing this, the following is a partial list:

1. The largest element is set to unity.
2. The length of the mode vector is set to unity.
3. A particular, physically significant, element is set to unity.
4. The modal mass is set to unity such that $\{\phi\}_m^T [M]\{\phi\}_m = \tilde{M}_{mm} = 1$

The first three of these are useful when the mode shapes are to be plotted. The last of the scaling schemes is implemented in NonStaD and will be explained presently. As a reminder that the mode shapes are some sort of normalized version of the displacements $\{\hat{u}\}$, the notation

$$\{\phi\} \equiv \text{normalized } \{\hat{u}\}$$

will be used.

Consider two arbitrary, non-null vectors $\{v\}_1$ and $\{v\}_2$. For the square matrix $[A]$ to be *positive definite*, we must have that the triple product

$$\{v\}_1^T [A]\{v\}_1 = \text{constant}$$

be greater than zero. If the matrix $[A]$ is symmetric, we also have that

$$\{v\}_1^T [A]\{v\}_2 = \{v\}_2^T [A]\{v\}_1$$

We will use these two important results to establish some properties of the mode shapes.

Each mode shape will satisfy the equation of motion, that is, when substituted into Equation (4.13) they give

$$[K]\{\phi\}_i = \lambda_i [M]\{\phi\}_i$$

Pre-multiply this by the transpose of another mode shape $\{\phi\}_j$

$$\{\phi\}_j^T [K]\{\phi\}_i = \lambda_i \{\phi\}_j^T [M]\{\phi\}_i$$

Now write the equation for the jth mode and pre-multiply this by the transpose of the ith mode; that is,

$$\{\phi\}_i^T [\, K \,]\{\phi\}_j = \lambda_j \{\phi\}_i^T [\, M \,]\{\phi\}_j$$

Subtract these, and since the mass and stiffness matrices are symmetric, then obtain

$$0 = (\lambda_i - \lambda_j)\{\phi\}_i^T [\, M \,]\{\phi\}_j$$

We chose the mode shapes to be at two different natural frequencies, therefore $\lambda_i \neq \lambda_j$ resulting in

$$\{\phi\}_i^T [\, M \,]\{\phi\}_j = 0$$

This is a statement of the *orthogonality* property of the mode shapes with respect to the mass matrix. By analogy to vector algebra, it means that the eigenvectors are perpendicular (orthogonal) to each other, and their vector dot product is therefore zero. It is emphasized, however, that in the present case we have a weighting factor $[\, M \,]$. In a similar manner, it can be seen that

$$\{\phi\}_i^T [\, K \,]\{\phi\}_j = 0$$

also. There are cases of *repeated roots*; that is, the system has different modes at the same frequency. The above development only shows that these modes are orthogonal to all other modes but not necessarily to each other. Actually, the eigenvectors are not unique and a linear combination of them may also satisfy the equations of motion. In these circumstances we will prescribe that the mode shapes associated with repeated roots be orthogonal to each other.

Modal Mass and Stiffness Matrices

If we set $i = j = m$ in the previous analysis, then the two mode shapes we are dealing with are the same, and therefore the triple product is equal to some nonzero constant. That is,

$$\{\phi\}_m^T [\, M \,]\{\phi\}_m = \tilde{M}_{mm}$$
$$\{\phi\}_m^T [\, K \,]\{\phi\}_m = \tilde{K}_{mm} = \lambda_m \tilde{M}_{mm} = \omega_m^2 \tilde{M}_{mm}$$

where $\tilde{M}_{mm}$ and $\tilde{K}_{mm}$ are called the *modal mass* and *modal stiffness* of the mth mode, respectively.

These relations show that the mass and stiffness matrix can be converted to a single constant, one for each mode, by multiplying by the mode shapes. Thus, construct the square matrix $[\, \Phi \,]$ whose columns are the normalized mode shape vectors as

$$[\, \Phi \,] \equiv \left[\left\{ \begin{array}{c} \phi_1 \\ \phi_2 \\ \vdots \\ \phi_N \end{array} \right\}_1 \left\{ \begin{array}{c} \phi_1 \\ \phi_2 \\ \vdots \\ \phi_N \end{array} \right\}_2 \cdots \left\{ \begin{array}{c} \phi_1 \\ \phi_2 \\ \vdots \\ \phi_N \end{array} \right\}_N \right]$$

The matrix $[\ \Phi\ ]$ is referred to as the *modal matrix*. It is a fully populated matrix of order $[N \times N]$ and typically is not symmetric.

The orthogonal properties of the mode shapes and the definition of the modal mass and stiffness can now be expressed in matrix form as

$$[\ \Phi\ ]^T[\ M\ ][\ \Phi\ ] = \lceil\ \tilde{M}\ \rfloor, \qquad [\ \Phi\ ]^T[\ K\ ][\ \Phi\ ] = \lceil\ \tilde{K}\ \rfloor$$

where $\lceil\ \tilde{M}\ \rfloor$ and $\lceil\ \tilde{K}\ \rfloor$ are diagonal matrices of order $[N \times N]$.

Example 4.23: The simple system shown in Figure 4.22 is modeled with two degrees of freedom. Determine the eigenvalues and eigenvectors associated with this system if the governing equations of motion are

$$\begin{bmatrix} 4 & -2 \\ -2 & 6 \end{bmatrix} \begin{Bmatrix} u_1 \\ u_2 \end{Bmatrix} + \begin{bmatrix} 1 & 0 \\ 0 & 2 \end{bmatrix} \begin{Bmatrix} \ddot{u}_1 \\ \ddot{u}_2 \end{Bmatrix} = \begin{Bmatrix} P_1 \\ P_2 \end{Bmatrix}$$

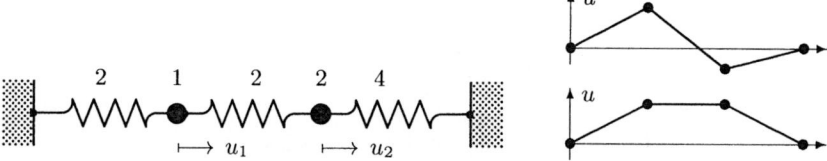

Figure 4.22: Two-degree-of-freedom system.

The spectral form of these equations for free vibration leads to the system

$$\begin{bmatrix} 4-\lambda & -2 \\ -2 & 6-2\lambda \end{bmatrix} \begin{Bmatrix} \hat{u}_1 \\ \hat{u}_2 \end{Bmatrix} = \begin{Bmatrix} 0 \\ 0 \end{Bmatrix}, \qquad \lambda = \omega^2$$

The determinant must be zero for a nontrivial solution; thus on multiplying out and rearranging, we get

$$(4-\lambda)(6-2\lambda) - 4 = 0 \quad \text{or} \quad \lambda^2 - 7\lambda + 10 = 0$$

Since this is quadratic, then the roots are

$$\lambda_{1,2} = \frac{7 \pm \sqrt{49-40}}{2} = \frac{7}{2} \pm \frac{3}{2} = 2,\ 5$$

Thus the ordered eigenvalues are $\lambda_1 = 2$ and $\lambda_2 = 5$. The two natural frequencies are
$$\omega_1 = \sqrt{\lambda_1} = \sqrt{2} \quad \text{and} \quad \omega_2 = \sqrt{\lambda_2} = \sqrt{5}$$

The mode shape for the first mode is obtained by substituting λ_1 into the original system to give

$$\begin{bmatrix} 4-2 & -2 \\ -2 & 6-4 \end{bmatrix} \begin{Bmatrix} \hat{u}_1 \\ \hat{u}_2 \end{Bmatrix}_1 = \begin{Bmatrix} 0 \\ 0 \end{Bmatrix}$$

and these become, when written out separately,

$$2\hat{u}_1 - 2\hat{u}_2 = 0$$
$$-2\hat{u}_1 + 2\hat{u}_2 = 0$$

From both equations we have that $\hat{u}_1 = \hat{u}_2$, thus the first mode shape is

$$\{\hat{u}\}_1 = \hat{u}_1\{\phi\}_1 = \hat{u}_1 \begin{Bmatrix} 1 \\ 1 \end{Bmatrix}$$

where the magnitude of $\hat{u}_1$ is arbitrary. Similarly, for the second mode we get after substituting for λ_2

$$\begin{bmatrix} 4-5 & -2 \\ -2 & 6-10 \end{bmatrix} \begin{Bmatrix} \hat{u}_1 \\ \hat{u}_2 \end{Bmatrix}_2 = \begin{Bmatrix} 0 \\ 0 \end{Bmatrix}$$

giving as separate equations

$$-\hat{u}_1 - 2\hat{u}_2 = 0$$
$$-2\hat{u}_1 - 4\hat{u}_2 = 0$$

Both of these equations give $\hat{u}_2 = -\frac{1}{2}\hat{u}_1$. Thus the second mode shape is

$$\{\hat{u}\}_2 = \hat{u}_1\{\phi\}_2 = \hat{u}_1 \begin{Bmatrix} 1 \\ -\frac{1}{2} \end{Bmatrix}$$

Again, this has been normalized to the first displacement.

These mode shapes are shown plotted in Figure 4.22. Strictly speaking, we should only plot the values of $\hat{u}_1$ and $\hat{u}_2$, but since the orthogonality properties are not affected by augmenting the eigenvectors with zeros, then a clearer picture is obtained by incorporating the zero displacement at the attachment points.

Example 4.24: Show that the eigenvectors of the previous example are orthogonal.

We will consider only the normalized forms

$$\{\phi\}_1 = \begin{Bmatrix} 1 \\ 1 \end{Bmatrix} \qquad \text{and} \qquad \{\phi\}_2 = \begin{Bmatrix} 1 \\ -\frac{1}{2} \end{Bmatrix}$$

First note that these are not orthogonal in the simple vector dot product sense, since

$$\begin{Bmatrix} 1 \\ 1 \end{Bmatrix}^T \begin{Bmatrix} 1 \\ -\frac{1}{2} \end{Bmatrix} = 1 - \frac{1}{2} \neq 0$$

The eigenvectors are orthogonal *with respect to* the mass and stiffness matrices.

For the mass matrix, we have

$$\begin{Bmatrix} 1 \\ 1 \end{Bmatrix}^T \begin{bmatrix} 1 & 0 \\ 0 & 2 \end{bmatrix} \begin{Bmatrix} 1 \\ -\frac{1}{2} \end{Bmatrix} = \begin{Bmatrix} 1 \\ 1 \end{Bmatrix}^T \begin{Bmatrix} 1 \\ -1 \end{Bmatrix} = 0$$

And for the stiffness matrix

$$\begin{Bmatrix} 1 \\ 1 \end{Bmatrix}^T \begin{bmatrix} 4 & -2 \\ -2 & 6 \end{bmatrix} \begin{Bmatrix} 1 \\ -\frac{1}{2} \end{Bmatrix} = \begin{Bmatrix} 1 \\ 1 \end{Bmatrix}^T \begin{Bmatrix} 5 \\ -5 \end{Bmatrix} = 0$$

Example 4.25:   Determine the modal mass and modal stiffness for the system of Figure 4.22.

Recalling that

$$\{\phi\}_1 = \begin{Bmatrix} 1 \\ 1 \end{Bmatrix} \qquad \text{and} \qquad \{\phi\}_2 = \begin{Bmatrix} 1 \\ -\frac{1}{2} \end{Bmatrix}$$

then the generalized mass for the first mode is

$$\tilde{M}_{11} = \begin{Bmatrix} 1 \\ 1 \end{Bmatrix}^T \begin{bmatrix} 1 & 0 \\ 0 & 2 \end{bmatrix} \begin{Bmatrix} 1 \\ 1 \end{Bmatrix} = 3$$

and for the second mode

$$\tilde{M}_{22} = \begin{Bmatrix} 1 \\ -\frac{1}{2} \end{Bmatrix}^T \begin{bmatrix} 1 & 0 \\ 0 & 2 \end{bmatrix} \begin{Bmatrix} 1 \\ -\frac{1}{2} \end{Bmatrix} = \frac{3}{2}$$

Note that there is only one mass for each mode. The modal stiffnesses are

$$\tilde{K}_{11} = \begin{Bmatrix} 1 \\ 1 \end{Bmatrix}^T \begin{bmatrix} 4 & -2 \\ -2 & 6 \end{bmatrix} \begin{Bmatrix} 1 \\ 1 \end{Bmatrix} = 6, \quad \tilde{K}_{22} = \begin{Bmatrix} 1 \\ -\frac{1}{2} \end{Bmatrix}^T \begin{bmatrix} 4 & -2 \\ -2 & 6 \end{bmatrix} \begin{Bmatrix} 1 \\ -\frac{1}{2} \end{Bmatrix} = \frac{15}{2}$$

It is useful to note that these results for the stiffness could also be obtained by using the relationship involving the resonant frequency, that is,

$$\tilde{K}_{11} = \omega_1^2 \tilde{M}_{11} = 2 \times 3 = 6 \qquad \text{and} \qquad \tilde{K}_{22} = \omega_2^2 \tilde{M}_{ii} = 5 \times 3/2 = 15/2$$

The exact same results are also obtained when the full modal matrix is used. First establish the modal matrix as

$$[\ \Phi\ ] = \begin{bmatrix} 1 & 1 \\ 1 & -\frac{1}{2} \end{bmatrix}$$

Now pre- and post-multiply the mass and stiffness matrices by this modal matrix to get for the mass

$$\lceil\ \tilde{M}\ \rfloor = \begin{bmatrix} 1 & 1 \\ 1 & -\frac{1}{2} \end{bmatrix} \begin{bmatrix} 1 & 0 \\ 0 & 2 \end{bmatrix} \begin{bmatrix} 1 & 1 \\ 1 & -\frac{1}{2} \end{bmatrix} = \begin{bmatrix} 3 & 0 \\ 0 & \frac{3}{2} \end{bmatrix}$$

and for the stiffness

$$\lceil\ \tilde{K}\ \rfloor = \begin{bmatrix} 1 & 1 \\ 1 & -\frac{1}{2} \end{bmatrix} \begin{bmatrix} 4 & -2 \\ -2 & 6 \end{bmatrix} \begin{bmatrix} 1 & 1 \\ 1 & -\frac{1}{2} \end{bmatrix} = \begin{bmatrix} 6 & 0 \\ 0 & \frac{15}{2} \end{bmatrix}$$

These results emphasize the diagonal nature of the modal mass and stiffness matrices. They also show that if the mode shapes are normalized in a different manner then different numerical values will be obtained for the modal stiffness and mass.

Transformation to Principal Coordinates

Consider a system described by the following coupled equations of motion:

$$\begin{bmatrix} k_{11} & k_{12} \\ k_{12} & k_{22} \end{bmatrix} \begin{Bmatrix} u_1 \\ u_2 \end{Bmatrix} + \begin{bmatrix} c_1 & 0 \\ 0 & c_2 \end{bmatrix} \begin{Bmatrix} \dot{u}_1 \\ \dot{u}_2 \end{Bmatrix} + \begin{bmatrix} m_1 & 0 \\ 0 & m_2 \end{bmatrix} \begin{Bmatrix} \ddot{u}_1 \\ \ddot{u}_2 \end{Bmatrix} = \begin{Bmatrix} P_1 \\ P_2 \end{Bmatrix}$$

In these equations, the coupling is due to the fact that the stiffness matrix is not diagonal. This is called *elastic* or *static* coupling. When the mass matrix is not diagonal, the coupling is termed *inertial* or *dynamic* coupling. If we obtained diagonal mass and stiffness matrices simultaneously, then the system would be uncoupled and each equation would be similar to that of a single-degree-of-freedom system. These could then be solved independently of each other. Such a transformation will be shown here.

Consider the transformation of the displacements to new values by the equation

$$\{u\} = [\; \Phi \;]\{\eta\}$$

where $[\; \Phi \;]$ is the modal matrix. Here $\{\eta\}$ are called *principal coordinates* or *normal coordinates*. The equations of motion in terms of these new coordinates are

$$[\;K\;][\; \Phi \;]\{\eta\} + [\;C\;][\; \Phi \;]\{\dot{\eta}\} + [\;M\;][\; \Phi \;]\{\ddot{\eta}\} = \{P\}$$

Pre-multiply this by the transpose of the modal matrix to get

$$[\; \Phi \;]^T[\;K\;][\; \Phi \;]\{\eta\} + [\; \Phi \;]^T[\;C\;][\; \Phi \;]\{\dot{\eta}\} + [\; \Phi \;]^T[\;M\;][\; \Phi \;]\{\ddot{\eta}\} = [\; \Phi \;]^T\{P\}$$

Because of the orthogonal properties of the mode shapes and the definition of the modal mass and stiffness, the terms associated with the mass and stiffness are diagonalized. This, however, is not true of the damping matrix. A simplifying assumption usually introduced at this stage is that the distribution of damping throughout the structure is *proportional*. That is,

$$[\;C\;] = \alpha[\;M\;] + \beta[\;K\;]$$

where α and β are constants; this is usually referred to as *Rayleigh damping*. Clearly, $[\;C\;]$ will be diagonalized so that the equations of motion become

$$\lceil\; \tilde{K} \;\rfloor\{\eta\} + \lceil\; \tilde{C} \;\rfloor\{\dot{\eta}\} + \lceil\; \tilde{M} \;\rfloor\{\ddot{\eta}\} = [\; \Phi \;]^T\{P\}$$

This represents N uncoupled equations of the form

$$\tilde{K}_{mm}\eta_m + \tilde{C}_{mm}\dot{\eta}_m + \tilde{M}_{mm}\ddot{\eta}_m = \{\phi\}_m^T\{P\} = \tilde{P}_m$$

where $\{\phi\}_m$ is the mth mode shape, $\tilde{M}_{mm}$, $\tilde{C}_{mm}$, and $\tilde{K}_{mm}$ are the mth modal mass, damping, and stiffness, respectively. Each equation above is the equation of motion for a single-degree-of-freedom system, and since $\tilde{K}_{mm} = \omega_m^2 \tilde{M}_{mm}$, can be written as

$$\omega_m^2 \eta_m + 2\zeta_m \omega_m \dot{\eta}_m + \ddot{\eta}_m = \frac{\tilde{P}_m}{\tilde{M}_{mm}} = \frac{\{\phi\}_m^T \{P\}}{\{\phi\}_m^T [M] \{\phi\}_m} \tag{4.14}$$

where the damping ratio is defined as

$$\zeta_m = \frac{\tilde{C}_{mm}}{2\sqrt{\tilde{K}_{mm} \tilde{M}_{mm}}} = \frac{\tilde{C}_{mm}}{2\omega_m \tilde{M}_{mm}} = \frac{\alpha + \beta \omega_{mm}^2}{2\omega_m}$$

We see that although the damping matrix $[C]$ is "constant," the effect it has on the response is different for each mode. Specifically, for mass proportional damping ($\beta = 0$), the higher modes have very little damping.

Following on from earlier in the chapter, the uncoupled equations of motion can be integrated directly to give the generalized response as

$$\eta_m(t) = \left[\eta_m(0) \cos \omega_m t + \frac{\zeta_m \omega_m \eta_m(0) + \dot{\eta}_m(0)}{\omega_d} \sin \omega_m t \right] e^{-\zeta \omega_m t}$$

$$+ \frac{1}{\omega_m \tilde{M}_{mm}} \int_0^t \tilde{P}_m(\tau) e^{-\zeta \omega_m(t-\tau)} \sin[\omega_m(t-\tau)] \, d\tau$$

where $\omega_d \equiv \omega_m \sqrt{1 - \zeta_m^2}$. This is Duhamel's integral for each mode. For simple forcing functions (for example, stepped loading) the integration may be performed analytically, but generally it must be done numerically.

Once time responses for all the η_m are obtained, the solution in terms of the original coordinates can be obtained by simply transforming back to the physical coordinates according to

$$\{u\} = [\Phi]\{\eta\}$$

In general, for an N-degree-of-freedom system, the responses are

$$\begin{Bmatrix} u_1 \\ u_2 \\ \vdots \\ u_N \end{Bmatrix} = \begin{Bmatrix} \phi_1 \\ \phi_2 \\ \vdots \\ \phi_N \end{Bmatrix}_1 \eta_1(t) + \begin{Bmatrix} \phi_1 \\ \phi_2 \\ \vdots \\ \phi_N \end{Bmatrix}_2 \eta_2(t) + \cdots \begin{Bmatrix} \phi_1 \\ \phi_2 \\ \vdots \\ \phi_N \end{Bmatrix}_M \eta_M(t)$$

This can also be written as

$$\{u(t)\} = \sum_m^M \{\phi\}_m \eta_m(t)$$

This is a fundamental relation in the dynamics of structures. It shows that the response of any complicated system can be conceived as the superposition of the responses of the natural vibration modes. Furthermore, the summation need not extend to N (the system size) but can be truncated at $M < N$. The justification is that if the loading $\{P\}(t)$ does not have a frequency content high enough to excite the higher modes, then it is not necessary to include these higher modes. Indeed, for many practical problems (such as earthquake analysis of buildings) M is significantly less than N and generally on the order of 1 to 5.

Example 4.26: Consider again the simple system described by the equations of motion

$$\begin{bmatrix} 4 & -2 \\ -2 & 6 \end{bmatrix} \begin{Bmatrix} u_1 \\ u_2 \end{Bmatrix} + \begin{bmatrix} 1 & 0 \\ 0 & 2 \end{bmatrix} \begin{Bmatrix} \ddot{u}_1 \\ \ddot{u}_2 \end{Bmatrix} = \begin{Bmatrix} P_1 \\ P_2 \end{Bmatrix}$$

Obtain the equations of motion in principal coordinates. Also, determine the free vibration characteristics of the system.

We have already shown that the modal matrix $[\ \Phi\ ]$ is assembled as

$$[\ \Phi\ ] = \begin{bmatrix} 1 & 1 \\ 1 & -\frac{1}{2} \end{bmatrix}$$

The coordinate transformation is therefore given by

$$\begin{Bmatrix} u_1 \\ u_2 \end{Bmatrix} = \begin{bmatrix} 1 & 1 \\ 1 & -\frac{1}{2} \end{bmatrix} \begin{Bmatrix} \eta_1 \\ \eta_2 \end{Bmatrix}$$

Applying this to the equation of motion and pre-multiplying by $[\ \Phi\ ]^T$ gives

$$[\ \Phi\ ]^T \begin{bmatrix} 4 & -2 \\ -2 & 6 \end{bmatrix} [\ \Phi\ ] \begin{Bmatrix} \eta_1 \\ \eta_2 \end{Bmatrix} + [\ \Phi\ ]^T \begin{bmatrix} 1 & 0 \\ 0 & 2 \end{bmatrix} [\ \Phi\ ] \begin{Bmatrix} \ddot{\eta}_1 \\ \ddot{\eta}_2 \end{Bmatrix} = [\ \Phi\ ]^T \begin{Bmatrix} P_1 \\ P_2 \end{Bmatrix}$$

Multiplying out, get

$$\begin{bmatrix} 6 & 0 \\ 0 & 15/2 \end{bmatrix} \begin{Bmatrix} \eta_2 \\ \eta_2 \end{Bmatrix} + \begin{bmatrix} 3 & 0 \\ 0 & 3/2 \end{bmatrix} \begin{Bmatrix} \ddot{\eta}_1 \\ \ddot{\eta}_2 \end{Bmatrix} = \begin{bmatrix} 1 & 1 \\ 1 & -1/2 \end{bmatrix} \begin{Bmatrix} P_1 \\ P_2 \end{Bmatrix}$$

Note how the mass and stiffness matrices have been diagonalized. The equations of motion can be separated as

$$6\eta_1 + 3\ddot{\eta}_1 = P_1 + P_2$$
$$\tfrac{15}{2}\eta_2 + \tfrac{3}{2}\ddot{\eta}_2 = P_1 - \tfrac{1}{2}P_2$$

It is worth observing that while the coordinates are uncoupled, all the applied forces now act at each generalized node as shown in Figure 4.23.

Figure 4.23: The modal representation of the undamped 2-DoF system.

Example 4.27: Consider a special case of the last example when $P_2 = 0$ and P_1 is a stepped loading of magnitude P_o. Obtain the solution if the system is initially at rest.

The differential equations after $t = 0$ are simplified to

$$\ddot{\eta}_1 + 2\eta_1 = \tfrac{1}{3}P_o$$
$$\ddot{\eta}_2 + 5\eta_2 = \tfrac{2}{3}P_o$$

Integrate this using the Duhamel's integral. The initial conditions are that $\eta(0)$ and $\dot{\eta}(0)$ are zero for both modes, leading to the modal responses

$$\eta_1(t) = \tfrac{1}{6}P_o[1 - \cos\omega_1 t], \qquad \omega_1 = \sqrt{2}$$
$$\eta_2(t) = \tfrac{4}{45}P_o[1 - \cos\omega_2 t], \qquad \omega_1 = \sqrt{5}$$

Performing the transformation back to physical coordinates gives the total response as

$$u_1(t) = \eta_1(t) + \eta_2(t) = \tfrac{1}{6}P_o[1 - \cos(\omega_1 t)] + \tfrac{4}{45}P_o[1 - \cos(\omega_2 t)]$$
$$u_2(t) = \eta_1(t) - \tfrac{1}{2}\eta_2(t) = \tfrac{1}{6}P_o[1 - \cos(\omega_1 t)] - \tfrac{2}{45}P_o[1 - \tfrac{1}{2}\cos(\omega_2 t)]$$

This is shown plotted in Figure 4.24. Note that the response for $u_1(t)$ appears somewhat random even though it is made up of only two sinusoids.

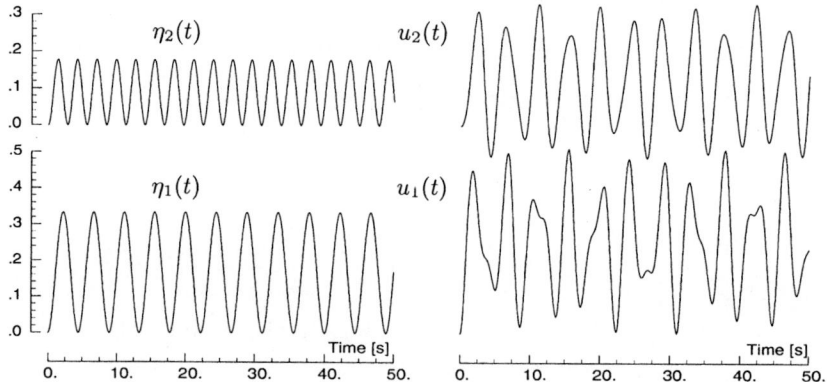

Figure 4.24: Transient response of the 2-DoF system.

Example 4.28: With reference to the last two examples, what set of applied forces will excite only the second mode of vibration?

If the applied force for the first mode is zero, then there will be no response in that mode. For the present system we have that

$$\tilde{P}_1 = P_1 + P_2$$

Hence by choosing $P_2(t) = -P_1(t)$, then this mode is not excited. That is, if equal but opposite forces are applied at the two masses, then only the second mode is excited.

Damped Free Vibration Modes

We now consider the effect of damping on the vibration modes. The essential idea to emerge is the distinction between proportional and nonproportional damping.

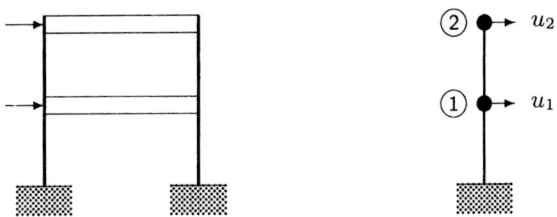

Figure 4.25: Two-degree-of-freedom system.

Consider the simple structure shown in Figure 4.25 and which has the reduced equations of motion

$$\begin{bmatrix} k_1 + k_2 & -k_2 \\ -k_2 & k_2 + k_3 \end{bmatrix} \begin{Bmatrix} u_1 \\ u_2 \end{Bmatrix} + \begin{bmatrix} c_1 & 0 \\ 0 & c_2 \end{bmatrix} \begin{Bmatrix} \dot{u}_1 \\ \dot{u}_2 \end{Bmatrix} + \begin{bmatrix} m_1 & 0 \\ 0 & m_2 \end{bmatrix} \begin{Bmatrix} \ddot{u}_1 \\ \ddot{u}_2 \end{Bmatrix} = \begin{Bmatrix} P_1 \\ P_2 \end{Bmatrix}$$

For free vibration we ask the question if the motion can be of the form

$$\begin{Bmatrix} u_1 \\ u_2 \end{Bmatrix}(t) = \begin{Bmatrix} \hat{u}_1 \\ \hat{u}_2 \end{Bmatrix} e^{i\mu t}$$

On substituting we get

$$\left[\begin{bmatrix} k_1 + k_2 & -k_2 \\ -k_2 & k_2 + k_3 \end{bmatrix} + i\mu \begin{bmatrix} c_1 & 0 \\ 0 & c_2 \end{bmatrix} - \mu^2 \begin{bmatrix} m_1 & 0 \\ 0 & m_2 \end{bmatrix} \right] \begin{Bmatrix} \hat{u}_1 \\ \hat{u}_2 \end{Bmatrix} = \begin{Bmatrix} 0 \\ 0 \end{Bmatrix}$$

The determinant must be zero for a nontrivial solution; thus on multiplying out and rearranging, we get the characteristic equation to determine μ. In the present case, this is quadratic, and therefore there are two roots and associated shapes

$$(\mu, \{\phi\})_1 , \qquad (\mu, \{\phi\})_2$$

In general, the roots and shapes are complex functions.

As was pointed out before, a simplifying assumption usually introduced is that the distribution of damping throughout the structure is proportional. That is,

$$[C] = \alpha[M] + \beta[K]$$

where α and β are constants. We will not do that here; instead, in the following examples, we will consider the special case when the governing equations of motion are

$$\begin{bmatrix} 4 & -2 \\ -2 & 6 \end{bmatrix} \begin{Bmatrix} u_1 \\ u_2 \end{Bmatrix} + c \begin{bmatrix} 1 & 0 \\ 0 & a \end{bmatrix} \begin{Bmatrix} \dot{u}_1 \\ \dot{u}_2 \end{Bmatrix} + \begin{bmatrix} 1 & 0 \\ 0 & 2 \end{bmatrix} \begin{Bmatrix} \ddot{u}_1 \\ \ddot{u}_2 \end{Bmatrix} = \begin{Bmatrix} P_1 \\ P_2 \end{Bmatrix}$$

When the coefficient a has the value 2, the damping matrix is proportional, otherwise it is nonproportional.

Example 4.29: The simple frame shown in Figure 4.25 is modeled with two degrees of freedom. Determine the eigenvalues and eigenvectors associated with the proportionally damped system with $a = 2$.

For free vibrations, the spectral form of equations are

$$\left[\begin{bmatrix} 4 & -2 \\ -2 & 6 \end{bmatrix} + i\mu c \begin{bmatrix} 1 & 0 \\ 0 & 2 \end{bmatrix} - \mu^2 \begin{bmatrix} 1 & 0 \\ 0 & 2 \end{bmatrix} \right] \begin{Bmatrix} \hat{u}_1 \\ \hat{u}_2 \end{Bmatrix} = \begin{Bmatrix} 0 \\ 0 \end{Bmatrix}$$

This leads to the system

$$\begin{bmatrix} 4 + ic\mu - \mu^2 & -2 \\ -2 & 6 + 2ic\mu - 2\mu^2 \end{bmatrix} \begin{Bmatrix} \hat{u}_1 \\ \hat{u}_2 \end{Bmatrix} = \begin{Bmatrix} 0 \\ 0 \end{Bmatrix}$$

The determinant must be zero for a nontrivial solution; thus on multiplying out and rearranging, we get

$$\mu^4 - 2ic\mu^3 - 7\mu^2 - c^2\mu^2 + 7ic\mu + 10 = 0$$

This has the factorization

$$[\mu^2 - ic\mu - 2][\mu^2 - ic\mu - 5] = 0$$

which leads to the two roots

$$\mu_1 = \tfrac{1}{2}ic \pm \sqrt{2 - \tfrac{1}{4}c^2}, \qquad \mu_2 = \tfrac{1}{2}ic \pm \sqrt{5 - \tfrac{1}{4}c^2}$$

These roots are plotted in Figure 4.26 for different values of c. Note that each mode has a different critical value of damping, for example, $C_{c1} = 2\sqrt{2}$, $C_{c2} = 2\sqrt{5}$. This indicates that it is possible to simultaneously have some modes overdamped while others are underdamped.

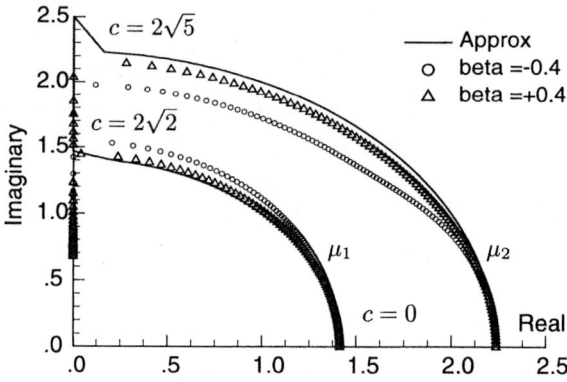

Figure 4.26: Roots for damped free vibration.

The mode shape for a given mode is obtained by substituting μ into the original system to give

$$\hat{u}_2 = \tfrac{1}{2}[4 + ic\mu - \mu^2]\hat{u}_1$$

Doing this for both modes leads to

$$\left\{ \begin{matrix} \hat{u}_1 \\ \hat{u}_2 \end{matrix} \right\}_1 = \hat{u}_1 \left\{ \begin{matrix} 1 \\ 1 \end{matrix} \right\}, \qquad \left\{ \begin{matrix} \hat{u}_1 \\ \hat{u}_2 \end{matrix} \right\}_2 = \hat{u}_1 \left\{ \begin{matrix} 1 \\ -\frac{1}{2} \end{matrix} \right\}$$

These mode shapes are identical to those of the undamped case. Note that this result is true irrespective of the amount of damping.

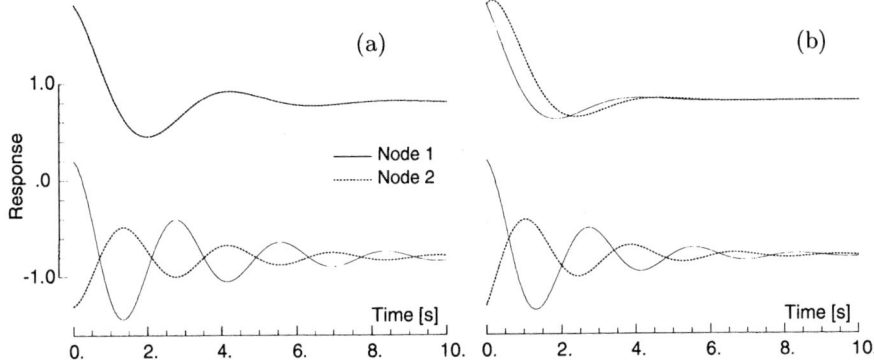

Figure 4.27: Free vibration damped responses. (a) Proportional damping. (b) Non-proportional damping.

Example 4.30: Determine the eigenvalues and eigenvectors associated with the nonproportionally damped system with $a = 2(1 + \beta)$.

Note that by setting $\beta = 0$ we get the proportionally damped situation. The spectral form for the governing equations of motion are

$$\left[\begin{bmatrix} 4 & -2 \\ -2 & 6 \end{bmatrix} + i\mu c \begin{bmatrix} 1 & 0 \\ 0 & a \end{bmatrix} - \mu^2 \begin{bmatrix} 1 & 0 \\ 0 & 2 \end{bmatrix} \right] \left\{ \begin{matrix} \hat{u}_1 \\ \hat{u}_2 \end{matrix} \right\} = \left\{ \begin{matrix} 0 \\ 0 \end{matrix} \right\}$$

and this leads to the system

$$\begin{bmatrix} 4 + ic\mu - \mu^2 & -2 \\ -2 & 6 + ica\mu - 2\mu^2 \end{bmatrix} \left\{ \begin{matrix} \hat{u}_1 \\ \hat{u}_2 \end{matrix} \right\} = \left\{ \begin{matrix} 0 \\ 0 \end{matrix} \right\}$$

The determinant must be zero for a nontrivial solution; thus on multiplying out and rearranging, we get

$$\mu^4 - ic\mu^3(2 + \beta) - 7\mu^2 - \mu^2 c^2(1 + \beta) + 3ic\mu(7 + 4\beta) + 10 = 0$$

The roots are not obvious but it has the approximate factorization

$$[\mu^2 - ic\mu(1 + \tfrac{2}{3}\beta) - 2][\mu^2 - ic\mu(1 + \tfrac{1}{3}\beta) - 5] = 0$$

This leads to the two approximate roots

$$\mu_1 = \tfrac{1}{2}ic(1 + \tfrac{2}{3}\beta) \pm \sqrt{2 - \tfrac{1}{4}c^2(1 + \tfrac{2}{3}\beta)^2}, \qquad \mu_2 = \tfrac{1}{2}ic(1 + \tfrac{1}{3}\beta) \pm \sqrt{5 - \tfrac{1}{4}c^2(1 + \tfrac{1}{3}\beta)^2}$$

These roots are also shown plotted in Figure 4.26 for values of $\beta = \pm 0.4$. The comparison with the exact values shows that the approximation is reasonable especially for small values of damping. This being so, we can rewrite the approximations as

$$\mu_1 = \tfrac{1}{2}i\bar{c}_1 \pm \sqrt{2 - \tfrac{1}{4}\bar{c}_1^2}\,, \quad \mu_2 = \tfrac{1}{2}i\bar{c}_2 \pm \sqrt{5 - \tfrac{1}{4}\bar{c}_2^2}\,, \quad \bar{c}_1 \equiv c(1 + \tfrac{2}{3}\beta)\,, \; \bar{c}_2 \equiv c(1 + \tfrac{1}{3}\beta)$$

which shows that the behavior of the roots is essentially that of the proportional case except that the damping is increased or decreased by β.

For the purpose of our discussions of the effect of β on the mode shapes, we can make the further approximations

$$\mu_1 \approx \sqrt{2} + \tfrac{1}{2}ic(1 + \tfrac{2}{3}\beta)\,, \quad \mu_2 \approx \sqrt{5} + \tfrac{1}{3}ic(1 + \tfrac{1}{3}\beta)$$

This is the vertical sections of the plots in Figure 4.26. The mode shape for a given mode is obtained by substituting μ into the original system to give

$$\hat{u}_2 = \tfrac{1}{2}[4 + ic\mu - \mu^2]\hat{u}_1$$

Doing this for both modes leads approximately to

$$\left\{ \begin{matrix} \hat{u}_1 \\ \hat{u}_2 \end{matrix} \right\}_1 = \hat{u}_1 \left\{ \begin{matrix} 1 \\ 1 - \tfrac{2}{3}ic\sqrt{2}\beta \end{matrix} \right\}, \quad \left\{ \begin{matrix} \hat{u}_1 \\ \hat{u}_2 \end{matrix} \right\}_2 = \hat{u}_1 \left\{ \begin{matrix} 1 \\ -\tfrac{1}{2} - \tfrac{1}{6}ic\sqrt{5}\beta \end{matrix} \right\}$$

Only if the deviation from proportionality, β, is zero are these mode shapes the same as for the undamped case.

The effect of β is to make the mode shapes complex. Hence, they have a slightly different interpretation than for the proportionally damped case. We will see their meaning by looking at the time traces in the next example.

Example 4.31:   Plot the displaced shape of the structure as a function of time.

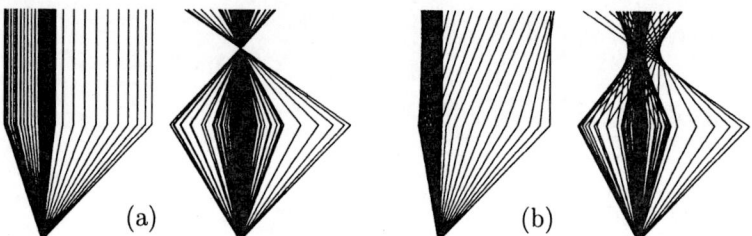

Figure 4.28: Free vibration mode shapes. (a) Proportionally damped. (a) Nonproportionally damped.

We recover the actual time histories by noting that

$$u(t) = \hat{u}e^{i\mu t} = (u_R + iu_I)e^{i(\mu_R + i\mu_I)t} = (u_R + iu_I)(\cos\mu_R t + i\sin\mu_R t)e^{-\mu_I t}$$
$$= (u_R\cos\mu_R t - u_I\sin\mu_R t)e^{-\mu_I t} + i(u_R\sin\mu_R t + u_I\cos\mu_R t)e^{-\mu_I t}$$

and now we just use the real part. For the first mode, we get

$$u_1(t) = u_o[\cos\sqrt{2}t]e^{-ct/2}\,, \quad u_2(t) = u_o[\cos\sqrt{2}t + \tfrac{2}{3}c\sqrt{2}\beta\sin\sqrt{2}t]e^{-ct/2}$$

and for the second mode, we get

$$u_1(t) = u_o[\cos \sqrt{5}t]e^{-ct/3}, \qquad u_2(t) = u_o[-\tfrac{1}{2}\cos \sqrt{5}t + \tfrac{1}{6}c\sqrt{5}\beta \sin \sqrt{5}t]e^{-ct/3}$$

These are shown plotted in Figure 4.27. We see for the proportional case ($\beta = 0$) that the shape remains the same over time — only the amplitude changes. For the nonproportional case, both the amplitude and the shape changes because there is a phase lag between the components. Consequently, we do not get sharp modal lines — they seem to wander in time. This is illustrated in Figure 4.28.

4.5   Relation of Wave Responses to Vibrations

When a structure experiences a localized disturbance over a short period of time, an impulse, the energy propagates throughout the structure as waves. When the structure is relatively large so that the waves do not have many interactions with the boundaries, then the analysis must be done using wave analysis methods. Some of these techniques are developed in Reference [23]. Our interest here, however, is to show the connection between a wave response and a vibration analysis; that is, when the time scale is such that there are many reflections.

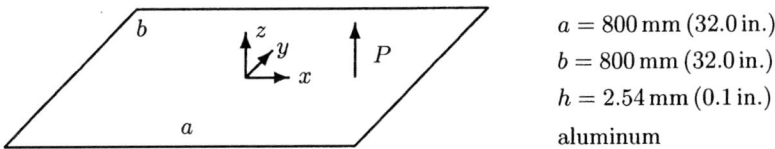

$a = 800\,\text{mm } (32.0\,\text{in.})$

$b = 800\,\text{mm } (32.0\,\text{in.})$

$h = 2.54\,\text{mm } (0.1\,\text{in.})$

aluminum

Figure 4.29: Simply supported plate with transverse impact loading.

Consider the plate shown in Figure 4.29, which is simply supported on all sides. We will impact it at the center, $x = 0$, $y = 0$.

The short-term responses are shown in Figure 4.30 for a central impact. At the impact site, we see an impulse-type response followed by some oscillatory behavior. It can be shown [23] that the velocity response at the impact site of a large plate is given by

$$\frac{dw}{dt} = \frac{1}{8\sqrt{\rho h D}}P(t)$$

That is, the velocity history has the same shape as the impact force history. The very early response at the central location thus illustrates the forcing history.

Elementary wave propagation, as for example a longitudinal wave in a long rod, is an example of a *nondispersive* wave, which propagates at a constant speed without change of its shape. The shape is retained even after reflection, and therefore reflections from various boundaries are easily identified simply by their arrival times (speed × distance) and their shape.

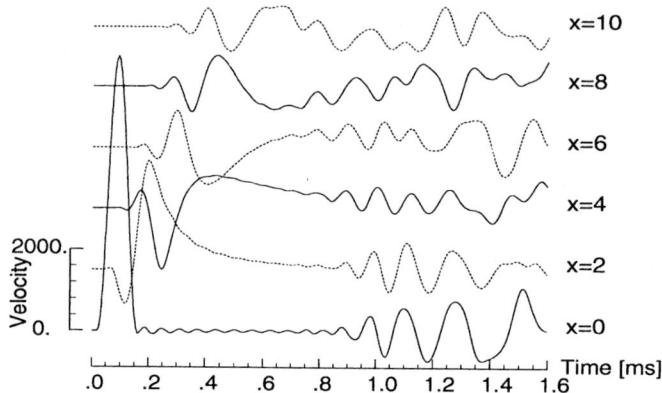

Figure 4.30: Short-term velocity responses due to a central impact.

Wave propagation in beams and plates are *dispersive*, meaning they have a spectrum of propagating speeds and change their shape. It is not our intention here to delve deeply into such waves, but only to describe the observed responses. For $x = 0$ at a time of about 0.8 ms, the observed oscillations are the responses due to the reflected wave. The eye can follow these, so that at $x = 10$ they arrive at the earlier time of about 0.6 ms. This evolution into an oscillatory behavior can now be observed in the head of the outgoing wave: at $x = 0$ we see no zeros, at $x = 2$ we see perhaps one zero, at $x = 4$ we see three zeros, by $x = 10$ we can just about discern six zeros. The reflections, having traveled on the order of 800 mm (32 in.), therefore exhibit the most zeros.

Since our interest is in the connection to structural vibrations, we now look at a smaller plate that is simply supported on all sides with $a = 400$ mm (16 in.), $b = 200$ mm (8 in.). We will impact it at some points along $y = 0$.

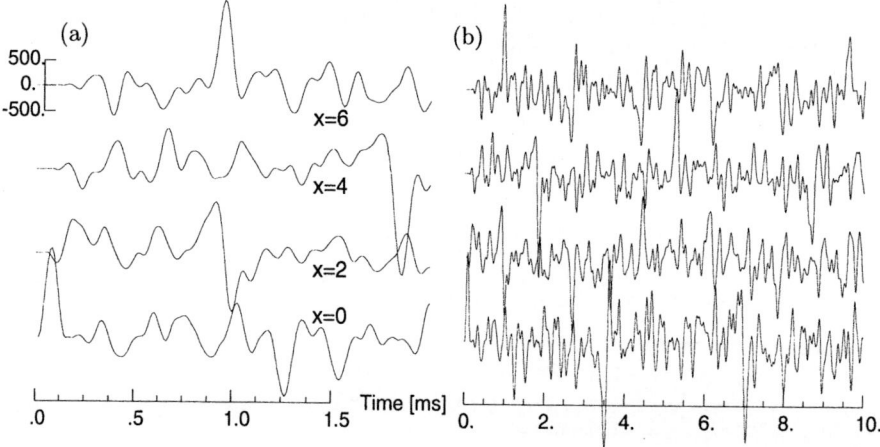

Figure 4.31: Velocity responses due to a central impact. (a) Short-term responses. (b) Long-term responses.

The short- and long-term responses are shown in Figure 4.31. The short term response is somewhat similar to that of Figure 4.30 but with earlier arrivals of reflections due to the shorter b side. When we look at the long-term response, we see the superposition of very many reflections giving an almost random look to the response.

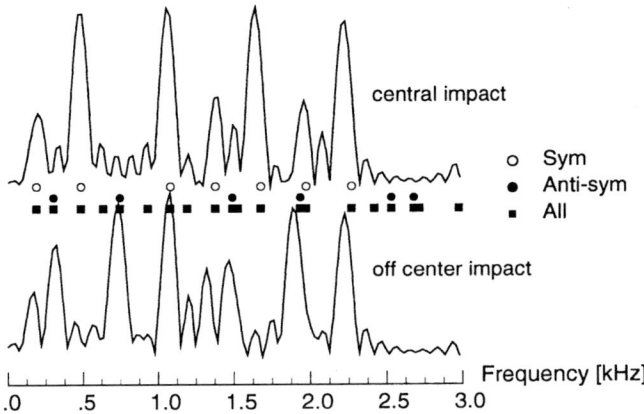

Figure 4.32: Frequency response due to a central and off-center impacts.

However, when we look at the response in the frequency domain as in Figure 4.32, we see a definite structure. In fact, we see the emergence of resonance-type behavior associated with the spectral peaks. What is happening is that the many reflections are superposing, some constructively, some destructively, thereby enhancing some frequency components and diminishing others. In the limit of very long time we would have very sharp spectral peaks corresponding to the modal responses. Thus the complex wave response is decomposing itself into a collection of modal responses.

Also shown on the figure are the resonance frequencies as obtained from an eigenanalysis. We see that there are fewer spectral peaks than resonances; this comes about because the central impact will excite only some of the modes — the doubly symmetric ones in this case. The off-center impact excites some of the nonsymmetric ones but not all of them.

4.6 Parametric Excitations

The previous sections considered the vibration of systems whose governing equations have constant coefficients. There are important situations, however, that give rise to equations with variable coefficients; examples of such problems are the compression of a piston rod in an engine and the effect of rotating machinery on a building. It will also arise in the nonlinear analysis of structures where there are interaction effects between different loadings; this is our main reason for considering the problem. A good introduction is given in Reference [56].

Types of Parametric Excitation

We begin our discussion with the simple system shown in Figure 4.33(a): a concentrated mass is attached to the end of a light flexible bar whose effective length is constrained by the position of the bushing.

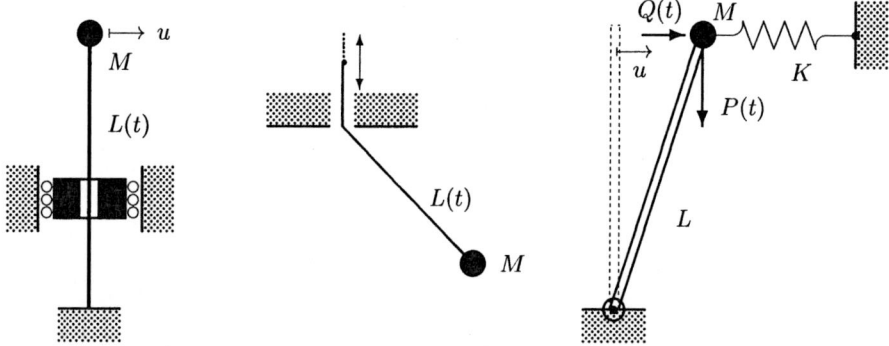

Figure 4.33: Examples of parametric excitation. (a) Stiffness change. (b) Mass change. (c) Loading.

The flexural stiffness is $3EI/L^3$ (think of the bar as a single beam element cantilevered at one end) and the equation of motion for free vibration is therefore

$$M\ddot{u} + \frac{3EI}{L^3}u = 0$$

Now let the bushing move along the bar such that

$$L = L_o + A\sin\omega t$$

We see that the stiffness also changes in time and the equation of motion is

$$M\ddot{u} + \frac{3EI}{[L_o + A\sin\omega t]^3}u = 0$$

The oscillations of the mass can no longer be called free, since they occur under the time-dependent external force associated with the periodic change of stiffness. On the other hand, they cannot be called forced, since the external force is not a driving force and occurs on the left-hand side of the equation. These oscillations, which occur for a fixed type of change of the parameters of a system (in this case the stiffness), are said to be *parametrically* excited.

A simple example of an inertia-based parametric excitation is shown in Figure 4.33(b): the length of the pendulum string is given a periodic change. This is also the action of a playground swing; the oscillations are excited by the periodic squatting of the person standing on the swing, which causes a periodic change in the center of gravity of the system.

Our third example, shown in Figure 4.33(c), is associated with an external load. The equation of motion, obtained by summing the moments about the base, is given by

$$M\ddot{u} + [K - P(t)/L]u = Q(t)$$

We see that the two forces appear in different positions in the equation; $Q(t)$ appears on the right-hand side as a usual applied load, whereas $P(t)$ appears on the left-hand side. This load is referred to as parametric loading, and has the distinctive feature that the static behavior can cause instability as we will see in later chapters.

The equation of motion of the type of systems we are interested in can be described by

$$M\ddot{u} + P(t)u + Ku = Q(t)$$

Although this system is linear, it is still very difficult to solve and we will need some mathematical developments (known as Floquet's theory) in order to progress with an analysis. We will look at the case where $Q = 0$ and $P(t)$ varies harmonically as

$$P(t) = P_o + \hat{P}\cos\omega t$$

This gives a system with periodic coefficients.

Solution Structure: Floquet's Theory

Because of our interest in vibrations, we are interested in establishing the existence of periodic solutions. Rewrite the equation as

$$\ddot{u} + \bar{p}(t)u = 0, \qquad \bar{p}(t) \equiv \omega_o^2[(1 - \frac{P_o}{K}) - \frac{\hat{P}}{K}\cos\omega t], \qquad \omega_o = \sqrt{\frac{K}{M}}$$

where $\bar{p}(t)$ is a periodic function of period T, that is, $\bar{p}(t+T) = \bar{p}(t)$. Just because $\bar{p}(t)$ is periodic does not mean that $u(t)$ is periodic; indeed, in most cases it is not. The system is second order, hence it has two fundamental solutions, which we write as

$$u(t) = Au_1(t) + Bu_2(t)$$

Because of the periodic coefficient, we know that $u_i(t+T)$ are also solutions (although not necessarily periodic). We will now establish the form of the functions $u_1(t)$ and $u_2(t)$. The results are taken from References [38, 82].

The time-shifted solution functions can be written as a linear combination of the fundamental solutions

$$u_1(t+T) = a_1u_1(t) + a_2u_2(t), \qquad u_2(t+T) = b_1u_1(t) + b_2u_2(t)$$

The total solution is then

$$
\begin{aligned}
u(t+T) &= Au_1(t+T) + Bu_2(t+T) \\
&= A[a_1u_1(t) + a_2u_2(t)] + B[b_1u_1(t) + b_2u_2(t)] \\
&= [Aa_1 + Bb_1]u_1(t) + [Aa_2 + Bb_2]u_2(t)
\end{aligned}
$$

We now ask: under what circumstance is the time-shifted solution similar to the original solution? That is,

$$
u(t+T) = \mu u(t) \quad \text{or} \quad [Aa_1 + Bb_1]u_1(t) + [Aa_2 + Bb_2]u_2(t) = \mu[Au_1(t) + Bu_2(t)]
$$

where μ is some constant. Equating coefficients leads to the eigenvalue problem

$$
\begin{bmatrix} a_1 & b_1 \\ a_2 & b_2 \end{bmatrix} \left\{ \begin{array}{c} A \\ B \end{array} \right\} - \mu \left\{ \begin{array}{c} A \\ B \end{array} \right\} = 0
$$

The values of μ_i are obtained from the quadratic equation

$$
\mu^2 - (a_1 + b_2)\mu + (a_1b_2 - a_2b_1) = 0
$$

The importance of this result is the possibility of special values of μ implying the existence of periodic solutions. That is,

$$
u(t + mT) = \mu u(t + (m-1)T) = \mu^2 u(t + (m-2)T) = \cdots = \mu^m u(t)
$$

Hence, if μ is the mth root of unity (with m being a positive integer), then $u(t)$ has periodicity of mT. Some particular values of μ are

$$
\begin{aligned}
m = 1: &\qquad \mu = 1 \\
m = 2: &\qquad \mu = \pm 1 \\
m = 3: &\qquad \mu = (e^{i2\pi})^{1/3} = 1, \tfrac{1}{2}(-1 \pm \sqrt{3}) \\
m = 4: &\qquad \mu = \pm 1, \pm i
\end{aligned}
\tag{4.15}
$$

We see there are many possibilities for periodic solutions of which period T is only one.

In general, μ_i are complex numbers so represent them as

$$
\mu_i = e^{i\lambda_i T}
$$

and we can therefore write the solution at time $t + T$ as

$$
u(t+T) = \mu_1 Au_1(t) + \mu_2 Bu_2(t) = Ae^{i\lambda_1 T}u_1(t) + Be^{i\lambda_2 T}u_2(t)
$$

Accordingly, a typical fundamental solution is written as

$$
u_i(t+T) = e^{i\lambda_i T}u_i(t)
$$

Multiply both sides by $e^{-i\lambda_i(t+T)}$, which leads to

$$u_i(t+T)e^{-i\lambda_i(t+T)} = e^{i\lambda_i T}u_i(t)e^{-i\lambda_i(t+T)} = u_i(t)e^{-i\lambda_i t}$$

This gives us the important conclusion that the functions

$$p_i(t) \equiv e^{-i\lambda_i t}u_i(t)$$

are periodic functions of period T. Consequently, we can write the general solution in the form

$$u(t) = Ae^{i\lambda_1 t}e^{-i\lambda_1 t}u_1(t) + Be^{i\lambda_1 t}e^{-i\lambda_1 t}u_2(t) = Ae^{i\lambda_1 t}p_1(t) + Be^{i\lambda_2 t}p_2(t) \quad (4.16)$$

Depending on the values of λ_i, the total solution could be periodic or aperiodic. As a simple illustration, the fundamental solutions

$$u_1(t) = e^{-2t}\sin 3t, \qquad u_2(t) = e^{-3t}\cos 2t$$

give a response that is not periodic.

The periodic functions $p_i(t)$ are called Mathieu functions and can be found tabulated in Reference [1].

Approximate Solutions

Now that we know the structure of the solution, we can attempt to get an approximate solution; the method to be used is explained in additional detail in References [38, 82]. First, re-write the governing equation in the form

$$\ddot{u} + [\alpha - \beta\cos\omega t]u = 0, \qquad \alpha \equiv \omega_o^2\left(1 - \frac{P_o}{K}\right), \quad \beta \equiv \omega_o^2\frac{\hat{P}}{K}$$

Since $p_i(t)$ are periodic with the same period as the forcing term, then let the solution be represented with the Fourier expansion

$$u(t) = e^{i\lambda t}\sum_{n=-\infty}^{\infty} b_n e^{in\omega t}$$

Substitute this into the governing equation and regroup using the complex relation for the cosine term

$$\cos\omega t = \tfrac{1}{2}(e^{i\omega t} - e^{i\omega t})$$

to get

$$\sum_{n=-\infty}^{\infty} \left\{ \tfrac{1}{2}\beta b_{n-1} + [-(\lambda + n\omega)^2 + \alpha]b_n + \tfrac{1}{2}\beta b_{n+1} \right\} e^{i(\lambda + n\omega)t} = 0$$

This must be true for all time; hence, the term inside the braces must be zero leading to the recurrence relation

$$\tfrac{1}{2}\beta b_{n-1} + [-(\lambda + n\omega)^2 + \alpha]b_n + \tfrac{1}{2}\beta b_{n+1} = 0$$

This is an homogeneous set of equations, and to get a nontrivial solution we set the determinant to zero. This, then, specifies the value λ for a given set of α and β. With λ so determined, we can then solve for b_n in terms of b_0. Finally, b_0 can be determined from the initial conditions.

Example 4.32: Use a three-term approximation to investigate the motion of a system under periodic loading when released from an initial displacement.

Approximate the system with just the three middle terms, which leads to the set of equations

$$\begin{bmatrix} -(\lambda-\omega)^2+\alpha & \tfrac{1}{2}\beta & 0 \\ \tfrac{1}{2}\beta & -(\lambda-0)^2+\alpha & \tfrac{1}{2}\beta \\ 0 & \tfrac{1}{2}\beta & -(\lambda+\omega)^2+\alpha \end{bmatrix} \begin{Bmatrix} b_{-1} \\ b_0 \\ b_1 \end{Bmatrix} = 0$$

We get a nontrivial solution only when the determinant is zero. Represent the determinant as

$$\Delta = \Delta(\alpha, \beta; \lambda, \omega) = \Delta(\bar{\alpha}, \bar{\beta}; \bar{\lambda})$$

where the parameters are normalized as

$$\bar{\alpha} = \alpha/\omega^2, \qquad \bar{\beta} = \beta/\omega^2, \qquad \bar{\lambda} = \lambda/\omega$$

Using a root solver, we can determine the values of $\bar{\lambda}$, which, for given values of $\bar{\alpha}$ and $\bar{\beta}$, make the determinant zero. We seek only real values of $\bar{\lambda}$ and the results are shown in Figure 4.34.

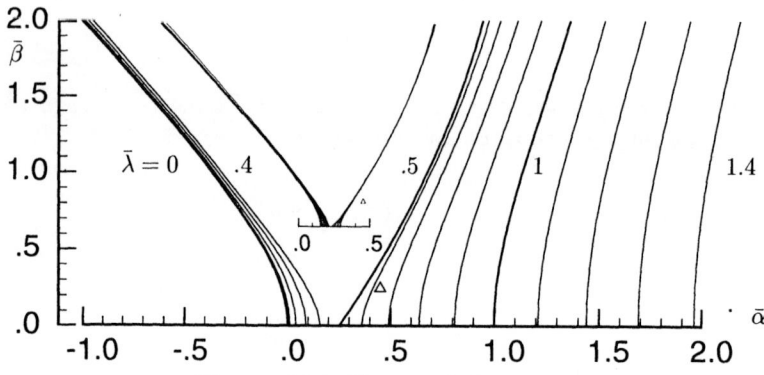

Figure 4.34: Characteristic exponent.

A curious feature of this plot is the absence of any parameter values in the wedges just above $\bar{\alpha} = 0.25$ and to the left of zero. The former is confirmed by the inset, which is for $\bar{\lambda}$ ranging from 0.4 to 0.5 in steps of 0.01: the contours are either close to the 0.4 line or close to the 0.5 line. Actually, the values of λ are complex in these regions, leading to unstable solutions. We will consider this in greater detail in Chapter 7.

We solve for b_{-1} and b_1 in terms of b_0 as

$$b_{-1} = \frac{-\frac{1}{2}\beta b_0}{[-(\lambda - \omega)^2 + \alpha]}, \qquad b_1 = \frac{-\frac{1}{2}\beta b_0}{[-(\lambda + \omega)^2 + \alpha]}$$

Imposing the initial condition that $u(t=0) = u_o$ gives

$$b_0 = u_o / \left[\frac{-\frac{1}{2}\beta}{[-(\lambda - \omega)^2 + \alpha]} + 1 + \frac{-\frac{1}{2}\beta}{[-(\lambda + \omega)^2 + \alpha]} \right]$$

Our final solution is

$$u(t) = b_0 \left\{ \frac{-\frac{1}{2}\beta}{[-(\lambda - \omega)^2 + \alpha]} e^{i(\lambda - \omega)t} + e^{i(\lambda + 0)t} + \frac{-\frac{1}{2}\beta b_0}{[-(\lambda + \omega)^2 + \alpha]} e^{i(\lambda + \omega)t} \right\}$$

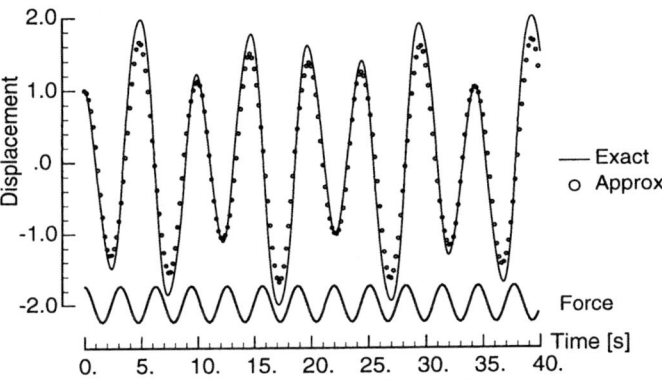

Figure 4.35: Comparison for displacement response.

Figure 4.35 shows a reconstruction and a comparison with the numerically integrated solution. The parameters used were

$$\omega = 2, \qquad \alpha = 1.8, \qquad \beta = 1.0$$

leading to the graph parameters

$$\bar{\alpha} = 0.45, \qquad \bar{\beta} = 0.25, \qquad \bar{\lambda} = 0.64, \qquad \lambda = 1.28$$

This is indicated with a triangle in Figure 4.34. The approximate solution captures the main characteristics of the response. What is especially worth noting is that the response does not bear any obvious period relation to the forcing history.

Systems of Equations

Consider a system of second-order linear equations described by

$$\ddot{w}_i + \sum_j a_{ij}\dot{w}_j + \sum_j b_{ij}w_j = p_i$$

where the coefficients a_{ij} and b_{ij} could be functions of time. Introduce the new variable $v_i \equiv \dot{w}_i$, then the system can be written as

$$\frac{d}{dt} \left\{ \begin{matrix} v \\ w \end{matrix} \right\}_i = \left[\begin{matrix} -\sum a_{ij}v_j & -\sum b_{ij}w_j \\ v_i & 0 \end{matrix} \right] + \left\{ \begin{matrix} p \\ 0 \end{matrix} \right\}_i$$

Using $\{u\} = \{v, w\}^T$ as the state vector, we then have the relation in matrix notation as

$$\{\dot{u}\} = [A(t)]\{u\} + \{P\}$$

where the elements of the $[N \times N]$ matrix $[A]$ could be functions of time. We will consider some properties of this general system.

Let $\{\psi\}_1, \{\psi\}_2, \cdots, \{\psi\}_N$, be linearly independent solutions, then the matrix

$$[\Psi] = [\{\psi\}_1, \{\psi\}_2, \cdots, \{\psi\}_N]$$

is called the *fundamental matrix* and we can write

$$[\dot{\Psi}(t)] = [A(t)][\Psi(t)]$$

This is equivalent to our original system. When the matrix $[A(t)]$ is a function of time, it is usually impossible to construct an explicit fundamental matrix. However, it is very useful in organizing some of the properties of the solutions and that is our main purpose in using it.

A general solution of the equations can be written in terms of the fundamental matrix as

$$\{u(t)\} = [\Psi(t)]\{c\}$$

where the array $\{c\}$ contains constants associated with the initial conditions. We conclude that any two fundamental matrices are related through

$$[\Psi(t)]_2 = [\Psi(t)]_1 [C]$$

since each column of $[\Psi]_2$ is a linear combination of each column of $[\Psi]_1$. Noting that

$$\{u(0)\} = [\Psi(0)]\{c\} \qquad \text{giving} \qquad \{c\} = [\Psi(0)]^{-1}\{u(0)\}$$

from which we get that the solution of the initial value problem is

$$\{u(t)\} = [\Psi(t)][\Psi(0)]^{-1}\{u(0)\}$$

which shows the dependence of the solution on the initial conditions.

For the inhomogeneous system with $\{P\} \neq 0$, we can write the solution as

$$\{u(t)\} = \{u_p(t)\} + \{\psi_c(t)\}$$

where $\{u_p(t)\}$ is the particular solution and $\{\psi_c(t)\}$ is the complementary solution of the associated homogeneous system. The solution of the inhomogeneous system can be written as

$$\{u(t)\} = [\Psi(t)][\Psi(0)]^{-1}\{u(0)\} + [\Psi(t)]\int_0^t [\Psi(\tau)]^{-1}\{P(\tau)\}\,d\tau$$

The second term is the particular solution, while the first is the complementary solution. This is a particular form of Duhamel's integral introduced earlier in this chapter.

Let the Wronskian (determinant) be defined as

$$W \equiv \det\|[\ \Psi\ ]\| = \det\|\{\psi\}_1, \{\psi\}_2, \cdots, \{\psi\}_N\|$$

then we conclude from the above that either

$$W(t) = \det\|[\ \Psi\ ]\| = 0 \qquad \text{or} \qquad W(t) = \det\|[\ \Psi\ ]\| \neq 0 \qquad \text{for all time}$$

The first occurs if the solutions are linearly dependent, and the second implies that $[\ \Psi\ ]$ is a fundamental matrix. The behavior of this determinant plays an important role in the behavior of the system, and therefore it is of value to know some of its properties. For example, it can be shown (Reference [38] gives a nice proof) that

$$\frac{dW}{dt} = \text{Tr}[\ A\ ]W(t) \qquad \text{or} \qquad \frac{\dot{W}}{W} = \text{Tr}[\ A\ ] \tag{4.17}$$

where $\text{Tr}[\ A\ ]$ means the sum of the elements of the diagonal of $[\ A\ ]$. This relation can be integrated to give

$$W(t) = W(0)\exp\{\int_0^t \text{Tr}[A(\tau)]\,d\tau\} \tag{4.18}$$

as a useful alternative form.

Example 4.33:   Determine the fundamental matrix for the system of equations

$$\begin{Bmatrix} \dot{u}_1 \\ \dot{u}_2 \end{Bmatrix} = \begin{bmatrix} 1 & 1 \\ 0 & f(t) \end{bmatrix}\begin{Bmatrix} u_1 \\ u_2 \end{Bmatrix}, \qquad f(t) = \frac{\sin t}{2 - \cos t}$$

The second equation gives

$$[2 - \cos t]\dot{u}_2 = [\sin t]u_2 \qquad \text{or} \qquad g(t)\,\dot{u}_2 = \dot{g}(t)\,u_2$$

from which we conclude that

$$u_2(t) = b[2 - \cos t]$$

where b is some constant. Substituting for u_2 into the first governing equation gives the inhomogeneous equation

$$\dot{u}_1 = u_1 + b[2 - \cos t]$$

whose solution is
$$u_1(t) = ae^t + b[-2 + \tfrac{1}{2}\cos t - \tfrac{1}{2}\sin t]$$
We can combine both into the solution representation
$$\left\{\begin{array}{c} u_1 \\ u_2 \end{array}\right\} = \left[\begin{array}{cc} e^t & -2 + \tfrac{1}{2}\cos t - \tfrac{1}{2}\sin t \\ 0 & 2 - \cos t \end{array}\right] \left\{\begin{array}{c} a \\ b \end{array}\right\}$$
We identify the square matrix as the fundamental solution.

The Wronskian is
$$W = \det\|[\ \Psi\ ]\| = e^t[2 - \cos t]$$
and therefore
$$\frac{\dot{W}}{W} = \frac{e^t[2 - \cos t + \sin t]}{e^t[2 - \cos t]} = 1 + \frac{\sin t}{2 - \cos t}$$
which is clearly the trace of $[A(t)]$. We also see that $W(t)$ never goes through zero.

A common occurrence is where the matrix $[A(t)]$ is a periodic function of time with minimal period T. That is, we have
$$\{\dot{u}(t)\} = [A(t)]\{u(t)\}, \qquad [A(T + t)] = [A(t)]$$
As we showed earlier, just because the coefficient matrix is periodic in time does not mean that the solutions are also periodic.

Using the fundamental matrix representation, we can write
$$[\dot{\Psi}(t)] = [A(t)][\Psi(t)]$$
At the later time we also have
$$[\dot{\Psi}(T + t)] = [A(T + t)][\Psi(T + t)] = [A(t)][\Psi(T + t)]$$
This shows that $[\Psi(T + t)]$ is also a fundamental matrix and therefore we have
$$[\Psi(T + t)] = [\Psi(t)][\ C\ ]$$
Evaluating this at $t = 0$ allows
$$[\Psi(T + 0)] = [\Psi(0)][\ C\ ] \qquad \text{or} \qquad [\ C\ ] = [\Psi(0)]^{-1}[\Psi(T)]$$
Let $(\mu_i, \{\phi\}_i)$ be the eigenpairs of the matrix $[\ C\ ]$, that is,
$$[\ C\ ]\{\phi\}_i = \mu_i\{\phi\}_i \qquad\qquad\qquad (4.19)$$
We therefore can write the initial conditions as $\{u(0)\} = \sum_k a_k\{\phi\}_k$. The solution at arbitrary time is
$$\{u(T + t)\} = [\Psi(T + t)]\{u(0)\} = [\Psi(t)][\ C\ ]\sum_k a_k\{\phi\}_k$$

and using Equation (4.19) becomes

$$\{u(T+t)\} = [\Psi(t)] \sum_k a_k \mu_k \{\phi\}_k$$

Let $\mu_k = e^{i\lambda_k T}$ where λ_k are called Floquet's exponents. Furthermore, let

$$\{u(t)\}_k \equiv [\Psi(t)] a_k \{\phi\}_k \quad \text{so that} \quad \{u(T+t)\}_k = [\Psi(t)] a_k e^{+i\lambda_k T} \{\phi\}_k$$

Now introduce

$$\{p(t)\}_k \equiv e^{-i\lambda_k t} \{u(t)\}_k$$

then

$$
\begin{aligned}
\{p(T+t)\}_k &= e^{-i\lambda_k(T+t)}\{u(T+t)\}_k = e^{-i\lambda_k(T+t)}[\Psi(t)] a_k e^{+i\lambda_k T} \{\phi\}_k \\
&= e^{-i\lambda_k t}\{u(t)\}_k = \{p(t)\}_k
\end{aligned}
$$

Therefore each of the components of $\{p(t)\}_k$ are periodic functions with period T. We now have the representation of the solution as

$$\{u(t)\} = \sum_k \{u(t)\}_k = \sum_k A_k e^{+i\lambda_k t} \{p(t)\}_k \qquad (4.20)$$

This is the result equivalent to Equation (4.16).

We also have the interesting result

$$W(T) = \mu_1 \mu_2 \cdots \mu_N = \exp\{ \int_0^T \mathrm{Tr}[\ A\ ] ds \} \qquad (4.21)$$

We will find a useful application of this formula in the pulsating beam problem of Chapter 7.

Problems

4.1 Consider the forced frequency response of an undamped system initially at rest.

- Show that the response is

$$u(t) = -\frac{\omega_o \omega P_o/K}{[\omega_o^2 - \omega^2]} \sin(\omega_o t) + \frac{P_o/K}{[\omega_o^2 - \omega^2]} \sin(\omega t)$$

- Plot this function for a value of ω close to ω_o and observe the "beating" effect.
- Consider when the forcing frequency is near resonance, that is, $\omega - \omega_o = 2\Delta\omega$ is small. Show that the early time response is

$$u(t) = \frac{\omega_o^2 P_o/K}{\Delta\omega[\omega_o + \omega]} \sin(\Delta\omega t) \cos[(\omega_o + \omega)t/2] \approx \frac{\omega_o^2 P_o/K}{[\omega_o + \omega]} t \cos[(\omega_o + \omega)t/2]$$

- Plot this function and compare to the above result. Observe the linear increase in amplitude with time.
- What happens in the limit as $\Delta\omega \longrightarrow 0$?

4.2 Consider the free vibration of a free/free rod of length L.
- Show that the characteristic equation is

$$k \sin kL = 0, \qquad k \equiv \omega \sqrt{\rho A / EA}$$

- Determine the first two natural frequencies.
- Sketch the first two mode shapes.

4.3 Consider the free longitudinal vibrations of a rocket: The effective length is 20 ft, net cross-sectional area 10 in.^{2}, unloaded structural weight 200 lb, and material aluminum. Making whatever assumptions seem reasonable,
- Determine the first three natural frequencies for the empty rocket.
- Determine the first three natural frequencies when the rocket carries a distributed load of 100 lb.

4.4 A fixed/fixed rod has a sinusoidal forcing function applied to its mid-point as shown.
- Determine the response against frequency and plot.
- Compare the frequencies of the spectral peaks to the natural frequencies; comment.
- Plot the response shape at the three frequencies: $\frac{1}{2}\omega_1$, ω_1, $\frac{3}{2}\omega_1$, where ω_1 is close to the first natural frequency.

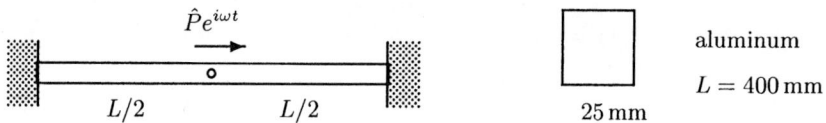

4.5 A rod, fixed at one end and having a concentrated mass at the other, is vibrating longitudinally.
- Use the Ritz method to estimate the first and second natural frequencies of vibration.
- Sketch the dependence of the frequencies on the mass ratio $m_c/\rho AL$.
- Superpose the exact results for the cases of no mass and when both ends are fixed.

$$\text{no mass:} \quad \omega_n = \frac{(n + \frac{1}{2})\pi}{L} \sqrt{\frac{EA}{\rho A}}, \qquad\qquad \text{fixed:} \quad \omega_n = \frac{(n)\pi}{L} \sqrt{\frac{EA}{\rho A}}$$

4.6 Consider the free vibration of a cc-ss-cc-ss rectangular plate.
- Use the analytical solution to determine some of the resonance frequencies.
- Use the Ritz method to estimate the fundamental frequency of free vibration.
- Compare both results with the analytical solution for a fixed-fixed beam.

4.7 A three-story building frame is to be modeled as a shear building with the following description

$$
\begin{bmatrix} 1800 & -800 & 0 \\ -800 & 1400 & -600 \\ 0 & -600 & 600 \end{bmatrix} \begin{Bmatrix} u_1 \\ u_2 \\ u_3 \end{Bmatrix} + \begin{bmatrix} 3 & 0 & 0 \\ 0 & 3 & 0 \\ 0 & 0 & 2 \end{bmatrix} \begin{Bmatrix} \ddot{u}_1 \\ \ddot{u}_2 \\ \ddot{u}_3 \end{Bmatrix} = \begin{Bmatrix} P_1 \\ P_2 \\ P_3 \end{Bmatrix}
$$

The location 1 is closest to the ground.
- Determine the natural frequencies and mode shapes.
- Sketch the mode shapes.

4.8 Consider the strain energy and kinetic energy in principal coordinates.
- Show that the expression for the strain energy takes on the following simple form: $2U = w_1^2 \eta_1^2 + w_1^2 \eta_1^2 + \cdots$.
- Show that the expression for the kinetic energy takes on the following simple form: $2T = \dot{\eta}_1^2 + \dot{\eta}_1^2 + \cdots$.

4.9 Consider the free vibrations of a cantilever beam.
- Estimate the fundamental frequency using the single Ritz function $g(x) = [x^2]$.
- Now use the static deflection under its own (uniformly distributed) weight as the Ritz function to estimate the fundamental frequency.
- Compare both results to the exact

$$
\omega_n = \frac{3.516}{L^2} \sqrt{\frac{EI}{\rho A}}
$$

4.10 Consider an infinite plate.
- Show that the response to a rectangular pulse of duration τ is

$$
w(r,t) = \frac{P}{4\pi b \rho h} \left[t H(\frac{r^2}{4bt}) - (t-\tau) H(\frac{r^2}{4b(t-\tau)}) \right], \quad b = \sqrt{D/\rho h}
$$

$$
H(x) \equiv \frac{\pi}{2} - Si(x) - \sin(x) + xCi(x)
$$

where Si and Ci are the sine and cosine integrals, respectively.
- Do an FEM analysis to confirm this result.
— Reference [29], p. 243

4.11 Determine the resonances of a circular plate with simply/supported boundary conditions.
- Do an FEM analysis to confirm the results.
— Reference [45].

5
Nonlinear Dynamics

This chapter considers the effect of nonlinearities on the dynamic response of structures. As with the nonlinear static analysis, an incremental/iterative approach is developed; here, however, we must also discretize the time. The computer methods established will be applicable to most dynamics problems; in the analysis, however, the emphasis is on vibrations, and we consider the two cases of free vibration and forced vibration. For structural applications, the dominant behavior is linear elastic, therefore we treat the nonlinearities as perturbations on this behavior. This will allow us to effect some simplified solutions.

A characteristic of nonlinear dynamic systems is that a small change in initial conditions can give quite different steady-state behavior. Additionally, nonlinear equations can exhibit a wide range of complex behaviors for a relatively small range of parameter changes. By complex behavior we include seemingly random or "chaotic" behavior from the system even though the solutions remain deterministic in terms of their initial values and there is no random input. Consequently, prediction becomes impossible.

As an illustration of chaos, Figure 5.1 shows phase plane plots (position against velocity) for the Lorenz equations for two slightly different initial conditions. The two solutions are completely different in the detail; nonetheless, they exhibit a similarity in that the two solutions jump back and forth about between two orbits. These orbits are called *strange attractors* and act to give organization to the chaotic behavior.

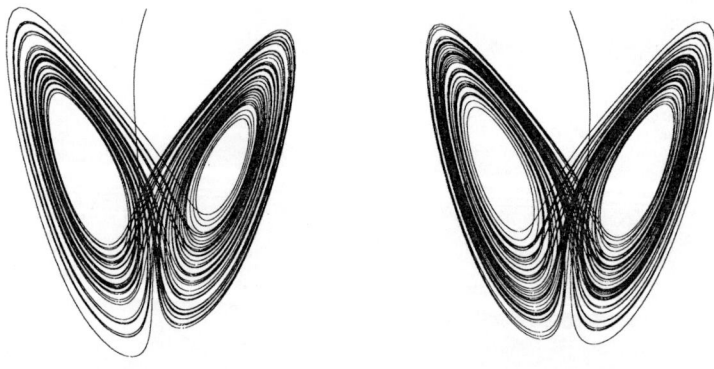

Figure 5.1: Phase plane plot of the Lorenz equations.

5.1   Free Vibration

We begin with one of the simpler nonlinear dynamics problems, namely, that of free vibration. This case, nonetheless, illustrates the difficulties encountered with nonlinear problems as well as showing many of the ways the solutions differ from their linear counterpart.

In order to effect a solution, we first need to impose some global restrictions. It seems reasonable to assume that, since there is no damping, we look for motions that are periodic. That is, we seek solutions for which $u(t + T) = u(t)$. Since the solutions are periodic on T, then they have a Fourier series representation in the form

$$u(t) = a_0 + a_1 \cos(\omega_f t) + a_2 \cos(2\omega_f t) + a_3 \cos(3\omega_f t) + \cdots, \qquad \omega_f = \frac{2\pi}{T}$$

In this, the coefficients a_n and the period T (hence ω_f) are unknown. That we anticipate multiple harmonics in the solution means that we expect the solution to be periodic but not harmonic. This method is sometimes called the method of *harmonic balance*, although the way we will use it is really as part of a perturbation solution. Reference [60] has a concise discussion of some of the other solution methods available for solving nonlinear problems.

Symmetric Return Force

The equation we wish to solve is

$$M\ddot{u} + K(1 + \alpha u^2)u = 0 \qquad \text{or} \qquad \ddot{u} + \omega_o^2(1 + \alpha u^2)u = 0$$

where $\omega_o \equiv \sqrt{K/M}$ is the natural frequency of the linear system and α can be either positive or negative.

For the symmetric return force, motions on the half period will be the minus image, that is, $u(t + \frac{1}{2}T) = -u(t)$. Therefore, we must only have odd terms in our expansion leaving as the solution

$$u(t) = A[\cos(\omega_f t) + \epsilon \cos(3\omega_f t) + \cdots], \qquad \omega_f = \frac{2\pi}{T}$$

The amplitude ratios are replaced with ϵ on the assumption that the various higher harmonic contributions are small. The other terms in the governing equation are

$$\begin{aligned}
\ddot{u} &= -\omega_f^2 A[\cos(\omega_f t) + 9\epsilon \cos(3\omega_f t) + \cdots] \\
u^3 &= A^3[\cos^3(\omega_f t) + 3\epsilon \cos^2(\omega_f t)\cos(3\omega_f t) + \cdots]
\end{aligned}$$

Noting that

$$\cos^3(\omega_f t) = \tfrac{1}{4}\cos(3\omega_f t) + \tfrac{3}{4}\cos(\omega_f t)$$

gives the governing equation as

$$-\omega_f^2 A[\cos(\omega_f t) + 9\epsilon\cos(3\omega_f t) + \cdots] + \omega_o^2 A[\cos(\omega_f t) + \epsilon\cos(3\omega_f t) + \cdots]$$
$$+\alpha\omega_o^2 A^3[\tfrac{1}{4}\cos(3\omega_f t) + \tfrac{3}{4}\cos(\omega_f t) + \cdots] \quad = 0$$

This must be true for all time, and since the cosine functions are linearly independent, we get

$$\cos(1\omega_f t): \qquad -\omega_f^2 A + \omega_o^2 A + \alpha\omega_o^2 A^3\tfrac{3}{4} = 0$$
$$\cos(3\omega_f t): \qquad -\omega_f^2 A9\epsilon + \omega_o^2 A\epsilon + \alpha\omega_o^2 A^3\tfrac{1}{4} = 0$$

The first of these gives

$$\omega_f = \omega_o\sqrt{1 + \alpha\tfrac{3}{4}A^2}$$

which indicates an amplitude-dependent frequency. Using this in the second equation gives

$$\epsilon = \frac{\alpha\omega_o^2 A^2\tfrac{1}{4}}{9\omega_f^2 - \omega_o^2} = \frac{\alpha\omega_o^2 A^2\tfrac{1}{4}}{9\omega_o^2 + 9\omega_o^2\tfrac{3}{4}\alpha A^2 - \omega_o^2} = \frac{\alpha A^2\tfrac{1}{4}}{8 + \tfrac{27}{4}\alpha A^2} \approx \frac{\alpha}{32}A^2$$

Our complete solution may now be written as

$$u(t) = A[\cos(\omega_f t) + \frac{\alpha}{32}A^2\cos(3\omega_f t) + \cdots], \qquad \omega_f = \omega_o\sqrt{1 + \alpha\tfrac{3}{4}A^2}$$

A similar process, although more cumbersome, can be used to determine the higher harmonic contributions.

This solution exhibits a characteristic that is very different from the corresponding linear solution; that is, it has an amplitude-dependent frequency. As a consequence, the principle of superposition does not hold and all solution methods based on this principle (such as Fourier analysis or Duhamel's integral) are no longer directly applicable. Furthermore, conclusions arising from such analyses (modal matrix, etc.) no longer apply.

Example 5.1: Determine the free motion response of a system with $M = 1$, $K = 1$, and $\alpha = 0.4$ when it is released from an initial position of $u(0) = 1$.
   Imposing the initial condition, we get

$$u(0) = 1 = A[1 + \frac{\alpha}{32}A^2]$$

from which we can determine that $A \approx 1$ (the other solutions give imaginary A, which are rejected on physical grounds).

The solution is compared to the numerically generated response and the linear response in Figure 5.2. The value of α used corresponds to $\omega_f/\omega_o \approx 1.14$. It is seen from the figure that this is about the ratio of the periods.

What may seem curious is why there is no apparent amplitude effect. This occurs because the initial displacement is matched at $t = 0$ and since the solution is periodic, then it must be matched at every nT. Consequently, only the phase shift is noticeable.

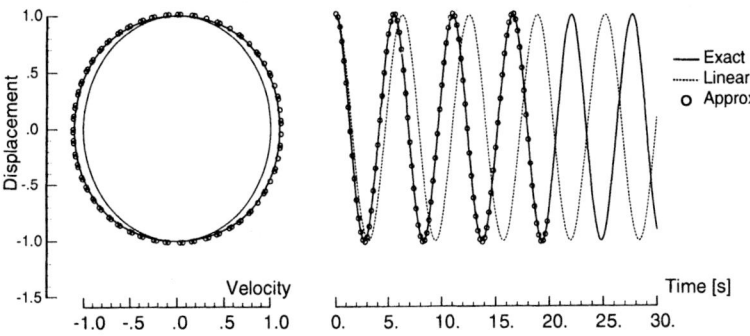

Figure 5.2: Free vibration response for a symmetric return spring.

Nonsymmetric Return Force

The equation we wish to solve is

$$M\ddot{u} + K(1 + \beta u)u = 0 \qquad \text{or} \qquad \ddot{u} + \omega_o^2(1 + \beta u)u = 0$$

where again $\omega_o \equiv \sqrt{K/M}$ is the natural frequency of the linear system.

For the nonsymmetric return force we must keep all terms in the expansion, leaving as our solution

$$u(t) = A_0 + A[\cos(\omega_f t) + \epsilon \cos(2\omega_f t) + \cdots], \qquad \omega_f = \frac{2\pi}{T}$$

Again, the amplitude ratios are replaced with ϵ on the assumption that the various higher harmonic contributions are small. Working as before gives the governing equation as

$$-\omega_f^2 A[\cos(\omega_f t) + 4\epsilon\cos(2\omega_f t) + \cdots] \quad + \quad \omega_o^2[A_o + A\{\cos(\omega_f t) + \epsilon\cos(2\omega_f t) + \cdots\}]$$
$$+\beta\omega_o^2\left[A_o^2 + 2A_o A\cos(\omega_f t) \quad + \quad A^2[\tfrac{1}{2} + \tfrac{1}{2}\cos(2\omega_f t) + \cdots]\right] = 0$$

This must be true for all time, and since the cosine functions are linearly independent, we get

$$\cos(0\omega_f t): \qquad \omega_o^2 A_o + \beta\omega_o^2 A_o^2 + \beta\omega_o^2 A^2 \tfrac{1}{2} = 0$$
$$\cos(1\omega_f t): \qquad -\omega_f^2 A + \omega_o^2 A + \beta\omega_o^2 2A_o A = 0$$
$$\cos(2\omega_f t): \qquad -\omega_f^2 A4\epsilon + \omega_o^2 A\epsilon + \beta\omega_o^2 A^2 \tfrac{1}{2} = 0$$

The first of these gives

$$A_o = \frac{-1 \pm \sqrt{1 - 2\beta^2 A^2}}{2\beta} \approx -\tfrac{1}{2}\beta A^2$$

indicating a nonlinear negative off-set. Combine this with the second equation to get

$$\omega_f = \omega_o\sqrt{1 + 2\beta A_o} = \omega_o[1 - 2\beta^2 A^2]^{1/4} \approx \omega_o\sqrt{1 - \tfrac{1}{2}\beta^2 A^2}$$

Again, we get an amplitude dependent-frequency — this time, however, it results in a period elongation. Using this in the third equation gives

$$\epsilon = \frac{\beta \omega_o^2 A \frac{1}{2}}{4\omega_f^2 - \omega_o^2} \approx \frac{1}{6}\beta A$$

Our complete solution is therefore

$$u(t) = -\tfrac{1}{2}\beta A^2 + A[\cos(\omega_f t) + \tfrac{1}{6}\beta A \cos(2\omega_f t) + \cdots], \qquad \omega_f = \omega_o[1 - 2\beta^2 A^2]^{1/4}$$

A similar process can be used to determine the higher harmonic contributions.

Within the limits of our approximation, this solution indicates that as the initial amplitude is increased, the frequency can go to zero and for an amplitude slightly larger, there would be a static instability. This is an example where slightly different initial conditions give very different responses. Actually, this example illustrates the type of behavior that would be exhibited by the truss structure of Figure 3.2. The instability observed would correspond to "snap-through" where the truss ends up on the negative side. We reconsider this again later in this chapter.

Example 5.2:   Determine the free motion response of a system with $M = 1$, $K = 1$, and $\beta = 0.4$ when it is released from an initial position of $u(0) = 1$.

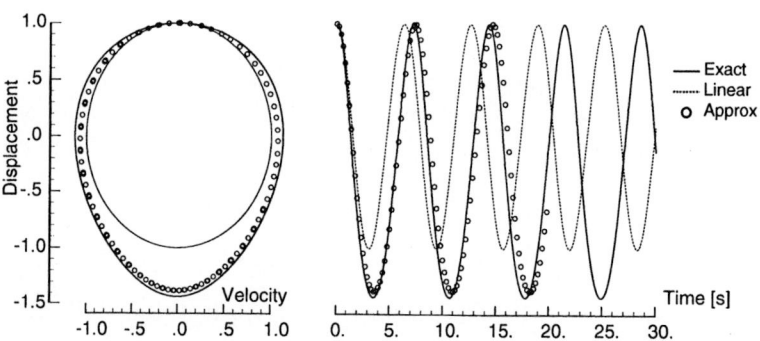

Figure 5.3: Free vibration response for a nonsymmetric return spring.

Imposing the initial condition, we get

$$u(0) = 1 = -\tfrac{1}{2}\beta A^2 + A + \tfrac{1}{6}\beta A^2 \qquad \text{or} \qquad \tfrac{1}{3}\beta A^2 - A + 1 = 0$$

from which we can determine that $A \approx 1.19$ and $A \approx 6.3$. The latter corresponds to an imaginary frequency ω_f, hence we take the former value.

The solution is compared to the numerically generated response and the linear response in Figure 5.3. We see the period elongation relative to the linear response as expected. A value of $\beta = 0.4$ corresponds to an off-set of -0.20. This closely matches the value in the figure — note that the difference seen is twice the off-set value.

Again the positive peaks agree because the initial displacement is matched at $t = 0$ and since the solution is periodic, then it must be matched at every nT.

Frictional Contact

One of the early examples of frictional contact studied was by Rayleigh in connection with the oscillations of a violin string. His analysis led to the equation

$$M\ddot{v} - \mu[1 - \tfrac{1}{3}\beta\dot{v}^2]\dot{v} + Kv = 0$$

where the nonlinearity is associated with the "damping" term. Before discussing this further, we will put it into the standard form of a *Van der Pol equation* by differentiation and making the substitution $\dot{v} \longrightarrow u$ to give

$$\ddot{u} - \epsilon[1 - \beta u^2]\dot{u} + \omega_o^2 u = 0$$

The sign of the damping contribution depends on the relative magnitude of the displacement: for $|u| < 1/\sqrt{\beta}$ it is negative and the system receives energy, while for $|u| > 1/\sqrt{\beta}$ it is positive and the system dissipates energy. Thus on a complete cycle, energy is both gained and lost; for this reason, the responses of these systems are often referred to as *self-excited oscillations*, although it is somewhat of a misnomer since there is an energy source eternal to the system (for example, the rotating drum in the introduction, or the energy to the violin string coming from the bow).

We will obtain a solution based on the assumption that ϵ is small. First consider the linear case:

$$\ddot{u} - \epsilon\dot{u} + \omega_o^2 u = 0$$

Because this has constant coefficients, seek solutions of the form $e^{i\mu t}$, which leads to the roots

$$\mu_{1,2} = -\tfrac{1}{2}i\epsilon \pm \tfrac{1}{2}\sqrt{4\omega_o^2 - \epsilon^2} \approx -\tfrac{1}{2}i\epsilon \pm \omega_o^2$$

The solution is therefore

$$u(t) = e^{\epsilon t/2}\Big[A\cos\omega_o t + B\sin\omega_o t\Big]$$

For a given set of initial conditions, this solution corresponds to spirals away from the origin. That is, the solution is unstable. This is as expected, since the friction corresponds to negative damping. However, when we return to the nonlinear equation, it shows that, when the amplitude exceeds a certain value, the friction becomes dissipative and, consequently, we should not get an infinite amplitude. Actually what happens is that the solution finds a *limit cycle* — an orbit in the phase plane that is constant. Limit cycles are equilibrium motions (in contrast to equilibrium points where the system is at rest) where the system is performing a periodic motion. Neighboring paths are not closed but spiral into (stable) or away (unstable) from the limit cycle. In the present case they are spiraling into the limit cycle. Limit cycles are an important feature of nonlinear systems (linear systems with constant coefficients cannot exhibit limit cycles) and we will discuss them in greater detail later.

Once the system has achieved the limit cycle, then the solution can be represented as

$$u(t) = A\cos(\omega_f t) + \cdots , \qquad \omega_f = \frac{2\pi}{T}$$

We now wish to determine A and ω_f. Using $x_1 = u$ and $x_2 = \dot{u}$, we can rewrite our system as

$$\dot{x}_1 = x_2 , \qquad \dot{x}_2 = \epsilon[1 - \beta x_1^2]x_2 - \omega_o^2 x_1$$

We are expecting the limit cycle to be nearly circular in the phase plane, so introduce the polar coordinates

$$r^2 = \omega_o^2 x_1^2 + x_2^2 , \qquad r\cos\theta = \omega_o x_1 , \qquad r\sin\theta = x_2$$

Differentiate the radius,

$$r\dot{r} = \omega_o^2 x_1\dot{x}_1 + x_2\dot{x}_2 = \epsilon[1 - \beta x_1^2]x_2^2$$

and substitute for x_1 and x_2, to get

$$\dot{r} = \epsilon[1 - \gamma r^2 \cos^2\theta]r\sin^2\theta , \qquad \gamma = \beta/\omega_o^2$$

This shows that the radius is nearly constant (since its rate of change is of order ϵ) and can be represented by

$$r(t) = A + O(\epsilon)$$

In a similar manner, we get

$$\dot{\theta} = -\omega_o + \epsilon[1 - \gamma r^2 \cos^2\theta]\sin\theta\cos\theta \approx -\omega_o + O(\epsilon)$$

and the frequency is predominantly that of the linear system without any friction.

It remains to determine A. Since the cycles repeat, the total change in r must be zero leading to

$$\int_0^T \dot{r}\,dt = 0 = \int_{2\pi}^0 \frac{\dot{r}}{\dot{\theta}}d\theta \approx \frac{-1}{\omega_o}\int_{2\pi}^0 \epsilon[1 - \gamma A^2\cos^2\theta]A\sin^2\theta d\theta \quad \Rightarrow \quad 0 = [1 - \gamma A^2\tfrac{1}{4}]$$

From this we get

$$A = \frac{2}{\omega_o\sqrt{\beta}} \qquad \text{and} \qquad u(t) = \frac{2}{\omega_o\sqrt{\beta}}\cos\omega_o t$$

This solution resembles the solution for a linear oscillator without friction. There is a significant difference, however; for the linear oscillator, the amplitude depends on the initial conditions — the bigger the initial energy, the bigger the radius. Here, the radius depends on the system parameters only, and is the same irrespective of how the motion is initiated.

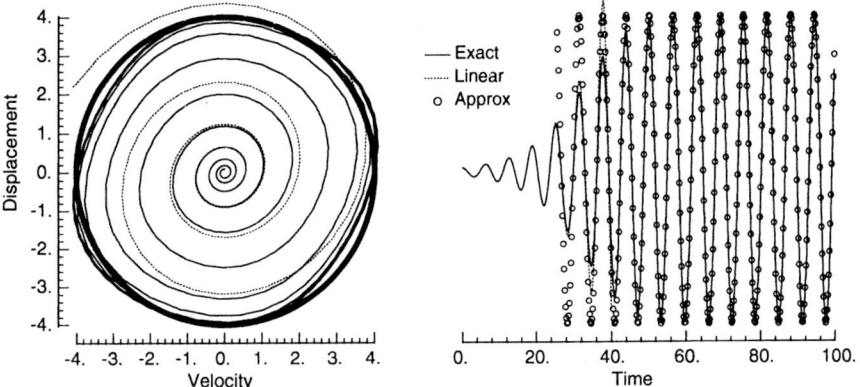

Figure 5.4: Free vibration response for the Van der Pol equation.

Example 5.3:   Determine the free motion response of a system with $M = 1$, $K = 1$, $\mu = 0.2$, and $\beta = 0.25$ when it is released from an initial position of $u(0) = 0.1$.

The linear and approximate limit cycle solutions are compared to the numerically generated response in Figure 5.4. We clearly see how the linear part of the true solution grows as an outward spiral until it reaches the limit cycle. (Note that because displacement is plotted on the vertical axis, the time axis is into the page and therefore the spirals are clockwise as assumed in the above derivations.) The radius depends on the strength of the nonlinearity as $1/\sqrt{\beta}$. Thus, when $\beta = 0$, we get the linear solution, which corresponds to a spiral going to infinity. The phase of the two solutions match very well, and the approximation estimates the radius of the limit cycle quite well.

When the initial conditions are such that $u(0) > 4.0$, the solution spirals inward toward the limit cycle.

5.2   Forced Response

We saw that a linear system can exhibit resonance behavior when forced harmonically. We therefore divide our nonlinear approximate analysis into two situations: first when the excitation is far from the linear resonance, and second when it is close to the linear resonance. In both cases, we only consider the symmetric return case for which the governing equation is

$$M\ddot{u} + K(1 + \alpha u^2)u = P(t) = \hat{P}\cos\omega t$$

and ω is the forcing frequency.

Forced Response Away from Resonance

Begin by rewriting the equation as

$$\ddot{u} + \omega_o^2(1 + \alpha u^2)u = \tilde{P}\cos\omega t$$

We will get an approximate solution by way of a perturbation method. To that end, consider ϵ to be small and let

$$u(t, \epsilon) \approx u_0(t) + u_1(t)\epsilon + u_2(t)\epsilon^2 + \cdots, \qquad \epsilon \equiv \alpha\sqrt{K/M} \equiv \alpha\omega_o^2$$

Substitute these into the governing equation to get

$$[\ddot{u}_0 + \ddot{u}_1\epsilon + \ddot{u}_2\epsilon^2 + \cdots] + \omega_o^2[u_0 + u_1\epsilon + u_2\epsilon^2 + \cdots] + \epsilon[u_0^3 + 3u_0u_1\epsilon + \cdots] = \tilde{P}\cos\omega t$$

Regroup in equal powers of ϵ and impose the condition that the equation must be satisfied (that is, be equal to zero) for equal powers of ϵ. This gives

$$\epsilon^0: \qquad \ddot{u}_0 + \omega_o^2 u_0 = \tilde{P}\cos\omega t$$

$$\epsilon^1: \qquad \ddot{u}_1 + \omega_o^2 u_1 = -u_0^3$$

$$\epsilon^2: \qquad \ddot{u}_2 + \omega_o^2 u_2 = -3u_0u_1$$

and so on. We are interested in solutions with period $T = 2\pi/\omega$ because they are related to the forcing frequency. However, we suppose that the forcing frequency is not an integer fraction of the natural frequency, $\omega \neq \omega_o/n$. The major term $u_0(t)$ is a periodic solution of the linearized equation. Hence, we are restricted to finding solutions which are close to the linearized result. The method will not expose other results.

Integrate the first equation to get

$$u_0(t) = A\cos\omega_o t + B\sin\omega_o t + \frac{\tilde{P}}{\omega_o^2 - \omega^2}\cos\omega t$$

If ω_o is not an integer multiple of ω, then the above is not periodic on $2\pi/\omega$ and we must have $A = 0$ and $B = 0$, leaving us with

$$u_0(t) = \frac{\tilde{P}}{\omega_o^2 - \omega^2}\cos\omega t$$

The second equation becomes

$$\ddot{u}_1 + \omega_o^2 u_1 = \left(\frac{\tilde{P}}{\omega_o^2 - \omega^2}\right)^3 \cos^3\omega t = \frac{\tilde{P}^3}{(\omega_o^2 - \omega^2)^3}\left[\tfrac{3}{4}\cos(\omega t) + \tfrac{1}{4}\cos(3\omega t)\right]$$

Integrating, and retaining only those solutions with the required periodicity, gives

$$u_1(t) = -\frac{3}{4}\frac{\tilde{P}^3}{(\omega_o^2 - \omega^2)^4}\cos(\omega t) - \frac{1}{4}\frac{\tilde{P}^3}{(\omega_o^2 - \omega^2)^3(\omega_o^2 - 9\omega^2)}\cos(3\omega t)$$

Our second approximation now becomes

$$u(t) = \frac{\tilde{P}}{\omega_o^2 - \omega^2}\cos\omega t - \frac{\alpha\omega_o^2}{(\omega_o^2 - \omega^2)}\left[\tfrac{3}{4}\cos\omega t + \tfrac{1}{4}\cos 3\omega\frac{(\omega_o^2 - \omega^2)}{(\omega_o^2 - 9\omega^2)}\right]\frac{\tilde{P}^3}{(\omega_o^2 - \omega^2)^3}$$

The series continues with $\cos 5\omega t$, $\cos 7\omega t$, and so on.

The method fails when $\omega_o/\omega = 1, 3, 5, \cdots$; this is averted by the integer condition, but nonetheless for a nearby frequency it indicates the series would have difficulty converging, hence we have a "near" resonance. That is, the solution shows that we get large responses at or near

$$\omega = \omega_o \quad \text{or} \quad \omega = \frac{1}{3}\omega_o \quad \text{or} \quad \omega = \frac{1}{5}\omega_o \quad \text{or} \quad \cdots$$

These are "resonances" due to the higher harmonics in u^3 feeding back into the linear system. These are called *subharmonic resonances*.

Example 5.4: Analyze the case when $M = 1$, $K = 1$, and $\alpha = 0.4$.

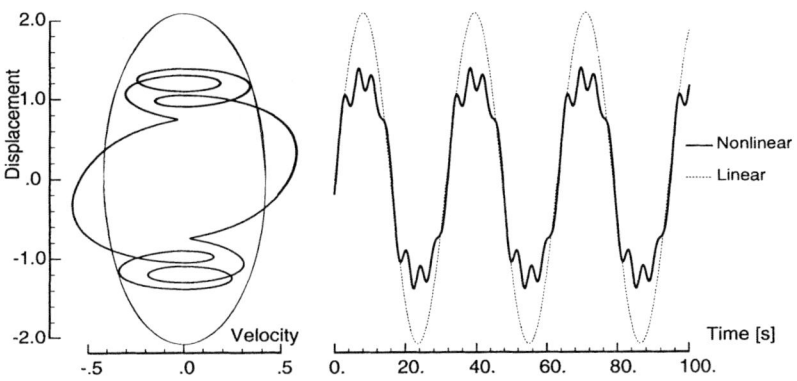

Figure 5.5: Subharmonic resonances.

Figure 5.5 shows an example of forcing the system at $\omega = \omega_o/5$. It is clear that a higher ω_o response is being generated; the time traces show the higher harmonic rider on the forcing frequency, the phase-plane plot shows small "loop-the-loop" patterns. This behavior would not happen in a linear system.

The results shown are for the steady state response. This was achieved by having $C = 0.1$ with the initial conditions $u_o = -0.2$, $\dot{u}_o = 0.48$. If the initial conditions are not set correctly, then the initial response will always show the presence of the resonance frequency. However, for linear systems they die down due to the damping.

Forced Motion Near Resonance

Now consider the situation where the forcing frequency is close to the linear natural frequency.

The equation we are considering is rewritten as

$$\ddot{u} + \omega_o^2 u + \epsilon u^3 = \tilde{P}\cos\omega t, \qquad \epsilon \equiv \omega_o^2 \alpha$$

Let the forcing frequency be near resonance such that

$$\omega_o^2 = (1 + \epsilon\beta)\omega^2$$

where β is a small parameter. Also, let the force be small, that is, $\tilde{P} = \epsilon\gamma$. Our equation now becomes

$$\ddot{u} + \omega^2 u = \epsilon[\gamma\cos\omega t - \beta\omega^2 u - u^3]$$

As in the last section, consider ϵ to be small and let

$$u(t, \epsilon) \approx u_0(t) + u_1(t)\epsilon + u_2(t)\epsilon^2 + \cdots$$

Substitute these into the governing equation and group equal powers of ϵ to get

$$\epsilon^0 : \quad \ddot{u}_0 + \omega^2 u_0 = 0$$
$$\epsilon^1 : \quad \ddot{u}_1 + \omega^2 u_1 = \gamma\cos\omega t - \beta\omega^2 u_0 - u_0^3$$

Integrate the first equation to get

$$u_0(t) = A\cos\omega t + B\sin\omega t$$

Since there is no damping, we can take $B = 0$
    The second equation now becomes

$$\ddot{u}_1 + \omega^2 u_1 = \gamma\cos\omega t - \beta\omega^2 A\cos\omega t - A^3[\tfrac{3}{4}\cos(\omega t) + \tfrac{1}{4}\cos(3\omega t)]$$

Note that the particular solution for $\ddot{u} + \omega^2 u = a\cos\omega t$ is $u(t) = (at/2\omega)\sin\omega t$, which introduces a *secular* term, that is, a term increasing in time. Hence we must have the associated coefficient be zero, leading to

$$\gamma - \beta\omega^2 A - \tfrac{3}{4}A^3 = 0$$

Substituting for β and γ gives

$$A(\omega_o^2 - \omega^2) + \tfrac{3}{4}\alpha\omega_o^2 A^3 - \tilde{P} = 0$$

The three solutions for A are shown in Figure 5.6(a). Consider the situation with $\tilde{P} > 0$: for $\omega > 1.5$, there are three real solutions; for $\omega < 1.5$, there is one real solution and two complex solutions. The case $\tilde{P} = 0$ is similar except that for $\omega < 1.5$, two of the solutions are purely imaginary. Note that the only physically viable solutions are those with real only components. These are shown as magnitudes in Figure 5.6(b) for three different load values.
    Note that even if $\omega^2 = \omega_o^2$, then

$$A = \left[\frac{\tilde{P}}{\frac{3}{4}\alpha\omega_o^2}\right]^{1/3}$$

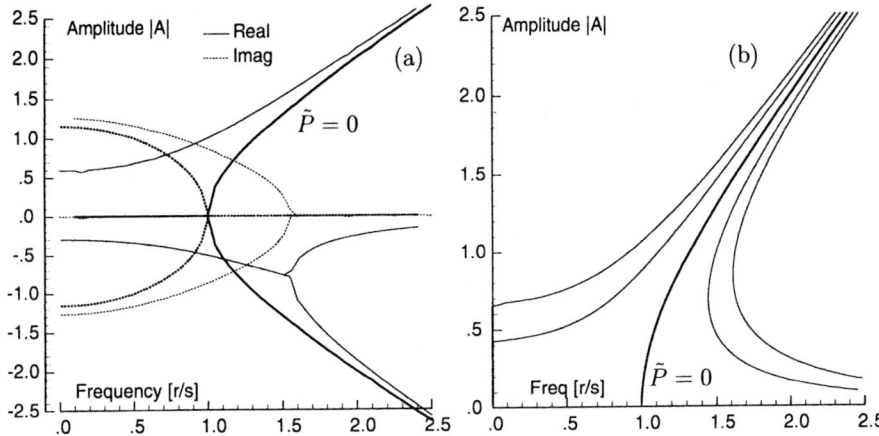

Figure 5.6: Variation of amplitude $A(\omega)$ near resonance for different load values. (a) Three complex roots. (b) Real only solutions.

showing that the amplitude is finite. It is only when $\alpha = 0$ (the linear case) that the amplitude becomes infinite. This interesting result shows that nonlinearities in the system would prevent the infinite displacement associated with linear resonance phenomena. Another special case is when $\tilde{P} = 0$, the free vibration case. We get

$$A = 0\,, \pm\sqrt{\frac{\omega^2 - \omega_o^2}{\frac{3}{4}\alpha\omega_o^2}}$$

This is the same relation as derived in the earlier section on free vibration. Here it corresponds to the centerline of the forced frequency response curve in Figure 5.6(b).

Effect of Damping

The effect of damping in linear systems is to decrease the peak amplitude at resonance. It is of interest to know if damping has a similar effect on the nonlinear peaks.

The equation we are considering is rewritten as

$$\ddot{u} + \eta\dot{u} + \omega_o^2 u + \epsilon u^3 = \tilde{P}\cos\omega t\,, \qquad \epsilon \equiv \omega_o^2\alpha\,, \quad \eta \equiv c/M$$

where we have introduced a linear viscous damping term. Let the forcing frequency be near resonance such that

$$\omega_o^2 = (1 + \epsilon\beta)\omega^2$$

where β is a small parameter. Also, let the force and damping be small, that is, $\tilde{P} = \epsilon\gamma$, $\eta = \epsilon\kappa$. Our equation now becomes

$$\ddot{u} + \omega^2 u = \epsilon[\gamma\cos\omega t - \kappa\dot{u} - \beta\omega^2 u - u^3]$$

As in the last section, we consider ϵ to be small and let

$$u(t, \epsilon) \approx u_0(t) + u_1(t)\epsilon + u_2(t)\epsilon^2 + \cdots$$

Substitute these into the governing equation and group equal powers of ϵ to get

$$\epsilon^0 : \qquad \ddot{u}_0 + \omega^2 u_0 = 0$$
$$\epsilon^1 : \qquad \ddot{u}_1 + \omega^2 u_1 = \gamma \cos \omega t - \kappa \dot{u}_o - \beta \omega^2 u_0 - u_0^3$$

Integrate the first equation to get

$$u_0(t) = A \cos \omega t + B \sin \omega t$$

Since there is damping, we cannot take $B = 0$.

Using relations such as

$$\cos^3 x = \tfrac{3}{4} \cos x + \tfrac{1}{4} \cos 3x, \qquad \sin^2 x \cos x = [1 - \cos^2 x] \cos x = \tfrac{3}{4} \cos x - \tfrac{1}{4} \cos 3x$$

the second equation now becomes

$$\begin{aligned}
\ddot{u}_1 + \omega^2 u_1 \quad = \quad & \left[\kappa \omega A - B[\beta \omega + \tfrac{3}{4}(A^2 + B^2)] \right] \sin \omega t \\
& + \left[-\kappa \omega B - A[\beta \omega + \tfrac{3}{4}(A^2 + B^2)] + \gamma \right] \cos \omega t \\
& + \left[-\tfrac{1}{4}B^3 - \tfrac{3}{4}A^2 B \right] \sin 3\omega t + \left[-\tfrac{1}{4}A^3 + \tfrac{3}{4}AB^2 \right] \cos 3\omega t
\end{aligned}$$

As noted before, the particular solution for equations such as $\ddot{u} + \omega^2 u = a \cos \omega t$ is $u(t) = (at/2\omega) \sin \omega t$, which introduces a secular term. Hence, we must have the associated coefficient be zero, leading to

$$\begin{aligned}
0 \quad &= \quad \kappa \omega A - B[\beta \omega^2 + \tfrac{3}{4}(A^2 + B^2)] \\
0 \quad &= \quad \kappa \omega B + A[\beta \omega^2 + \tfrac{3}{4}(A^2 + B^2)] - \gamma
\end{aligned}$$

Substituting for β, κ, and γ gives

$$\begin{aligned}
\eta \omega A - B[(\omega_o^2 - \omega^2) + \tfrac{3}{4}\omega_o^2 \alpha (A^2 + B^2)] \quad &= \quad 0 \\
\eta \omega B + A[(\omega_o^2 - \omega^2) + \tfrac{3}{4}\omega_o^2 \alpha (A^2 + B^2)] \quad &= \quad \tilde{P}
\end{aligned} \qquad (5.1)$$

If there is no damping, we get $B = 0$, and the second equation becomes the result previously obtained.

Squaring both equations and adding, we get the equation for the amplitude as

$$\eta^2 \omega^2 C^2 + C^2 [(\omega_o^2 - \omega^2) + \tfrac{3}{4}\omega_o^2 \alpha C^2]^2 = \tilde{P}^2, \qquad C^2 \equiv A^2 + B^2$$

The solution for C is shown in Figure 5.7 for different excitation levels. Parenthetically, the simplest way to make this plot is to solve for ω^2 (which is only quadratic) for different values of real-only amplitude.

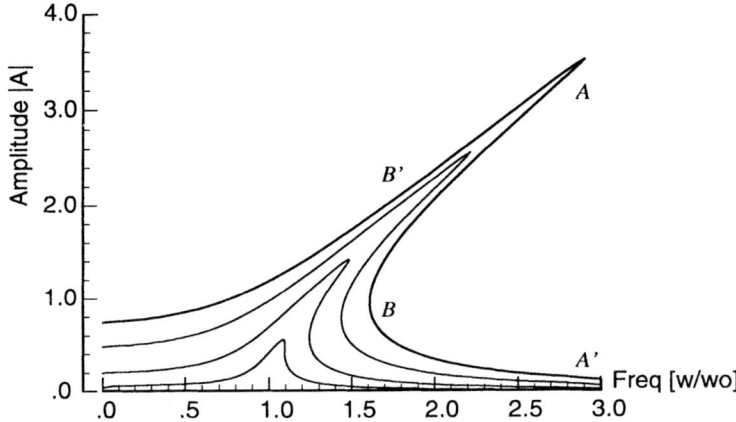

Figure 5.7: Effect of damping on the amplitude response near resonance.

The major point to note about the plot is that, for a single excitation level, there is only one solution curve but it folds on itself. Consider the following thought experiment to help explain the plots. Set P at a low value and slowly scan through ω from $\omega_1 < \omega_o$ to $\omega_2 > \omega_o$ and back again. As ω is increased, in the vicinity of $\omega \approx \omega_o$, we will see a peak response but it gets smaller on further increase of ω. The return response for decreasing ω is similar. These are familiar behaviors from the damped linear theory. Now increase P and again scan through ω. Just past point A, the amplitude will show a sudden decrease corresponding to a jump to point A'. Thereafter, it will show a steady decrease. On the return journey, it will get to point B before suddenly jumping to B' with a larger amplitude. These jump behaviors are characteristic of an instability — we will therefore take this up in greater detail in the next two chapters.

Example 5.5: Use numerical integration methods to investigate the dependence of the forced nonlinear response on the initial conditions

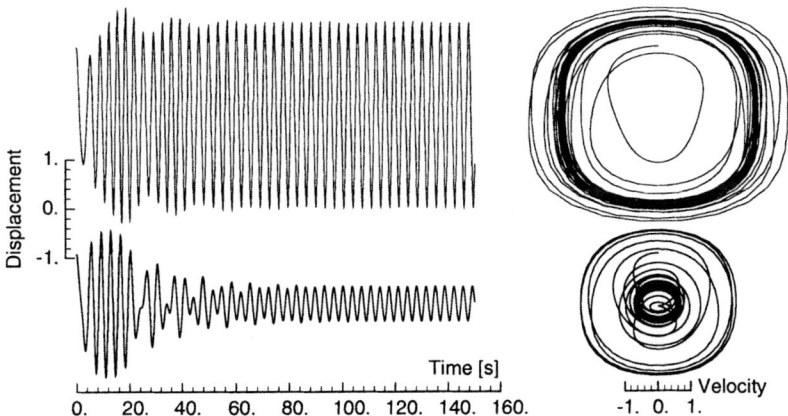

Figure 5.8: Time and phase plane plots for two slightly different initial conditions.

Owing to the folding of the response plot in Figure 5.7, there is a multiplicity of amplitudes for a given excitation frequency. The effect of this is illustrated in Figure 5.8, which shows two slightly different initial conditions (under the same forcing conditions) converging to two very different steady-state behaviors. Figure 5.7 shows that three amplitudes are possible for a given frequency (beyond B) and yet the simulations exhibit only two amplitudes. The reason is that the middle amplitude solution is unstable.

The phase-plane plots are also shown in Figure 5.8. In both cases, the beginning point is clearly seen. The heavier regions are the steady-states achieved. This dependence of the final steady state solution on the values of the initial conditions is a hallmark of nonlinear systems. In the linear case, there is a nice separation of the two.

5.3   Dynamic Equilibrium with Large Rotations

In the case of thin-walled structures, we must take into account the effect of large rotations and the possibility that joint masses may be sizable in extent. As will be seen, the equations of motion governing the rotation are nonlinear even for a single mass.

Rotational Motion

Consider a body rotating about a fixed axis (as in Figure 3.13) with angular velocity $\hat{\omega}$. The origin is on the rotation axis. Each point not on the axis moves in a circle in a plane normal to the axis with velocity and acceleration

$$\dot{\hat{r}} = \hat{\omega} \times \hat{r}$$
$$\ddot{\hat{r}} = \dot{\hat{\omega}} \times \hat{r} + \hat{\omega} \times (\hat{\omega} \times \hat{r})$$

Consider now a rotation about a fixed point O rather than an axis. Euler's theorem says that any finite motion of the body is equivalent to a rotation of the body about some particular axis through this point.

During a time Δt the body will rotate through an angle $\Delta\phi$ about the unique rotation axis. As Δt approaches zero, the ratio $\Delta\phi/\Delta t$ becomes the magnitude of the angular velocity $\hat{\omega}$. The direction of $\hat{\omega}$ is along the rotation axis.

For a general motion of a body about a fixed point, the rotation axis is not a line fixed in the body. We can represent the velocity and acceleration as above, the difference between the general motion and the one about a fixed axis lies in the angular acceleration $\dot{\hat{\omega}}$ term: when the axis is fixed, the vector $\dot{\hat{\omega}}$ is directed along the rotation axis and represents the rate of change of $\hat{\omega}$; in the general case it reflects the change in direction of $\hat{\omega}$ as well as its change in magnitude and it is not directed along the rotation axis.

Finite rotations are not commutative; infinitesimal rotations do, however, obey the parallelogram law of vector addition. It follows that the angular velocities may be added vectorally, that is,

$$\hat{\omega} = \dot{\hat{\phi}} = \hat{\omega}_1 + \hat{\omega}_2$$

Note that for finite rotations $\omega \neq \dot{\phi}$.

Kinetics

Consider a rigid body with axes (x, y, z) attached with origin at the mass center G. The angular velocity of the reference frame is therefore $\hat{\omega}$. When the body rotates, the inertia changes when referred to the inertial frame of reference. We will therefore find it more convenient to refer the governing equations to a set of body fixed coordinates.

The angular momentum of a small element of mass $dm = \rho \, dV$ is

$$d\hat{H} = \hat{r} \times \hat{v} \, dm = \hat{r} \times (\bar{v} + \hat{\omega} \times \hat{r}) \, dm$$

But $\int \hat{r} \times \bar{v} \, dm = 0$ since the origin is the mass center, hence, we have

$$d\hat{H} = \hat{r} \times (\hat{\omega} \times \hat{r}) dm \qquad \text{or} \qquad \hat{H} = \int \hat{r} \times (\hat{\omega} \times \hat{r}) \rho \, dxdydz$$

The explicit form for the components of $d\hat{H}$ are

$$\begin{aligned}
dH_x &= [\quad (y^2 + z^2)\omega_x & -xy\omega_y & -xz\omega_z] \, dm \\
dH_y &= [\quad -yx\omega_x & (z^2 + x^2)\omega_y & -yz\omega_z] \, dm \\
dH_z &= [\quad -zx\omega_x & -zy\omega_y & (x^2 + y^2)\omega_z] \, dm
\end{aligned}$$

Now let

$$I_{xx} \equiv \int_V (y^2 + z^2) \, dm , \qquad I_{xy} \equiv \int_V (yz) \, dm , \qquad \text{and so on}$$

where I_{ij} are the mass moments of inertia. The angular momentum expression now becomes

$$\begin{aligned}
H_x &= [\quad I_{xx}\omega_x & -I_{xy}\omega_y & -I_{xz}\omega_z] \\
H_y &= [\quad -I_{yx}\omega_x & I_{yy}\omega_y & -I_{yz}\omega_z] \\
H_z &= [\quad -I_{zx}\omega_x & -I_{zy}\omega_y & I_{zz}\omega_z]
\end{aligned}$$

This lends itself to the matrix forms

$$\begin{Bmatrix} H_x \\ H_y \\ H_z \end{Bmatrix} = \begin{bmatrix} I_{xx} & -I_{xy} & -I_{xz} \\ -I_{yx} & I_{yy} & -I_{yz} \\ -I_{zx} & -I_{zy} & I_{zz} \end{bmatrix} \begin{Bmatrix} \omega_x \\ \omega_y \\ \omega_z \end{Bmatrix} \qquad \text{or} \qquad \{H\} = [\, J \,]\{\omega\}$$

The quantity $[\, J \,]$ is a second-order tensor and therefore transforms analogously to stress. It has principal values and directions. This relation shows that the

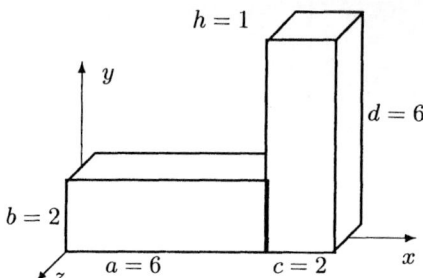

Figure 5.9: Geometry of composite object.

direction of the angular velocity and angular momentum are generally not the same.

Example 5.6: Determine the moments of inertia for the angle bracket shown in Figure 5.9.

Some useful formulas for moments of inertias about the mass center are

Block $[a \times b \times c]$: $I_{xx} = \frac{1}{12}m(b^2 + c^2)$, $I_{yy} = \frac{1}{12}m(c^2 + a^2)$, $I_{zz} = \frac{1}{12}m(a^2 + b^2)$

Cylinder $[L \times R]$: $I_{xx} = I_{yy} = \frac{1}{12}m(L^2 + 3R^2)$, $I_{zz} = \frac{1}{2}mR^2$

Sphere $[R]$: $I_{xx} = I_{yy} = I_{zz} = \frac{2}{5}mR^2$

The approach to general bodies is to conceive of them as made of simpler parts and then to use the parallel axis theorem to add them with respect to a common point. This theorem says that the inertia matrix about a set of axes at $\bar{x}$ can be written in terms of inertias about a set of parallel axes at the center of mass $\bar{x}^G$ using the formulas

$$I_{xx} = I_{xx}^G + m(x^G - x)^2, \qquad I_{xy} = I_{xy}^G - m(x^G - x)(y^G - y), \qquad \text{and so on}$$

where m is the mass of the body.

For the given problem, we must first determine the mass center for the composite object. Choosing the origin as shown, we have for the first moments

$$
\begin{aligned}
M_y &= \rho(abh)a/2 + \rho(cdh)(a + c/2) = \rho(abh + cdh)c_x \\
M_x &= \rho(abh)b/2 + \rho(cdh)d/2 = \rho(abh + cdh)c_y
\end{aligned}
$$

Hence the centroid is at

$$c_x = \frac{(ab)a + (cd)(2a + c)}{2(ab + cd)} = \frac{69}{14} = 4.93, \qquad c_y = \frac{(ab)b + (cd)d}{2(ab + cd)} = \frac{19}{7} = 2.71$$

The second moments are

$$I_{xx} = \frac{1}{12}\rho(abh)(b^2 + h^2) + \rho(abh)(b/2)^2 + \frac{1}{12}\rho(cdh)(d^2 + h^2) + \rho(cdh)(d/2)^2 = \rho299.7$$

$$I_{yy} = \frac{1}{12}\rho(abh)(a^2 + h^2) + \rho(abh)(a/2)^2 + \frac{1}{12}\rho(cdh)(c^2 + h^2) + \rho(cdh)(a + c/2)^2 = \rho934.3$$

$$I_{xy} = 0 - \rho(abh)(a/2)(b/2) + 0 - \rho(cdh)(a + c/2)(d/2) = -\rho484$$

The parallel axis theorem could be used to relocate this inertia to the center of mass.

Equations of Motion

For an absolute reference frame, the change of momentum is related to the externally applied moments according to

$$\sum \hat{M} = \frac{d\hat{H}}{dt}$$

where the terms are taken about either a fixed point O or about the mass center G. When $\hat{H}$ is expressed relative to a moving coordinate system, then

$$\sum \hat{M} = \frac{d\hat{H}}{dt}\Big|_{xyz} + \hat{\Omega} \times \hat{H}$$

This is written in component form as

$$\begin{aligned}
\sum M_x &= \dot{H}_x - H_y\Omega_z + H_z\Omega_y \\
\sum M_y &= \dot{H}_y - H_z\Omega_x + H_x\Omega_z \\
\sum M_z &= \dot{H}_z - H_x\Omega_y + H_y\Omega_x
\end{aligned}$$

Let the reference axes coincide with the principal axes of inertia, then I_{xy} and the other off diagonal terms are zero and we have $\hat{\Omega} = \hat{\omega}$ leading to

$$\begin{aligned}
I_{xx}\dot{\omega}_x + (I_{zz} - I_{yy})\omega_y\omega_z &= M_{ex} \\
I_{yy}\dot{\omega}_y + (I_{xx} - I_{zz})\omega_z\omega_x &= M_{ey} \\
I_{zz}\dot{\omega}_z + (I_{yy} - I_{xx})\omega_x\omega_y &= M_{ez}
\end{aligned} \tag{5.2}$$

These equations are known as *Euler's relations*. Note that they are nonlinear, and this can give rise to some unexpected phenomena. In the special case when all principal values are the same, the Euler relations uncouple and become

$$I_{xx}\dot{\omega}_x = M_{ex}, \qquad I_{yy}\dot{\omega}_y = M_{ey}, \qquad I_{zz}\dot{\omega}_z = M_{ez}$$

These are linear and are the equations usually employed in a linear small displacement analysis.

The orientation of the rigid body is not simply the integral of the angular velocities; indeed we must use the transformation developed in Chapter 3. That is, during an incremental scheme, we use the above equations to determine estimates of the angular velocities from which increments in the orientation is then

$$\Delta\phi = \omega\Delta t$$

and the orientation of the triad vectors are then updated as

$$\{v'\} = \left[[\ I\ ] + \frac{\sin\phi}{\Delta\phi}[S(\Delta\phi)] + \frac{1 - \cos\Delta\phi}{\Delta\phi^2}[S(\Delta\phi)]^2 \right]\{v\} = [R(\Delta\phi)]\{v\}$$

If desired, the small angle approximation

$$\{v'\} \approx \left[[\ I\ ] + [S(\Delta\phi)] + [S(\Delta\phi)]^2 \right] \{v\}$$

can be used.

Example 5.7: Consider a rigid block that experiences moment impulses in such a way that one of the rotational velocities is significantly larger than the other two. Determine the subsequent free motion.

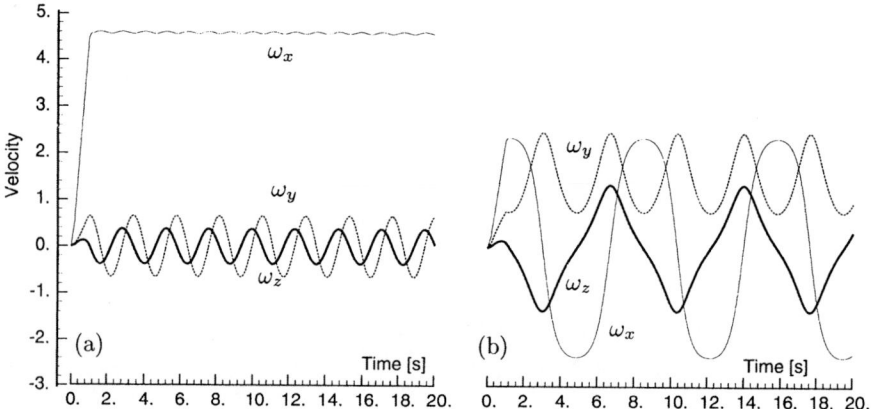

Figure 5.10: Rotational speeds due to moment impulse. (a) Impulse ratio is [5:1:1]. (b) Impulse ratio is [1:5:1].

Let ω_x be the dominant velocity, then we can approximate the response as

$$\omega_x \approx \omega_o + \xi_x, \qquad \omega_y \approx \xi_y, \qquad \omega_z \approx \xi_z$$

where ξ_i are small terms. Substitute this approximation into the Euler's relations and neglecting nonlinear terms we obtain the free response equations as

$$
\begin{aligned}
I_{xx}\dot{\xi}_x &= 0 \\
I_{yy}\dot{\xi}_y + (I_{xx} - I_{zz})\omega_o\xi_z &= 0 \\
I_{zz}\dot{\xi}_z + (I_{yy} - I_{xx})\omega_o\xi_y &= 0
\end{aligned}
$$

The first equation shows that ξ_x, and hence ω_x, will remain essentially constant.

The second and third equations are coupled first-order equations that can be written as

$$\frac{d}{dt}\left\{ \begin{array}{c} I_{yy}\xi_y \\ I_{zz}\xi_z \end{array} \right\} = \left[\begin{array}{cc} 0 & (I_{xx} - I_{zz})\omega_o \\ (I_{yy} - I_{xx})\omega_o & 0 \end{array} \right] \left\{ \begin{array}{c} \xi_y \\ \xi_z \end{array} \right\}$$

These can be uncoupled to give second-order equations such as

$$I_{yy}\ddot{\xi}_y + (I_{xx} - I_{zz})(I_{yy} - I_{xx})\omega_o^2\xi_y = 0$$

which we recognize as the equation of a linear oscillator. Note, however, that the "stiffness" term can be positive or negative depending on the relative magnitudes of the inertias.

We will deal with the first-order equations. Since the equations have constant coefficients, we seek solutions of the form

$$\left\{ \begin{array}{c} \xi_y \\ \xi_z \end{array} \right\}(t) = \left\{ \begin{array}{c} \xi_y \\ \xi_z \end{array} \right\}_o e^{i\mu t}$$

On substituting, this leads to the homogeneous equation

$$\left[\begin{array}{cc} I_{yy}i\mu & (I_{xx} - I_{zz})\omega_o \\ (I_{yy} - I_{xx})\omega_o & I_{zz}i\mu \end{array} \right] \left\{ \begin{array}{c} \xi_y \\ \xi_z \end{array} \right\}_o = 0$$

To have a nontrivial solution, the determinant must be zero; this will determine the allowable values for μ. Multiply out to get

$$\mu = \pm\omega_o \sqrt{\frac{(I_{xx} - I_{yy})(I_{xx} - I_{zz})}{I_{yy}I_{zz}}}$$

As long as $I_{xx} > I_{yy}, I_{zz}$ or $I_{xx} < I_{yy}, I_{zz}$ then μ is real and we have oscillatory behavior for ξ_y, ξ_z and similarly for ω_y, ω_z. The frequency of the oscillation is dependent on the magnitude of ω_o. This is shown in Figure 5.10(a) for the parameters

$$I_{xx} = 1, \qquad I_{yy} = 2, \qquad I_{zz} = 3$$

and the ratio of the major impulse to the other two is $5 : 1$. From the figure we get that

$$\mu = \pm\omega_o/\sqrt{3} \approx \pm 4.6/\sqrt{3} = 2.66 = \omega$$

This corresponds to a period of $T = 2\pi/\omega = 2.36$, which is approximately as shown in the figure.

It is worth noting that if I_{xx} is neither the maximum nor minimum inertia, then μ is imaginary leading to exponentially increasing solutions. As shown in Figure 5.10(b) this does not actually occur. What takes place is that, as predicted, ω_y and ω_z do indeed increase, but at a certain stage they are no longer small as assumed in the analysis above and we have a fully nonlinear situation. As shown in the previous section, the nonlinearities will prevent an infinite value occurring.

5.4   Nonlinear Dynamics and Chaotic Motions

The previous sections highlighted some of the differences between the vibration of linear and nonlinear systems. This section attempts to place these ideas in a broader dynamics context.

A characteristic of nonlinear dynamic systems is that a change in initial conditions can give quite different steady-state behavior. Nonetheless, certain aspects of the long-run behavior of such systems show systematic characteristics with a certain degree of independence of the initial conditions and parameter values. This is what we want to look at in this section.

Phase Space Plots

When the differential equations are not explicitly dependent on time, they are called *autonomous*; while if time does appear explicitly, they are called *nonautonomous*. Equations with a time-dependent forcing term are examples of nonautonomous systems. However, such a system can always be converted to an autonomous one by increasing the order of the system. For example, the driven pendulum

$$ML\frac{d^2\theta}{dt^2} + C\frac{d\theta}{dt} + Mg\sin\theta = \hat{P}\sin(\omega t)$$

can be converted to the system

$$\dot{x}_1 = x_2$$
$$\dot{x}_2 = -cx_2 - \omega_o^2\sin x_1 + \tilde{P}\sin x_3$$
$$\dot{x}_3 = \omega$$

where we have introduced three new variables: $x_1 = \theta$, $x_2 = d\theta/dt$, $x_3 = \omega t$; and the parameters $c = C/ML$, and $\omega_o^2 = g/L$.

The *phase space* of a dynamical system is a mathematical space with orthogonal coordinates representing each of the variables necessary to specify the instantaneous state of the system. A particle moving in one dimension is described by its position and velocity. A particle moving in three dimensions would have a six-dimensional phase space with three position and three velocity coordinates. If a forcing term is present, then the dimension of the phase space is increased by one.

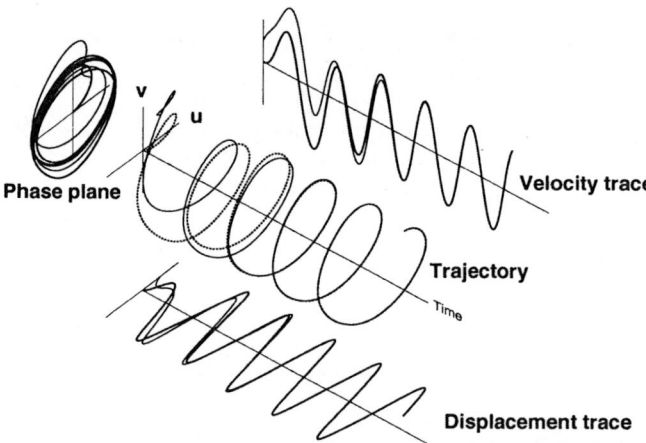

Figure 5.11: Phase space plots for a driven pendulum.

The path in phase space followed by a system as it evolves in time is called an *orbit* or *trajectory*. An example of trajectories in three-dimensional phase space is shown in Figure 5.11 for the pendulum. The two-dimensional projections of the three-dimensional plot onto the different planes are collectively called phase-plane plots; two we recognize as the displacement and velocity traces, the third (which shows the motion without time) is usually called the phase-plane plot. These plots show the state of the pendulum during the initial transient period as well as its long-term behavior. In this case, each trajectory starts at a different initial condition but limits to the same closed orbit. This closed orbit is an *attractor* and is called a *limit cycle*.

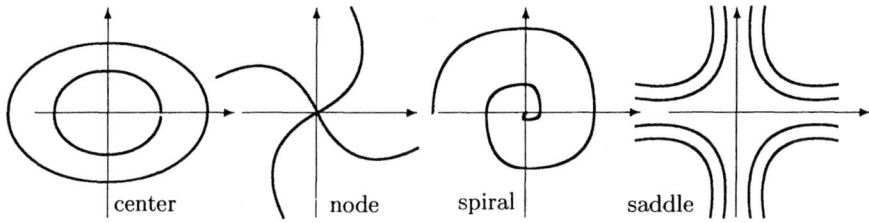

Figure 5.12: Classification of some equilibrium points.

While the transient portion of the motion is usually very complex, sometimes the long-term behavior exhibits a degree of regularity as above. In these cases, we can ignore time and the position/velocity projection contains the information we require. In the cases we will be looking at, we will generally not plot the initial transient and focus only on the long-term behavior.

Some special phase plane plots are shown in Figure 5.12. The center is always stable. The spiral is stable when the trajectories go toward the origin and is typified by a damped oscillator. The stable node (focus) is an overdamped oscillator. The saddle point is unstable.

Chaotic Motions

The central characteristic of chaotic systems is that its motion does not repeat its past behavior even approximately. Despite this lack of regularity, they do follow deterministic equations. A chaotic system resembles a stochastic system (a system subjected to random external excitations); however, the source of the irregularity is quite different. For chaotic motions, the irregularity is part of the intrinsic dynamics of the system, not unpredictable outside influences.

The phase-plane behavior of a chaotic system is shown in Figure 5.13(a), where the individually computed state points (not connected) are shown. Apparently, the solution is "all over the place." Its central characteristic is that the system does not repeat its past behavior even approximately.

Necessary conditions for chaotic motion [6] are that the system have at least three degrees of freedom (or two with a forcing term) and have a nonlinearity that couples at least two of the variables. In the case of the pendulum, we see

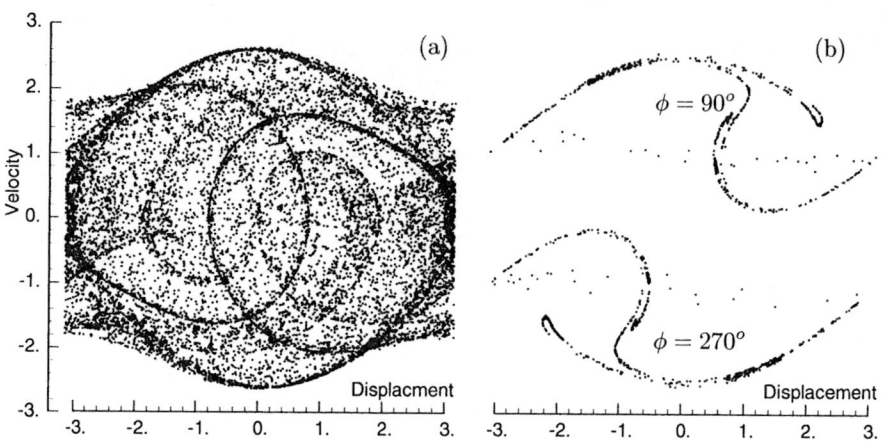

Figure 5.13: Chaotic motion and two Poincare sections.

that the nonlinearity $\sin x_1$ couples x_1 and x_2. For the dynamical systems that have a great number of degrees of freedom, we see that chaotic motion is always quite possible.

The search for periodicities can be assisted by plotting Poincare maps appropriate to the period being looked for: with a forcing frequency of ω, then $2\pi/\omega$ for the forcing frequency, $4\pi/\omega$ for the doubling frequency, and so on. It is constructed by viewing the phase space diagram stroboscopically in such a way that the motion is observed periodically. The long-term behavior in Figure 5.11, for example, would appear as a single dot. If a motion whose natural frequency is ω is strobed at a frequency of 2ω, the Poincare section would have two points. The initial transients are discarded when making the Poincare sections.

When we apply this procedure to chaotic output, we find that there is some degree of underlying structure as shown in Figure 5.13(b). The shape of the Poincare sections varies with the phase at which they are taken — the figure shows sampling at two different phases. There are no fixed points observed in the chaotic regime; instead the set of Poincare section points plays a similar role. The set, to a large extent, is independent of the initial conditions in that it has the property that any point once in the set generates a sequence of first returns all of which lie in the set. This is described as a *strange attractor*.

Bifurcation Diagrams

Consider the simple example of a linear oscillator with a displacement-dependent "gain" G,

$$M\ddot{u} + Ku = Gu + P(t)$$

The free vibration $e^{i\mu t}$ response has the roots

$$-\mu^2 M + K = G \qquad \text{or} \qquad \mu = \sqrt{\frac{K-G}{M}}$$

Clearly the nature of the response changes when G becomes equal to and exceeds K; that is, the behavior changes from being oscillatory to being exponentially increasing. This is an example of a *static* instability.

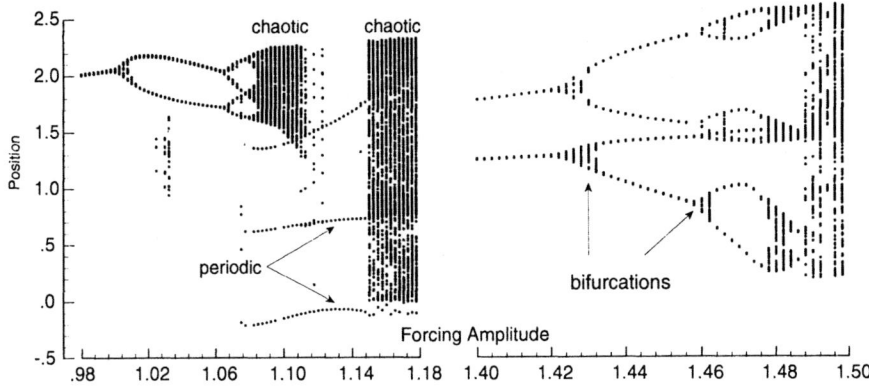

Figure 5.14: Bifurcation plots for the driven pendulum.

In statics and dynamics, a change in the number of solutions to a differential equation can occur as a parameter is varied; this is called a *bifurcation*. This passage from one set of responses to another often occurs very suddenly or "catastrophically." A static example is the buckling of a column: initially there is only axial compression, but after the bifurcation there is both axial and transverse displacements.

Figure 5.14 shows a bifurcation plot for the nonlinear driven pendulum as the amplitude parameter is changed. Note that it alternately goes through regions of periodic behavior and chaos. The regions of periodic behavior exhibit a number of bifurcations. To get these plots, we set the forcing amplitude and them sample the long-term steady state-response as a Poincare plot. Thus the dots on each vertical segment (at a given amplitude) required a complete time trace analysis.

We associate limit and bifurcation points with instabilities of the system and leave their fuller discussion until the next two chapters; the next example, however, illustrates some of the dynamics issues.

Example 5.8: A rigid pendulum rotates about a vertical axis with constant angular velocity Ω and is elastically constrained by a spring as shown in Figure 5.15. The massless spring is assumed to rotate with the mass about the vertical axis. Investigate the bifurcation behavior as the rotational speed is changed.

The mass has a velocity component $L\dot{\theta}$ about the pendulum axis, and a component $\Omega L \sin \theta$ about the axle, hence the kinetic energy is

$$T = \tfrac{1}{2}M[L\dot{\theta}]^2 + \tfrac{1}{2}M[\Omega L \sin \theta]^2$$

The total potential is given by

$$\Pi = U + V = \tfrac{1}{2}K[L \sin \theta]^2 + MgL[1 - \cos \theta]$$

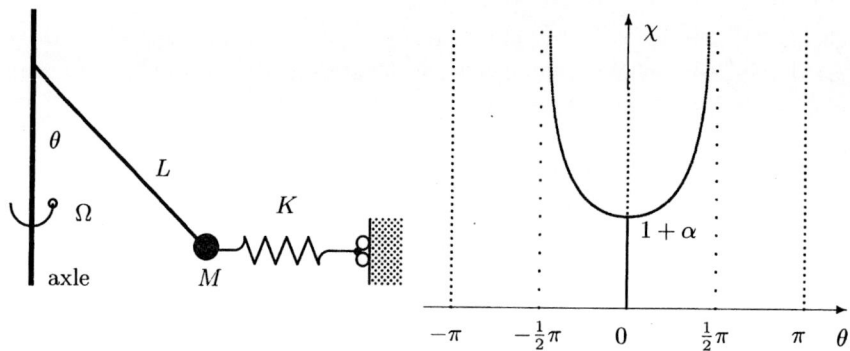

Figure 5.15: Static equilibrium paths.

Substituting these into Lagrange's equation

$$\mathcal{F}_\theta = \frac{d}{dt}\{\frac{\partial T}{\partial \dot\theta}\} - \{\frac{\partial T}{\partial \theta}\} + \{\frac{\partial (U+V)}{\partial \theta}\} = 0$$

leads to

$$\ddot\theta = \frac{g}{L}[\chi \cos\theta - 1 - \alpha \cos\theta]\sin\theta \equiv \mathcal{F}_s(\theta;\chi), \qquad \chi \equiv \frac{L\Omega^2}{g}, \qquad \alpha \equiv \frac{KL}{Mg}$$

We will identify χ as a "loading" parameter in the sense that we are interested in the change of solutions as χ is increased.

The static equilibrium paths ($\ddot\theta = 0$) are given by

$$\mathcal{F}_s(\theta;\chi) = \frac{g}{L}[\chi \cos\theta - 1 - \alpha \cos\theta]\sin\theta = 0$$

This has two sets of solutions

$$\sin\theta = 0, \qquad \chi \cos\theta - 1 - \alpha \cos\theta = 0$$

The first is independent of χ and gives

$$\theta = 0, \ \pm\pi, \ \pm 2\pi, \dots$$

This is called the *fundamental path*, and corresponds to when the pendulum is in the vertical alignment. The other solution gives

$$\chi = \frac{1 + \alpha \cos\theta}{\cos\theta}$$

Both solutions are shown plotted in Figure 5.15. The two solutions intersect at $\theta = 0$ when $\chi = 1 + \alpha$ leading to a bifurcation or two possible equilibrium states. Bifurcations are an important aspect of stability analysis, and we consider them in greater detail in Chapters 6 and 7.

To draw the phase-plane plots, we need to determine the velocity. Second-order systems always have a first integration, which can be obtained from

$$\ddot\theta = \frac{d}{dt}(\dot\theta) = \frac{d(\dot\theta)}{d\theta}\frac{d\theta}{dt} = \dot\theta\frac{d\dot\theta}{d\theta} = \frac{1}{2}\frac{d}{d\theta^2}(\dot\theta) = \mathcal{F}(\theta;\chi)$$

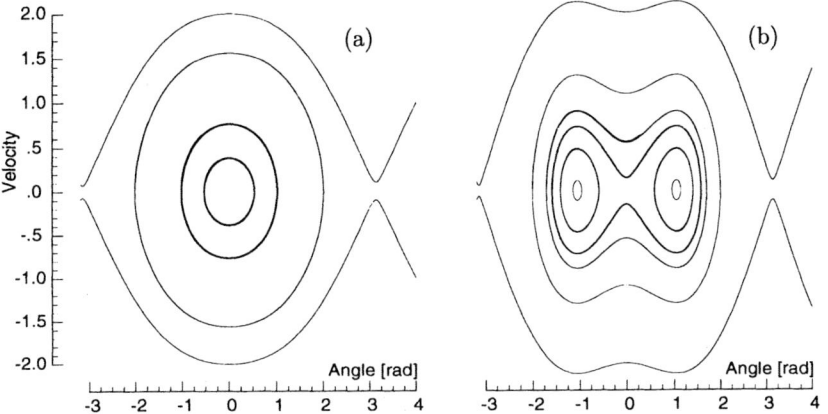

Figure 5.16: Contours of equal energy for two load levels. (a) $\chi = 0.5$. (b) $\chi = 2.0$.

Integrating both sides with respect to θ gives

$$\dot{\theta}^2 = \frac{2g}{L}[-\tfrac{1}{2}\chi \cos\theta + 1 + \tfrac{1}{2}\alpha \cos\theta]\cos\theta$$

The phase-plane plots are shown in Figure 5.16 for two values of χ. For $\chi < 1+\alpha$, there is a single center that is stable. For $\chi > 1 + \alpha$, there are two centers, both stable, which are separated by an unstable saddle point located at the fundamental path.

Small variations in the loading parameter near the bifurcation value can cause the system to flip-flop between having a single center and having two centers, with a resulting chaotic motion. This is demonstrated next.

Example 5.9: Use numerical methods to demonstrate the effect of the load level on the free vibration response.

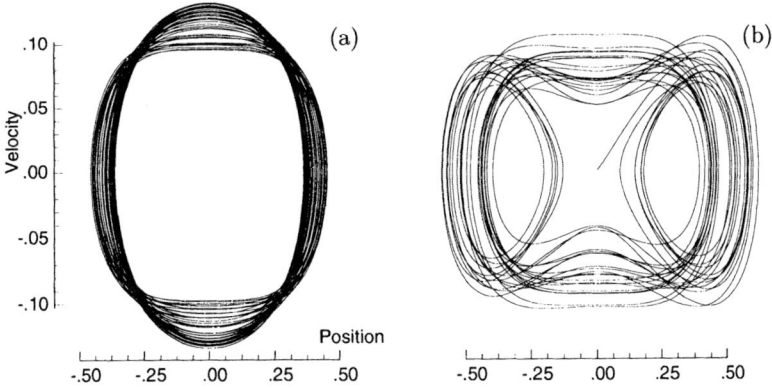

Figure 5.17: Effect of load level on the free vibration response. (a) $\chi_o = 1.95$. (b) $\chi_o = 2.05$.

We use the parameters $\alpha = 1$, $g/L = 1$, and take the load level as

$$\chi = \chi_o[1 + 0.05 \sin \omega t], \qquad \omega = 0.1$$

The static bifurcation occurs when $\chi = 1 + \alpha = 2$. This variation of the loading makes the problem a parametric excitation problem.

Figure 5.17 shows the phase-plane plots for two values of χ_o. Case (b) is initially unstable and therefore the final trajectories are independent of the initial velocities. Case (a), on the other hand, is stable and the radius of the trajectories does depend on the initial velocity; therefore, the initial velocity of Case (a) was adjusted so that the two phase-plane plots are about the same size.

For χ greater than the static bifurcation, the trajectories show a chaotic behavior. We can identify the two centers of Figure 5.16(b), but there also seems to be the single center of Figure 5.16(a). The response is such that it dwells periodically in orbits around each of the centers. The response seems to have three attractors.

5.5   Time Integration of Linear Systems

It is clear from the foregoing sections that the only feasible way to generate solutions for general nonlinear systems is by numerical methods. This section introduces some (direct) time integration methods for finding the dynamic response of structures; that is, the dynamic equilibrium equations are integrated directly in a step-by-step fashion. We operate with the full structural matrices in the general linear form

$$[\,K\,]\{u\} + [\,C\,]\{\dot{u}\} + [\,M\,]\{\ddot{u}\} = \{P\}$$

The nonlinear systems are treated in later sections,

The equations for direct integration are either *explicit* or *implicit*. In the explicit case, the equations of motion are written at the current time and as a result neither $\{u\}$ nor $\{\dot{u}\}$ at the current time is a function of the acceleration $\{\ddot{u}\}$ at the next time $t + \Delta t$. In the implicit case, on the other hand, the equations of motion are used at the next time, $t + \Delta t$. An example of both will be given.

Central Difference Method

To construct the central difference algorithm, we begin with finite difference expressions for the nodal velocities and accelerations at the current time t

$$\{\dot{u}\}_t = \frac{1}{2\Delta t}\{u_{t+\Delta t} - u_{t-\Delta t}\}$$

$$\{\ddot{u}\}_t = \frac{1}{\Delta t^2}\{u_{t+\Delta t} - 2u_t + u_{t-\Delta t}\} \tag{5.3}$$

Substitute these into the equations of motion written at time t to get

$$[\,K\,]\{u\}_t + [\,C\,]\frac{1}{2\Delta t}\{u_{t+\Delta t} - u_{t-\Delta t}\} + [\,M\,]\frac{1}{\Delta t^2}\{u_{t+\Delta t} - 2u_t + u_{t-\Delta t}\} = \{P\}_t$$

We can rearrange this equation so that only quantities evaluated at time $t + \Delta t$ are on the left-hand side

$$\left[\frac{1}{2\Delta t}C + \frac{1}{\Delta t^2}M \right] \{u\}_{t+\Delta t} = \{P\}_t \tag{5.4}$$

$$- \left[K - \frac{2}{\Delta t^2}M \right] \{u\}_t - \left[\frac{1}{\Delta t^2}M - \frac{1}{2\Delta t}C \right] \{u\}_{t-\Delta t}$$

This scheme is therefore explicit. Note that the stiffness is on the right-hand side in the effective load vector; therefore, these equations cannot recover the static solution in the limit of very slow loading. Furthermore, if the mass matrix is not positive definite (that is, if it has some zeros on the diagonal), then the scheme does not work because the square matrix on the left-hand side is not invertible in general.

The algorithm for the step-by-step solution operates as follows: We start at $t = 0$, initial conditions prescribe $\{u\}_0$ and $\{\dot{u}\}_0$, from these and the equation of motion, we find the acceleration $\{\ddot{u}\}_0$ if it is not prescribed. These equations also yield the displacements $\{u\}_{-\Delta t}$ needed to start the computations; that is, from the differences for the velocity and acceleration we get

$$\{u\}_{-\Delta t} = \{u\}_0 - (\Delta t)\{\dot{u}\}_0 + \tfrac{1}{2}(\Delta t)^2\{\ddot{u}\}_0$$

The set of Equations (5.3) and (5.4) are then used repeatedly; the equation of motion gives $\{u\}_{\Delta t}$, then the difference equations gives $\{\ddot{u}\}_{\Delta t}$ and $\{\dot{u}\}_{\Delta t}$, and then the process is repeated. We defer describing the full algorithm until we have considered the nonlinear case.

The solution is inexpensive if the mass and damping matrices are diagonal. This is a significant advantage. The computational cost, in general, is approximately

$$\text{cost} = \tfrac{1}{2}NB_m^2 + [2N(2B_m - 1) + N2B]q$$

where q is the number of time increments, B is the semi-bandwidth of the stiffness matrix, and B_m is the semi-bandwidth of the mass (and damping) matrix. When the mass matrix is diagonal, this reduces to

$$\text{cost} = 2NBq$$

The cost is linear in the number of time steps.

Example 5.10: Use the central difference scheme to determine the free response of the system
$$Ku + M\ddot{u} = P(t) = 0$$
with $K = 1$, $M = 1$, and $u(0) = 1$.

The recurrence relations become

$$\frac{M}{\Delta t^2}\{u\}_{t+\Delta t} = -[K - \frac{2M}{\Delta t^2}]\{u\}_t - \frac{M}{\Delta t^2}\{u\}_{t-\Delta t}$$

Substitute for the mass and stiffness and using a time step of $\Delta t = 1$ we get

$$\{u\}_{t+\Delta t} = \{u\}_t - \{u\}_{t-\Delta t}$$

We first determine the starter values from

$$\{u\}_{-\Delta t} = \{u\}_0 - (\Delta t)\{\dot{u}\}_0 + \tfrac{1}{2}(\Delta t)^2\{\ddot{u}\}_0 = 1 - 0 + \tfrac{1}{2}1 = \tfrac{3}{2}$$

where the acceleration was obtained from the equation of motion. The sequence of values is

$$u_1 = +1.0 - 0.5 = +0.5$$
$$u_2 = +0.5 - 1.0 = -0.5$$
$$u_3 = -0.5 - 0.5 = -1.0$$
$$u_4 = -1.0 + 0.5 = -0.5$$
$$u_5 = -0.5 + 1.0 = +0.5$$
$$u_6 = +0.5 + 0.5 = +1.0$$
$$u_7 = +1.0 - 0.5 = +0.5$$
$$u_8 = +0.5 - 1.0 = -0.5$$

The full sequence is shown plotted in Figure 5.18(a) along with the exact solution. Note that the amplitude is correct, but there is an increasing phase difference. This is an order $O(\Delta t^2)$ algorithm, hence halving the time step will decrease the error by a factor of 4.

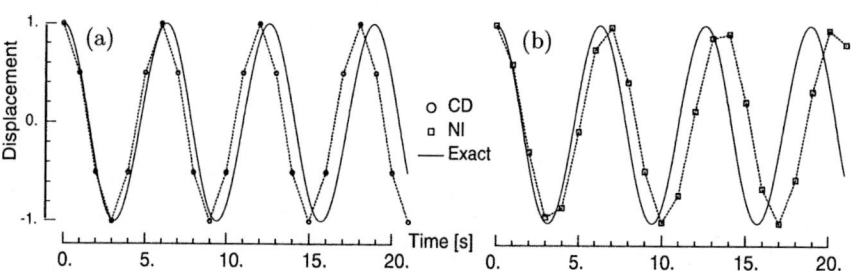

Figure 5.18: Time integration schemes. (a) Central difference (CD) scheme. (b) Newmark implicit (NI) scheme.

Constant Average Acceleration Method

We will now derive a different integration scheme by considering the equations of motion at the time step $t + \Delta t$. Assume that the acceleration is constant over the small time step Δt and given by its average value. That is,

$$\ddot{u}(t) \approx \tfrac{1}{2}(\ddot{u}_t + \ddot{u}_{t+\Delta t}) = \text{constant} = \alpha$$

Integrate this to give the velocity and displacement as

$$\begin{aligned} \dot{u}(t) &= \dot{u}_t + \alpha t \\ u(t) &= u_t + \dot{u}_t t + \tfrac{1}{2}\alpha t^2 \end{aligned}$$

We estimate the average acceleration by evaluating the displacement at time $t + \Delta t$ to give

$$\tfrac{1}{2}(\ddot{u}_t + \ddot{u}_{t+\Delta t}) = \alpha = \frac{2}{\Delta t^2}\{u_{t+\Delta t} - u_t - \dot{u}_t \Delta t\}$$

These equations can be rearranged to give difference formulas for the new acceleration and velocity (at time $t + \Delta t$) in terms of the new displacement as

$$\begin{aligned} \{\ddot{u}\}_{t+\Delta t} &= 2\{\alpha\} - \{\ddot{u}\}_t = \frac{4}{\Delta t^2}\{u_{t+\Delta t} - u_t\} - \frac{4}{\Delta t}\{\dot{u}\}_t - \{\ddot{u}\}_t \\ \{\dot{u}\}_{t+\Delta t} &= \{\alpha\}\Delta t + \{\dot{u}\}_t = \frac{2}{\Delta t}\{u_{t+\Delta t} - u_t\} - \{\dot{u}\}_t \end{aligned} \tag{5.5}$$

Substitute these into the equations of motion at the new time $t + \Delta t$ to obtain the implicit scheme

$$\begin{aligned} [\,K\,]\{u\}_{t+\Delta t} &+ [\,C\,]\left\{\frac{2}{\Delta t}\{u_{t+\Delta t} - u_t\} - \{\dot{u}\}_t\right\} \\ &+ [\,M\,]\left\{\frac{4}{\Delta t^2}\{u_{t+\Delta t} - u_t\} - \frac{4}{\Delta t}\{\dot{u}\}_t - \{\ddot{u}\}_t\right\} = \{P\}_{t+\Delta t} \end{aligned}$$

All terms that have been evaluated at time t are now shifted to the right-hand side. The rearranged equations of motion are then

$$\begin{aligned} \left[K + \frac{2}{\Delta t}C + \frac{4}{\Delta t^2}M\right]\{u\}_{t+\Delta t} &= \{P\}_{t+\Delta t} + [\,C\,]\left\{\frac{2}{\Delta t}u + \dot{u}\right\}_t \\ &+ [\,M\,]\left\{\frac{4}{\Delta t^2}u + \frac{4}{\Delta t}\dot{u} + \ddot{u}\right\}_t \end{aligned} \tag{5.6}$$

The new displacements are obtained by solving this system of simultaneous equations. This scheme is a special case of Newmark's method [7].

The algorithm operates as follows: We start at $t = 0$, initial conditions prescribe $\{u\}_0$ and $\{\dot{u}\}_0$. From these and the equations of motion (written at time $t = 0$) we find $\{\ddot{u}\}_0$ if it is not prescribed. Then the above system of equations are solved for the displacement $\{u\}_{\Delta t}$, from which estimates of the accelerations $\{\ddot{u}\}_{\Delta t}$ and the velocities $\{\dot{u}\}_{\Delta t}$ can also be obtained. These are used to obtain current values of the right-hand side. Then solving the equation of motion again yields $\{u\}_{2\Delta t}$, and so on. The solution procedure for $\{u\}_{t+\Delta t}$ is not trivial, but the coefficient matrix need be reduced to $[\,U\,]^T\lfloor\,D\,\rfloor[\,U\,]$ form only once if Δt and all of the system matrices do not change during the integration. Note that in the limit of large ΔT we recover the static solution. We again defer describing the full algorithm until we have considered the nonlinear case.

The computational cost is approximately

$$\text{cost} = \tfrac{1}{2}NB^2 + [2NB + 2N(2B_m - 1)]q$$

where the first term is the cost of the matrix decomposition. When the mass matrix is diagonal, this reduces to

$$\text{cost} = \tfrac{1}{2}NB^2 + 2NBq$$

Except for the cost of the initial decomposition, the total cost is the same as for the central difference method.

Example 5.11:   Use the constant acceleration scheme to determine the free response of the system

$$Ku + M\ddot{u} = P(t) = 0$$

with $K = 1$, $M = 1$, and $u(0) = 1$.

The recurrence relations become

$$[K + \frac{4M}{\Delta t^2}]\{u\}_{t+\Delta t} = M[\frac{4}{\Delta t^2}]\{u\}_t + \frac{4}{\Delta t}]\{\dot{u}\}_t + \{\ddot{u}\}_t]$$

Substitute for the mass and stiffness and using a time step of $\Delta t = 1$, we get

$$5\{u\}_{t+\Delta t} = \tfrac{1}{5}[4\{u\}_t + 4\{\dot{u}\}_t + \{\ddot{u}\}_t]$$

The updating is done as

$$\begin{aligned}
\{\ddot{u}\}_{t+\Delta t} &= 4\{u_{t+\Delta t} - u_t\} - 4\{\dot{u}\}_t - \{\ddot{u}\}_t \\
\{\dot{u}\}_{t+\Delta t} &= 2\{u_{t+\Delta t} - u_t\} - \{\dot{u}\}_t
\end{aligned}$$

Again the acceleration was obtained from the equation of motion.

The sequence of values is

$$\begin{aligned}
u_1 &= [4 \times 1.0 + 4 \times 0.0 - 1.0]/5 = 0.6 \\
u_2 &= [4 \times 0.6 + 4 \times -0.8 - 0.6]/5 = -0.28 \\
u_3 &= [4 \times -0.28 + 4 \times -0.96 + 0.28]/5 = -0.94 \\
u_4 &= [4 \times -0.94 + 4 \times -0.35 + 0.94]/5 = -0.84 \\
u_5 &= [4 \times -0.84 + 4 \times 0.54 + 0.84]/5 = -0.76 \\
u_6 &= [4 \times -0.76 + 4 \times 1.0 + 0.76]/5 = 0.75 \\
u_7 &= [4 \times 0.75 + 4 \times 0.66 - 0.75]/5 = 0.98 \\
u_8 &= [4 \times 0.98 + 4 \times -0.21 - 0.98]/5 = 0.42
\end{aligned}$$

The full sequence is shown plotted in Figure 5.18(b) along with the exact solution. Note that the amplitude is correct but there is an increasing phase difference. Just like the central difference method, this is an order $O(\Delta t^2)$ algorithm, hence halving the time step will decrease the error by a factor of 4.

Numerical Stability

Numerical integration schemes are susceptible to instabilities, a symptom of which is that the solution diverges at each time step. We now analyze both algorithms in this respect.

A point to note is that, if the system of equations are integrated directly, then, since we are dealing with linear systems, the same results are obtained if a modal transformation is first performed, the integration done numerically, and the physical responses reconstructed. Therefore, to study the accuracy of the direct integration methods, we may focus attention only on integrating a single modal equation. In this way, the only variables to be considered are Δt, ω_m, and ζ_m. Furthermore, because all equations are similar, we need only study the integration of a typical one given by

$$\ddot{u} + 2\zeta\omega\dot{u} + \omega^2 u = p$$

where ω is the modal frequency and ζ is the modal damping.

To investigate the central difference algorithm, we begin with finite difference expressions in time for modal velocities and accelerations at the current time t as given in Equation (5.3) and substitute these into the equations of motion, written at time t, to get

$$(1 + \zeta\omega\Delta t)u_{n+1} - (2 - \omega^2\Delta t^2)u_n + (1 - \zeta\omega\Delta t)u_{n-1} = p_n\Delta t^2$$

By appending the identity $(1 + \zeta\omega\Delta t)u_n = (1 + \zeta\omega\Delta t)u_n$, we can write the recurrence relation as

$$\begin{Bmatrix} u_{n+1} \\ u_n \end{Bmatrix} = \frac{1}{(1 + \zeta\xi)} \begin{bmatrix} 2 - \xi^2 & -1 + \zeta\xi \\ 1 + \zeta\xi & 0 \end{bmatrix} \begin{Bmatrix} u_n \\ u_{n-1} \end{Bmatrix} + \begin{Bmatrix} p_n\Delta t^2 \\ 0 \end{Bmatrix}$$

where $\xi \equiv \omega\Delta t$. We seek solutions of the homogeneous problem in the form

$$\{u\}_n = \{u\}_0 e^{i\lambda t} = \{u\}_0 e^{i\lambda n\Delta t} = \{u\}_0 (e^{i\lambda\Delta t})^n = \{u\}_0 \rho^n$$

Substitute this to give the characteristic equation associated with this difference equation as

$$\rho^2 - \rho\frac{(2 - \xi^2)}{(1 + \zeta\xi)} + \frac{(1 - \zeta\xi)}{(1 + \zeta\xi)} = 0$$

The roots are

$$\rho = \left[(2 - \xi^2) \pm \sqrt{\xi^4 - 4\xi^2(1 - \zeta^2)}\right]/(2 + 2\zeta\xi) = Ae^{i\phi}$$

The amplitude and phase are given by

$$A = \sqrt{\frac{1 - \zeta\omega\Delta}{1 + \zeta\omega\Delta t}}, \qquad \tan\phi = \frac{\pm\omega\Delta t\sqrt{4(1 - \zeta^2) - \omega^2\Delta t^2}}{(2 - \omega^2\Delta t^2)}$$

When damping is negligible, these reduce to

$$A = 1, \qquad \tan\phi = \frac{\pm\omega\Delta t\sqrt{1 - \omega^2\Delta t^2/4}}{(1 - \omega^2\Delta t^2/2)}$$

indicating there are no amplitude errors but there is a phase advance. A plot of the amplitude A and phase-related term ϕ/ξ are given in Figure 5.19. Note the abrupt change at $\xi = \omega\Delta t = 2$.

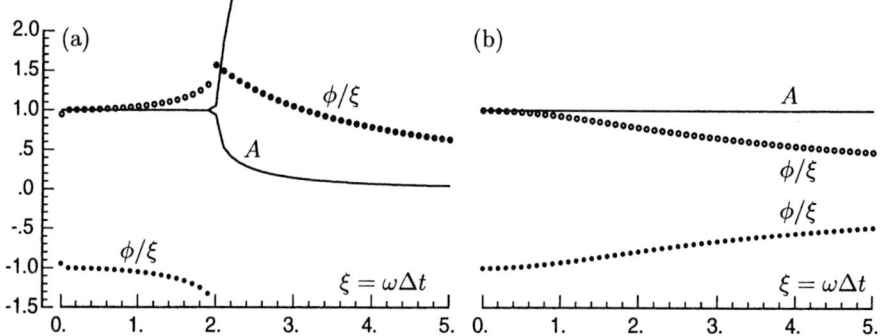

Figure 5.19: Amplitude and phase behaviors of the two algorithms. (a) Central difference. (b) Newmark implicit.

We need to have oscillating solutions (because of our second-order system), hence the radical must be negative. Thus, we require that

$$\xi^4 - 4\xi^2(1 - \zeta^2) < 0 \qquad \text{or} \qquad \xi = \omega\Delta t < 2\sqrt{1 - \zeta^2}$$

Hence the method is only conditionally stable, since it is possible for this criterion not to be satisfied in some circumstances. When there is no damping, this simplifies to

$$\xi = \omega\Delta t < 2 \qquad \text{or} \qquad \Delta t < \frac{2}{\omega} = \frac{T}{\pi}$$

where T is the period associated with the frequency ω. Thus, for numerical stability, the step size must be less than one-third the period. This seems easily achieved, since it is generally considered [12, 18] that the step size should be less than one-tenth the period for an accurate solution. As we will see for multiple degree of freedom systems, this is not so straightforward.

A similar analysis for the average acceleration method gives

$$\left\{ \begin{array}{c} u \\ \dot{u}\Delta t \\ \ddot{u}\Delta t^2 \end{array} \right\}_{n+1} = \frac{1}{D} \left[\begin{array}{ccc} 1 + \zeta\xi & 1 + \frac{1}{2}\zeta\xi & \frac{1}{4} \\ -\frac{1}{2}\xi^2 & 1 - \frac{1}{4}\xi^2 & \frac{1}{2} \\ -\xi^2\Delta t^2 & -(2\zeta\xi + \xi^2) & -\zeta\xi - \frac{1}{4}\xi^2 \end{array} \right] \left\{ \begin{array}{c} u \\ \dot{u}\Delta t \\ \ddot{u}\Delta t^2 \end{array} \right\}_{n}$$

where $D = 1 + \zeta\xi + \frac{1}{4}\xi^2$. We can obtain a characteristic equation for this system by assuming a solution of the form

$$\{u\} = \{C\}\rho^n$$

After substitution, this gives rise to an eigenvalue problem, the roots of which are

$$\rho = \frac{1 - \frac{1}{4}\xi^2 \pm i\xi}{1 + \frac{1}{4}\xi^2} = Ae^{i\phi}$$

for the no damping case. The magnitude and phase are given by

$$A = 1, \qquad \tan\phi = \frac{\xi}{1 - \frac{1}{4}\xi^2} = \frac{\omega\Delta t}{1 - \frac{1}{4}\omega^2\Delta t^2}$$

indicating there are no amplitude errors but there is a phase lag. A plot of the amplitude A and phase-related term ϕ/ξ are given in Figure 5.19(b). Note that both functions are monotonic over the region plotted.

We note that there is always only an imaginary part to $i\phi$, hence the solution will exhibit the desired oscillations. Consequently, we conclude that the system is *unconditionally stable*, gives no amplitude decay, but will exhibit a phase shift.

Example 5.12: Show how the phase and amplitude behaviors of the two algorithms manifest themselves for the simple oscillator.

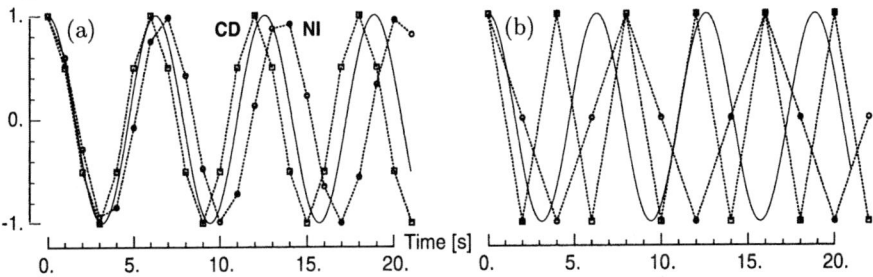

Figure 5.20: Reconstructions showing effects of phase shifts. (a) $\xi = 1.0$, (b) $\xi = 1.95$.

The results are shown in Figure 5.20 for two different values of $\xi = \omega\Delta t$. Both algorithms essentially maintain the proper amplitude; it is the phase that shows the most degradation — the two algorithms shift it in opposite directions.

The significant difference of the two algorithms is that the central difference scheme is only conditionally stable, but once $\xi < 2.0$, both algorithms give about the same accuracy.

Explicit Versus Implicit Schemes

The foregoing analysis shows that the computational cost for the explicit central difference scheme can be less than the implicit average acceleration method. Furthermore, the stability analysis shows that the criterion

$$\Delta t < \frac{2}{\omega} = \frac{T}{\pi}$$

for the explicit scheme will automatically be satisfied because of accuracy considerations. Hence, we could conclude that the explicit scheme is preferable.

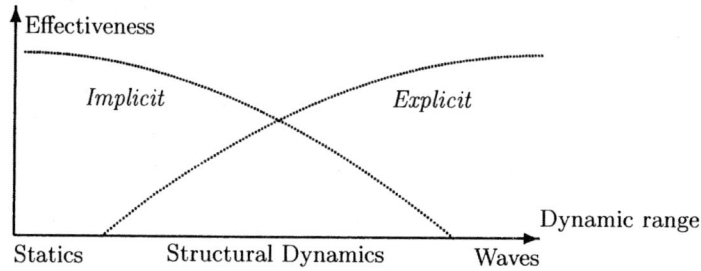

Figure 5.21: Overlap of effectiveness of implicit and explicit methods.

A very important factor was overlooked in the above: a multiple-degree-of-freedom system will have many modes and when direct integration is used, this is equivalent to integrating each mode with the same time step Δt. Therefore, the above stability criterion must be applied to the highest modal frequency of the system even if our interest is in the low-frequency response. In other words, if we energize the system in such a way as to excite only the lower frequencies, we must nonetheless choose an integration step corresponding to the highest possible mode.

The significance of this is that the matrix analyses of structures produce so-called *stiff equations*. In the present context, stiff equations characterize a structure whose highest natural vibration frequencies are much greater than the lowest. Especially stiff structures therefore include those with a very fine mesh, and a structure with near-rigid support members. If the conditionally stable algorithm is used for these structures, Δt must be very small, usually orders of magnitude smaller than for the implicit scheme.

In summary, explicit methods are conditionally stable and therefore require a small Δt but produce equations that are cheap to solve. The implicit methods are (generally) unconditionally stable and therefore allow a large Δt but produce equations that are more expensive to solve. The size of Δt is governed by considerations of accuracy rather than stability; that is, we can adjust the step size appropriate to the excitation force or the number of modes actually excited. The difference factor can be orders of magnitude and will invariably outweigh any disadvantage in having to decompose the system matrices. Based on these considerations, the implicit scheme is generally the method of choice for general structural dynamics problems that are linear and where the frequency content is relatively low. The explicit scheme is generally the method of choice for wave-type dynamics problems where the frequency content is relatively high. A schematic of this division is shown in Figure 5.21.

More discussions can be found in References [12]. Additional considerations come into play when we look at nonlinear systems and this is explored next.

5.6 Explicit Nonlinear Scheme

When inertia effects are significant in nonlinear problems, as in such problems as wave propagation, we have the choice of using either an explicit integration scheme or an implicit scheme. We will see that the explicit scheme is simpler to implement (primarily because it does not utilize the tangent stiffness matrix) and therefore is generally the more preferable of the two.

Nonlinear Formulation

Assume the solution is known at time t and we wish to find it at time $t + \Delta t$. The applied loads in this case include the inertia and damping forces. We begin with the dynamic equilibrium equation at time t

$$[\,M\,]\{\ddot{u}\}_t + [\,C\,]\{\dot{u}\}_t = \{P\}_t - \{F\}_t$$

where $\{F\}$ is the assembled vector of element nodal forces. The difference expressions for the nodal velocities and accelerations at the current time are

$$\{\dot{u}\}_t = \frac{1}{2\Delta t}\{u_{t+\Delta t} - u_{t-\Delta t}\}$$

$$\{\ddot{u}\}_t = \frac{1}{\Delta t^2}\{u_{t+\Delta t} - 2u_t + u_{t-\Delta t}\} \tag{5.7}$$

Substitute these into the equilibrium equation and rearrange so that only quantities evaluated at time $t + \Delta t$ are on the left-hand side

$$\left[\frac{1}{2\Delta t}C + \frac{1}{\Delta t^2}M\right]\{u\}_{t+\Delta t} = \{P\}_t - \{F\}_t \tag{5.8}$$

$$- \left[-\frac{2}{\Delta t^2}M\right]\{u\}_t - \left[\frac{1}{\Delta t^2}M - \frac{1}{2\Delta t}C\right]\{u\}_{t-\Delta t}$$

The initial conditions prescribe $\{u\}_0$ and $\{\dot{u}\}_0$, from these and the equation of motion, we find the acceleration $\{\ddot{u}\}_0$ if it is not prescribed. These equations also yield the displacements $\{u\}_{-\Delta t}$ needed to start the computations; that is, from the differences for the velocity and acceleration we get

$$\{u\}_{-\Delta t} = \{u\}_0 - (\Delta t)\{\dot{u}\}_0 + \tfrac{1}{2}(\Delta t)^2\{\ddot{u}\}_0$$

Generally, the mass and damping matrices are arranged to be diagonal and hence the solution for the displacements is a trivial computational effort. Furthermore, there are no equilibrium iteration loops as will be needed in the implicit schemes.

A very important feature of this scheme is that the structural stiffness matrix need not be formed [10]. That is, we can assemble the element nodal forces directly for each element. This is significant in terms of saving memory and also has the advantage that the assemblage effort can be easily distributed on a multi-processor computer. Furthermore, since the nodal forces $\{F\}$ are obtained directly for each element, there is flexibility in the types of forces that can be allowed. For example, Reference [83] uses a cohesive stress between element sides in order to model dynamic fracture of brittle materials. In this context, we can also model deformation dependent loading effects (such as follower forces or sliding friction) without changing the basic algorithm.

Nonlinear Algorithm

The basic algorithm for the central difference method can be stated as:

Step 1: Specify algorithm parameters such as number and size of time steps.

Step 2: Read the initial geometry and material.

Step 3: Initialize the triads.

Step 4: Assemble the effective inertia matrix as

$$[M_{eff}] \equiv \left[\frac{1}{2\Delta t}C + \frac{1}{\Delta t^2}M\right]$$

Step 5: Decompose the inertia matrix if necessary

$$[M_{eff}] = [\ U\ ]^T\lceil\ D\ \rfloor[\ U\ ]$$

Step 6: Specify the initial conditions for $\{u\}_0$ and $\{\dot{u}\}_0$. Obtain $\{\ddot{u}\}_0$ from the equations of motion.

Step 7: Read the load vector $\{P\}_t$. It may be necessary to interpolate this from non-equispaced values.

Step 8: *Begin loop over time increments:*

Step T.1: Assemble the nodal loads vector $\{F\}$.

Step T.2: Form the effective load vector

$$\{P_{eff}\}_t \equiv \{P\}_t - \{F\}_t - \left[-\frac{2}{\Delta t^2}M\right]\{u\}_t - \left[\frac{1}{\Delta t^2}M - \frac{1}{2\Delta t}C\right]\{u\}_{t-\Delta t}$$

Step T.3: Solve for the new displacement increments from

$$[\ U\ ]^T\lceil\ D\ \rfloor[\ U\ ]\{u\}_{t+\Delta t} = \{P_{eff}\}_t$$

Step T.4: Update the geometry, stresses and triads.

Step T.5: Store results for this time step.

Step 9: *End loop over time increments.*

Step 10: END

It is possible to enhance this algorithm by including automatic step changes, automatic testing for appropriate time step size, and so on.

Example 5.13: Use the explicit integration scheme to investigate the nonlinear dynamic response of a shallow arch uniformly loaded.

For illustrative purposes, we let the arch have a radius of 1651 mm (65 in.) and a span of 1270 mm (50 in.). This results in an initial height of the arch of 127 mm ($H = 5.0$ in.). We model the arch as a frame with elements and apply the load in a smoothed stepped fashion.

For low loads, the response is a transient vibration finally settling into a displaced position. However, as the load is increased, there is a critical value beyond which the behavior is quite different. Figure 5.22 shows the response for stepped loads of 425 and 435 that straddle the critical load.

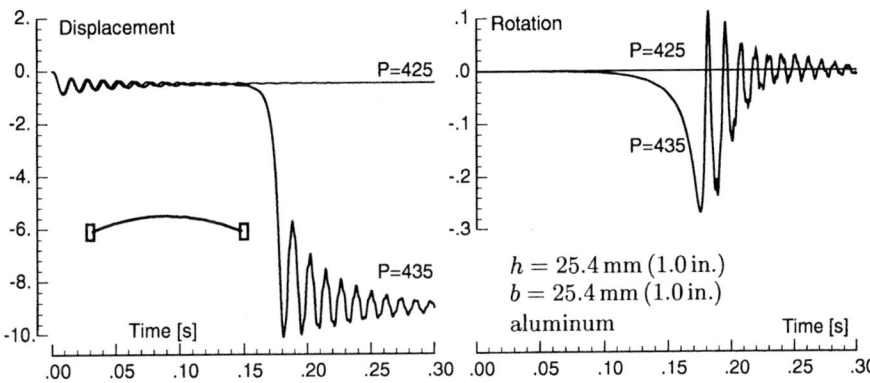

Figure 5.22: Dynamic response of the center point of a shallow arch. (a) Vertical displacement. (b) Rotation.

It is clear that the slightly higher load has caused a snap-through of the arch and the structure finds a new equilibrium position now curved downward. Damping was used to give an indication of the final settled value of the deflection, otherwise the oscillations can be large enough to cause the structure to snap through multiple times. In fact, it is interesting to note how energetic the snap-through has been.

What is quite curious about this problem is that the rotation at the center node is nonzero. To emphasize this point: the geometry is symmetric, the loading is symmetric, and therefore, we expect the deformation to be symmetric, and specifically that the rotation at the center be zero. As will be demonstrated in the next chapter, what is happening is that as the load gets close to the limit load, the tangent stiffness becomes singular. At that point there is a rigid body mode quite similar to the first buckling load which is antisymmetric. Being a rigid body mode, any slight disturbance in that direction is not resisted and a relatively large deflection can ensue. It is true that there are no applied loads in that direction but accumulated round-off errors serve the purpose and so we note the asymmetry. Figure 5.23(a) clearly shows the asymmetry during the snap-through.

To further justify our conjecture the problem was re-run, but this time with a very small asymmetric ping — two small forces acting in opposite directions located around the center point. Figure 5.23(b) shows the results as the strength of the ping is varied. A strong ping induces the jump sooner than a weak one. Indeed, we see the non-ping case as essentially being $Q \approx 10^{-5}$, which is on the order of the round-off error.

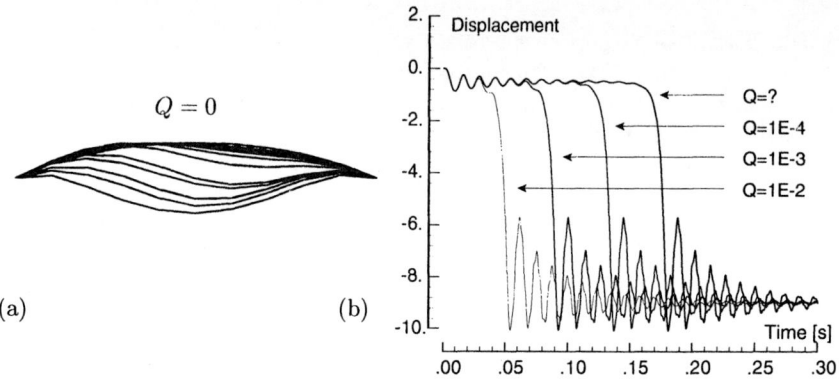

Figure 5.23: Effect of ping on the initiation of the snap-through. (a) Deformed shapes, (b) Vertical deflection.

A final point worth making is that if the primary load is made large enough, we can cause the snap-through without any appreciable rotation. This is in agreement with our conjecture, because the rotation accumulates only slowly while the primary load simply pushes the structure through. Furthermore, if we incorrectly assume symmetry and model only half of the arch, then the snap-through load obtained would be higher than the actual load.

Example 5.14:   Analyze the longitudinal impact of a rod by a sphere.

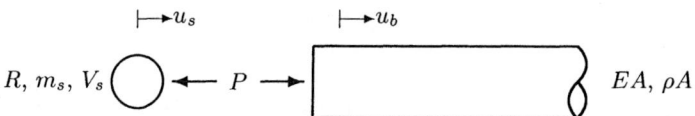

Figure 5.24: Impact of a rod with a spherical ball.

Consider a long bar impacted by a spherical ball traveling at velocity V_s. Let u_s be the motion of the center of mass of the ball and u_b be the motion of a representative point on the rod. The equation of motion of the impactor is

$$M_s \ddot{u}_s = -P$$

Assume that the impact generates a longitudinal plane wave that travels in the bar, then the velocity of our representative point is related to the force by [23]

$$\dot{u}_b = -\frac{c_o \sigma}{E} = \frac{c_o}{EA} P$$

Assume that this relation is valid even near the impact site. Let the contact between the ball and the flat end of the rod be described by the Hertzian contact law; that is, the force and relative indentation are given by [27]

$$P = K\alpha^{3/2}, \qquad \alpha = u_s - u_b, \qquad K = \tfrac{4}{3}\sqrt{R_s} \left(\frac{k_s k_b}{k_s + k_b} \right), \qquad k_i \equiv \frac{E_i}{1 - \nu_i^2}$$

This allows u_s and u_b to be obtained from the following differential equations:

$$\ddot{u}_s = -\frac{K}{M_s}(u_s - u_b)^{3/2}, \qquad \dot{u}_b = \frac{c_o K}{EA}(u_s - u_b)^{3/2}$$

with the initial conditions that at $t = 0$

$$u_s = 0, \qquad \dot{u}_s = V_s, \qquad u_b = 0$$

These equations can be programmed to allow calculation of the resulting histories.

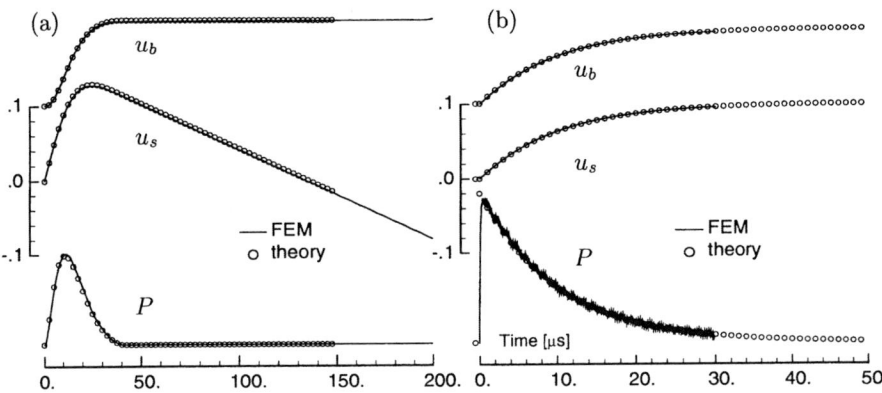

Figure 5.25: Responses for impacted rod with $V_s = 250\,\text{m/s}\,(10000\,\text{in./s})$. (a) Hertzian contact. (b) Flush contact.

The above differential equations can be combined into a single equation for the indentation as

$$M_s \ddot{\alpha} + K\alpha^{3/2} + C\alpha^{1/2}\dot{\alpha} = 0, \qquad C \equiv \frac{3c_o M_s K}{2EA}$$

with the initial conditions that at $t = 0$: $\alpha = 0$, $\dot{\alpha} = V_s$. Note that this has the form of a nonlinear spring/mass system with a nonlinear viscous damper — the apparent damping arises because the long rod is a conduit of energy out of the contact region.

The responses are shown in Figure 5.25(a). The force history is nonsymmetric. Note that the ball has a rebound velocity.

The comparison is with a rod element modeling. Incorporating the contact force within the explicit formulation is relatively straightforward because we need only modify the load vector according to

$$\{P\} - \{F\} \qquad \Longrightarrow \qquad \{P\} - \{F\} + \{F_c\}$$

where $\{F_c\}$ is the vector of contact forces computed from the current deformation state and the Hertzian contact law — in the present case there are only two nonzero components. The most significant difference between the theory and the FEM results would be the reflected waves; otherwise the comparison is almost identical.

Example 5.15:   Analyze the longitudinal impact of a rod by a small cylinder.

By way of contrast, consider the case where the impactor is cylindrical so that the contact is flush. The equations of motion are similar to the above but we impose the condition that the contacting surfaces have the same motion. That is,

$$\ddot{u}_s = \ddot{u}_b \qquad \text{or} \qquad -\frac{1}{M_s}P = \frac{c_o}{EA}\frac{dP}{dt}$$

This differential equation for $P(t)$ can be integrated to give

$$P = \frac{V_s EA}{c_o}e^{-\beta t}, \qquad u_s = u_b = \frac{V_s}{\beta}[1 - e^{-\beta t}], \qquad \beta \equiv \frac{EA}{c_o M_s}$$

What is interesting is that the maximum force occurs at $t = 0$, and it does not depend on the mass of the impactor. The reason is that the actual boundary condition is that of an imposed velocity, and this, of course, is independent of the mass. Where the mass has a big effect is on the time decay through the parameter β — the larger the mass, the flatter the force trace, and the longer the duration of the force. Thus two very long, flat ended rods, impacting collinearly, would produce a force history that is simply a step function that exists for a very long time.

The displacement and force histories are shown in Figure 5.25(b). The most significant difference in comparison to Figure 5.25(a) is the very sharp rise time. Indeed, for the FEM comparison it was necessary to refine the element size an order of magnitude. There are various ways of implementing this contact condition; the one chosen to present the results is in the form of a cohesive law where

$$\sigma_n = \gamma\sigma_c\frac{u}{\delta}\left[1 - \frac{u}{\delta}\right]^2, \qquad \sigma_c = E/10$$

and γ and δ are parameters. This has been used to model the cohesive debonding of materials; however, in the present case we only use it for the compressive behavior. Note that in problems like these, the dominant effects are associated with the dynamics of the impacting bodies, and not so much with the particulars of the contact laws. Consequently, this gives greater latitude in modeling the contact conditions.

5.7 Nonlinear Structural Dynamics

As indicated in Figure 5.21, our meaning of structural dynamics is the frequency range that begins close to zero and extends a moderate way, perhaps five to fifty of the modal frequencies. Since this encompasses nearly static problems, we see that we must use some sort of implicit integration scheme that utilizes the tangent stiffness.

Nonlinear Formulation

In our incremental solution, we assume the solution is known at time t and wish to find it at time $t + \Delta t$. As was done in the static case, we write the dynamic equilibrium equation at time $t + \Delta t$

$$[M]\{\ddot{u}\}_{t+\Delta t} + [C]\{\dot{u}\}_{t+\Delta t} = \{P\}_{t+\Delta t} - \{F\}_{t+\Delta t} \tag{5.9}$$

where $\{P\}$ are the externally applied loads and $\{F\}$ is the assembled element nodal forces. The element nodal loads are not known but can be approximated as

$$\{F\}_{t+\Delta t} \approx \{F\}_t + \{\Delta F\} = \{F\}_t + [K_T]\{\Delta u\}, \qquad [K_T] \equiv [\frac{\partial F}{\partial u}] \tag{5.10}$$

where $\{\Delta F\}$ is the increment in element nodal loads due to an increment in the displacements, and $[K_T]$ is the tangent stiffness matrix. In addition, we can approximate the accelerations and velocities using the finite differences of the constant acceleration method

$$
\begin{aligned}
\{u\}_{t+\Delta t} &= \{u\}_t + \{\Delta u\} \\
\{\dot{u}\}_{t+\Delta t} &= \{\dot{u}\}_t + \tfrac{1}{2}\Delta t \{\ddot{u}_t + \ddot{u}_{t+\Delta t}\} \\
\{\ddot{u}\}_{t+\Delta t} &= \frac{4}{\Delta t^2}(\{u\}_{t+\Delta t} - \{u\}_t) - \frac{4}{\Delta t}\{\dot{u}\}_t - \{\ddot{u}\}_t
\end{aligned}
\tag{5.11}
$$

Substituting these into the equilibrium equation allows the increment in displacements to be obtained by solving

$$\left[K_T + \frac{4}{\Delta t}C + \frac{4}{\Delta t^2}M\right]\{\Delta u\} = \{P\}_{t+\Delta t} - \{F\}_t \tag{5.12}$$

$$+ [C]\left\{\frac{2}{\Delta t}u + \dot{u}\right\}_t + [M]\left\{\frac{4}{\Delta t^2}u + \frac{4}{\Delta t}\dot{u} + \ddot{u}\right\}_t$$

Once the increment in displacement is obtained, then all variables can be updated. At this stage it is possible to move onto the next load level and repeat the above to get the next displacement increment.

It must be realized, however, that just as in the static case, if the estimates for the new displacements are substituted into the equilibrium relations of Equation (5.9), then this equation will not be satisfied because the displacements were obtained using the approximation of Equation (5.10). What we can do, however, is repeat the above process at the same load level until we get convergence. That is, noting

$$u_{t+\Delta t} = u_{t+\Delta t}^i + \Delta u_{t+\Delta t}^i, \qquad \{F\}_{t+\Delta t} \approx \{F\}_{t+\Delta t}^i + [K_T]^i\{\Delta u\}^i$$

and substituting along with Equation (5.11) into Equation (5.9), we obtain the iterative increments of displacement as

$$\left[K_T + \frac{4}{\Delta t}C + \frac{4}{\Delta t^2}M\right]^{i-1}\{\Delta u\}^i = \{P\}_{t+\Delta t} - \{F\}_{t+\Delta t}^{i-1} \tag{5.13}$$

$$+[\,C\,]\left\{\frac{-2}{\Delta t}(u_{t+\Delta t}^{i-1}-u_t)+\dot{u}\right\}_t$$

$$+[\,M\,]\left\{\frac{-4}{\Delta t^2}(u_{t+\Delta t}^{i-1}-u_t)+\frac{4}{\Delta t}\dot{u}+\ddot{u}\right\}_t$$

This is repeated using the updates

$$\{u\}_{t+\Delta t}^i=\{u\}_{t+\Delta t}^{i-1}+\{\Delta u\}^i,\quad [\,K\,]_{t+\Delta t}^i=[\,K_t\,](\{u\}_{t+\Delta t}^i),\quad \{F\}_{t+\Delta t}^i=\{F\}(\{u\}_{t+\Delta t}^i)$$

The iteration process is started (at each increment) using the starter values

$$\{u\}_{t+\Delta t}^0=\{u\}_t,\qquad [K_T]_{t+\Delta t}^0=[K_T]_t,\qquad \{F\}_{t+\Delta t}^0=\{F\}_t$$

This basic algorithm is known as the *full Newton-Raphson method*. It has the disadvantage that during each iteration, the tangent stiffness matrix must be formed and decomposed. For large systems, the cost of this can be prohibitive.

The modified Newton-Raphson method is basically as above except that the tangent stiffness is not updated during the iterations. This generally requires more iterations and sometimes is less stable but it is less computationally costly.

Nonlinear Algorithm

The basic algorithm for the full Newton-Raphson method can be stated as:

Step 1: Specify parameters of the algorithm such as tolerances and time step.

Step 2: Read the initial geometry and material properties.

Step 3: Initialize all triads.

Step 4: Assemble the mass and damping matrices $[\,M\,]$ and $[\,C\,]$.

Step 5: Specify the initial conditions for $\{u\}_0$ and $\{\dot{u}\}_0$. Obtain $\{\ddot{u}\}_0$ from the equations of motion.

Step 6: Read the load vector $\{P\}_t$. It may be necessary to interpolate this from non-equispaced values.

Step 7: *Begin loop over time increments:*

> **Step T.1:** Increment the applied load vector $\{P\}_{t+\Delta t}$.
>
> **Step T.2:** Initialize for equilibrium iterations
>
> $$u_{t+\Delta t}^0=u_t,\qquad K_{t+\Delta}^0=K_t,\qquad F_{t+\Delta t}^0=F_t$$
>
> **Step T.3:** *Begin loop over iterations:*
>
> > **Step I.1:** ITERATE:
> >
> > **Step I.2:** Assemble the element nodal force vector $\{F\}^i$.
> >
> > **Step I.3:** Form the effective load vector
> >
> > $$\{P_{eff}\}_{t+\Delta t}\;=\;\{P\}_{t+\Delta t}-\{F\}_{t+\Delta t}^{i-1}+[\,C\,]\left\{\frac{-2}{\Delta t}(u_{t+\Delta t}^{i-1}-u_t)+\dot{u}\right\}_t$$
> >
> > $$+\;[\,M\,]\left\{\frac{-4}{\Delta t^2}(u_{t+\Delta t}^{i-1}-u_t)+\frac{4}{\Delta t}\dot{u}+\ddot{u}\right\}_t$$

Step I.4: Assemble the linear stiffness matrix $[K_E]$.

Step I.5: Assemble the geometric stiffness matrix $[K_G]$.

Step I.6: Form the tangent stiffness matrix as

$$[K_T] = [K_E] + [K_G]$$

Step I.7: Decompose the effective stiffness to

$$\left[K_T + \frac{4}{\Delta t}C + \frac{4}{\Delta t^2}M\right] = [\ U\ ]^T \lceil\ D\ \rfloor[\ U\ ]$$

Step I.8: Solve for the new displacement increments from

$$[\ U\ ]^T \lceil\ D\ \rfloor[\ U\ ]\{\Delta u\}^i = \{P_{eff}\}_{t+\Delta t}$$

Step I.9: Update the displacements

$$\{u\}^i_{t+\Delta t} \quad = \quad \{u\}^{i-1}_{t+\Delta t} + \{\Delta u\}^i$$

Step I.10: Update geometry and triads.

Step I.11: Test for convergence.

if: $|\{\Delta u\}^i|/|\{u\}^i| \quad < \quad tol \qquad$ converged, goto UPDATE

if: $|\{\Delta u\}^i|/|\{u\}^i| \quad > \quad tol \qquad$ not converged, goto ITERATE

Step T.4: *End loop over iterations.*

Step T.5: UPDATE:

$$\{u\}_{t+\Delta t} \quad = \quad \{u\}^i_{t+\Delta t}$$
$$\{xyz\}_{t+\Delta t} \quad = \quad \{xyz\}^i_{t+\Delta t}$$

Step T.6: Compute orientation of global nodes.

Step T.7: Store results for this time step.

Step 8: *End loop over time increments.*

Step 9: END

It is possible to enhance this algorithm by including automatic step changes, automatic testing for appropriate time step size, and monitoring the spectral properties of the tangent stiffness. This algorithm is clearly more involved than the one for the central difference scheme.

Example 5.16: Analyze the response of a mass attached to a rod and allowed to swing under gravity loading.

Without energy losses, the mass should swing back and forth always achieving the same height.

Figure 5.27(a) shows the effect of the time step on the stability of the algorithm. Unlike the linear case, which just shows a slow deterioration, the nonlinear case exhibits catastrophic failure. Worse, there is no prior indication of imminent

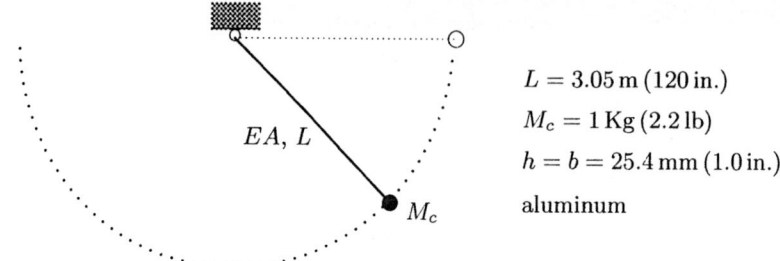

$$L = 3.05\,\text{m}\,(120\,\text{in.})$$
$$M_c = 1\,\text{Kg}\,(2.2\,\text{lb})$$
$$h = b = 25.4\,\text{mm}\,(1.0\,\text{in.})$$
aluminum

Figure 5.26: Pendulum with elastic link.

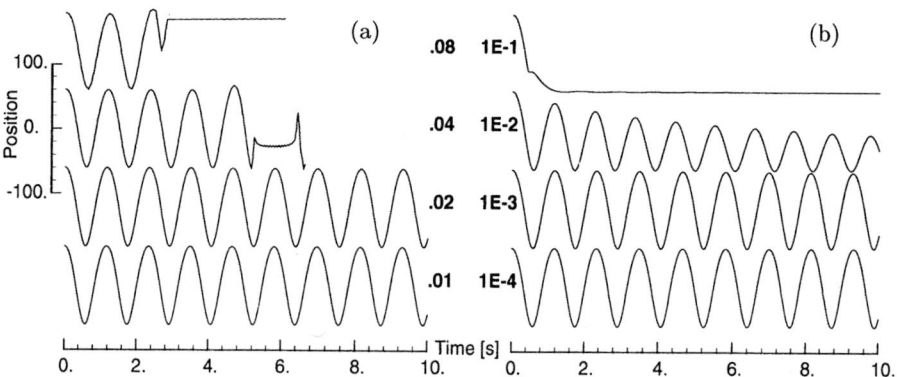

Figure 5.27: Pendulum responses. (a) Effect of time step. (b) Effect of convergence tolerance.

failure. This means that the two stable cases shown might actually fail at a later time.

Figure 5.27(b) shows the effect of the convergence tolerance. In dynamic problems, it is essential to have equilibrium checks with a fairly tight tolerance.

This pendulum problem illustrates that the implicit integration schemes, when applied to nonlinear problems, have a step size restriction much more restrictive than for the corresponding linear problem. Schemes for improving the implicit algorithm are discussed in References [20, 25, 41] but all require additional computations at the element level and therefore not readily adapted to general purpose finite element programs. The best that can be done at present is to always test convergence with respect to step size and tolerance.

Example 5.17: Investigate the choice of proportional damping on the dynamic response of structures.

We take the damping as proportional in the form

$$[\,C\,] = \alpha[\,M\,] + \beta[\,K\,]$$

The easiest way to see the effect of damping is to observe the attenuation behavior of a wave propagating in a structure. To this end, we will impact the rod shown in Figure 5.28(a) with a relatively sharp pulse and observe the dynamic response. These responses are shown in Figure 5.29(a) where the impacted end is monitored.

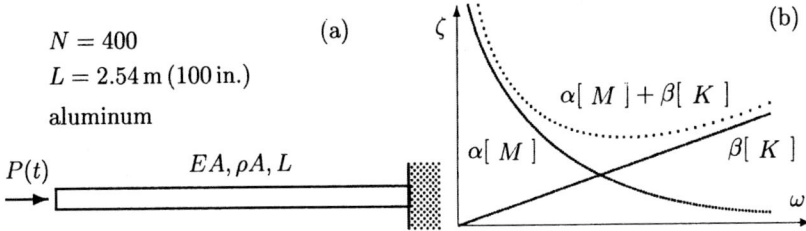

Figure 5.28: Impacted rod. (a) Geometry. (b) Modal damping.

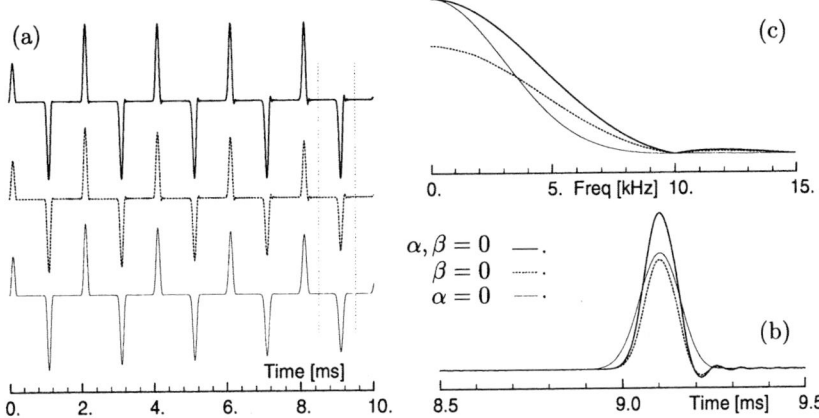

Figure 5.29: Impacted rod responses. (a) Time histories. (b) Expanded time trace. (c) Amplitude spectrum.

There is an attenuation of the response; a closer look at the response in the vicinity of $t = 9.0$ ms is shown in Figure 5.29(b) and indicates a definite difference between the two types of damping. The amplitude spectrum of these is shown in Figure 5.29(c) from which it is apparent that, when $\alpha \neq 0$ (mass proportional), the low frequencies are attenuated, whereas, when $\beta \neq 0$ (stiffness proportional), the high frequencies are attenuated.

To explain the frequency effect of the "constant" $[\,C\,]$ matrix, we return to the linear modal analysis of the previous chapter; by choosing the damping to be proportional, the modal matrix can diagonalize the damping matrix to give

$$\lceil\,\tilde{C}\,\rfloor = \alpha\lceil\,\tilde{M}\,\rfloor + \beta\lceil\,\tilde{K}\,\rfloor$$

Since $\tilde{K}_{mm} = \omega_m^2 \tilde{M}_{mm}$, we can introduce the modal damping ratio defined as

$$\zeta_m = \frac{\tilde{C}_{mm}}{2\sqrt{\tilde{K}_{mm}\tilde{M}_{mm}}} = \frac{\tilde{C}_{mm}}{2\omega_m\tilde{M}_{mm}} = \frac{\alpha + \beta\omega_{mm}^2}{2\omega_m}$$

This is shown in Figure 5.28(b). In broad terms, we expect a similar behavior in the nonlinear case.

These results have an important implication for our choice of time integration scheme. If we choose an explicit integration scheme, then the stiffness matrix is not readily available and we must use a mass proportional damping. Consequently, the high-frequency components will not be attenuated. If, in order to get some sort of high-frequency attenuation, $[\,C\,]$ is approximated somehow in terms of an initial stiffness matrix, then we loose the advantage of dealing with diagonal matrices and the computational cost increases significantly.

5.8   Dynamic Load Interactions

Structures and machines typically have a combination of different loads. For example, a building may have dead weight load of snow acting vertically and then a strong wind blows, or, an airplane may have the equilibrated loads of lift, cargo, drag, etc., and then has transient impact loads on landing.

For structures that behave linearly, the effects of these separate loads may be calculated separately and then simply superposed. This cannot be done when the systems are nonlinear.

Simple Example

We are familiar with the idea that superposition cannot be used for nonlinear cases. Thus, a load of $P = 2$, say, does not have the same effect as twice the load of $P = 1$. A more subtle nonlinear effect is the interaction effect between different components of load.

Consider the simple two-degree-of-freedom system

$$K_{11}u_1 + K_{12}u_2^{\alpha} = P_1$$
$$K_{21}u_1^{\alpha} + K_{22}u_2 = P_2$$

where the nonlinearity is taken only in the coupling terms. The relation between u_1 and P_1 is

$$P_1 = K_{11}u_1 + K_{12}\left[\frac{P_2 - K_{21}u_1^{\alpha}}{K_{22}}\right]^{\alpha}$$

Consider this as a one-degree-of-freedom system with P_2 as a parameter; then the tangent stiffness is given by

$$K_T = \frac{\partial P_1}{\partial u_1} = K_{11} - \alpha^2 \frac{K_{12}K_{21}}{k_{22}^{\alpha}} \left[P_2 - K_{21}u_1^{\alpha}\right]^{\alpha-1} u_1^{\alpha-1}$$

Contrast the linear case ($\alpha = 1$) with a nonlinear case ($\alpha = 2$)

$$\alpha = 1: \ K_T = K_{11} - \frac{K_{12}K_{21}}{k_{22}}, \qquad \alpha = 2: \ K_T = K_{11} - 4\frac{K_{12}K_{21}}{k_{22}^2}\left[P_2 - K_{21}u_1^2\right]u_1$$

In the linear case, the tangent stiffness is constant and independent of the second load. Consequently, resultant effects can be computed as simple superpositions.

For the nonlinear case, the tangent stiffness depends explicitly on the second load as well as on the current level of deformation. Consequently, effects that depend on the current tangent stiffness (such as vibrations) will be affected by the value of the second load.

Effect of Ping Loads

The general case of load interaction requires a full nonlinear analysis; however, to gain a sense of how loads can interact, we consider the special case of a two load system where one load is quasi-static and the other is a short-term impulse or "ping." This is shown schematically in Figure 5.30, and we chose it because it is amenable to a perturbation-type analysis.

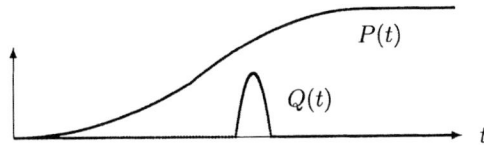

Figure 5.30: Two load histories.

Let us conceive of the total response and applied load as made up of two parts

$$\{u(t)\} = \{u_o(t)\} + \epsilon\{\xi(t)\}, \qquad \{P(t)\} = \{P_o(t)\} + \epsilon\{Q(t)\}$$

That is, there is the primary response $\{u_o\}$, which is due to $\{P_o\}$, and the smaller perturbation response $\{\xi\}$, which is due to the ping load $\{Q\}$. Substituting this into the equation of motion for the total response gives

$$[\,M\,]\{\ddot{u}_o + \epsilon\ddot{\xi}\} + [\,C\,]\{\dot{u}_o + \epsilon\dot{\xi}\} = \{P_o\} + \epsilon\{Q\} - \{F(u_o + \epsilon\xi)\}$$

Since $\epsilon\{\xi\}$ is a small perturbation, we can expand $\{F\}$ to give

$$\{F(u_o + \epsilon\xi)\} \approx \{F(u_o)\} + [\frac{\partial F}{\partial u_o}]\epsilon\{\xi\} + \ldots = \{F(u_o)\} + \epsilon[K_T(u_o)]\{\xi\} + \ldots$$

where $[K_T]$ is the tangent stiffness at the current state of deformation. Substituting this expression for $\{F\}$ into the equation of motion and grouping according to different powers of ϵ gives the two equations

$$\epsilon^0: \qquad [\,M\,]\{\ddot{u}_o\} + [\,C\,]\{\dot{u}_o\} = \{P_o\} - \{F(u_o)\}$$

$$\epsilon^1: \qquad [\,M\,]\{\ddot{\xi}\} + [\,C\,]\{\dot{\xi}\} + [K_T]\{\xi\} = \{Q\} \qquad (5.14)$$

We see that the response due to the ping is that of a linear system with a constant stiffness $[K_T]$. How the stiffness changes is governed by the first equation and for quasi-static problems reduces to

$$[\,M\,]\{\ddot{u}_o\} + [\,C\,]\{\dot{u}_o\} = \{P_o\} - \{F(u_o)\} \approx 0$$

This is the situation covered in Chapter 3. As the deformation unfolds, the tangent stiffness can change; indeed, it can even become singular — this is one of the situations covered in the next two chapters. Clearly, then, the spectral content of the ping free vibration response will give information about the current tangent stiffness $[K_T]$.

The above analysis has much in common with the example in Section 1.7 where the effect of residual stresses on wave propagation was investigated. In both cases, the initial stress changed the tangent stiffness/modulus and thus changed the dynamic response. For the in-plane problem, the effects were on the order of 1/1000 for stress levels comparable to the yield stress and hence is generally neglected. For structures, and in particular thin-walled structures, the interaction effects occur at a significantly lower stress level, and hence must be accounted for.

Example 5.18: Investigate the effect of an axial load on the dynamic response of a beam.

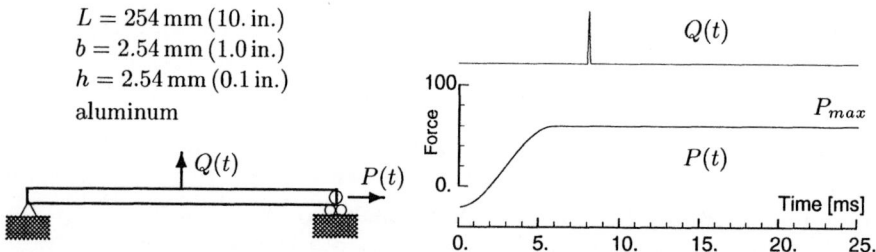

$L = 254\,\text{mm}\,(10.\,\text{in.})$
$b = 2.54\,\text{mm}\,(1.0\,\text{in.})$
$h = 2.54\,\text{mm}\,(0.1\,\text{in.})$
aluminum

Figure 5.31: Simultaneous load histories applied to a simply supported beam.

Consider the simply supported beam shown in Figure 5.31. We will apply two loads to this structure. The first is the axial load, which on its own will just cause elongation, and the beam behaves as a rod. The second load is transverse and on its own will simply cause a transient vibration of the beam. The two loads applied simultaneously will have a nonlinear interaction effect, which we now investigate. The two load histories are also shown in Figure 5.31.

The first load is initially zero and then increases to P_{max}; this is done over a time scale that causes very few transients. Three situations are considered: one when it is tensile, one when it is compressive, and one when it is zero.

The second load is comprised of a sharp transverse ping load when $P(t) = P_{max}$. In the linear case, this would just cause free vibration of the beam. The pulse is sharp enough to cause multiple modes whose frequencies are given by

$$f_n = \frac{\omega_n}{2\pi} = \frac{1}{2\pi}\left(\frac{n\pi}{L}\right)^2 \sqrt{\frac{EI}{\rho A}}$$

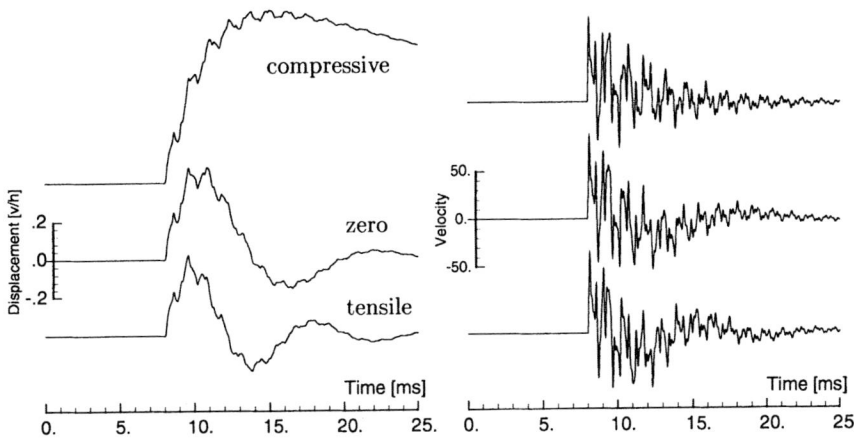

Figure 5.32: Transverse responses of a loaded beam.

Damping is used to bring this transient response to zero within a reasonable time.

When P is zero, the transverse displacement shows an increase due to the first ping, but it eventually comes back to zero. The velocity shows a typical multimode response. Significant differences are observed, however, after the axial load is applied. When the load is tensile, there is a slight decrease in the peak displacement but it eventually comes back to zero. When the load is compressive, the peak is greater but remains high for a much longer time.

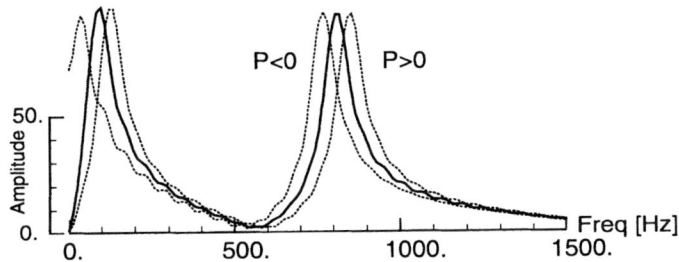

Figure 5.33: Amplitude spectrum of transverse velocity responses.

From a cursory glance, the velocity responses are similar, but we can get a deeper insight by looking at their spectral content as shown in Figure 5.33.

A tensile P causes an increase in the frequency and this coincides with our idea that an axial load increases the transverse stiffness as in a clothes line or a violin string. A compressive P causes a decrease in frequency as is clear for the second mode. The first mode is the most interesting; its frequency has decreased to near zero. This implies that there is a load value that could bring it to zero giving rise to a d.c. component — a component that monotonically increases in time. If we wait long enough, the transverse displacement will then be very large (definitely disproportionate to the linear case). This is the hallmark of an instability, indeed we are seeing a dynamic manifestation of static buckling. This simple beam case has static buckling loads [22] of

$$P_{cr} = EI(\frac{n\pi}{L})^2 \qquad \text{compressive}$$

The value of the applied P_{max} is $\pm 80\%$ of the lowest critical value confirming that we are close to buckling.

A more comprehensive dynamic view of instabilities is taken up in the next two chapters.

Problems

5.1 A mass M is attached to the mid-point of an elastic string fixed at both ends and of length $2L$.
- Obtain the equations of motion.
- Use the method of harmonic balance to estimate the frequency relation.
- Use numerical methods to integrate these equations.

5.2 Look at the design of an equal frequency shock mount.
                                                                    — Reference [56], p. 131

5.3 The set of equations considered by Lorenz in his discussion of the unpredictability (chaotic nature) of the weather were

$$\dot{x}_1 = 10(x_2 - x_1)\,, \qquad \dot{x}_2 = -x_2 + 28x_1 - x_1 x_3\,, \qquad \dot{x}_3 = -\tfrac{8}{3}x_3 + x_1 x_2$$

- Use numerical methods to integrate these equations.
- Show that they have a strange attractor.
- Replace the coefficient 28 with the parameter χ and investigate the possibility of bifurcations as χ is varied.   — Reference [26], p. 291

5.4 A shear building with walls under compressive loading has the following linearized equations of motion

$$\begin{bmatrix} 4 & -2 \\ -2 & 6 \end{bmatrix} \begin{Bmatrix} u_1 \\ u_2 \end{Bmatrix} - P \begin{bmatrix} 1 & -1 \\ -1 & 2 \end{bmatrix} \begin{Bmatrix} u_1 \\ u_2 \end{Bmatrix} + \begin{bmatrix} 1 & 0 \\ 0 & 2 \end{bmatrix} \begin{Bmatrix} \ddot{u}_1 \\ \ddot{u}_2 \end{Bmatrix} = \begin{Bmatrix} P_1 \\ P_2 \end{Bmatrix}$$

- Determine the mode shapes and frequencies as a function of P.
- Determine the modal mass and stiffness. How are they affected by P?
- Rewrite the system of equations in Principal (or Normal) coordinates.
- Discuss the effect of P on the coupling of the applied loads $\{P_1, P_2\}$.
- What combination of forces should be used if only the first mode is to be excited?
- Devise a set of FEM tests to confirm the results.

5.5 Use numerical time integration to show that the following systems have no periodic solutions:

- $\dot{x}_1 = x_2$ $\dot{x}_2 = 1 + x_1^2 - (1 - x_1)x_2$
- $\dot{x}_1 = -(1 - x_1)^3 + x_1 x_2^2$ $\dot{x}_2 = x_2 + x_2^3$
- $\dot{x}_1 = x_1^3 + 2x_1 x_2$ $\dot{x}_2 = -x_1^2 + x_2 - x_2^2 + x_2^3$
- $\dot{x}_1 = x_1$ $\dot{x}_2 = 1 + x_1 + x_2^2$
- $\dot{x}_1 = x_2$ $\dot{x}_2 = -1 - x_1^2$
- $\dot{x}_1 = 1 - x_1^3 + x_2^2$ $\dot{x}_2 = 2x_1 x_2$
- $\dot{x}_1 = x_2$ $\dot{x}_2 = (1 + x_1^2)x_2 + x^3$

                                                                    — Reference [38], p. 92

6
Stability of Structures

In the previous chapters, we investigated the equilibrium (both static and dynamic) of various structural systems. Now we wish to consider a very important property of systems in equilibrium, namely, their *stability*. Basically, we are interested in what happens to the structure when it is slightly disturbed from its equilibrium position: as shown in Figure 6.1, does it return to its equilibrium position, or does it depart even further? We term the former case *stable equilibrium* and the latter *unstable equilibrium* — clearly a load-carrying structure in a state of unstable equilibrium is unreliable and hazardous. Reference [32] gives a good background setting for the study of structural stability, whereas Reference[77] is an excellent compendium of examples and solutions. Mention should also be made of References [74, 75, 85], which place stability in a much broader context.

It is difficult to discuss stability without simultaneously discussing dynamics. We will therefore introduce a dynamic view of nominally static instability problems; fundamentally, this view considers instability to be synonymous with motion and large displacements, and thus requires a fully nonlinear dynamic perspective. We use the dynamic view to navigate the complex problem of mode jumping.

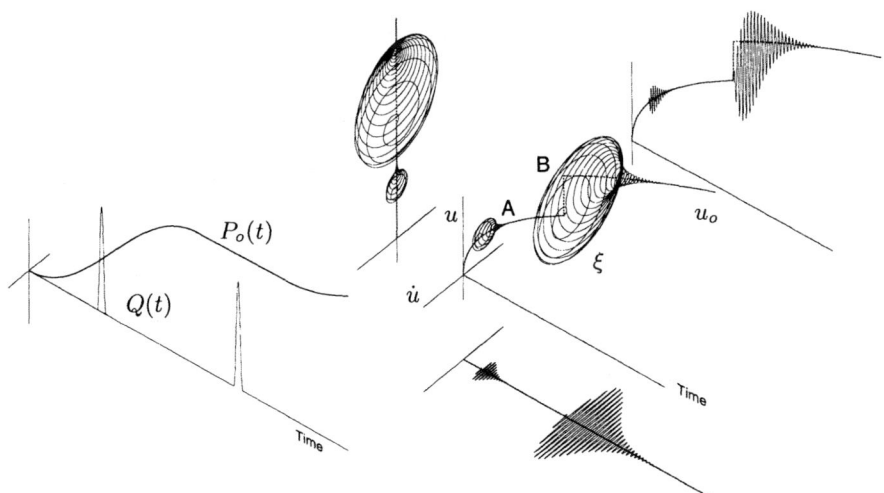

Figure 6.1: Dynamic view of a static instability.

6.1   Concepts of Stability

There are a number of definitions of stability and the one chosen should be appropriate for the phenomena being investigated. Our basic notion is that if a system is slightly disturbed from its equilibrium position and eventually returns to the original position after removal of the disturbance, then it is stable. Conversely, if it gives a disproportionate response, then it is unstable. There are other concepts of stability such as the original as proposed by Euler, which was phrased in terms of alternate equilibrium positions not "too far" from the configuration under discussion. We look at Euler's concept first.

Euler's Static Equilibrium View

To clarify our stability idea in an actual situation, consider the simple pinned structure shown in Figure 6.2(a). The initially vertical bar is assisted in remaining vertical by the action of the horizontal spring; the spring is unstretched when the bar is vertical. An equilibrium analysis in the undeformed state shows that the displacements are related to the forces through the following stiffness relations

$$Ku = Q, \qquad K_b v = -P \qquad\qquad (6.1)$$

where K_b is the axial stiffness of the bar. For the purpose of the later discussion, let the bar be very stiff, from which we conclude that the displacement is only horizontal, that is, $v = 0$.

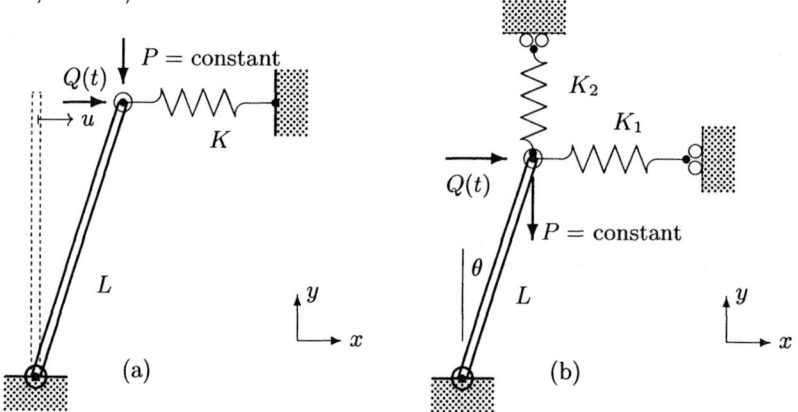

Figure 6.2: Disturbed equilibrium state of a pinned bar. (a) Small deflection model. (b) Large deflection model.

A key point in Euler's stability analysis is to look at equilibrium of the structure in its deformed state. In this first analysis, we will consider a linear analysis where all displacements are assumed to be small. Consider the situation when the bar has already displaced by an amount u as shown in Figure 6.2(a). Since it is in equilibrium, summing the moments about the base gives

$$-QL - Pu + KuL = 0 \qquad \text{or} \qquad [K - P/L]u = Q$$

It is worth noting at this stage the different roles played by the two forces: Q appears as a regular applied load on the right-hand side, but P appears on the left-hand side almost like a stiffness term. We refer to this as an initial stress term and is the same as the geometric stiffness of Chapter 3.

Clearly, we can solve for the displacements for any combinations of loads, thus

$$u = \frac{Q}{K - P/L}$$

These are shown plotted in Figure 6.3(a) for different values of Q. The obvious point of note is that in the vicinity of $P = KL$ the displacements are very large even for negligible Q. (Note that it is a consequence of the linear approximations that the infinity occurs — deflections in a real nonlinear situation would be finite.) Keep in mind that every point shown in the figure is an equilibrium position, and based on our discussion above we would declare that the structure is unstable near $P = KL$. One could imagine, therefore, a loading sequence that need only "skip over" this special value of load and we would not have to worry about stability. The situation is much subtler than this, and we need to refine our concepts of equilibrium and stability.

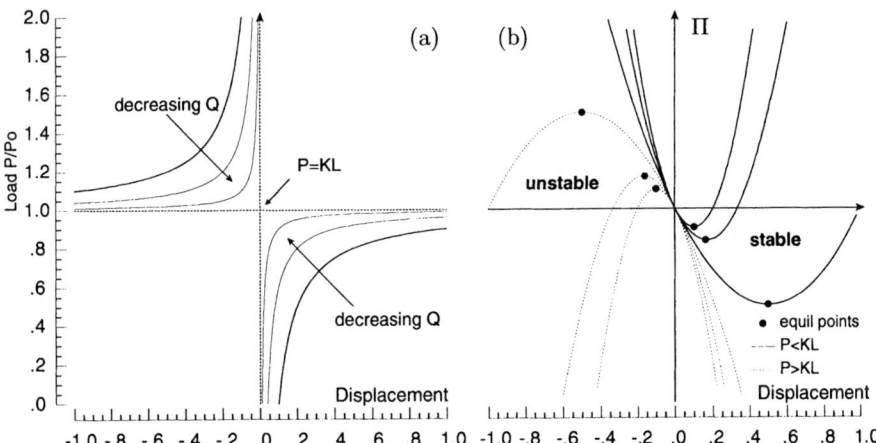

Figure 6.3: Behavior of the linear bar model. (a) Displacements for various values of loads. (b) Potential energy for different P values and constant Q.

The total potential for our simple system is comprised of the strain energy

$$U = \tfrac{1}{2}Ku^2$$

and the potentials of the loads

$$V = -Pv - Qu \approx -\tfrac{1}{2}(P/L)u^2 - Qu$$

The approximation is based on small deflections. The total potential is

$$\Pi = U + V = \tfrac{1}{2}Ku^2 - \tfrac{1}{2}(P/L)u^2 - Qu$$

This is shown plotted in Figure 6.3(b) for different values of P but the same value of Q. Note that most points on these plots are not equilibrium points — equilibrium is where

$$\mathcal{F} = \frac{\partial \Pi}{\partial u} = Ku - (P/L)u - Q = 0$$

and are shown as dots in the figure.

For $P < KL$, the plots show a series of valleys the bottoms of which correspond to the equilibrium positions. For $P > KL$, the plots show a series of peaks the top of which correspond to the equilibrium positions. Is there a difference in the nature of the two sets of equilibrium states?

Consider the following scenario: for $P < KL$, increment Q by a small amount ΔQ. This will move the system to a new equilibrium position. Now release ΔQ; since the system is no longer in equilibrium, a dynamic process ensues. That is, the additional potential is converted to kinetic energy. The system will pass through a valley and come to rest temporarily but again this will be at a nonequilibrium position. And the process repeats. Now let the system be "sluggish" so that on each cycle energy is lost, then eventually all the excess potential is dissipated and the system finally comes to rest in the valley at an equilibrium point. We therefore say that the points for which $P < KL$ are stable.

Now consider the same scenario but for the case $P > KL$. The increment ΔQ causes a decrease in potential; this conceivably could be restored on release leaving the system back at its original equilibrium position. However, because of our stipulation that there is dissipation, then this cannot happen. Hence, if the system comes to rest temporarily, it will not be an equilibrium position. Thus as the dynamic process ensues potential energy is continually lost but, since there is no valley in which to finally come to rest, the system continues the process until the displacement is very large. At this stage the system has collapsed. We therefore say that for any load for which $P > KL$ the system is unstable.

We have thus gone beyond declaring just the point $P = KL$ unstable and now see that all points beyond $P = KL$ are unstable. The special point $P = KL$ is called the critical value or sometimes a bifurcation point. In those cases when Q is small, the deflections in the region close by are small; we will take advantage of this to do a linearized eigenvalue analysis to determine these bifurcation points. In the literature (see Reference [77] for a very large collection of solutions) this is generally known as a *buckling analysis*. It is emphasized, however, that such an analysis determines only the critical points and says nothing of the postbuckling behavior.

Limit Points and Bifurcation Points

Consider the loaded column shown in Figure 6.2(b); this is similar to that of Figure 6.2(a) except for the additional spring and that the deflection can be large. We use the angle off the vertical, θ, as the independent variable and allow it to vary in the range from zero to π.

Since the frame member is rigid, we easily establish the relationship between the vertical and horizontal displacements as

$$u = L\sin\theta, \qquad v = L\cos\theta$$

The total potential energy is given by

$$
\begin{aligned}
\Pi &= \tfrac{1}{2}K_1 u^2 + \tfrac{1}{2}K_2 v^2 - Pv - Qu \\
    &= \tfrac{1}{2}KL^2\sin^2\theta + \tfrac{1}{2}K_2 L^2(1-\cos\theta)^2 - PL(1-\cos\theta) - QL\sin\theta
\end{aligned}
$$

The equilibrium configurations are found by setting

$$\mathcal{F} = \frac{\partial\Pi}{\partial\theta} = \left[K_1\cos\theta + K_2(1-\cos\theta) - \frac{P}{L}\sin\theta\right]\sin\theta - \frac{Q}{L}\cos\theta = 0$$

Rearranging gives

$$P = K_1\cos\theta + K_2(1-\cos\theta) - \frac{Q\cos\theta}{L\sin\theta}$$

where we view Q simply as a parameter.

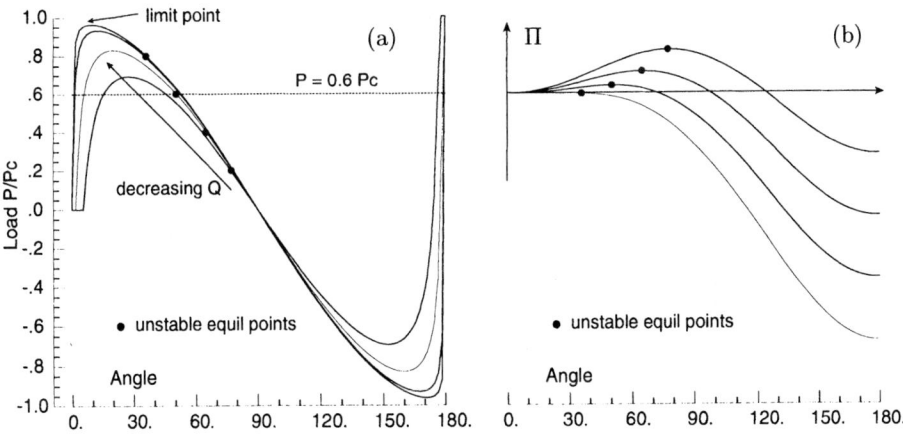

Figure 6.4: Behavior of the nonlinear bar model with $K_2 = 0$. (a) Rotation for various values of loads. (b) Potential energy for different P values and small constant Q.

Figure 6.4(a) shows a plot of the force P against angle for different values of Q when $K_2 = 0$. In each case, the force cannot exceed a particular maximum value and in the limit of small Q this corresponds to the critical value $P_c = K_1 L$. One interpretation of this figure is to view it as imposing the displacement and the force is the resulting reaction. Suppose, however, we wish to impose the force, then what is the interpretation of the decreasing force? More specifically, suppose we are at the peak value and we increment the load a small amount ΔP, what happens?

To answer this question, we need to look at the potential energy plotted in Figure 6.4(b). The total potential energy (when $K_2 = 0$) is given by

$$\Pi = \tfrac{1}{2} K_1 L^2 \sin^2 \theta - PL \cos \theta - QL \sin \theta$$

Again, it is emphasized that most points on the potential plots are nonequilibrium points. Consider the line corresponding to the load values $P = 0.6P_c$ in Figure 6.4(a). It intersects the equilibrium curve at three points, but only two of the points (near 0 and 180) are stable. In other words, all points immediately past the peak are unstable and therefore a load/angle combination in this range would cause a large displacement. The member would rotate until it found the second stable equilibrium point at a large angle.

The maximum load point is called a *limit point*, because the load cannot exceed this limiting value. The phenomenon of quickly jumping from one equilibrium configuration to another distant one is called *snap-through*.

In the previous developments we saw that $Q \longrightarrow 0$ is a special case. Let us now deal directly with this situation. The equilibrium relation is

$$\mathcal{F} = \frac{\partial \Pi}{\partial \theta} = \left[K_1 \cos \theta + K_2 (1 - \cos \theta) - \frac{P}{L} \sin \theta \right] \sin \theta = 0$$

This has two solutions. The first is

$$\sin \theta = 0 \qquad \text{or} \qquad \theta = 0, \pm \pi, \pm 2\pi, \ldots$$

These correspond to when the bar is in a vertical alignment. The other solution gives

$$P = K_1 \cos \theta + K_2 (1 - \cos \theta)$$

Both solutions are shown plotted in Figure 6.5 for different values of K_2/K_1. We see that the solutions intersect at $P/K_1 L = 1$, which is the critical value we previously identified. The presence of K_2 does not affect this critical value but it does change the shape of the equilibrium curve. We now investigate its stability.

The first variation of the potential energy is related to the equilibrium solution, while the second variation determines its stability. This is given by

$$K_T = \frac{\partial \mathcal{F}}{\partial \theta} = \frac{\partial^2 \Pi}{\partial \theta^2} = K_1 L^2 (2 \cos^2 \theta - 1) + K_2 L^2 (1 - 2 \cos^2 \theta + \cos \theta) - PL \cos \theta$$

For the solution $\theta = 0$, this becomes

$$K_T = \frac{\partial^2 \Pi}{\partial \theta^2} = K_1 L^2 - PL$$

Hence, when $P > K_1L$ this solution path is unstable. That is, the intersection with the second solution has caused this solution path to become unstable. On the other hand, for $\theta = \pm\pi$ this becomes

$$K_T = \frac{\partial^2 \Pi}{\partial \theta^2} = K_1L^2 - K_2L^2 + PL$$

When $K_2 = 0$, say, then the solution is stable as long as P is not reversed. If $K_2 > K_1$, then a minimum value of P is required to maintain stability otherwise the bar will snap back to the original upright position.

For the second solution with $P = K_1 \cos\theta + K_2(1 - \cos\theta)$, we get

$$K_T = \frac{\partial^2 \Pi}{\partial \theta^2} = (K_2 - K_1)L^2 \sin^2 \theta$$

Thus, so long as $K_2 > K_1$ the solution paths are stable. In particular, if $K_2 = 0$, the solution is always unstable, confirming what we saw in Figure 6.4(a). The complete picture is shown in Figure 6.5.

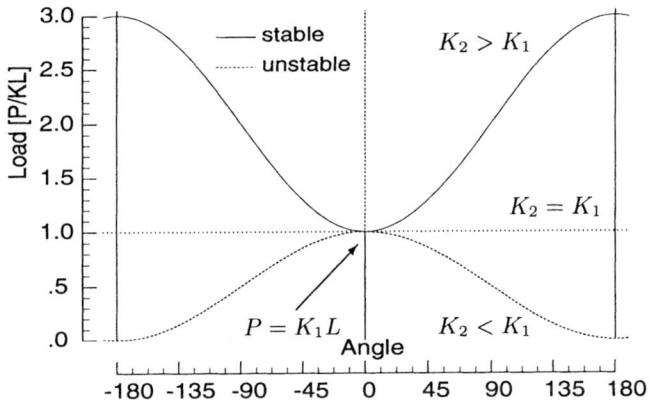

Figure 6.5: Bifurcation view.

We now have the following bifurcation view of the loading process: as P is increased, the solution ($\theta = 0$) is stable until the critical point $P = K_1L$ is reached; a further increment of load along this path makes the solution unstable — any disturbance will cause it to snap. Where it goes depends on the relative values of K_1 and K_2: for $K_2 > K_1$ there is a small increase in θ onto the stable path and the load can then continue to be increased. For $K_2 < K_1$, there is a very large increase in θ where the bar snaps to the inverted position. This snap, in reality, would involve a dynamic process.

A Dynamic View of Static Instability

It was difficult to discuss the Euler view of stability without simultaneously discussing dynamics. Thus it seems that dynamics is inseparable from our view of stability. In fact, a common intuitive notion of stability asks the simple question: if the structure is slightly perturbed, what happens to the ensuing dynamics? That is, if the structure returns to its current state, then it is stable; otherwise it is unstable and a dynamic process ensues. The purpose of this section is to pursue the implications of this simple notion of stability; later we revisit it within the context of modern computational methods, and a more comprehensive dynamic view will be developed in Chapter 7.

Fundamentally, the dynamic view considers instability to be synonymous with motion and large displacements, and thus requires a fully nonlinear dynamic perspective. In comparison to the static methods, we make two operational changes. First, the independent variable is time; all variables, including the load, are functions of time, and the complete state vector — both velocity and displacement — is computed and monitored. Secondly, we explicitly separate the loading associated with the fundamental path from that which initiates or interrogates the instability. There are very few papers devoted to a dynamic analysis of static instability; however, mention should be made of Reference [59], which introduced a combined static/dynamic analysis, and References [44, 85] give very clear discussions of the difference between the static and dynamic methods of stability analysis.

A schematic of our dynamic view is illustrated in Figure 6.1. A stable loading state (resulting from the slowly applied load $P(t)$) is illustrated by the segment A; a disturbance (in the form of a short duration ping load $Q(t)$) causes oscillations about the equilibrium path; these are temporary and the structure eventually comes back to the equilibrium path. At, or beyond, a critical point, however, a small disturbance will cause a significant dynamic process to ensue. Depending on the particular problem, a nearby equilibrium path may or may not be found. It is worth noting that the new equilibrium path may not be statically connected to the original one; that is, we could not devise a proportional loading sequence (where the ratio of all the loads is kept constant) to connect the two equilibrium states.

Consistent with our dynamic view of instability, let us conceive of the total response and applied load as made up of two parts

$$\{u(t)\} = \{u_o(t)\} + \epsilon\{\xi(t)\}, \qquad \{P(t)\} = \{P_o(t)\} + \epsilon\{Q(t)\}$$

That is, there is the primary response $\{u\}_o$, which is due to $\{P_o\}$, and the smaller perturbation response $\{\xi\}$, which is due to the ping load $\{Q\}$. As shown in Section 5.8 on dynamic load interactions, since ϵ is small, this leads to two equations

$$\epsilon^0: \qquad [M]\{\ddot{u}_o\} + [C]\{\dot{u}_o\} = \{P_o\} - \{F(u_o)\}$$

$$\epsilon^1: \qquad [M]\{\ddot{\xi}\} + [C]\{\dot{\xi}\} + [K_T]\{\xi\} = \{Q\}$$

The second equation, which is often referred to as the variational equation [44], shows that the response due to the ping is that of a linear system with a constant stiffness $[K_T]$.

How the stiffness changes is governed by the first equation and for quasistatic problems reduces to

$$[M]\{\ddot{u}_o\} + [C]\{\dot{u}_o\} = \{P_o\} - \{F(u_o)\} \approx 0$$

As we showed in Chapter 3, the incremental form of this relation becomes

$$[K_T]\{\Delta u\}_{t+\Delta t} = \{P\}_{t+\Delta t} - \{F\}_t$$

and so long as $[K_T]$ is nonsingular we can uniquely determine the deformation. As the deformation unfolds, the tangent stiffness can change, and in particular it can become singular — this is precisely the situation of interest here.

At a given load P_o, the tangent stiffness matrix is constant and therefore the small disturbance response is that of a linear system. After the ping is applied, the system is in free vibration governed by

$$[K_T]\{\xi\} + [C]\{\dot{\xi}\} + [M]\{\ddot{\xi}\} = \{0\}$$

We look for solutions of the form $\{\xi(t)\} = \{\phi\}e^{i\mu t}$. Substitute into the equation of motion to get

$$\Big[[K_T] + i\mu[C] - \mu^2[M]\Big]\{\phi\}e^{i\mu t} = 0$$

There can be nontrivial solutions only if the determinant is zero, which leads to a characteristic equation to determine the eigenvalues μ_i and eigenvectors $\{\phi\}_i$. The general solution is written as a combination of

$$\{\phi\}_1 e^{i\mu_1 t}, \quad \{\phi\}_2 e^{i\mu_2 t}, \quad \cdots, \quad \{\phi\}_N e^{i\mu_N t}$$

We state our stability criterion in terms of the properties of the eigenvalues μ_i. For the system to be asymptotically stable we want

$$\text{Im}\,[\mu] > 0 \quad \text{since} \quad \{\xi\}e^{i\mu t} = \{\phi\}e^{i(\mu_R + i\mu_I)t} = \{\phi\}e^{-\mu_I t}e^{i\mu_R t}$$

Thus, a negative imaginary component of μ_i would give an exponentially increasing function of time. If the criterion is not true for any one of the roots, then the system is unstable.

The static instability criterion of Euler is essentially the case of $\mu_1 = 0$; that is, both the real and imaginary parts are zero simultaneously. There are structural problems, however, where this criterion is insufficient. For example, for follower-force type problems (and related problems such as aeroelastic flutter) instability occurs when the real part of μ_1 is still positive. Such situations are usually referred to as dynamic (or kinetic) instabilities [85]. These situations are covered in the next chapter; the next example discusses a pseudo-static situation.

Example 6.1: Investigate the stability of a slightly out-of-round shaft rotating with constant angular speed Ω.

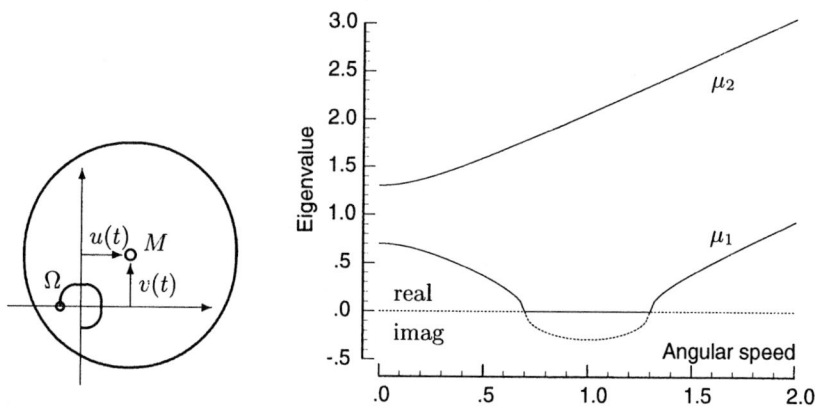

Figure 6.6: Rotating shaft with eccentricity.

Let the center of the shaft have displacements $u(t)$ and $v(t)$ about its original position. The slight out-of-roundness causes it to have different principal bending stiffnesses. Let these principal values be represented by the spring constants K_1 and K_2 with $K_1 < K_2$, then the strain energy is given by

$$U = \tfrac{1}{2}K_1 u^2 + \tfrac{1}{2}K_2 v^2$$

The kinetic energy has two components: that due to the motion about the center, and that due to the rotation,

$$T = \tfrac{1}{2}M\dot{u}^2 + \tfrac{1}{2}M\dot{v}^2 + \tfrac{1}{2}M[\Omega u]^2 + \tfrac{1}{2}M[\Omega v]^2$$

Following Section 4.3, the Coriolis acceleration is given by

$$2\,\hat{\omega} \times \dot{\hat{u}} = 2\Omega\,\hat{k} \times (\dot{u}\hat{\imath} + \dot{v}\hat{\jmath}) = 2\Omega(-\dot{v}\hat{\imath} + \dot{u}\hat{\jmath})$$

Therefore, the components of force acting on the system are

$$Q_u = 2M\Omega\dot{v}, \qquad Q_v = -2M\Omega\dot{u}$$

Substituting into Lagrange's equation gives

$$
\begin{aligned}
\mathcal{F}_u &= \frac{d}{dt}\{\frac{\partial T}{\partial \dot{u}}\} - \frac{\partial T}{\partial u} + \frac{\partial(U+V)}{\partial u} - Q_u = 0 = M\ddot{u} - M\Omega^2 u + K_1 u - 2M\Omega\dot{v} \\
\mathcal{F}_v &= \frac{d}{dt}\{\frac{\partial T}{\partial \dot{v}}\} - \frac{\partial T}{\partial v} + \frac{\partial(U+V)}{\partial v} - Q_v = 0 = M\ddot{v} - M\Omega^2 v + K_2 v + 2M\Omega\dot{u}
\end{aligned}
$$

This can be rearranged to the form

$$
\begin{bmatrix} \omega_1^2 & 0 \\ 0 & \omega_2^2 \end{bmatrix} \begin{Bmatrix} u \\ v \end{Bmatrix} - \Omega^2 \begin{bmatrix} 1 & 0 \\ 0 & 1 \end{bmatrix} \begin{Bmatrix} u \\ v \end{Bmatrix} + 2\Omega \begin{bmatrix} 0 & -1 \\ 1 & 0 \end{bmatrix} \begin{Bmatrix} \dot{u} \\ \dot{v} \end{Bmatrix} + \begin{bmatrix} 1 & 0 \\ 0 & 1 \end{bmatrix} \begin{Bmatrix} \ddot{u} \\ \ddot{v} \end{Bmatrix} = \begin{Bmatrix} 0 \\ 0 \end{Bmatrix}
$$

where $\omega_1^2 = K_1/M$ and $\omega_2^2 = K_2/M$.

We now investigate the stability of the system as a function of the angular speed. In this case, the angular speed plays a similar role to the load in the previous examples. First note that, because the system is linear, this is also the equation of motion for the perturbation. The tangent stiffness matrix is

$$[K_T] = \begin{bmatrix} \omega_1^2 - \Omega^2 & 0 \\ 0 & \omega_2^2 - \Omega^2 \end{bmatrix}$$

which shows the possibility of diagonal terms becoming zero and then negative. In particular, the determinant is zero when

$$\det[K_T] = [\omega_1^2 - \Omega^2][\omega_2^2 - \Omega^2] = 0 \qquad \text{or} \qquad \Omega = \omega_1, \qquad \Omega = \omega_2$$

There are two singular values. Note that the dynamic potential at the zero velocity point is

$$U - T = \tfrac{1}{2}K_1 u^2 + \tfrac{1}{2}K_2 v^2 - \tfrac{1}{2}M[\Omega u]^2 - \tfrac{1}{2}M[\Omega v]^2$$

and this is negative for all $\Omega > \omega_1$, irrespective of ω_2, implying that all speeds $\Omega > \omega_1$ are unstable. As we will see next, this conclusion is not correct.

A dynamic stability analysis begins by assuming free vibration solutions of the form

$$\begin{Bmatrix} \xi_1 \\ \xi_2 \end{Bmatrix}(t) = \begin{Bmatrix} \hat{\xi}_1 \\ \hat{\xi}_2 \end{Bmatrix} e^{i\mu t}$$

Substituting this into the governing equation gives the eigenvalue system

$$\begin{bmatrix} \omega_1^2 - \Omega^2 - \mu^2 & -2\Omega i\mu \\ 2\Omega i\mu & \omega_2^2 - \Omega^2 - \mu^2 \end{bmatrix} \begin{Bmatrix} \hat{\xi}_1 \\ \hat{\xi}_2 \end{Bmatrix} = \begin{Bmatrix} 0 \\ 0 \end{Bmatrix}$$

which leads to the characteristic equation

$$\mu^4 - \mu^2[\omega_1^2 + \omega_2^2 + 2\Omega^2][\omega_1^2 - \Omega^2][\omega_2^2 - 2\Omega^2] = 0$$

This is quadratic in μ^2, hence there are four roots appearing as $\pm$ pairs.

First look at the possibility of a static instability — this occurs when $\mu = 0$

$$[\omega_1^2 - \Omega^2][\omega_2^2 - 2\Omega^2] = 0 \qquad \text{or} \qquad \omega_1 = \Omega, \qquad \omega_2 = \Omega$$

which, of course, is the same result as above. The actual roots are given by

$$\mu_{1,2} = \left[\tfrac{1}{2}(\omega_1^2 + \omega_2^2) + \Omega^2 \pm \tfrac{1}{2}\sqrt{(\omega_1^2 - \omega_2^2)^2 + 8\Omega^2(\omega_1^2 + \omega_2^2)} \right]^{1/2}$$

which are shown plotted in Figure 6.6. It is only in the region $\omega_1 < \Omega < \omega_2$ that μ has a negative imaginary part and hence that the shaft is unstable. If the Coriolis force is ignored in the analysis, then the roots are given by

$$\mu_1 = \pm\tfrac{1}{2}\sqrt{\omega_1^2 - \Omega^2}, \qquad \mu_2 = \pm\tfrac{1}{2}\sqrt{\omega_2^2 - \Omega^2}$$

which shows all speeds $\Omega > \omega_1$ to be unstable. That is, the Coriolis force has stabilized the shaft for speeds $\Omega > \omega_2$; such a result could not be predicted by the static methods because the Coriolis force would never appear.

The special case $K_1 = K_2 = K$ (or $\omega_1 = \omega_2 = \omega_o$) is interesting because it gives a critical (where $\mu = 0$) speed of $\Omega = \omega_o = \sqrt{K/M}$ that is stable. Of course, any imperfection in the shaft (or loading) would destroy that.

General Comments

To summarize, motivated by dynamic heuristics, our view of static instability is that the structure is in a state of unstable equilibrium when the second derivative of the total potential energy is negative. The agent for the change in the above analysis was an applied load Q quite separate from the primary loading P, but note that we get a critical value even in the limit of $Q = 0$.

The agent Q is referred to as an *imperfection* since it can be thought of as a slightly mis-applied P, and since slight geometric imperfections have the same result. The structure (and the solution) is called perfect when $Q = 0$. Imperfect structures exhibit limit type instabilities while the perfect structure can also exhibit bifurcations.

In the remainder of this chapter we will concentrate on static instabilities and generally adopt the Euler approach of looking at equilibrium in the deformed configuration. We focus on monitoring the eigenvalues of the tangent stiffness matrix, our stability criterion being that if one of them goes to zero then the load is at a singular point. This rules out static problems that have a non-symmetric tangent stiffness — we leave some of these cases until the next chapter where a fully dynamic view is developed. The discussion of stability involves an analysis of the post-buckling behavior — this clearly needs our full nonlinear analysis techniques developed in the earlier chapters. In many situations it is sufficient just to know the first critical load and in the next few sections we develop methods for obtaining this information without doing a complete nonlinear post-buckling analysis. In the final sections, however, we navigate through a complete load/unload cycle for a structure undergoing large deformations and buckling.

6.2   Beams with Axial Forces

In this section, we take into account the effect an axial load has on a beam. The axial force may be either tensile or compressive, but compression is of more interest here because of its effect on stability. The key to the analysis is to look at equilibrium of the beam in a slightly displaced position.

We consider equilibrium in two steps. The first step corresponds to a linear analysis where we establish the forces and moments acting on the beam segment; the beam is still in its initial configuration. This is called the *pre-buckle analysis*. Then we displace the beam segment slightly leaving the loads intact as shown in Figure 6.7. This will give us the measure of the sensitivity of the equilibrium condition to changes in configuration. Since the change in configuration is only slight, we can retain all the usual assumptions of the elementary beam theory.

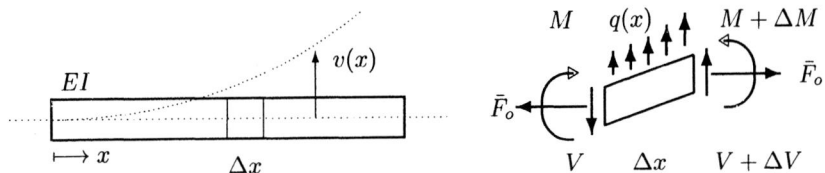

Figure 6.7: Equilibrium of small beam segment slightly deformed.

Governing Equations

Consider the Lagrangian strain in a beam bending in the $x - y$ plane

$$E_{xx} = \frac{\partial u}{\partial x^o} + \tfrac{1}{2}\left[\left(\frac{\partial u}{\partial x^o}\right)^2 + \left(\frac{\partial v}{\partial x^o}\right)^2\right]$$

where we are using the large displacement coordinate description of Chapter 1. The kinematics of thin-beam theory gives

$$u(x^o, y^o) \approx u(x^o) - y^o \frac{\partial v}{\partial x^o}, \qquad v(x^o, y^o) \approx v(x^o)$$

Substitute these into the strain/displacement relation to get

$$E_{xx} = \frac{\partial u}{\partial x^o} + \tfrac{1}{2}\left[\left(\frac{\partial u}{\partial x^o}\right)^2 + \left(\frac{\partial v}{\partial x^o}\right)^2\right] - y^o\frac{\partial^2 v}{\partial x^{o2}} + \tfrac{1}{2}\left[\left(-y^o\frac{\partial^2 v}{\partial x^{o2}}\right)^2\right]$$

$$\approx \frac{\partial u}{\partial x^o} - y^o\frac{\partial^2 v}{\partial x^{o2}} + \tfrac{1}{2}\left(\frac{\partial v}{\partial x^o}\right)^2$$

The approximation is based on the assumption that the transverse deflection is significantly larger than that of the axial displacement. We recognize the leading term as that of the linear theory of rods; the second as that of the linear flexure theory of beams; the third term, which is nonlinear, comes from the large deflection of the beam. Now that we have established the nonlinear contribution to the strain, we can use the small deflection notation that does not distinguish between the deformed and undeformed configurations.

The axial strain gives rise to a stress according to

$$\sigma_{xx} = E E_{xx}$$

The strain energy for a beam in uniaxial stress is

$$U = \tfrac{1}{2}\int_V \sigma_{xx} E_{xx}\, dV = \tfrac{1}{2}\int_V E E_{xx}^2\, dA\, dx$$

After substitution and integration we are left with the three terms

$$U = \tfrac{1}{2}\int_L\left[EI\left(\frac{d^2 v}{dx^2}\right)^2\right] dx + \tfrac{1}{2}\int_L\left[EA\left(\frac{\partial u}{\partial x}\right)^2\right] dx + \tfrac{1}{2}\int_L\left[\bar{F}_o\left(\frac{\partial v}{\partial x}\right)^2\right] dx$$

In this, we introduced the term

$$\bar{F}_o = EA\frac{\partial u}{\partial x}$$

which corresponds to an axial load. Strictly, this is deformation dependent as we showed in Chapter 3, but in the following it is assumed to be preexisting in the beam and therefore not subject to variation. Thus, the energy expression comprises the contribution for the elastic behavior plus terms associated with the axial loading.

The potential of the distributed applied loads is

$$V = -\int q_u u \, dx - \int qv \, dx$$

Using Hamilton's principle, and assuming sectional properties and axial force $\bar{F}_o$ are piecewise uniform, leads to two uncoupled governing equations [23]

$$EA\frac{\partial^2 u}{\partial x^2} - \rho A\frac{\partial^2 u}{\partial t^2} = q_u(x,t), \qquad EI\frac{\partial^4 v}{\partial x^4} - \bar{F}_o\frac{\partial^2 v}{\partial x^2} + \rho A\frac{\partial^2 v}{\partial t^2} = q(x,t)$$

The associated boundary conditions are specified using the terms

$$\{u, \ F = EA\frac{du}{dx}\}, \qquad \{v, \ V = -EI\frac{d^3 v}{dx^3} + \bar{F}_o\frac{dv}{dx}\}, \qquad \{\phi, \ M = +EI\frac{d^2 v}{dx^2}\}$$

To see the meaning of the resultants, look at the free body diagram in Figure 6.7. It is apparent that the axial force causes a bending action and therefore will enter the beam equilibrium equations. Thus, balance of force and moment on the small segment of Figure 6.7 gives (for small slopes and deflections), respectively,

$$\frac{dV}{dx} = -q, \qquad \frac{dM}{dx} + V - \frac{dv}{dx}\bar{F}_o = 0$$

All the relationships for the flexural quantities may now be collected as

$$\text{Displacement}: \qquad v = v(x)$$

$$\text{Slope}: \qquad \phi = \frac{dv}{dx}$$

$$\text{Moment}: \qquad M = +EI\frac{d^2 v}{dx^2}$$

$$\text{Shear}: \qquad V = -EI\frac{d^3 v}{dx^3} + \bar{F}_o\frac{dv}{dx}$$

$$\text{Loading}: \qquad q = +EI\frac{d^4 v}{dx^4} - \bar{F}_o\frac{d^2 v}{dx^2} \qquad (6.2)$$

We shall refer to these equations as the *coupled beam equations*. The only difference in comparison with the elementary beam equations is the addition of the $\bar{F}_o$

related terms in the expressions for the loading and shear. Thus with $\bar{F}_o$ treated as a parameter, it is seen from these that the transverse displacement $v(x)$ (via its derivatives) can be viewed as the fundamental unknown of interest.

For the purpose of integrating these equations, we treat the axial term as constant — it is known either by specification or through a separate linear analysis of the in-plane loads. In this way, the problem is reduced to being linear even though it takes the deformed configuration into account. As an example, to integrate the loading relation, we will assume that the transverse loading $q = q_o$ is constant, then integrating twice gives

$$EI\frac{d^2v}{dx^2} - \bar{F}_o v = c_1^* x + c_2^* + \tfrac{1}{2}q_o x^2$$

This is an inhomogeneous second-order differential equation. The complete solution comprises a solution to the homogeneous problem plus a particular solution. The character of the solution differs depending on whether $\bar{F}_o$ is tensile or compressive. The general solution for compressive loading (i.e., $\bar{F}_o = -P$) is

$$v(x) = c_1 \cos kx + c_2 \sin kx + c_3 x + c_4 + \frac{q_o}{2EIk^2}x^2, \qquad k^2 \equiv \frac{-\bar{F}_o}{EI} = \frac{P}{EI} \quad (6.3)$$

and for tensile loading (i.e., $\bar{F}_o = P$) is

$$v(x) = c_1 \cosh kx + c_2 \sinh kx + c_3 x + c_4 + \frac{q_o}{2EIk^2}x^2, \qquad k^2 \equiv \frac{\bar{F}_o}{EI} = \frac{P}{EI} \quad (6.4)$$

The constants of integration c_1, c_2, c_3, c_4 are evaluated by imposing particular boundary conditions. The shear distribution is given by the simple expression

$$V(x) = \pm k^2 EI\, c_3 - q_o x$$

with the $\pm$ corresponding to tensile and compressive axial loading, respectively. When the distributed loading is zero, the shear is constant and simply related to the constant c_3.

Example 6.2: Consider the cantilevered beam shown in Figure 6.8 with the combination of axial and transverse forces. Determine the deflected shape.

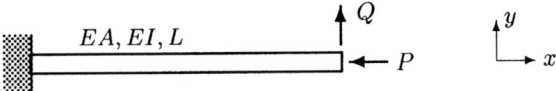

Figure 6.8: Cantilevered beam with end loads.

Let us first solve this problem in the manner of elementary beam statics (this will constitute the pre-buckle analysis). The axial behavior gives a displacement and force distribution of

$$u(x) = -\frac{Px}{EA}, \qquad F(x) = -P$$

That is, the axial analysis simply gives that the axial force is the same everywhere along the beam and is P compressive. Imposing the boundary conditions of zero transverse displacement and slope at one end, zero moment and an applied force of Q at the other, gives the deflected shape as

$$v(x) = \frac{Q}{6EI}[3xL^2 - x^3]$$

We have thus solved for the transverse displacements independent of the axial behavior. This is the nature of the linear uncoupled theory.

We will now solve the coupled beam equations. The distributed loading is zero and the axial force is compressive, therefore, the general deflected shape is

$$v(x) = c_1 \cos kx + c_2 \sin kx + c_3 x + c_4, \qquad k = \sqrt{P/EI}$$

We impose the following boundary conditions to determine the coefficients. At $x = 0$, we have

$$v = 0 \quad \Rightarrow \quad 0 = c_1 + c_4$$
$$\frac{dv}{dx} = 0 \quad \Rightarrow \quad 0 = c_2 k + c_3$$

At $x = L$, we have

$$M = EI\frac{d^2 v}{dx^2} = 0 \quad \Rightarrow \quad 0 = EI[-c_1 k^2 \cos kL - c_2 k^2 \sin kL]$$
$$V = -EI\frac{d^3 v}{dx^3} - P\frac{dv}{dx} = Q \quad \Rightarrow \quad Q = -EIk^2 c_3$$

Solving directly gives (noting that $P = k^2 EI$)

$$c_1 = -\frac{Q}{k^3 EI}\frac{\sin kL}{\cos kL}, \qquad c_2 = \frac{Q}{k^3 EI}, \qquad c_3 = -\frac{Q}{k^2 EI}, \qquad c_4 = \frac{Q}{k^3 EI}\frac{\sin kL}{\cos kL}$$

As a result, the deflected shape is determined to be

$$v(x) = \frac{Q}{k^3 EI}\frac{1}{\cos kL}[-\sin kL \cos kx + \cos kL \sin kx - kx \cos kL + \sin kL]$$

A couple of points to note about this solution. First, if Q is removed, then the beam returns to its initially straight position. Furthermore, this elastic behavior is linear — a doubling of Q results in a doubling of deflection. Second, the axial load appears in the solution in a complicated fashion; the deflection is not linear with respect to this load. This is another example of load interactions. A final point is that we can recover the uncoupled case (straight beam) by appropriately letting $P \to 0$. We now look at varying P.

Example 6.3: Determine the load interaction effect of the axial load on the stiffness relation for the cantilever beam.

Since we are eventually interested in stiffness-type relations, consider the above solution for a particular point. For convenience, choose the end point $x = L$, then

$$v(L) = \frac{Q}{k^3 EI}\left[\frac{\sin kL - kL \cos kL}{\cos kL}\right]$$

Rearrange this as a stiffness relation in the form

$$\frac{EI}{L^3}\left[\frac{k^3 L^3 \cos kL}{\sin kL - kL \cos kL}\right] v(L) = Q \qquad \text{or} \qquad [K_T]v_L = Q$$

Figure 6.9 shows a plot of the stiffness as a function of the initial axial load. The axial load is seen to have a profound effect on the stiffness. Indeed, there are multiple points where the stiffness is zero and other points where it is infinite (both positive and negative).

To show that these results are consistent with the straight beam, choose P small so that the parameter $\xi \equiv kL$ is also small, then

$$v(L) \approx \frac{Q}{k^3 EI}\left[\frac{\xi - \xi^3/6 \cdots - \xi(1 - \xi^2/2 \cdots)}{1 - \xi^2/6 \cdots}\right] = \frac{Q}{k^3 EI}\left[\frac{\xi^3/3 \cdots}{1 - \xi^2/6 \cdots}\right] = \frac{QL^3}{3EI}$$

This indeed is the uncoupled value obtained in the previous example.

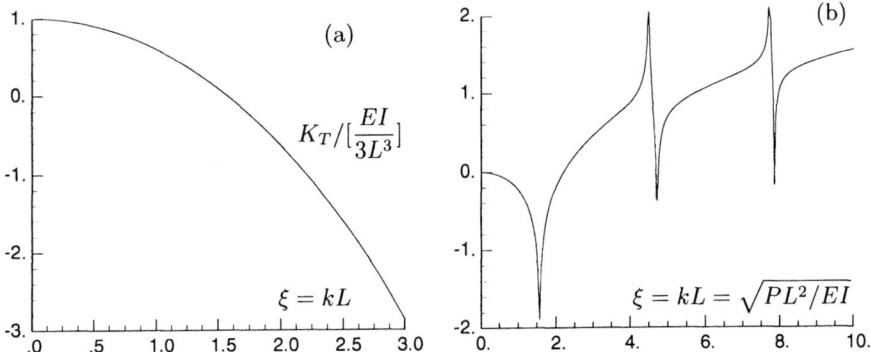

Figure 6.9: Stiffness as a function of axial load. (a) Linear plot over small range. (b) Log plot over large range.

The zero stiffness crossings in this example correspond to critical loads because at these points the beam has complete loss of stiffness and is in a state of neutral equilibrium. That is, even for a nominal Q the deflections are indicated to be very large. (An alternative viewpoint is that any specified position can be obtained with very little Q.) These critical points occur at

$$\cos kL = 0 \qquad \text{or} \qquad kL = \tfrac{1}{2}\pi, \tfrac{3}{2}\pi, \tfrac{5}{2}\pi, \cdots, \tfrac{1}{2}(n+1)\pi$$

On substituting for k in terms of the force, this gives

$$P = (n+1)^2 \pi^2 \frac{EI}{4L^2}$$

There are many critical loads. The minimum load to cause buckling is at $n = 0$, hence

$$P_{cr} = EI\left(\frac{\pi}{2L}\right)^2$$

An alternative way to write this relation is in terms of the critical stress. Let the moment of inertia of the beam cross-section be $I = \frac{1}{12}bh^3$, then

$$\sigma_{cr} = \frac{P_{cr}}{bh} = \alpha E[\frac{h}{L}]^2, \qquad \alpha = \frac{\pi^2}{8}$$

The ratio L/h is called the *slenderness ratio*. We see from this relation the profound effect the slenderness of the beam has on the failure stress: typically, the yield stress is related to the modulus as $\sigma_Y \approx E/200$ and therefore, for beams with $L > 15h$, the beam will buckle collapse before it yields. Furthermore, each doubling of the beam length reduces the critical load by a factor of four.

Effect of Load and Geometric Imperfections

A perfect beam/column loaded perfectly along its axis would not buckle. For buckling to occur we need an agent; in the previous developments, we considered the transverse load Q as the agent for causing the instability. Furthermore, we saw that even in the limit of Q going to zero we still could ascertain the critical axial loads.

A related idea is the role played by geometric imperfections and load eccentricities. Essentially, the effect of both is to give the beam a "nudge" in the transverse direction and thus set it in motion in the transverse direction even though the loads may be only axial. That is, they too can act as agents for the instability.

Example 6.4:   Determine the effect of the load eccentricity on the deformation of the pinned/pinned beam shown in Figure 6.10.

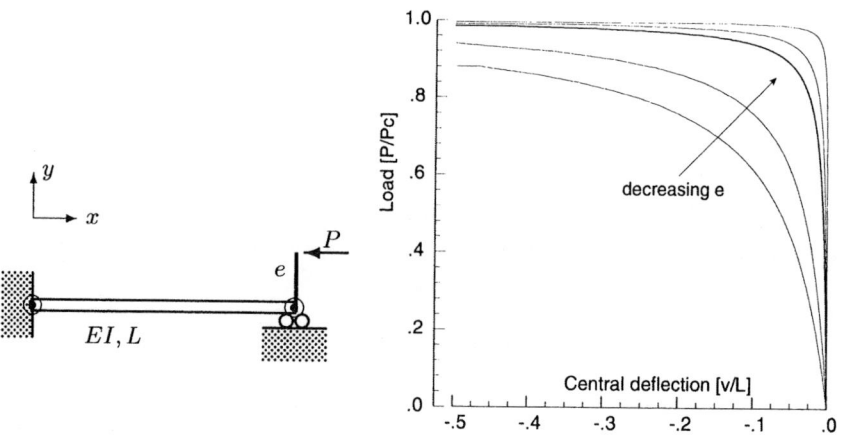

Figure 6.10: Pinned/pinned beam with eccentric load. The eccentricities range from $L/10$ to $L/1000$.

Let the load be applied a distance e off the axis of the beam. This can be thought of as a geometric imperfection (the beam is not straight with respect to the load point and the support) or as a load imperfection (the actual load at the end of the perfect beam is a combination axial load and moment). We will take this latter view. By inspection get that the axial load is $\bar{F}_o = -P$, and the boundary conditions are

at $x = 0$: $v = 0 \quad \Rightarrow \quad 0 = c_1 + c_4$

$$M = EI\frac{d^2v}{dx^2} = 0 \quad \Rightarrow \quad 0 = -c_1 k^2$$

at $x = L$:
$$v = 0 \quad \Rightarrow \quad 0 = c_1 \cos kL + c_2 \sin kL + c_3 L + c_4$$

$$M = EI\frac{d^2v}{dx^2} = Pe \quad \Rightarrow \quad \frac{Pe}{EI} = -c_1 k^2 \cos kL - c_2 k^2 \sin kL$$

Note that the additional moment term enters the equations in that same way as Q previously. We can solve for all the coefficients to get

$$v(x) = \frac{-Pe}{EIk^2}\left[\frac{\sin kx}{\sin kL} - \frac{x}{L}\right] = -e\left[\frac{\sin kx}{\sin kL} - \frac{x}{L}\right], \qquad k = \sqrt{\frac{P}{EI}}$$

Clearly, whenever

$$\sin kL = 0 \qquad \text{or} \qquad kL = \pi, 2\pi, 3\pi, \ldots \qquad \text{or} \qquad P = \frac{EI}{L^2}[\pi^2, 4\pi^2, 9\pi^2, \ldots]$$

we get a very large deflection somewhere along the beam. A plot of the mid-beam deflection for different values of eccentricity is shown in Figure 6.10. In the limit as e goes to zero, the plot limits to a bifurcation.

We get the same effect if we consider an initial curvature (or geometric imperfection) given, say, by $v_o(x) = e \sin(\pi x/L)$.

Instability as an Eigenvalue Problem

In a buckling analysis we seek only the critical values of load and possibly the corresponding deflection shapes. This section shows that we can get this information from an eigenvalue analysis without explicitly considering the agents for the instability.

To see this, recall that there are four constants of integration that are to be determined from the boundary conditions. Thus, we always set up a system of simultaneous relations as

$$[A(P)]\{c\} = \{B(Q)\}$$

where $[A(P)]$ means the system of equations depends on the axial loading P, the vector of coefficients is $\{c\} = \{c_1, c_2, c_3, c_4\}$, and $\{B(Q)\}$ are the specific form of boundary conditions that may depend on the transverse loads Q. To solve for c_1, say, we could use Cramer's rule, but for this to be valid we must have $\det[\,A\,] \neq 0$. Note, however, that it is possible for $\det[\,A\,]$ to be zero and at precisely those values Cramer's rule breaks down and there is not a unique connection between the coefficients (i.e., the deflection shape) and the boundary conditions. This means that one or more of the equations are linearly dependent and therefore the solution is not unique. In other words, there is a second solution $\{c^*\}$ that also satisfies the equations

$$[A(P)]\{c^*\} = \{B(Q)\}$$

Let the difference of the two solutions be $\{\phi\} = \{c\} - \{c^*\}$; it satisfies the homogeneous equations

$$[A(P)]\{\phi\} = \{0\}$$

In general, this admits only the trivial solution $\{\phi\} = \{0\}$, but when $\det[\,A\,]$ is zero, there are other solutions.

From the above, it is clear that we need only consider the homogeneous problem in order to determine the critical loads. That is, we consider the problem

$$[A(\lambda)]\{\phi\} = \{0\}$$

where λ is some parameter (in our case the critical load), special values of which cause the determinant of $[A(\lambda)]$ to be zero. This is known as an *eigenvalue problem*; actually, since $[\,A\,]$ contains transcendental functions, it is usually referred to as a transcendental eigenvalue problem. We obtain the special values of λ (called *eigenvalues*) by setting the determinant to zero. Corresponding to each eigenvalue, we can find a solution for $\{\phi\}$, this is called an *eigenvector*. This terminology is the same as introduced during the discussion of vibrations in Chapter 4.

Example 6.5:   Determine the buckling loads for the pinned/pinned beam shown in Figure 6.11.

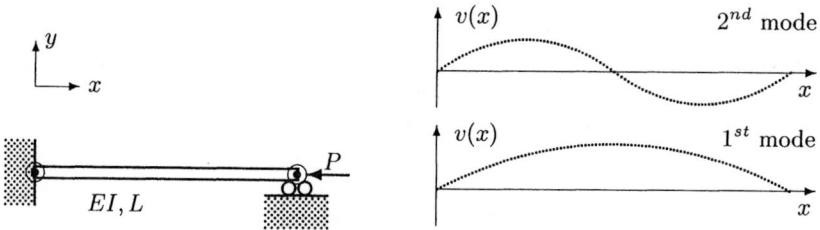

Figure 6.11: Mode shapes for pinned/pinned beam.

Except for P, the problem does not state any other applied loads, therefore, by inspection get that $\bar{F}_o = -P$. The boundary conditions are

at $x = 0$:
$$v = 0 \quad \Rightarrow \quad 0 = c_1 + c_4$$
$$M = EI\frac{d^2v}{dx^2} = 0 \quad \Rightarrow \quad 0 = -c_1 k^2$$

at $x = L$:
$$v = 0 \quad \Rightarrow \quad 0 = c_1 \cos kL + c_2 \sin kL + c_3 L + c_4$$
$$M = EI\frac{d^2v}{dx^2} = 0 \quad \Rightarrow \quad 0 = -c_1 k^2 \cos kL - c_2 k^2 \sin kL$$

where $k = \sqrt{P/EI}$. These equations are quite similar to those of the eccentric load problem except that the moment is zero. These four equations can be put into matrix form as

$$
\begin{bmatrix}
1 & 0 & 0 & 1 \\
-k^2 & 0 & 0 & 0 \\
\cos kL & \sin kL & L & 1 \\
-k^2 \cos kL & -k^2 \sin kL & 0 & 0
\end{bmatrix}
\begin{Bmatrix}
c_1 \\ c_2 \\ c_3 \\ c_4
\end{Bmatrix} = 0
$$

We now inquire if the determinant of this matrix can be zero. On multiplying out get

$$
\det = k^4 L \sin kL = 0
$$

There are many values of k when this equation is satisfied. The obvious one of $k = 0$ corresponds to the trivial case of zero axial load. The other possibilities are

$$
kL = \pi,\, 2\pi,\, 3\pi,\, \cdots,\, n\pi
$$

On substituting for k in terms of the force, this gives

$$
P_{cr} = n^2 \pi^2 \frac{EI}{L^2}
$$

There are many critical loads, and corresponding to each there is a different deflected shape. To determine these shapes, let us reconsider the relation among the coefficients. At the special values of $kL = n\pi$, the matrix for determining $\{c\}$ reduces to

$$
\begin{bmatrix}
1 & 0 & 0 & 1 \\
-k^2 & 0 & 0 & 0 \\
1 & 0 & L & 1 \\
-k^2 & 0 & 0 & 0
\end{bmatrix}
\begin{Bmatrix}
c_1 \\ c_2 \\ c_3 \\ c_4
\end{Bmatrix} = 0
$$

From this we get

$$
c_1 = 0,\qquad c_3 = 0,\qquad c_4 = 0
$$

But we cannot determine c_2. In other words, the equations are satisfied for any value of c_2. Hence the mode shape is

$$
v(x) = c_2 \sin kx,\qquad k = n\pi/L
$$

The first two mode shapes are shown in Figure 6.11. Note that, since c_2 is unknown, we have determined the *shape* of the deflection but not the actual deflection.

Example 6.6: Determine the effect of an elastic foundation on the buckling loads of a simply supported beam.

The governing equation for a beam on an elastic foundation is

$$
EI\frac{d^4 v}{dx^4} - \bar{F}_o \frac{d^2 v}{dx^2} + Kv = 0
$$

Since this has constant coefficients, we seek solutions of the form $v(x) = Ae^{ikx}$, which leads to the characteristic equation

$$
k^4 + \tilde{P}k^2 - \beta^4 = 0,\qquad \beta^4 \equiv \frac{(-K)}{EI},\qquad \tilde{P} \equiv \frac{P_o}{EI}
$$

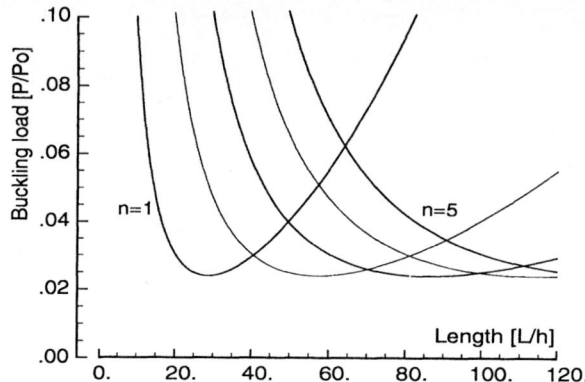

Figure 6.12: Buckling loads for a pinned/pinned beam on an elastic foundation.

from which we obtain the four spectrum relations

$$k_{1,3}(P) = \pm\sqrt{\sqrt{\beta^4 + \tfrac{1}{4}\tilde{P}^2} - \tfrac{1}{2}\tilde{P}} \equiv \pm\alpha, \quad k_{2,4}(P) = \pm i\sqrt{\sqrt{\beta^4 + \tfrac{1}{4}\tilde{P}^2} + \tfrac{1}{2}\tilde{P}} \equiv \pm\bar{\alpha}$$

Thus, the general solution is represented by

$$v(x) = \mathbf{A}e^{-i\alpha x} + \mathbf{B}e^{-\bar{\alpha}x} + \mathbf{C}e^{+i\alpha x} + \mathbf{D}e^{+\bar{\alpha}x} \tag{6.5}$$

or alternatively,

$$v(x) = c_1 \cos(\alpha x) + c_2 \sin(\alpha x) + c_3 \cosh(\bar{\alpha}x) + c_4 \sinh(\bar{\alpha}x)$$

when α and $\bar{\alpha}$ are likely to be real only.

For the simply supported boundary conditions, we have that at $x = 0$

$$v = 0 \quad \Rightarrow \quad \hat{v} = 0 = c_1 + c_3$$

$$M = 0 \quad \Rightarrow \quad \frac{d^2\hat{v}}{dx^2} = 0 = -\alpha^2 c_1 + \bar{\alpha}^2 c_3 = 0$$

This leads to $c_1 = 0$ and $c_3 = 0$. At $x = L$, we have

$$v = 0 \quad \Rightarrow \quad \hat{v} = 0 = c_1 S + c_3 S_h$$

$$M_{xx} = 0 \quad \Rightarrow \quad \frac{d^2\hat{v}}{dx^2} = 0 = -\alpha^2 c_1 S + \bar{\alpha}^2 c_3 S_h = 0$$

where $S \equiv \sin(\alpha L)$ and $S_h \equiv \sinh(\bar{\alpha}L)$. These two equations lead to the system

$$\begin{bmatrix} \sin\alpha L & \sinh\bar{\alpha}L \\ -\alpha^2 \sin\alpha L & -\bar{\alpha}^2 \sinh\bar{\alpha}L \end{bmatrix} \begin{Bmatrix} c_2 \\ c_4 \end{Bmatrix} = 0$$

The characteristic equation is obtained by setting the determinant equal to zero and gives

$$(\alpha^2 + \bar{\alpha}^2) \sin(\alpha L) \sinh(\bar{\alpha}L) = 0$$

Since $\sinh(\bar{\alpha}L)$ is always positive, this equation has the solutions

$$\alpha L = n\pi \quad \Rightarrow \quad \sqrt{\tfrac{1}{2}\sqrt{4\beta^4 + \tilde{P}^2} - \tfrac{1}{2}\tilde{P}} = \frac{n\pi}{L}$$

Expanding and rearranging gives

$$P_{cr} = EI\left[(\frac{n\pi}{L})^2 + \frac{K}{EI}(\frac{L}{n\pi})^2\right]$$

The corresponding mode shapes are given by

$$v_n(x) = c_2 \sin(\alpha x) = c_2 \sin(\frac{n\pi x}{L})$$

These mode shapes are the same as for the buckling of a simply supported beam.

The effect of the elastic foundation is to always increase the buckling load. In so doing, it actually changes the effective length of the beam. To see this, the variation of buckling load with beam length is shown in Figure 6.12. It is noted that for a beam of given length, the lowest buckling load does not necessarily coincide with the simplest mode shape. As the length increases, it is a higher and higher mode that buckles and the limiting buckling load is independent of length. We can get this overall minimum buckling by setting

$$\frac{\partial P_{cr}}{\partial L} = 0 = EI\left[-\frac{2}{L}(\frac{n\pi}{L})^2 + \frac{K}{EI}\frac{2}{L}(\frac{L}{n\pi})^2\right]$$

from which $(n\pi/L)^2 = \sqrt{K/EI}$ giving

$$P_{min} = 2\sqrt{K\,EI}\,, \qquad L_{\text{eff}} = \pi\sqrt{EI/K}$$

As the length is increased, the beam will always buckle in a shape that is multiples of L_{eff}.

Free Vibration with Axial Force

From the foregoing, we know an in-plane load can cause a stiffening or softening of a structure. Not surprisingly, therefore, is that it also affects the vibration characteristics. In this chapter, we are specifically interested in buckling and therefore wish to consider the vibration characteristics as the applied load approaches the critical value. As intimated in the introductory discussion of stability, the vibrational characteristics contain information about the stability state.

Consider the case when there is an elastic foundation; the governing equation is

$$EI\frac{\partial^4 v}{\partial x^4} - \bar{F}_o(t)\frac{\partial^2 v}{\partial x^2} + Kv + \rho A\frac{\partial^2 v}{\partial t^2} = 0$$

As written, this is a case where the coefficients are variable; solutions to these equations are very difficult to find as we saw in Section 4.6 on parametric excitation, therefore we leave this general case to the next chapter. To simplify matters, we will consider only the case when the applied load $P(t)$ is quasi-static, that is, $\bar{F}_o(t) = P(t) = P_o$.

The spectral form of the differential equation for $\hat{v}$ becomes

$$\frac{d^4\hat{v}}{dx^4} - \tilde{P}\frac{d^2\hat{v}}{dx^2} - \beta^4\hat{v} = 0\,, \qquad \beta^4 \equiv \frac{(\rho A\omega^2 - K)}{EI}\,, \qquad \tilde{P} \equiv \frac{\bar{F}_o}{EI} = \frac{P_o}{EI}$$

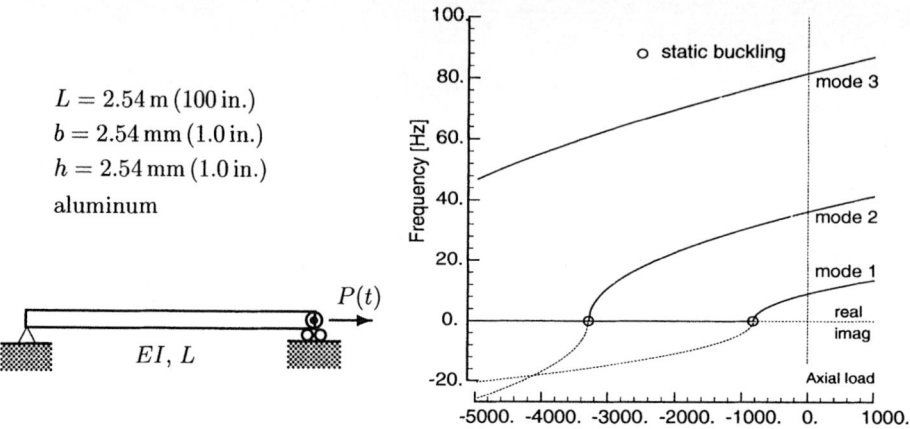

$L = 2.54\,\mathrm{m}\,(100\,\mathrm{in.})$
$b = 2.54\,\mathrm{mm}\,(1.0\,\mathrm{in.})$
$h = 2.54\,\mathrm{mm}\,(1.0\,\mathrm{in.})$
aluminum

$P(t)$

EI, L

Figure 6.13: Beam with quasi-static axial load and its effect on the vibration characteristics.

Since this has constant coefficients, we seek solutions of the form $\hat{v} = A e^{ikx}$, which leads to the characteristic equation

$$k^4 + \tilde{P}k^2 - \beta^4 = 0$$

from which we obtain the four spectrum relations

$$k_{1,3}(\omega) = \pm\sqrt{\tfrac{1}{2}\sqrt{4\beta^4 + \tilde{P}^2} - \tfrac{1}{2}\tilde{P}} \equiv \pm\alpha, \qquad k_{2,4}(\omega) = \pm i\sqrt{\tfrac{1}{2}\sqrt{4\beta^4 + \tilde{P}^2} + \tfrac{1}{2}\tilde{P}} \equiv \pm\bar{\alpha}$$

Thus, the general solution is represented by

$$\hat{v}(x) = \mathbf{A} e^{-i\alpha x} + \mathbf{B} e^{-\bar{\alpha}x} + \mathbf{C} e^{+i\alpha x} + \mathbf{D} e^{+\bar{\alpha}x} \qquad (6.6)$$

As noted in Chapter 4, in analyzing problems without damping (where α and $\bar{\alpha}$ are likely to be real only), we will find it more convenient to use the solution form

$$\hat{v}(x) = c_1 \cos(\alpha x) + c_2 \sin(\alpha x) + c_3 \cosh(\bar{\alpha}x) + c_4 \sinh(\bar{\alpha}x)$$

Both forms, of course, will lead to the same answer.

Example 6.7: Determine the resonance frequencies and mode shapes for a simply supported beam with an axial load.

The solution follows almost identically to that for the beam on the elastic foundation. At $x = 0$, we have

$$v = 0 \quad \Rightarrow \quad \hat{v} = 0 = c_1 + c_3$$

$$M = 0 \quad \Rightarrow \quad \frac{d^2\hat{v}}{dx^2} = 0 = -\alpha^2 c_1 + \bar{\alpha}^2 c_3 = 0$$

This leads to $c_1 = 0$ and $c_3 = 0$. At $x = L$, we have

$$v = 0 \quad \Rightarrow \quad \hat{v} = 0 = c_1 S + c_3 S_h$$

$$M_{xx} = 0 \quad \Rightarrow \quad \frac{d^2\hat{v}}{dx^2} = 0 = -\alpha^2 c_1 S + \bar{\alpha}^2 c_3 S_h = 0$$

where $S \equiv \sin(\alpha L)$ and $S_h \equiv \sinh(\bar{\alpha}L)$. These two equations lead to the system

$$
\begin{bmatrix} \sin \alpha L & \sinh \bar{\alpha} L \\ -\alpha^2 \sin \alpha L & -\bar{\alpha}^2 \sinh \bar{\alpha} L \end{bmatrix} \begin{Bmatrix} c_2 \\ c_4 \end{Bmatrix} = 0
$$

The characteristic equation is obtained by setting the determinant equal to zero and gives

$$
(\alpha^2 + \bar{\alpha}^2) \sin(\alpha L) \sinh(\bar{\alpha} L) = 0
$$

This has the solutions

$$
\alpha L = n\pi \qquad \Rightarrow \qquad \sqrt{\tfrac{1}{2}\sqrt{4\beta^4 + \tilde{P}^2} - \tfrac{1}{2}\tilde{P}} = n\pi/L
$$

Expanding and rearranging gives

$$
\omega_n = \sqrt{\frac{EI}{\rho A}} \sqrt{(\frac{n\pi}{L})^2 \left[(\frac{n\pi}{L})^2 + \frac{P_o}{EI}\right] + \frac{K}{EI}}
$$

The corresponding mode shapes are given by

$$
\hat{v}_n(x) = c_2 \sin(\alpha x) = c_2 \sin(\frac{n\pi x}{L})
$$

These mode shapes are the same as for the buckling of a simply supported beam.

The variation of resonance frequency with axial load is shown in Figure 6.13 for $K = 0$. We see that the axial tension increases the frequency, while an axial compression decreases it. In fact, the frequency can be driven to zero when

$$
P_o = P_c = EI(\frac{n\pi}{L})^2
$$

These are the collection of static buckling loads.

At the first static buckling load, the vibrational frequency is zero ($\lambda_1 = \omega^2 = 0$ in our stability analysis) but the mode shape is that of a single half-sine. At a load slightly more compressive, the mode shape stays the same but ($\lambda_1 < 0$) and the beam is unstable.

Example 6.8:   Investigate the effect of a varying axial force on the vibration characteristics of a beam.

Consider the case when the internal force distribution is symmetric and applied to a simply supported beam. The force and corresponding applied load are given by

$$
\bar{F}_o(x) = P_o \sin(3\pi x/L), \qquad q(x) = -\frac{d\bar{F}_o}{dx} = -\frac{3\pi}{L} P_o \cos(3\pi x/L)
$$

These are shown plotted in Figure 6.14(a). This distributed load was approximated over 24 elements and Figure 6.14(b) shows the results of the vibration eigenanalysis. There is no essential difference from the previous example except that the antisymmetric mode is relatively insensitive to the symmetric internal load.

We get the simple theory solution by using a Ritz approximation. The strain energy, in general, is

$$
U = \tfrac{1}{2} \int_L \left[EI(\frac{d^2v}{dx^2})^2\right] dx + \tfrac{1}{2} \int_L \left[\bar{F}_o(\frac{\partial v}{\partial x})^2\right] dx
$$

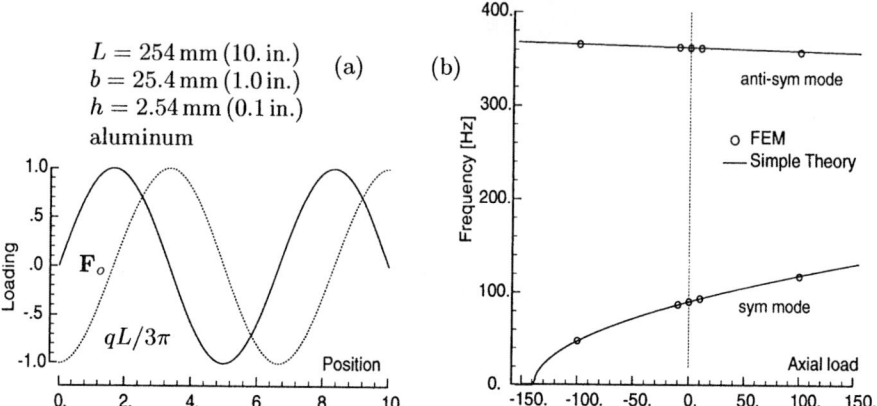

Figure 6.14: Effect of varying axial stress on the vibration. (a) Applied loading and internal force distribution. (b) Resonance frequency as a function of the load magnitude.

Now assume that the mode shapes can be approximated by

$$v(x) = \tilde{v}\sin(\xi_n x), \qquad \xi_n \equiv n\pi/L$$

where $n = 1$ is a symmetric mode and $n = 2$ is antisymmetric. Substituting this and the expression for $\bar{F}_o(x)$ into the energy expression, and integrating leads to

$$U = \tfrac{1}{2}\tilde{v}^2 EI\xi_n^4 L/2 + \tfrac{1}{2}\tilde{v}^2 \xi_n^2 P_o C_n L/2, \qquad C_n = 2.97, \; -0.30$$

The kinetic energy is handled in a similar way, leading to

$$T = \tfrac{1}{2}\int \rho A \dot{v}^2 \, dx \qquad \Rightarrow \qquad T = \tfrac{1}{2}\rho A \dot{\tilde{v}}^2 L/2$$

Assuming harmonic motion, and obtaining the stiffness and mass in the usual way, leads to the eigenvalue problem

$$\left[EI\xi_n^4 + \xi_n^2 P_o C_n - \rho A \omega^2\right]\tilde{v} = 0 \qquad \text{or} \qquad \omega_n = \sqrt{\frac{EI\xi_n^4 + \xi_n^2 P_o C_n}{\rho A}}$$

This is shown plotted as the full line in Figure 6.14(b). That the FEM results and this simple theory agree so well is not surprising, since both are based on a similar Ritz approximation. Indeed, this is the reason why in this book most of the comparisons with the FEM results use a method other than the Ritz method.

Local Buckling and Post-Buckling Behavior

An eigenanalysis gives the critical load, but it does not indicate what happens after that. In our discussion of the elastica in Section 3.1, Figure 3.8 indicates that some structural components continue to maintain their load-bearing capability and even increase it. We will now look closer at this.

We must distinguish between load-bearing capability and stiffness. A compressed column looses stiffness as the load is increased, at the critical point the

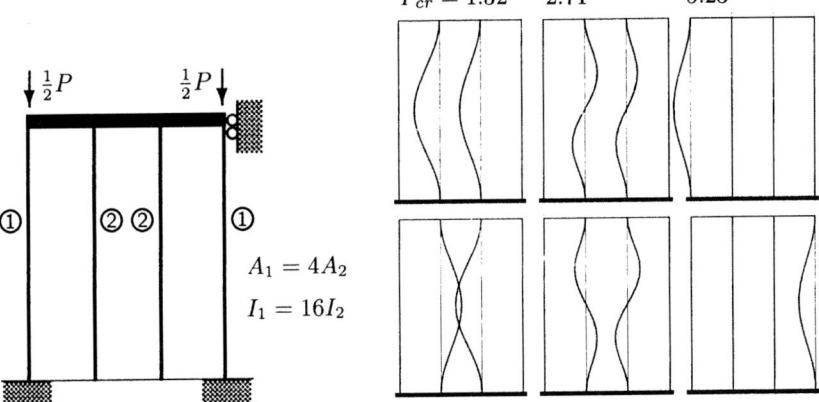

Figure 6.15: A constrained frame structure and its first six buckled shapes.

stiffness is zero but its load-bearing capability is still the applied load. Therefore, a simple approach to post-buckling analysis is to assume that only the critical members are affected, that they maintain their critical load after becoming critical, but do not share in any further increases in load.

This simple post-buckling analysis works best when there is *local buckling*, that is, the buckling mode is confined to a single member or portion of the structure. An example of local buckling is shown in Figure 6.15 where the inside and outside members have cross-sections $[a \times a]$ and $[2a \times 2a]$, respectively. Each member buckles as if it is a standalone beam with fixed/fixed boundary conditions. Note that the rigid horizontal bars plus the upper right constraint serves to isolate each column — if this constraint is removed, then the lowest buckling mode is a global one where all members tilt to the side.

Example 6.9: Estimate the post-buckling behavior of the frame shown in Figure 6.15.

Assume the members behave as clamped/clamped columns so that the critical load is $P_{cr} = 4EI(n\pi/L)^2$, then the ratio of critical loads are $P_1/P_2 = 16$ and the inner members become critical first. Actually, the inner members have two buckling loads before the outer ones become critical.

Just at the critical point $P_2 = 4EI_2(\pi/L)^2 = P_o$, the load is shared among all members in proportion to their areas so that

$$P = 2P_1 + 2P_2 = 2(4+1)P_o = 10P_o, \qquad \frac{\Delta L}{L} = \frac{P_2}{EA_2} = \frac{P_o}{EA_2} \equiv \epsilon_o$$

After this point, the outer members carry all the increments in load until they too become critical. Then

$$P = 2P_{1cr} + 2P_{2cr} = 2(16+1)P_o = 34P_o, \qquad \frac{\Delta L}{L} = \frac{P_1}{EA_1} = \frac{16P_o}{4EA_2} = 4\epsilon_o$$

Note that it is assumed that the middle members deform the same as the outer members. This is consistent with their loss of stiffness while maintaining their load-bearing ability.

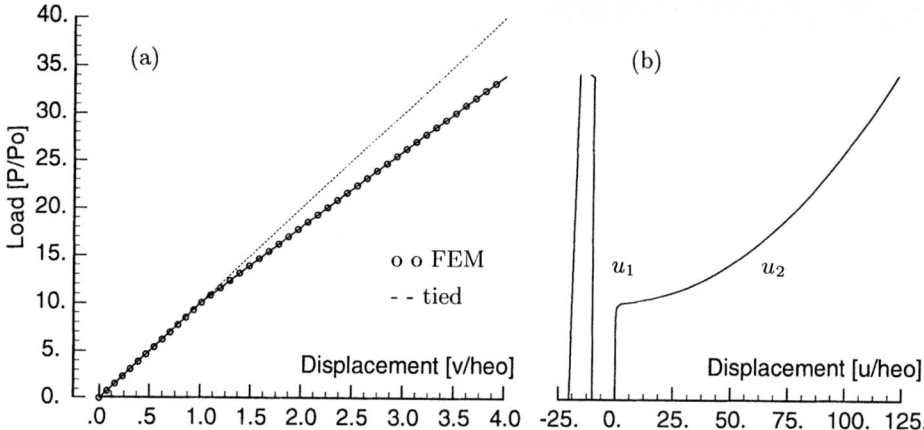

Figure 6.16: Post-buckling behavior of the frame. (a) End shortening. (b) Large deformation FEM results for transverse deflections.

The plot of load against shortening is shown in Figure 6.16(a). The overall stiffness is hardly affected, and the ultimate load achieved is more than three times that of the first buckling.

Let us now tie the inner members to the outer ones at their centers. This will prevent the first buckling mode from occurring, but will not affect the second mode. Our estimate of the first critical point is then

$$P = 2P_1 + 2P_2 = 2(4+1)4P_o = 40P_o \,, \qquad \frac{\Delta L}{L} = \frac{P_2}{EA_2} = \frac{4P_o}{EA_2} = 4\epsilon_o$$

and the second buckling is

$$P = 2P_{1cr} + 2P_{2cr} = 2(16+4)P_o = 40P_0 \,, \qquad \frac{\Delta L}{L} = \frac{P_1}{EA_1} = \frac{16P_o}{4EA_2} = 4\epsilon_o$$

In this particular instance, tying the members together caused the outer members to buckle at the same time as the inner ones. In this case, we have a global buckling.

Example 6.10:   Do a large deformation analysis of the frame shown in Figure 6.15 and compare with the simple theory.

Small transverse loads $(Q = P \times 10^{-5})$ are applied to the center of each member to act as the agent for the buckling. These results are shown plotted in Figure 6.16(b). Note the very large transverse deflection of the inner member relative to that of the outer member. In fact, the outer member hardly moves in the transverse direction until the collapse load is reached.

Also shown in Figure 6.16(a) are the large deformation results for the end shortening. The comparison with the simple theory is very good.

6.3   Buckling of Plates

Our development for the buckling analysis of plates follows along lines very similar to that for beams. That is, we include second-order effects in the strain displacement relations and then use Hamilton's principle to obtain the governing equations and the associated boundary conditions.

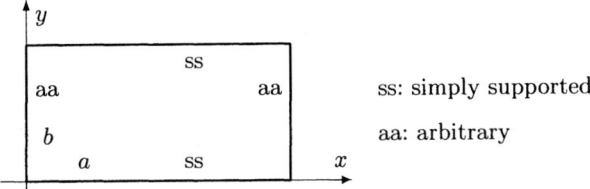

Figure 6.17: Rectangular plate of size $[a \times b]$ and the type of boundary conditions.

Governing Equations

Consider the x-component of the Lagrangian strain in a plate in flexure

$$E_{xx} = \frac{\partial u}{\partial x^o} + \frac{1}{2}\left[\left(\frac{\partial u}{\partial x^o}\right)^2 + \left(\frac{\partial v}{\partial x^o}\right)^2 + \left(\frac{\partial w}{\partial x^o}\right)^2\right]$$

where we are using the large displacement coordinate description of Chapter 1. Thin-plate theory gives approximations for the in-plane displacements as

$$u(x^o, y^o, z^o) \approx u(x^o, y^o) - z^o\frac{\partial w}{\partial x^o}, \qquad v(x^o, y^o, z^o) \approx v(x^o, y^o) - z^o\frac{\partial w}{\partial y^o}$$

and the out-of-plane deflection as $w(x^o, y^o, z^o) \approx w(x^o, y^o)$. Substitute these into the strain energy expression to get

$$
\begin{aligned}
E_{xx} &= \frac{\partial u}{\partial x^o} + \frac{1}{2}\left[\left(\frac{\partial u}{\partial x^o}\right)^2 + \left(\frac{\partial v}{\partial x^o}\right)^2 + \left(\frac{\partial w}{\partial x^o}\right)^2\right] \\
&\quad - z^o\frac{\partial^2 w}{\partial x^{o2}} + \frac{1}{2}\left[\left(-z^o\frac{\partial^2 w}{\partial x^{o2}}\right)^2 + \left(-z^o\frac{\partial^2 w}{\partial x^{o2}}\right)^2\right] \\
&\approx \frac{\partial u}{\partial x^o} - z^o\frac{\partial^2 w}{\partial x^{o2}} + \frac{1}{2}\left(\frac{\partial w}{\partial x^o}\right)^2
\end{aligned}
$$

The approximation is based on the assumption that the out-of-plane deflection is significantly larger than the in-plane displacements. We recognize the leading term as that of the linear membrane theory of plates, the second as that of the linear flexure theory of plates; the third term, which is nonlinear, comes from the large deflection of the plate. We have similar expressions for the other in-plane components of strain

$$E_{yy} = \frac{\partial v}{\partial y^o} - z^o \frac{\partial^2 w}{\partial y^{o2}} + \frac{1}{2}\left(\frac{\partial w}{\partial y^o}\right)^2, \qquad 2E_{xy} = \frac{\partial u}{\partial y^o} + \frac{\partial v}{\partial x^o} - 2z^o \frac{\partial^2 w}{\partial x^o \partial y^o} + \frac{\partial w}{\partial x^o} \frac{\partial w}{\partial y^o}$$

Now that we have established the nonlinear contribution to the strains we can revert back to the notation that does not distinguish between the deformed and undeformed configurations.

All these strains give rise to stresses according to

$$\sigma_{xx} = \frac{E}{1-\nu^2}[E_{xx} + \nu E_{yy}]$$

$$\sigma_{yy} = \frac{E}{1-\nu^2}[E_{yy} + \nu E_{xx}]$$

$$\sigma_{xy} = G2E_{xy}$$

The strain energy for a plate in plane stress is

$$U = \frac{1}{2}\int_V [\sigma_{xx}E_{xx} + \sigma_{yy}E_{yy} + \sigma_{xy}2E_{xy}]\, dV$$

After substitution and integration, we get

$$\begin{aligned}
U = \; &\frac{1}{2}D\int_A \left[(\nabla^2 w)^2 + 2(1-\nu)\left[(\frac{\partial^2 w}{\partial x \partial y})^2 - (\frac{\partial^2 w}{\partial x^2})(\frac{\partial^2 w}{\partial y^2})\right]\right] dxdy \\
&+ \frac{1}{2}\int_A \left[E^* h\left[(\frac{\partial u}{\partial x})^2 + (\frac{\partial v}{\partial y})^2 + 2\nu \frac{\partial u}{\partial x}\frac{\partial v}{\partial y}\right] + Gh\left(\frac{\partial u}{\partial y} + \frac{\partial v}{\partial x}\right)^2\right] dx\, dy \\
&+ \frac{1}{2}\int_A \left[\bar{N}_{xx}(\frac{\partial w}{\partial x})^2 + 2\bar{N}_{xy}(\frac{\partial w}{\partial x})(\frac{\partial w}{\partial y}) + \bar{N}_{yy}(\frac{\partial w}{\partial y})^2 + Kw^2\right] dxdy
\end{aligned}$$

where we have added the effect of an elastic foundation. In this, we grouped some terms according to

$$\bar{N}_{xx} = \frac{Eh}{1-\nu^2}\left(\frac{\partial u}{\partial x} + \nu\frac{\partial v}{\partial y}\right)$$

$$\bar{N}_{yy} = \frac{Eh}{1-\nu^2}\left(\frac{\partial v}{\partial y} + \nu\frac{\partial u}{\partial x}\right)$$

$$\bar{N}_{xy} = Gh\left(\frac{\partial u}{\partial y} + \frac{\partial v}{\partial x}\right)$$

which correspond to in-plane loads. Strictly, these are deformation dependent, but in the following they are assumed to be pre-existing in the plate and therefore not subject to variation. Thus the energy expression comprises the contribution for the linear behavior plus three new terms associated with the in-plane loading and the elastic foundation.

Using Hamilton's principle, this leads to the uncoupled governing equations for the membrane and flexural behaviors. The latter equation is [63]

$$D\nabla^2\nabla^2 w - \bar{N}_{xx}\frac{\partial^2 w}{\partial x^2} - 2\bar{N}_{xy}\frac{\partial^2 w}{\partial x\partial y} - \bar{N}_{yy}\frac{\partial^2 w}{\partial y^2} + Kw + \rho h\frac{\partial^2 w}{\partial t^2} = q(x,y,t) \quad (6.7)$$

As in Chapter 2, the associated boundary conditions are found to be

$$\left\{ w \quad \text{or} \quad V_{xz} = -D\left[\frac{\partial^3 w}{\partial x^3} + (2-\nu)\frac{\partial^3 w}{\partial x\partial y^2}\right] + \bar{N}_{xx}\frac{\partial^2 w}{\partial x^2} \right\}$$

$$\left\{ \frac{\partial w}{\partial x} \quad \text{or} \quad M_{xx} = D\left[\frac{\partial^2 w}{\partial x^2} + \nu\frac{\partial^2 w}{\partial y^2}\right] \right\} \quad (6.8)$$

For the purpose of integrating the governing equations, we treat the in-plane terms as constant. They are known either by specification or through a separate linear analysis of the in-plane loads. In this way, the problem is reduced to being linear even though it takes large deflections into account.

Solution Structure for Rectangular Plates

Following the method introduced in Chapter 2 and also used in Chapter 4, we consider solutions of the form

$$w(x,y) = \sum_m \tilde{w}_m(x)e^{i\xi_m y}, \qquad \xi_m = \frac{m\pi}{b}$$

where b is the width of the plate. We will determine the coefficients $\tilde{w}_m$ by requiring that the governing differential equations be satisfied. The differential equation for $\tilde{w}$ becomes

$$(\frac{d^2}{dx^2} - \xi^2)(\frac{d^2}{dx^2} - \xi^2)\tilde{w}_m - \beta^4\tilde{w}_m - \frac{1}{D}[\bar{N}_{xx}\frac{d^2}{dx^2} + 2i\xi\bar{N}_{xy}\frac{d}{dx} - \bar{N}_{yy}\xi^2]\tilde{w} = 0$$

with $\beta^4 \equiv (-K)/D$. This has constant coefficients, hence, e^{-ikx} is a kernel solution. The spectrum relations are then

$$k^4 + 2k^2\xi^2 + \xi^4 - \beta^4 + \frac{1}{D}[\bar{N}_{xx}k^2 + 2\bar{N}_{xy}k\xi + \bar{N}_{yy}\xi^2] = 0$$

Because of the odd power of k occurring in this characteristic equation, the roots do not appear as $\pm$ pairs. Thus, the general solution is represented by

$$w(x,y) = \sum_m \left[\mathbf{A}e^{-ik_1 x} + \mathbf{B}e^{-ik_2 x} + \mathbf{C}e^{-ik_3 x} + \mathbf{D}e^{-ik_4 x}\right] e^{i\xi_m y} \quad (6.9)$$

In this way, the solution structure is quite similar to that already used in Chapter 2.

Example 6.11: Determine the buckling loads and mode shapes for a simply supported rectangular plate of size $[a \times b]$ with axial compression in the x-direction. When the in-plane shear force is zero, the roots are determined to be

$$k_{1,3} = \pm \left[\sqrt{\beta^4 + \xi^2(\bar{N}_{xx} - \bar{N}_{yy})\frac{1}{D} + \frac{1}{4D^2}\bar{N}_{xx}^2} - \xi^2 - \frac{1}{2D}\bar{N}_{xx} \right]^{1/2} \equiv \pm \alpha_m$$

$$k_{2,4} = \pm i \left[\sqrt{\beta^4 + \xi^2(\bar{N}_{xx} - \bar{N}_{yy})\frac{1}{D} + \frac{1}{4D^2}\bar{N}_{xx}^2} + \xi^2 + \frac{1}{2D}\bar{N}_{xx} \right]^{1/2} \equiv \pm i\bar{\alpha}_m$$

Thus, the general solution is represented by

$$w(x, y) = \sum_m \left[\mathbf{A}e^{-i\alpha_m x} + \mathbf{B}e^{-\bar{\alpha}_m x} + \mathbf{C}e^{+i\alpha_m x} + \mathbf{D}e^{+\bar{\alpha}_m x} \right] e^{i\xi_m y} \qquad (6.10)$$

which is identical to that of Equation (2.20). Sometimes, we will find it easier to use the solution form

$$w(x, y) = \sum_m \left[c_1 \cos(\alpha_m x) + c_2 \sin(\alpha_m x) + c_3 \cosh(\bar{\alpha}_m x) + c_4 \sinh(\bar{\alpha}_m x) \right] \begin{Bmatrix} \cos(\xi_m y) \\ \sin(\xi_m y) \end{Bmatrix}$$

Looking at the boundary conditions at $y = 0$ and $y = b$, we see that

$$w = 0, \qquad \frac{\partial^2 w}{\partial y^2} = 0$$

always. That is, this is true for each m term and implies that this particular solution can solve only those problems with simply supported lateral sides.

Choose only the $\sin(\xi_m y)$ terms. We need only concentrate on the boundary conditions at $x = 0$ and $x = a$, the lateral boundary conditions are automatically satisfied. The solution is precisely the same as for the simple plate with the same boundary conditions obtained in Chapter 2. Hence, we have the same characteristic equation

$$(\alpha_m^2 + \bar{\alpha}_m^2) \sin(\alpha_m a) \sinh(\bar{\alpha}_m a) = 0$$

This has the solutions

$$\alpha_m a = n\pi \qquad \Rightarrow \qquad \sqrt{\beta^4 + \xi^2(\bar{N}_{xx} - \bar{N}_{yy}) + \frac{1}{4D^2}\bar{N}_{xx}^2} - \xi^2 - \frac{1}{2D}\bar{N}_{xx} = (\frac{n\pi}{a})^2$$

Expanding and rearranging gives

$$(\frac{n\pi}{a})^2 \bar{N}_{xx} + (\frac{m\pi}{a})^2 \bar{N}_{yy} = -D \left[\left[(\frac{n\pi}{a})^2 + (\frac{m\pi}{b})^2 \right]^2 + \frac{K}{D} \right]$$

The corresponding mode shapes are given by

$$\tilde{w}_{mn}(x, y) = c_2 \sin(\alpha_m x) \sin(\xi_m y) = c_2 \sin(\frac{n\pi x}{a}) \sin(\frac{n\pi y}{b})$$

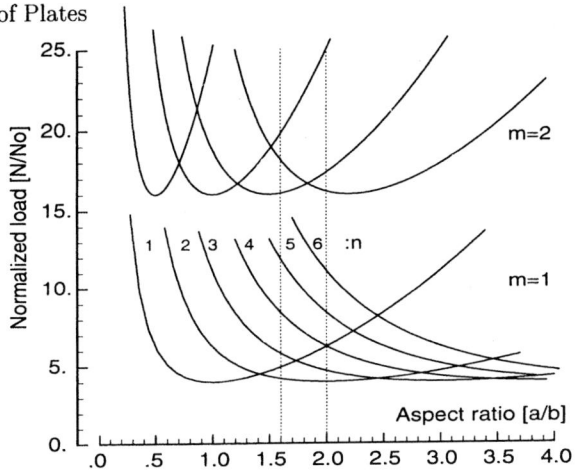

Figure 6.18: Buckling loads as a function of aspect ratio a/b.

In the special case of proportional loading ($N_{yy} = \gamma N_{xx}$) we have

$$\bar{N}_{xx} = -D(\frac{a}{n\pi})^2 \left[\left[(\frac{n\pi}{a})^2 + (\frac{m\pi}{b})^2 \right]^2 + \frac{K}{D} \right] / \left[1 + \gamma(\frac{ma}{nb})^2 \right]$$

It is clear that a biaxial compressive stress is more critical than a uniaxial stress. On the other hand, a tensile $\bar{N}_{yy}$ has a stiffening effect. The expression for the uniaxial case ($N_{yy} = 0$) is

$$\bar{N}_{xx} = -D(\frac{a}{n\pi})^2 \left[\left[(\frac{n\pi}{a})^2 + (\frac{m\pi}{b})^2 \right]^2 + \frac{K}{D} \right]$$

These buckling loads are shown plotted in Figure 6.18 for different values of aspect ratio a/b with $K = 0$. (The loads are normalized to $\bar{N}_o = D\pi^2/b^2$.) Note that the minimum loads correspond to the sine waves in the shortest dimension. Furthermore, for a given aspect ratio the sequence of modes does not necessarily correspond to going from simple to complex. As indicated for $a/b = 1.6$, for example, we get the (mn) sequence (2,1), (1,1), (3,1), (4,1), (3,2), (4,2), (2,2). Two buckling modes coincide when the aspect ratio is given by $a/b = \sqrt{n(n+1)}$, or $\sqrt{2} = 1.41$, $\sqrt{6} = 2.45$, $\sqrt{12} = 3.46$, and so on.

Furthermore, as the plate length increases, we get a minimum buckling load that is independent of the length. Referring to Figure 6.18, we get the minimum critical load by setting $m = 1$ and differentiating with respect to a (keeping b constant) to get

$$\frac{\partial \bar{N}_{xx}}{\partial a} = 0 \quad \Longrightarrow \quad \frac{a}{n} = b \quad \Longrightarrow \quad \bar{N}_{min} = -4D(\frac{\pi}{b})^2$$

The buckled (half) wavelength is the width of the plate. This behavior of a limiting wavelength is quite similar to that for a beam on an elastic foundation. Indeed, the $\xi_m = (m\pi/b)$ term enters the solution in the same way as K; consequently, we can think of the higher modes as being associated with higher orders of elastic constraint [23].

Example 6.12: Determine the buckled shapes for a rectangular plate simply supported on all sides with a uniform in-plane shear stress.

The differential equation for the plate with shear loads becomes

$$D\nabla^2\nabla^2 w - 2\bar{N}_{xy}\frac{\partial^2 w}{\partial x \partial y} + \rho h\frac{\partial^2 w}{\partial t^2} = q(x,y)$$

This problem was first solved in Reference [68] where inertia effects were also taken into consideration. We basically follow the same procedure but ignore the inertia effects. Consider solutions of the form

$$w(x,y,t) = \tilde{w}_m(y)e^{ikx/c}e^{i\omega t}$$

where $b = 2c$ is the width of the plate. We will determine the coefficients $\tilde{w}_m$ by requiring that the governing differential equations be satisfied. The differential equation for $\tilde{w}$ becomes

$$(\frac{d^2}{dz^2} - k^2)(\frac{d^2}{dz^2} - k^2)\tilde{w}_m - \beta^4\tilde{w}_m - 2ik\bar{N}_{xy}\frac{d\tilde{w}}{dz} = 0, \qquad \beta^4 \equiv \frac{(\rho h\omega^2 - K)b^4}{D}$$

with $z \equiv y/c$. This has constant coefficients, hence, $e^{i\lambda z}$ is a kernel solution. The spectrum relations are then

$$\lambda^4 + 2\lambda^2 k^2 + 2Ak\lambda + k^4 - B = 0, \qquad A \equiv \frac{c^2\bar{N}_{xy}}{D} \quad B \equiv \frac{(\rho h\omega^2 - K)c^4}{D}$$

The coefficients A and B are associated with the membrane load and inertia, respectively; in the following, we take $B = 0$. Thus, the general solution is represented by

$$\tilde{w}(z = y/b) = \mathbf{A}e^{i\lambda_1 z} + \mathbf{B}e^{i\lambda_2 z} + \mathbf{C}e^{i\lambda_3 z} + \mathbf{D}e^{i\lambda_4 z} \qquad (6.11)$$

For simply supported conditions we can use

$$\tilde{w}(z) = \mathbf{A}\sin(\lambda_1 z) + \mathbf{B}\sin(\lambda_2 z) + \mathbf{C}\sin(\lambda_3 z) + \mathbf{D}\sin(\lambda_4 z)$$

The boundary conditions are

$$z = \pm 1: \qquad \tilde{w} = 0$$
$$z = \pm 1: \qquad \frac{d^2\tilde{w}}{dz^2} = 0$$

After setting up the homogeneous system of equations we get a nontrivial solution only if the determinant is zero. This leads to

$$(\lambda_1^2 - \lambda_2^2)(\lambda_3^2 - \lambda_4^2)\sin(\lambda_1 - \lambda_3)\sin(\lambda_2 - \lambda_4) = (\lambda_1^2 - \lambda_3^2)(\lambda_2^2 - \lambda_4^2)\sin(\lambda_1 - \lambda_2)\sin(\lambda_3 - \lambda_4)$$

This complicated equation must be solved numerically.

The solution process to find the critical load has three steps. First we need to find λ_i as a function of the wavenumber k. This is shown in Figure 6.19(a) for a value of $A = 1.5$. Not all of these values of λ_i will satisfy the determinant relation associated with the boundary conditions. For each value of A we determine those roots that do satisfy the relation; the values of wavenumber where this happens is

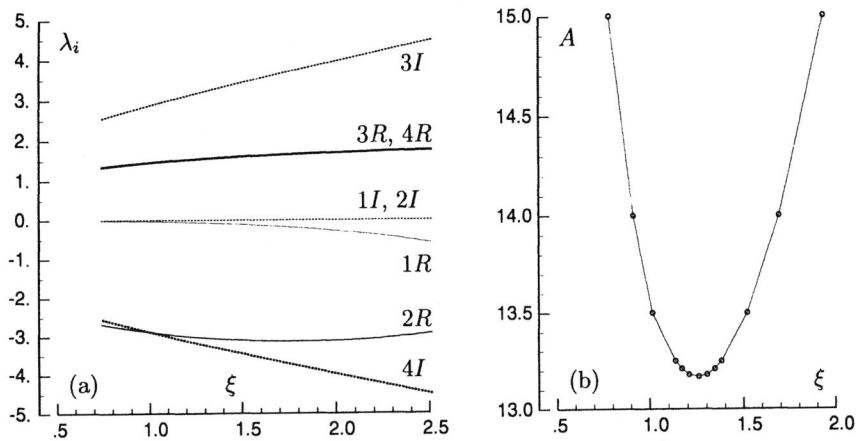

Figure 6.19: Steps in the solution for the plate with shear. (a) Roots λ_i for $A = 1.5$. (b) Values of coefficient A and wavenumber k that satisfy the boundary conditions.

shown in Figure 6.19(b). The critical value of load occurs where A is a minimum and is given by

$$A = 13.17 = \frac{c^2}{D}\bar{N}_{xy} = \frac{b^2}{4D}\bar{N}_{xy}, \qquad k = 1.25$$

Finally, with this value of A and k, we have the corresponding roots of

$$\lambda_1 = -0.076 + i0, \quad \lambda_2 = -2.854 + i0, \quad \lambda_3 = 1.465 + i3.06, \quad \lambda_4 = 1.465 - i3.06$$

These are then used to obtain the mode shape shown in Figure 6.20.

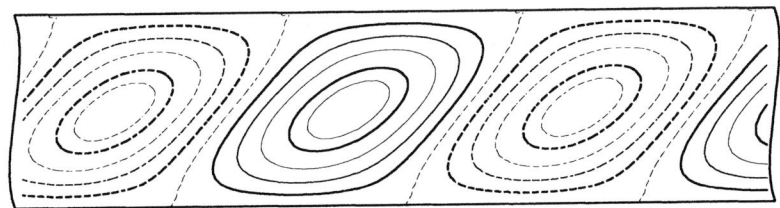

Figure 6.20: Mode shape at critical load.

Although the plate is infinitely long, there is a definite wavelength to the buckled shape. Clearly, this is connected to the width of the plate as being the characteristic length — a result we also found for the normal in-plane loads.

Reference [77] solves the problem of nearly square plates under shear loading. It then proposes the relation

$$\bar{N}_{xy} = [5.35 + 4(\frac{a}{b})^2]\frac{\pi^2 D}{b^2}$$

as a reasonable approximation for the minimum critical load for plates with arbitrary aspect ratio. This underestimates the exact solution derived above by about 6%.

Example 6.13: Compare the buckling performance of steel and aluminum plates.

Take the simply supported plate as an example. The stress at the critical load is

$$\sigma_{xx} = \frac{\bar{N}_{min}}{h} = -\frac{Eh^2}{3(1-\nu^2)}\left(\frac{\pi}{b}\right)^2$$

where h is the plate thickness. The density and modulus of steel are approximately three times that of aluminum. Therefore, for equal weight plates of same size $[a \times b]$, an aluminum plate is approximately three times the thickness of the steel plate, and consequently, the critical stress is three times that of steel. The critical load itself is nine times that for steel. From this comparison, we can see the importance of aluminum in light-weight thin-walled structures.

Example 6.14: Discuss the effect of thermoelastic stresses on the buckling of plates.

In the uncoupled theory of thermoelasticity, the effect of the temperature is to cause a volume change; if this is constrained, then stresses are generated. If these stresses are of the membrane type, then they need to be considered in the buckling analysis. There are no new essential difficulties introduced by the temperature. Thus, a buckling analysis of the problem in Figure 2.5, say, would simply use the computed elastic stresses in the buckling equation.

Reference [15] considers thermoelastic problems in general as well as some thermoelastic stability problems.

Example 6.15: A thin-walled structure is made from flat plates and frame reinforcers as shown in Figure 6.21. Estimate the buckling loads.

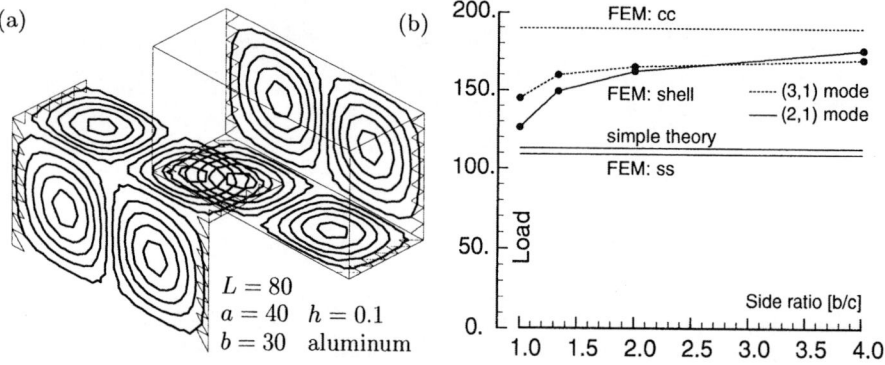

Figure 6.21: A 3-D folded plate shell structure. (a) First mode out-of-plane buckled contours. (b) Buckling load as a function of the short side.

A reasonable analytical approximation to 3-D folded plate structures of the type shown in Figure 6.21 is to treat each face separately as a plate with simply supported conditions on the two long edges. The end conditions are a little more problematic — these depend quite sensitively on the nature of the applied loading, the fixities, and local reinforcement, but some reasonable approximation for these end conditions can be made. We wish to look further into this.

The figure shows the out-of-plane deflection contours for the lowest mode and indeed, one of the striking aspects is the resemblance they bear to the behavior of simple plates. In particular, the long edge seems to rotate without offering a bending resistance, and therefore resembles the simply supported boundary condition. Note that the reinforcement was such that both ends behaved nearly as simply supported.

Figure 6.21 shows what happens as the side aspect ratio is changed. The simple theory, of course, predicts the same buckling load because the large side is not changing. The FEM results, however, show a significant change with the load increasing. The results are bounded by the flat plate results for simply-supported and clamped boundary conditions. What is happening is that the smaller the top plate becomes, the more of a constraint it imposes on the large side plate.

Effect of Membrane Loads on Plate Vibration

Consider a plate with the in-plate loadings $\bar{N}_{xx}$, $\bar{N}_{yy}$, $\bar{N}_{xy}$ and resting on an elastic foundation of stiffness K. The spectral form of Equation (6.7) is

$$D\nabla^2\nabla^2\tilde{w} - \bar{N}_{xx}\frac{\partial^2\tilde{w}}{\partial x^2} - 2\bar{N}_{xy}\frac{\partial^2\tilde{w}}{\partial x\partial y} - \bar{N}_{yy}\frac{\partial^2\tilde{w}}{\partial y^2} - (\rho h\omega^2 - i\eta h\omega - K)\tilde{w} = \tilde{q}(x,y)$$

Again, consider solutions of the form

$$\hat{w}(x,y) = \sum_m \tilde{w}_m(x)e^{i\xi_m y}, \qquad \xi_m = \frac{m\pi}{b}$$

where b is the width of the plate. We will determine the coefficients $\tilde{w}_m$ by requiring that the governing differential equations be satisfied. The differential equation for $\tilde{w}$ becomes

$$(\frac{d^2}{dx^2} - \xi^2)(\frac{d^2}{dx^2} - \xi^2)\tilde{w}_m - \beta^4\tilde{w}_m - \frac{1}{D}[\bar{N}_{xx}\frac{d^2}{dx^2} - 2i\xi\bar{N}_{xy}\frac{d}{dx} - \bar{N}_{yy}\xi^2]\tilde{w} = 0$$

with

$$\beta^4 \equiv \frac{(\rho h\omega^2 - i\eta h\omega - K)}{D}$$

The differential equation has constant coefficients, hence, e^{-ikx} is a kernel solution. The spectrum relations are then

$$k^4 + 2k^2\xi^2 + \xi^4 - \beta^4 + \frac{1}{D}[\bar{N}_{xx}k^2 + 2\bar{N}_{xy}k\xi + \bar{N}_{yy}\xi^2] = 0$$

Because of the uneven power of k occurring in this characteristic equation, the roots do not appear as $\pm$ pairs. As shown in the example of the sheared plate, this leads to awkwardness in the solution, hence, we will take the shear in-plane

force as zero. The solution then is very similar to that of Equation (6.10) but
with the above definition of β. That is, the general solution is represented by

$$\hat{w}(x,y) = \sum_m \left[\mathbf{A}e^{-i\alpha_m x} + \mathbf{B}e^{-\bar{\alpha}_m x} + \mathbf{C}e^{+i\alpha_m x} + \mathbf{D}e^{+\bar{\alpha}_m x} \right] e^{i\xi_m y} \qquad (6.12)$$

Again, in analyzing problems without damping, we will find it easier to use the
solution form

$$\hat{w}(x,y) = \sum_m \left[c_1 \cos(\alpha_m x) + c_2 \sin(\alpha_m x) + c_3 \cosh(\bar{\alpha}_m x) + c_4 \sinh(\bar{\alpha}_m x) \right] f(\xi_m y)$$

with $f(\xi_m y)$ being either $\cos(\xi_m y)$ or $\sin(\xi_m y)$.

Example 6.16: Determine the resonance frequencies and mode shapes for a
simply supported rectangular plate of size $[a \times b]$ with an axial compression in the
x-direction.

Choose only the $\sin(\xi_m y)$ terms. We need only concentrate on the boundary
conditions at $x = 0$ and $x = a$, since the lateral boundary conditions are automatically satisfied.

The solution is precisely the same as for the simple plate with the same boundary conditions. Hence, we get the same characteristic equation

$$(\alpha_m^2 + \bar{\alpha}_m^2) \sin(\alpha_m a) \sinh(\bar{\alpha}_m a) = 0$$

This has the solutions

$$\alpha_m a = n\pi \quad \Rightarrow \quad \sqrt{\beta^4 + \xi^2(\bar{N}_{xx} - \bar{N}_{yy})/D + \tfrac{1}{4}\bar{N}_{xx}^2/D^2} - \xi^2 - \tfrac{1}{2}\bar{N}_{xx}/D = (\frac{n\pi}{a})^2$$

Expanding and rearranging gives

$$\omega_{mn} = \sqrt{\frac{D}{\rho h}} \left[\left[(\frac{n\pi}{a})^2 + (\frac{m\pi}{b})^2 \right]^2 + (\frac{n\pi}{a})^2 \frac{\bar{N}_{xx}}{D} + (\frac{m\pi}{b})^2 \frac{\bar{N}_{yy}}{D} + K \right]^{1/2}$$

The corresponding mode shapes are given by

$$\tilde{w}_{mn}(x,y) = c_2 \sin(\alpha_m x) \sin(\xi_m y) = c_2 \sin(\frac{n\pi x}{a}) \sin(\frac{m\pi y}{b})$$

These mode shapes are the same as for the simple plate.

We see that $\bar{N}_{xx}$, $\bar{N}_{yy}$, and K all have similar effects on the frequency. The
variation of resonance frequency with membrane load is shown in Figure 6.22 for
two plate aspect ratios and $\bar{N}_{yy} = 0$, $K = 0$. In these plots, the normalizations
on the frequency and load are, respectively,

$$\omega_{mn}^* = \frac{\omega_{mn}}{\omega_o}, \qquad \bar{N}_{xx}^* = \frac{\bar{N}_{xx}}{N_o} \qquad \text{where} \qquad \omega_o = \frac{1}{a^2}\sqrt{\frac{D}{\rho h}}, \qquad N_o = \frac{D}{a^2}$$

We see that an in-plane tension increases the frequency, while an in-plane compression decreases it, and that a mode keeps the same shape irrespective of load.
The frequency can be driven to zero when

$$\bar{N}_{xx} = -D(\frac{a}{n\pi})^2 \left[(\frac{n\pi}{a})^2 + (\frac{m\pi}{b})^2 \right]^2$$

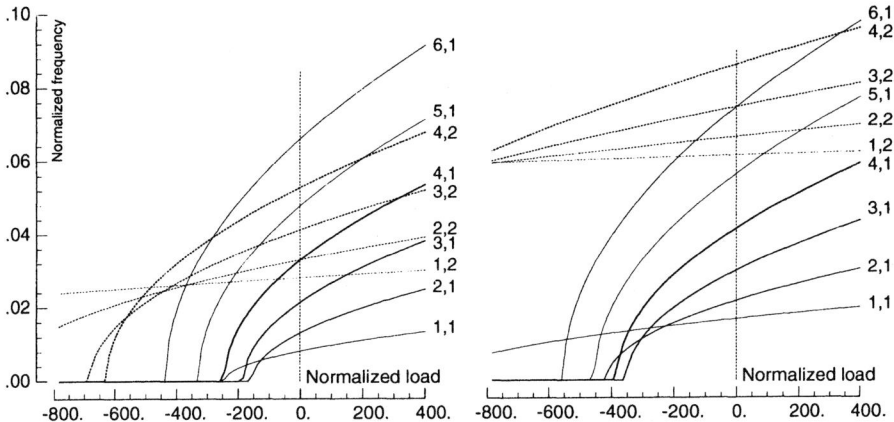

Figure 6.22: Variation of resonance with membrane load for plates with different aspect ratios. (a) $a/b = 2$, (b) $a/b = 3$.

These are precisely the collection of static buckling loads.

It is interesting to observe how the mode shapes interchange their order in going from a vibration with no membrane load to a vibration with the first critical value, say. That is, a vibration eigenanalysis done at the first buckling load will have a different sequence of mode shapes than one done at zero load. Furthermore, the sequence of mode shapes for the eigen-buckling problem is not the same sequence as for the eigen-vibration problem. In particular, the mode shape for the lowest vibration frequency at zero load does not coincide with the shape for the lowest static buckling load.

We will return to more discussion of these mode shapes later in the chapter.

6.4   Matrix Formulation for Buckling

As shown in the previous sections, the strain energies for both the beam and plate have three contributions. The first is the axial or in-plane elastic contribution, the second is the flexural elastic contribution, and the third is the second-order contribution from the in-plane loading. We have already established matrix representations for the first two, and we now turn our attention to the second-order contribution. First, however, we attempt to draw the connection between an eigenanalysis of stability and the incremental analysis of large deflections presented in the Chapter 3.

Relation of Eigenanalyses to Nonlinear Incremental Analyses

Recall that for our incremental formulation of the corotational scheme, at each time or load step, we must solve the equilibrium relation

$$\left[\sum_m [T_m^n][\bar{k}_E^n + \bar{k}_G^n]_m [T_m^n] \right] \{\Delta u\}^{n+1} = \{P\}^{n+1} - \{F\}^n$$

where the superscript n refers to the time steps t_0, t_1, t_2, $\ldots t_n$. Let us focus on the first two steps of this process

step 1: $\left[\sum_m [T_m^0][\bar{k}_E^0 + 0]_m [T_m^0] \right] \{\Delta u\}^1 = \{P\}^1 - 0$

update: $\{x\}^1 = \{x\}^0 + \{\Delta u\}^1$, $\{F\}^1$, $[\bar{k}_E]^1$, $[\bar{k}_G]^1$, $[\,T\,]^1$

step 2: $\left[\sum_m [T_m^1][\bar{k}_E^0 + \bar{k}_G^1]_m [T_m^1] \right] \{\Delta u\}^2 = \{P\}^2 - \{F\}^1$

For the first time step, the element nodal forces are zero, and, consequently, so also is the geometric stiffness. The elastic stiffness in local coordinates is always unchanged, the geometric stiffness in local coordinates changes because the axial and membrane loads change during the deformation.

To help make the following ideas concrete, consider specifically a truss structure with a load distribution $\chi\{P\}$; that is, $\{P\}$ specifies the distribution of loads while χ is a scalar load multiplier (this is called *proportional loading*). The first load increment causes an axial force in each member that is proportional to χP, that is,

$$\bar{F}_{om} \propto \chi P^1 \qquad \Longrightarrow \qquad [\bar{k}_G(\bar{F}_o)]_m = \chi [\bar{k}_G^*(P^1)]_m$$

Furthermore, let P^1 be a small increment so that the change of geometry is negligible. The second step now becomes

step 2: $\left[\sum_m [T_m^0][\bar{k}_E^0 + \chi \bar{k}_G^{*0}(P^1)]_m [T_m^0] \right] \{\Delta u\}^2 = \{P\}^2 - \{F\}^1$

or

step 2: $\left[K_E^0 + \chi K_G^{*0} \right] \{\Delta u\}^2 = \{P\}^2 - \{F\}^1$

For the incremental analysis, we could now proceed to obtain $\{\Delta u\}^2$ and continue to the next step. For the eigenanalysis of stability, however, we stop and ask the question: Is it possible to have two or more solutions to the above equations? We are asking if it is possible to have two equilibrium states for a given set of loads. If it is possible, then there is a second solution $\{\Delta u^*\}$ that also satisfies the equations

$$[K_E + \chi K_G^{*0}]\{\Delta u^*\} = \{P\}^2 - \{F\}^1$$

Let the difference of the two solutions be $\{\phi\} = \{\Delta u\}^2 - \{\Delta u^*\}$; it satisfies the homogeneous equations

$$[K_E + \chi K_G^{*0}]\{\phi\} = \{0\}$$

In general, this admits only the trivial solution $\{\phi\} = 0$. There are circumstances, however, when χ has special values (called *eigenvalues*) that are nontrivial solutions. We obtain the eigenvalues χ by setting the determinant to zero.

Finding these solutions is known as an *eigenanalysis*; in contrast to our earlier discussion with beams, this is called a linear eigenanalysis because the eigenvalue parameter χ appears linearly in the equations. Corresponding to each eigenvalue, we can find a solution for $\{\phi\}$, this is called an *eigenvector*.

In the above discussion, we assumed P^1 was small so that we could associate both $[K_E]$ and $[K_G]$ with the undeformed configuration. Consequently, our eigenanalysis is rooted in the small deformation theory of structures, and therefore is not applicable to instability states (such as limit points) that are often associated with large deflections. We will call this small deflection analysis a linear eigenanalysis or a *buckling* analysis. However, we can modify the above discussion so that instead of discussing times t_0 and t_1 we could discuss times t_n and t_{n+1}. In other words, it is possible to do a linear eigenanalysis at each stage of a nonlinear deformation process — the value of critical load then obtained is the load change from the current state that will cause buckling.

As we saw in Chapter 3 for incremental loading, we can use a rather poor geometric stiffness and still get good results because of the Newton-Raphson equilibrium iterations. This is not so here because the eigenanalysis goes to the maximum load in one step; consequently, there is a greater need for an accurate geometric stiffness. We could alleviate the problem somewhat, as indicated in the previous paragraph, by doing the eigenanalysis at an elevated load.

A final point worth nothing is that once the stiffness matrices are assembled, NonStaD uses the subspace iteration scheme [22] to solve the eigenvalue problem just as it does for the vibration eigenvalue problem.

Example 6.17: Show the connection between a buckling eigenanalysis and a vibration eigenanalysis at the elevated load.

The vibration eigenvalue problem is

$$[K_T]\{\phi\}_i - \omega_i^2 [M]\{\phi\}_i = 0$$

At the singular point, $\omega_1 = 0$, we have

$$[K_T]\{\phi\}_1 = 0$$

Now consider the case when the structural response is only slightly nonlinear, then we can represent the tangent stiffness in the form

$$[K_T] \approx [K_E] + \chi[K_G]$$

where $[K_E]$ is the elastic stiffness, $[K_G]$ is the geometric stiffness, and χ is a loading factor. At the singular point, we therefore have

$$[K_T]\{\phi\}_1 = [K_E + \chi K_G]\{\phi\}_1 = 0$$

This is the eigenvalue problem for the buckling of the structure, and we conclude that the vibration mode shape is the same as the first buckling mode shape. This result is consistent with Figure 6.22.

Information about the spectral properties of $[K_T]$ at an elevated load are not always readily available; the above shows that, in some circumstances, we can at least estimate the shape for the lowest mode by doing a linear buckling eigen-analysis.

Geometric Stiffness for Frames

For illustrative purposes, we will first look at the plane behavior of a frame and truss where we allow a combination of axial and bending effects. The degrees of freedom at each node are $\{u\} = \{u, v, \phi\}^T$. The strain energy due to the axial loading is

$$U_G = \tfrac{1}{2} \int_L \left[\bar{F}_o (\frac{\partial v}{\partial x})^2 \right] dx$$

for both trusses and frames and depends only on the transverse deflection.

For the truss, let the transverse deflection be represented by the linear shape functions

$$\bar{v}(\bar{x}) \;=\; [1 - \frac{\bar{x}}{L}]\bar{v}_1 + [\frac{\bar{x}}{L}]\bar{v}_2$$

We substitute these into the strain energy expression and integrate. This leads to

$$2U_G = \frac{\bar{F}_o}{L_o}[\bar{v}_1 - \bar{v}_2]^2$$

The contributions to the stiffness are obtained from

$$[k_{Gij}] = \frac{\partial^2 U_G}{\partial u_i \partial u_j} \quad \Rightarrow \quad [\bar{k}_G] = \frac{\bar{F}_o}{L} \begin{bmatrix} 0 & 0 & 0 & 0 & 0 & 0 \\ 0 & 1 & 0 & 0 & -1 & 0 \\ 0 & 0 & 0 & 0 & 0 & 0 \\ 0 & 0 & 0 & 0 & 0 & 0 \\ 0 & -1 & 0 & 0 & 1 & 0 \\ 0 & 0 & 0 & 0 & 0 & 0 \end{bmatrix}$$

This is precisely the same geometric stiffness matrix obtained in Chapter 4 using the corotational approach for trusses.

For the frame member, we could let the transverse deflection be represented as above, then of course we would get the exact same geometric stiffness for the frame as for the truss. Such a stiffness would be called *inconsistent* because it does not use the same shape functions as used in establishing the elastic stiffness matrix. Instead, let the deflected shape be represented by the beam shape functions

$$\bar{v}(\bar{x}) = \left[1 - 3(\frac{\bar{x}}{L})^2 + 2(\frac{\bar{x}}{L})^3 \right] \bar{v}_1 + (\frac{\bar{x}}{L}) \left[1 - 2(\frac{\bar{x}}{L}) + (\frac{\bar{x}}{L})^2 \right] L\phi_{1z}$$
$$+ (\frac{\bar{x}}{L})^2 \left[3 - 2(\frac{\bar{x}}{L}) \right] \bar{v}_2 + (\frac{\bar{x}}{L})^2 \left[-1 + (\frac{\bar{x}}{L}) \right] L\phi_{2z}$$

We substitute these into the strain energy expression and integrate. After differentiation as above, this leads to

$$[\bar{k}_G] = \frac{\bar{F}_o}{30L}
\begin{bmatrix}
0 & 0 & 0 & 0 & 0 & 0 \\
0 & 36 & 3L & 0 & -36 & 3L \\
0 & 3L & 4L^2 & 0 & -3L & -L^2 \\
0 & 0 & 0 & 0 & 0 & 0 \\
0 & -36 & -3L & 0 & 36 & -3L \\
0 & 3L & -L^2 & 0 & -3L & 4L^2
\end{bmatrix}
\tag{6.13}$$

This is not the same as obtained above for the truss or as obtained in Chapter 4 for the corotational schemes for frames. The question therefore arises as to differences caused by the different choice of interpolating or shape functions. This is answered in the following example.

Example 6.18: Compare the performance of the consistent and inconsistent geometric matrices for the frame element.

Figure 6.23 shows the first six buckling loads for a simply supported beam as the number of elements is increased. The mesh density is given as N where the number of elements is 2^N.

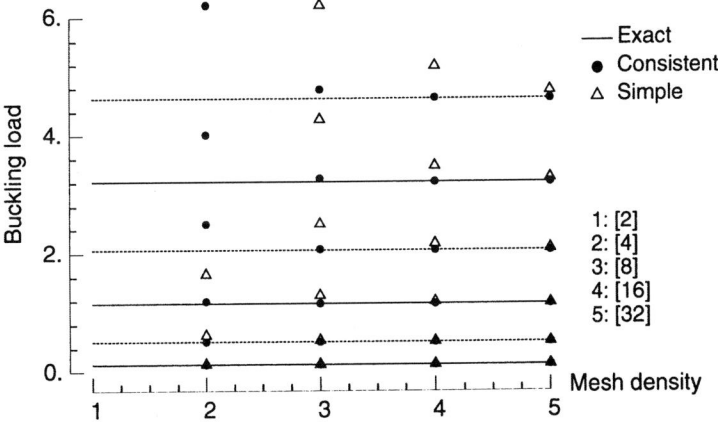

Figure 6.23: Comparison of convergence of two geometric stiffnesses for a beam.

The very important point to note is that both formulations exhibit convergence to the exact result and therefore both are acceptable. Clearly, however, the consistent formulation shows the better rate of convergence. Thus, all other things being equal, the consistent formulation is the more preferable of the two.

Sometimes it happens that certain factors, such as geometry or the need to model the mass distribution accurately, dictates the use of a small element and in those cases the inconsistent formulation might be slightly preferable because it is less computationally expensive to formulate and assemble.

Geometric Stiffness for Plates in Flexure

First consider the effect of the in-plane or membrane behavior of the plate. The strain energy due to the membrane loading is

$$U_G = \tfrac{1}{2} \int_A \left[\bar{N}_{xx}(\frac{\partial w}{\partial x})^2 + 2\bar{N}_{xy}(\frac{\partial w}{\partial x})(\frac{\partial w}{\partial y}) + \bar{N}_{yy}(\frac{\partial w}{\partial y})^2 \right] dxdy$$

This depends only on the transverse deflection. Let this transverse deflection be represented by the linear shape functions

$$w(x,y) = \sum_j h_j(x,y)w_j$$

The interpolations are those of the three-noded triangle. Note that the only degrees of freedom participating at each node are $\{u\} = \{u,\ v,\ w\}^T$. We substitute these into the strain energy expression and integrate to get

$$2U_G = \left[\bar{N}_{xx}(\frac{1}{2A}\sum_j b_j w_j)^2 + 2\bar{N}_{xy}(\frac{1}{2A}\sum_j b_j w_j)(\frac{1}{2A}\sum_j c_j w_j) + \bar{N}_{yy}(\frac{1}{2A}\sum_j c_j w_j)^2 \right]A$$

The contributions to the stiffness are obtained from

$$[k_{ij}] = \frac{\partial^2 U}{\partial u_i \partial u_j}$$

which leads to the $[3 \times 3]$ submatrix of the $[9 \times 9]$ element stiffness submatrix

$$[\bar{k}_G]_{ij} = \frac{\bar{N}_{xx}}{4A}\begin{bmatrix} 0 & 0 & 0 \\ 0 & 0 & 0 \\ 0 & 0 & b_i b_j \end{bmatrix} + \frac{\bar{N}_{xy}}{4A}\begin{bmatrix} 0 & 0 & 0 \\ 0 & 0 & 0 \\ 0 & 0 & b_i c_j + b_j c_i \end{bmatrix} + \frac{\bar{N}_{yy}}{4A}\begin{bmatrix} 0 & 0 & 0 \\ 0 & 0 & 0 \\ 0 & 0 & c_i c_j \end{bmatrix}$$

The contributions associated with the rotational DoF are zero. This is precisely the same geometric stiffness matrix obtained in Chapter 4 using the corotational approach if we convert the stresses to nodal forces.

We saw with the frame member that performance can be improved if better quality shape functions are used. The DKT element does not have a clean representation of its shape functions, so it is difficult to develop a consistent geometric stiffness to go with the DKT elastic stiffness. We will illustrate the procedure instead with an inconsistent representation but one that should perform better than that just derived.

Keeping in mind that the demands on the shape functions for formulating the geometric stiffness is less severe than for the elastic stiffness (because of the different orders of derivatives in the energy expression), one good possibility is to take the displacements in the form

$$
\begin{aligned}
w(x,y) = \quad & c_1 h_1 + c_2 h_2 + c_3 h_3 \\
+ \quad & c_4(h_1 h_2^2 + h) + c_5(h_2 h_3^2 + h) + c_6(h_3 h_1^2 + h) \\
+ \quad & c_7(h_1^2 h_2 + h) + c_8(h_2^2 h_3 + h) + c_9(h_3^2 h_1 + h) \qquad (6.14)
\end{aligned}
$$

where h_i are the triangle area coordinates and $2h \equiv h_1 h_2 h_3$. In this way, the 10 terms of the Pascal triangle of Chapter 2 are distributed among the nine coefficients. If we now determine the nine coefficients in terms of the nine nodal degrees of freedom, then we get the associated shape functions as

$$
\begin{aligned}
N_1 &= h_1 + (h_1^2 h_2 - h_1 h_2^2) + (h_1^2 h_3 - h_1 h_3^2) \\
N_2 &= -y_{12}(h_1^2 h_2 + h) + y_{31}(h_1^2 h_3 + h) \\
N_3 &= -x_{21}(h_1^2 h_2 + h) + x_{13}(h_1^2 h_3 + h)
\end{aligned}
\tag{6.15}
$$

The other six are obtained by permutation. With this formulation, let the displacements be represented by the participating degrees of freedom

$$
\{w(x,y)\} = [\,N\,]\{u\}, \qquad \{u\} = \{w_1, \phi_{x1}, \phi_{y1}; w_2, \phi_{x2}, \phi_{y2}; w_3, \phi_{x3}, \phi_{y3}\}
$$

After substituting into the energy expression and minimizing, we get the geometric stiffness as

$$
[\,\bar{k}_G\,] = \bar{N}_x[k_{Gx}] + \bar{N}_y[k_{Gy}] + \bar{N}_{xy}[k_{Gxy}]
$$

where

$$
[k_{Gx}] = \int [\frac{\partial N}{\partial x}][\frac{\partial N}{\partial x}]^T dA, \quad [k_{Gy}] = \int [\frac{\partial N}{\partial y}][\frac{\partial N}{\partial y}]^T dA, \quad [k_{Gxy}] = \int [\frac{\partial N}{\partial x}][\frac{\partial N}{\partial y}]^T dA
$$

These expressions are too lengthy to repeat here. In any event, we are slightly more interested in seeing how the simpler version fares.

Example 6.19: Use a convergence study to compare the effectiveness of the two geometric stiffness formulations for plates.

The generic mesh is shown in Figure 6.24. The other meshes are obtained by dividing this uniformly. We will use the simply supported plate as the test case.

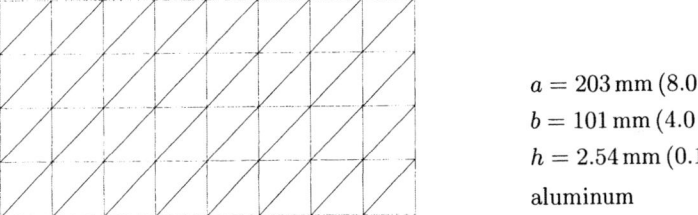

$a = 203\,\text{mm}\ (8.0\,\text{in.})$

$b = 101\,\text{mm}\ (4.0\,\text{in.})$

$h = 2.54\,\text{mm}\ (0.1\,\text{in.})$

aluminum

Figure 6.24: Generic $[4 \times 8]$ mesh.

The buckling loads are shown in Figure 6.25. In each case, the mesh represented the complete plate and subspace iteration was used to determine the eigenvalues. Just as for the frame member, we note that both formulations converge to the exact solution. In this case, however, the refined model is only marginally better than the simple membrane model.

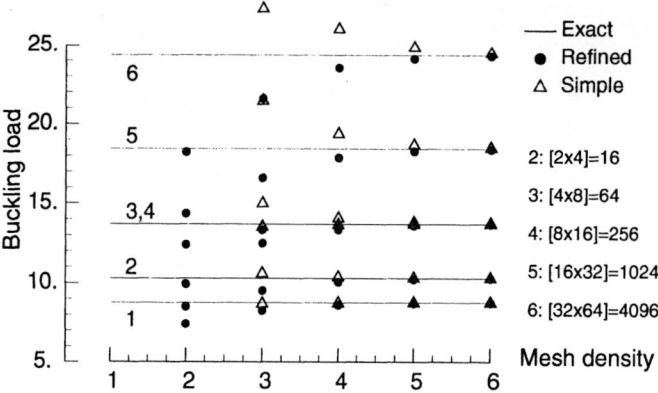

Figure 6.25: Convergence study for the buckling loads of a simply supported plate.

The rate of convergence with respect to element size is relatively slow. But since both models converge from opposite sides, this convergence rate could be improved by using an inconsistent model based on some average of the two, or, as indicated by the beam results, by using a truly consistent model.

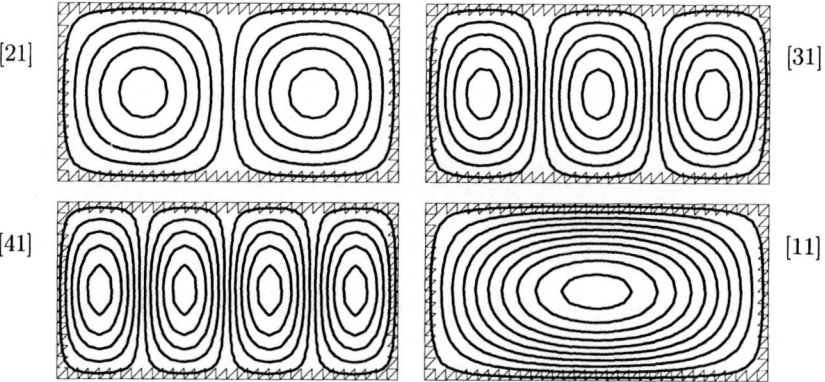

Figure 6.26: Contours of the first four buckled modes shapes. The sequence is (2,1), (3,1), (4,1), (1,1) with modes (4,1) and (1,1) having repeated roots.

The figure shows the first six modes, and, as predicted by the plots in Figure 6.18, roots 3 and 4 are repeated roots. We note that these roots appear as repeated roots in the numerical analysis only as the mesh is refined; for a coarser mesh they are distinct.

Contours of the first four modes are shown in Figure 6.26. The sequence is the top row first, followed by the second row; these shapes are in agreement with the predictions of Figure 6.18. Note that even though roots 3 and 4 are repeated roots their mode shapes are significantly different.

Example 6.20: Determine the buckling loads and mode shapes for Stein's plate.

Stein [69] did a thorough experimental study of the buckling and post-buckling behavior of plates. The parameters of the plate are interesting because they exhibit a mode jumping phenomenon. We will reconsider this problem again later.

This plate is of size $[25.36 \times 4.71 \times 0.39\,\text{mm}^3]$ and made of aluminum. The boundary conditions are that the long sides are simply supported, while the other two sides are clamped. This aspect ratio is such as to generate many close modes and therefore is a good challenge as a convergence test.

We first obtain the exact solution. The general solution is taken from Equation (6.10) as

$$w(x,y) = \sum_m \left[c_1 \cos(k_1 x) + c_2 \sin(k_1 x) + c_3 \cos(k_2 x) + c_4 \sin(k_2 x) \right] \sin(\xi_m y)$$

The simply supported boundary conditions are automatically satisfied, hence we need only consider the end constraints. Looking at the boundary conditions for $x = 0$, we have

$$\text{at } x = 0: \quad w = 0 \implies 0 = c_1 + c_3$$
$$\frac{\partial w}{\partial x} = 0 \implies 0 = k_1 c_2 + k_2 c_4$$

The solution is now written as

$$w(x,y) = \sum_m \left[c_1 [\cos(k_1 x) - \cos(k_2 x)] + c_2 [\sin(k_1 x) - \frac{k_1}{k_2} \sin(k_2 x)] \right] \sin(\xi_m y)$$

We also have for the other boundary condition

$$\text{at } x = a: \quad w = 0 \implies 0 = c_1(C_1 - C_2) + c_2(S_1 - \frac{k_1}{k_2} S_2)$$
$$\frac{\partial w}{\partial x} = 0 \implies 0 = c_1(-k_1 C_1 + k_2 C_2) + c_2(-k_1 S_1 - k_1 S_2)$$

where $C_1 \equiv \cos(k_1 a)$ and so on. These homogeneous equations have a nontrivial solution only if the determinant is zero; this leads to

$$\det = k_1(C_1 - C_2)^2 + k_1(S_1 - \frac{k_2}{k_1} S_2)(S_1 - \frac{k_1}{k_2} S_2)$$

The simplest way to obtain the critical load is to put this expression in a do-loop where the load is changed. There are multiple load values corresponding to the buckling loads. These are the exact values given in Figure 6.27.

The figure shows a comparison of the exact values with the finite element results. There is good convergence for both FEM formulations. It must be pointed out, however, that because so many of the roots are close, extra tightness on the convergence tolerance was needed to get the good results. There is not much to choose between the two formulations, hence, in future, we will use the simpler of the two.

In-Plane Stiffening

The emphasis in the preceding sections has been on the out-of-plane stiffening due to in-plane loads. For 3-D shell problems, which have all degrees of freedom

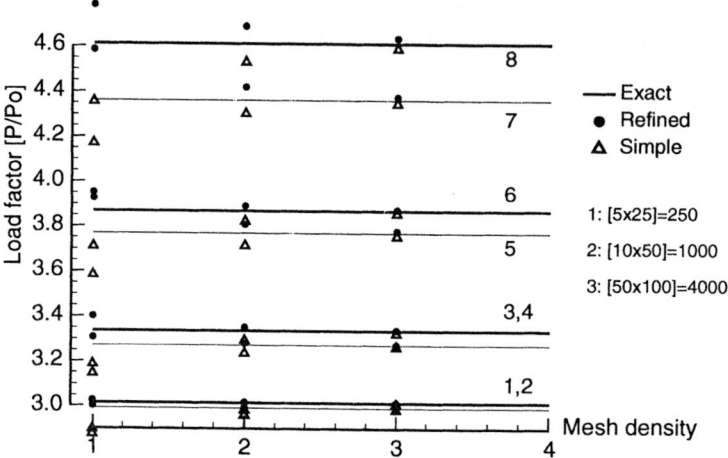

Figure 6.27: Buckling loads for Stein's plate ($P_o = 1000$).

participating, we can also get an in-plane stiffening effect. Our derivation follows a similar procedure as used for the flexure of plates.

Consider the x-component of the Lagrangian strain in a plate stretching only in its plane

$$E_{xx} = \frac{\partial u}{\partial x^o} + \frac{1}{2}\left[\left(\frac{\partial u}{\partial x^o}\right)^2 + \left(\frac{\partial v}{\partial x^o}\right)^2\right]$$

$$E_{yy} = \frac{\partial v}{\partial y^o} + \frac{1}{2}\left[\left(\frac{\partial v}{\partial y^o}\right)^2 + \left(\frac{\partial u}{\partial y^o}\right)^2\right]$$

$$2E_{xy} = \frac{\partial u}{\partial y^o} + \frac{\partial v}{\partial x^o} + \left[\frac{\partial u}{\partial x^o}\frac{\partial u}{\partial y^o} + \frac{\partial v}{\partial x^o}\frac{\partial v}{\partial y^o}\right]$$

These strains give rise to stresses according to

$$\sigma_{xx} = \frac{E}{1-\nu^2}\left[E_{xx} + \nu E_{yy}\right]$$

$$\sigma_{yy} = \frac{E}{1-\nu^2}\left[E_{yy} + \nu E_{xx}\right]$$

$$\sigma_{xy} = G2E_{xy}$$

The strain energy for a plate in plane stress is

$$U = \frac{1}{2}\int_V \left[\sigma_{xx}E_{xx} + \sigma_{yy}E_{yy} + \sigma_{xy}2E_{xy}\right]\, dV$$

After substitution, dropping higher products in the nonlinear terms, and integrating through the thickness, we get

$$U = \frac{1}{2}\int_A \left[E^*h\left[\left(\frac{\partial u}{\partial x}\right)^2 + \left(\frac{\partial v}{\partial y}\right)^2 + 2\nu\frac{\partial u}{\partial x}\frac{\partial v}{\partial y}\right] + Gh\left(\frac{\partial u}{\partial y} + \frac{\partial v}{\partial x}\right)^2\right]\, dx\, dy$$

$$+ \frac{1}{2}\int_A \left[\bar{N}_{xx}(\frac{\partial u}{\partial x})^2 + 2\bar{N}_{xy}(\frac{\partial u}{\partial x})(\frac{\partial v}{\partial y}) + 2\bar{N}_{xy}(\frac{\partial u}{\partial y})(\frac{\partial v}{\partial x}) + \bar{N}_{yy}(\frac{\partial v}{\partial y})^2\right]\, dx\, dy$$

Thus, the energy expression comprises the contribution for the linear behavior plus three new terms associated with the nonlinear in-plane loading. In this we grouped some terms according to

$$\bar{N}_{xx} = E^* h \left(\frac{\partial u}{\partial x} + \nu \frac{\partial v}{\partial y} \right)$$

$$\bar{N}_{yy} = E^* h \left(\frac{\partial v}{\partial y} + \nu \frac{\partial u}{\partial x} \right)$$

$$\bar{N}_{xy} = Gh \left(\frac{\partial u}{\partial y} + \frac{\partial v}{\partial x} \right)$$

which correspond to an approximation of the in-plane loads. Strictly, these are deformation dependent, but in the following they are assumed to be pre-existing in the plate and therefore not subject to variation.

In the energy expression, we have terms such as

$$E^* h \left(\frac{\partial u}{\partial x} \right)^2 , \qquad \bar{N}_{xx} \left(\frac{\partial u}{\partial x} \right)^2$$

which have a common deformation. Hence in the governing equations (after applying our variational principle) we would have a common coefficient $(E^* + \sigma_{xx})h$. In all practical situations, the modulus is larger than the stresses by several orders of magnitude, and we conclude that the in-plane stiffening effect is negligible compared to the elastic stiffness. Actually, the analysis just performed is essentially that of the example in Section 1.7 where initial stresses are shown to affect stress wave propagation.

For completeness, and for comparison with results from the corotational scheme of Chapter 3, we will evaluate the geometric stiffness for the in-plane stiffening. Let the in-plane displacements be represented by the linear shape functions

$$u(x,y) = \sum_j h_j(x,y) u_j , \qquad v(x,y) = \sum_j h_j(x,y) v_j$$

The interpolations are those of the three-noded triangle. Note that the only degrees of freedom participating at each node are $\{u\} = \{u, v\}^T$. We substitute these into the strain energy expression and integrate. The contributions to the stiffness are obtained from

$$[k_{ij}] = \frac{\partial^2 U}{\partial u_i \partial u_j}$$

which leads to the [3 × 3] submatrix

$$[\bar{k}_G]_{ij} = \frac{\bar{N}_{xx}}{4A} \begin{bmatrix} b_i b_j & 0 & 0 \\ 0 & 0 & 0 \\ 0 & 0 & 0 \end{bmatrix} + \frac{\bar{N}_{xy}}{4A} \begin{bmatrix} 0 & b_i c_j + b_i c_j & 0 \\ b_i c_j + b_i c_j & 0 & 0 \\ 0 & 0 & 0 \end{bmatrix} + \frac{\bar{N}_{yy}}{4A} \begin{bmatrix} 0 & 0 & 0 \\ 0 & c_i c_j & 0 \\ 0 & 0 & 0 \end{bmatrix}$$

This is not precisely the same geometric stiffness matrix obtained in Chapter 3 using the corotational approach even if we convert the stresses to nodal forces.

It is, however, the geometric stiffness corresponding to the membrane behavior of the total Lagrangian formulation. That is,

$$[\bar{k}_N] = [B_N]^T [\sigma^K][B_N] V^o$$

where

$$[\sigma^K] = \begin{bmatrix} \sigma^K_{xx} & \sigma^K_{xy} & 0 & 0 \\ \sigma^K_{xy} & \sigma^K_{yy} & 0 & 0 \\ 0 & 0 & \sigma^K_{xx} & \sigma^K_{xy} \\ 0 & 0 & \sigma^K_{xy} & \sigma^K_{yy} \end{bmatrix}, \quad [B_N] = \frac{1}{2A}\begin{bmatrix} b_1 & 0 & b_2 & 0 & b_3 & 0 \\ c_1 & 0 & c_2 & 0 & c_3 & 0 \\ 0 & b_1 & 0 & b_2 & 0 & b_3 \\ 0 & c_1 & 0 & c_2 & 0 & c_3 \end{bmatrix}$$

It is worth noting that we could do a similar procedure for frames while accounting for the effect on the out-of-plane stiffness. For the plane frame, we then get

$$[\bar{k}_G] = \frac{\bar{F}_o}{L}\begin{bmatrix} 1 & 0 & 0 & -1 & 0 & 0 \\ 0 & 1 & 0 & 0 & -1 & 0 \\ 0 & 0 & 0 & 0 & 0 & 0 \\ -1 & 0 & 0 & 1 & 0 & 0 \\ 0 & -1 & 0 & 0 & 1 & 0 \\ 0 & 0 & 0 & 0 & 0 & 0 \end{bmatrix}$$

which looks quite similar to the earlier result. There is one significant difference, however; this matrix is isotropic, that is, it has the same entries for all orientations.

6.5   Static View of Stability of Discrete Systems

We now discuss some of the previous stability results in a more general context for discrete systems. In particular, we wish to explore the meaning of instability when the deformations are nonlinear. Many of the ideas are developed in more depth in the paper by Allman [3] and the book by Crisfield [20].

Proportional Loading Along the Primary Path

We concentrate on the quasi-static primary loading path. In our decomposition, we assume this is a quasi-static path and hence governed by

$$[M]\{\ddot{u}_o\} + [C]\{\dot{u}_o\} = \{P_o\} - \{F(u_o)\} \approx 0$$

In the subsequent analysis, we drop the subscript on the displacements. The total potential energy of our general nonlinear system is

$$\Pi(u, \chi) = U(u) - \chi\{u\}^T\{P\}$$

where $U(u)$ is the strain energy that is a function only of the discrete displacement vector $\{u\}$, $\{P\}$ is a fixed load vector, and χ is a scalar load multiplier. Since the load vector never changes, we say that this loading is *proportional*.

Consider small changes in the total potential due to small changes in displacement (with χ fixed)

$$\delta\Pi(u,\chi) = \{\frac{\partial\Pi}{\partial u}\}\{\delta u\} + \frac{1}{2}\{\delta u\}^T[\frac{\partial^2\Pi}{\partial u\partial u}]\{\delta u\} + \cdots$$

We have that

$$\{\frac{\partial\Pi}{\partial u}\} = \{\frac{\partial U}{\partial u}\} - \chi\{P\} \equiv \{\mathcal{F}\}(u,\chi), \qquad [\frac{\partial^2\Pi}{\partial u\partial u}] = [\frac{\partial^2 U}{\partial u\partial u}] = [\frac{\partial\mathcal{F}}{\partial u}] \equiv [K_T]$$

where $[K_T]$ is the tangent stiffness. The variation of the potential is therefore given by

$$\delta\Pi(u,\chi) = \{\mathcal{F}\}^T\{\delta u\} + \frac{1}{2}\{\delta u\}^T[K_T]\{\delta u\} + \cdots$$

For an equilibrium loading path, the energy changes should be stationary. That is, the first variation should be zero irrespective of $\{\delta u\}$. Hence we have that

$$\{\frac{\partial\Pi}{\partial u}\} = \{\mathcal{F}\}(u,\chi) = \{\frac{\partial U}{\partial u}\} - \chi\{P\} = \{F\} - \chi\{P\} = 0$$

This is the equation that defines the equilibrium path; this path can be viewed as a continuous curve in $(\{u\}^T, \chi)$ space. In this, only the nodal force vector $\{F\}$ is a function of the displacements.

Consider a loading history along a sequence of equilibrium states. In particular, consider two equilibrium states, A and B, small $\{\Delta u\}$ and $\Delta\chi$ apart, as shown in Figure 6.28, then we have

$$\begin{aligned}\{\mathcal{F}\}(u,\chi)|_B &= \{\mathcal{F}\}(u_A + \Delta u, \chi_A + \Delta\chi)\\ &= \{\mathcal{F}\}(u,\chi)|_A + [\frac{\partial\mathcal{F}}{\partial u}]|_A\{\Delta u\} + \{\frac{\partial\mathcal{F}}{\partial\chi}\}|_A\Delta\chi + \cdots\end{aligned}$$

Since A and B are equilibrium states, then $\{\mathcal{F}\}|_A$ and $\{\mathcal{F}\}|_B$ are both zero giving

$$\{\frac{\partial\mathcal{F}}{\partial u}\}|_A\{\Delta u\} + [\frac{\partial\mathcal{F}}{\partial\chi}]|_A\Delta\chi = [\frac{\partial F}{\partial u}]|_A\{\Delta u\} - \Delta\chi\{P\} = [K_T]\{\Delta u\} - \Delta\chi\{P\} \approx 0$$

The approximation is because we are neglecting the higher-order terms. We will refer to this as the loading equation. Provided that $\det||K_T|| \neq 0$ we get

$$\{\Delta u\} = \Delta\chi[K_T]^{-1}\{P\}$$

This is the standard tangential solution used in a nonlinear analysis and described in Chapter 3. Strictly speaking, this is exact only when the loading path is linear, and some method such as Newton-Raphson iterations are needed to insure that a sequence of such increments do not deviate too much from the actual equilibrium path.

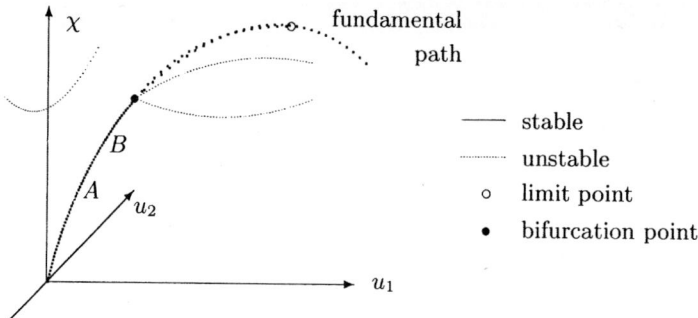

Figure 6.28: Examples of limit and bifurcation singular points occurring along a fundamental loading path.

Limit and Bifurcation Singular Points

For stable equilibrium, the small changes of energy should be positive for any small perturbation $\{\delta u\}$ about the equilibrium point, hence we require that

$$\Pi(u + \delta u, \chi) - \Pi(u, \chi) > 0 \qquad \text{or} \qquad \{\delta u\}^T [K_T]\{\delta u\} > 0 \qquad \text{for all } \{\delta u\}$$

since $\{\mathcal{F}\}^T\{\delta u\} = 0$ through equilibrium. For the above to be true, we require that $[K_T]$ be positive definite.

There are two situations of interest to us here. First, when

$$\{\delta u\}^T [K_T]\{\delta u\} < 0 \qquad \text{for some } \{\delta u\}$$

then $[K_T]$ is not positive definite and it will have at least one negative eigenvalue. It is therefore unstable. The other case is when

$$\{\delta u\}^T [K_T]\{\delta u\} = 0 \qquad \text{for some } \{\delta u\}$$

which is a neutral equilibrium state and $[K_T]$ has a zero eigenvalue. Consequently,

$$\det \|[K_T]\| = 0$$

A full investigation of the nature of its equilibrium requires the use of higher-order terms in the expansion of the potential; this we will consider in the next subsection.

For the neutral equilibrium case, we cannot find a unique $\{\Delta u\}$ using the loading equation and we have a singular point. This singular point can be either a *limit point* or a *bifurcation point*. To see the distinction between these two, we must look at the spectral properties of the tangent stiffness.

Consider the free vibration of the structure when loaded near a critical point; that is, let $\{\Delta u\} = \{\xi\}e^{i\omega t}$ leading to

$$[K_T]\{\xi\} - \omega^2 [M]\{\xi\} = 0$$

It is therefore sufficient for us to introduce the eigenvalues λ_m and eigenvectors $\{\phi\}_m$ of the tangent stiffness such that

$$[K_T]\{\phi\}_m = \lambda_m [\,M\,]\{\phi\}_m, \qquad \lambda_m = \omega_m^2$$

and let the eigenvectors be normalized such that

$$\{\phi\}_i^T [\,M\,]\{\phi\}_j = \delta_{ij}, \qquad \{\phi\}_m^T [K_T]\{\phi\}_m = \lambda_m$$

When the solution is following the path from a stable state, the lowest eigenvalue, λ_1, is zero at the singular point. Hence, we have that

$$[K_T]\{\phi\}_1 = 0$$

Now multiply the transpose of the loading equation by $\{\phi\}_1$ to get

$$(\{\Delta u\}^T [K_T] - \Delta\chi\{P\}^T)\{\phi\}_1 = 0 \qquad \text{giving} \qquad \Delta\chi\{P\}^T\{\phi\}_1 = 0$$

This relation is used to distinguish between limit and bifurcation points as follows:

$$\begin{aligned}
\text{limit point:} &\qquad \Delta\chi = 0, &\qquad \{P\}^T\{\phi\}_1 \neq 0 \\
\text{bifurcation point:} &\qquad \Delta\chi \neq 0, &\qquad \{P\}^T\{\phi\}_1 = 0 &\qquad (6.16)
\end{aligned}$$

Figure 6.28 shows examples of the two singular points. As shown in the figure, a 2-D plot of u_2 against load shows little or no motion until the singular point is reached, then it has two possible paths to take. In the simple example illustrated, the paths can be concave up (stable), concave down (unstable), or asymmetric (one stable, one unstable).

We can get further insight into these singular points by a modal representation of the displacement increment in terms of the eigenvectors

$$\{\Delta u\} = \eta_1\{\phi\}_1 + \eta_2\{\phi\}_2 + \cdots = \sum \eta_m\{\phi\}_m$$

Now substitute this into the loading equation, multiply the resulting equation by $\{\phi\}_m$, then making use of the orthogonality properties leads to

$$\eta_m\lambda_m - \Delta\chi\{P\}^T\{\phi\}_m = 0 \qquad \text{or} \qquad \eta_m = \frac{1}{\lambda_m}\Delta\chi\{P\}^T\{\phi\}_m$$

We therefore have the general modal representation for the displacement increment

$$\{\Delta u\} = \eta_1\{\phi\}_1 + \Delta\chi \sum_{m=2} \frac{1}{\lambda_m}\Big[\{P\}^T\{\phi\}_m\Big]\{\phi\}_m = \eta_1\{\phi\}_1 + \Delta\chi\{v\}$$

Note that $\{\phi\}_1^T\{v\} = 0$. For the limit point where $\Delta\chi = 0$ and assuming $\lambda_m \neq 0$, we get

$$\{\Delta u\} = \eta_1\{\phi\}_1$$

The displacement increment has the shape of the first eigenmode. The displacement increment for the bifurcation point, on the other hand, depends on all the modes. We look more closely at these shapes in the next section; the next example problem looks at the relation between load and singular points.

Example 6.21: Analyze the stability of the simple truss structure shown in Figure 6.29

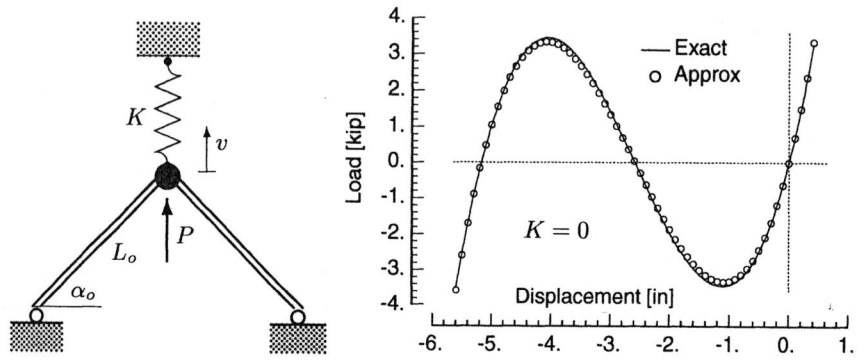

Figure 6.29: A pinned truss and the approximate representation of its force/displacement behavior.

We have already considered this problem in Chapter 3, but here we will look at it from the point of view of loading along the primary path. We established that the axial displacement in the member is

$$\bar{u} = \sqrt{L_o^2 + 2vL_o \sin\alpha_o + v^2} - L_o$$

The axial strain is $\bar{\epsilon} = \bar{u}/L_o$, hence the potential for the problem is

$$\Pi = \tfrac{1}{2}Kv^2 + 2\tfrac{1}{2}EA\bar{\epsilon}^2 - Pv = \tfrac{1}{2}Kv^2 + EA\left[\sqrt{1 + 2\frac{v}{L_o}\sin\alpha_o + (\frac{v}{L_o})^2} - 1\right]^2 - Pv$$

The equilibrium path is

$$\mathcal{F}_v = \frac{\partial\Pi}{\partial v} = Kv + 2EA\left[\sin\alpha_o + \frac{v}{L_o}\right]\left[1 - \frac{1}{\sqrt{1 + 2\frac{v}{L_o}\sin\alpha_o(\frac{v}{L_o})^2}}\right] - P = 0$$

or simply

$$F - P = 0$$

We are now in a position to determine P (and hence F) as a function of v.

The member force term is too complicated for our present purpose, so we will approximate it with a polynomial. Noting that the three zero-load points occur when v is 0, $-L_o \sin\alpha_o$, and $-2L_o \sin\alpha_o$, respectively, then a good approximation is

$$P = Kv + EA\sin^3\alpha_o\left[\bar{v}(1 + \bar{v})(2 + \bar{v})\right], \qquad \bar{v} \equiv \frac{v}{L_o \sin\alpha_o}$$

A comparison of the approximate and true static behavior of the truss is shown in Figure 6.29 for when $K = 0$. Clearly the approximation is good over the whole range of loads. The approximate form for the potential is now

$$\Pi = \tfrac{1}{2}Kv^2 + \beta\left[\bar{v}^2 + \bar{v}^3 + \tfrac{1}{4}\bar{v}^4\right]L_o \sin\alpha_o - Pv, \quad \beta \equiv EA\sin^3\alpha_o, \quad \bar{v} \equiv \frac{v}{L_o \sin\alpha_o}$$

This is shown plotted in Figure 6.30 for different values of P. The role of the linear spring K is simply to make the two minima for $P = 0$ to be different.

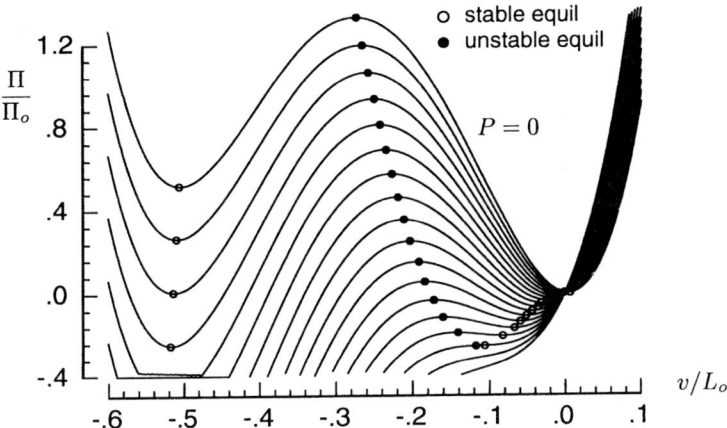

Figure 6.30: Total potential for different values of applied load.

For each low value of P there are three equilibrium points — two valleys and a peak corresponding to stable and unstable equilibrium points, respectively. The second stable equilibrium point is in the snap-through position. As P is increased downward, we have a stable decreasing equilibrium path indicated by the open circles. This reaches a minimum where the stable and the unstable equilibrium points come together and are annihilated. When they meet, the structure is unstable and the load is at the critical value. The only equilibrium point for a higher load corresponds to the snap-through position.

Limit point instabilities are sometimes also referred to as bifurcation instabilities [26] because the system goes from having three solutions (equilibrium positions) to having only one when the load passes through the critical value. Reference [75] refers to it as a *fold catastrophe* and the load parameter as the *unfolding* parameter. They also show that the potential requires just the cubic term for determinancy,

$$\Pi = av^3 + \chi v + \text{higher-order terms}$$

That is, it takes the same (topological) form whether or not higher-order terms are retained, and that the linear term is sufficient for its unfolding.

Classification of Singular Points

We now consider the higher-order terms in the expansion. Again, consider a loading history along a sequence of equilibrium states, and in particular, consider two nearby equilibrium states A and B, such that

$$
\begin{aligned}
\{\mathcal{F}\}(u,\chi)|_B &= \{\mathcal{F}\}(u_A + \Delta u, \chi_A + \Delta\chi) \\
&= \{\mathcal{F}\}(u,\chi)|_A + [\frac{\partial \mathcal{F}}{\partial u}]|_A \{\Delta u\} + \{\frac{\partial \mathcal{F}}{\partial \chi}\}|_A \Delta\chi \\
&\quad + \frac{1}{2}\{\Delta u\}^T [\frac{\partial^2 \mathcal{F}}{\partial u \partial u}]|_A \{\Delta u\} + \{\Delta u\}^T [\frac{\partial^2 \mathcal{F}}{\partial u \partial \chi}]|_A \Delta\chi + \frac{1}{2}[\frac{\partial^2 \mathcal{F}}{\partial \chi^2}]|_A \Delta\chi^2 + \cdots
\end{aligned}
$$

For proportional loading,

$$
\{\mathcal{F}\} = \{\frac{\partial U}{\partial u}\} - \chi\{P\} = \{F\} - \chi\{P\}
$$

giving that

$$
[\frac{\partial \mathcal{F}}{\partial u}] = [\frac{\partial F}{\partial u}] = [K_T], \qquad [\frac{\partial^2 \mathcal{F}}{\partial u \partial u}] = [\frac{\partial K_T}{\partial u}], \qquad [\frac{\partial^2 \mathcal{F}}{\partial u \partial \chi}] = 0, \qquad [\frac{\partial^2 \mathcal{F}}{\partial \chi^2}] = 0
$$

Also, consider a loading path such that

$$
\begin{aligned}
\{\Delta u\} &= \{\frac{\partial u}{\partial s}\}\Delta s + \frac{1}{2}\{\frac{\partial^2 u}{\partial s^2}\}\Delta s^2 = \{u'\}\Delta s + \frac{1}{2}\{u''\}\Delta s^2 \\
\Delta\chi &= \frac{\partial \chi}{\partial s}\Delta s + \frac{1}{2}\frac{\partial^2 \chi}{\partial s^2}\Delta s^2 = \chi'\Delta s + \frac{1}{2}\chi''\Delta s^2
\end{aligned}
$$

where s is the "arc-length" along the path. Substituting these into the equilibrium expression then leads to

$$
\begin{aligned}
\{\mathcal{F}\}(u,\chi)|_B &= \{\mathcal{F}\}(u,\chi)|_A \\
&\quad + \left([\frac{\partial \mathcal{F}}{\partial u}]\{u'\} + \{\frac{\partial \mathcal{F}}{\partial \chi}\}\chi'\right)\Big|_A \Delta s \\
&\quad + \frac{1}{2}\left([\frac{\partial \mathcal{F}}{\partial u}]\{u''\} + \{\frac{\partial \mathcal{F}}{\partial \chi}\}\chi'' + \{u'\}^T[\frac{\partial^2 \mathcal{F}}{\partial u \partial u}]\{u'\}\right)\Big|_A \Delta s^2 + \cdots
\end{aligned}
$$

Because the loading path is an equilibrium path, each term in this sequence must be zero. The second term gives

$$
[\frac{\partial \mathcal{F}}{\partial u}]\{u'\} + \{\frac{\partial \mathcal{F}}{\partial \chi}\}\chi' = 0 \qquad \Longrightarrow \qquad [K_T]\{u'\} - \{P\}\chi' = 0
$$

which is very similar to the loading equation. As developed in the previous subsection, we can get a modal representation of the displacement increment as

$$
\{u'\} = \eta_1\{\phi\}_1 + \chi'\{v\}, \qquad \{v\} = \sum_{m=2} \frac{1}{\lambda_m}[\{P\}^T\{\phi\}_m]\{\phi\}_m
$$

Substitution of this into the third term and pre-multiplying by $\{\phi\}_1$ leads to

$$B_1\eta_1^2 + 2B_2\eta_1\chi' + B_3\chi'^2 + B_4\chi'' = 0$$

where

$$B_1 = \{\phi\}_1^T[[\frac{\partial^2\mathcal{F}}{\partial u \partial u}]\{\phi\}_1]\{\phi\}_1 = \{\phi\}_1^T[[\frac{\partial K_T}{\partial u}]\{\phi\}_1]\{\phi\}_1$$

$$B_2 = \{\phi\}_1^T[[\frac{\partial^2\mathcal{F}}{\partial u \partial u}]\{v\}]\{\phi\}_1 = \{\phi\}_1^T[[\frac{\partial K_T}{\partial u}]\{v\}]\{\phi\}_1$$

$$B_3 = \{\phi\}_1^T[[\frac{\partial^2\mathcal{F}}{\partial u \partial u}]\{v\}]\{v\} = \{\phi\}_1^T[[\frac{\partial K_T}{\partial u}]\{v\}]\{v\}$$

$$B_4 = \{\phi\}_1^T[\frac{\partial\mathcal{F}}{\partial u}]\{v\} + \{\phi\}_1^T\{\frac{\partial\mathcal{F}}{\partial\chi}\} = -\{\phi\}_1^T\{P\}$$

Letting $\Delta\bar{s} = \eta_1\Delta s$ and $\Delta\chi = \chi'\Delta s$, we can write the above as

$$B_1\Delta\bar{s}^2 + 2B_2\Delta\bar{s}\Delta\chi + B_3\Delta\chi^2 + B_4\Delta^2\chi = 0 \qquad (6.17)$$

This is the equation we will use to classify the equilibrium states and the types of singular points occurring is structural analyses.

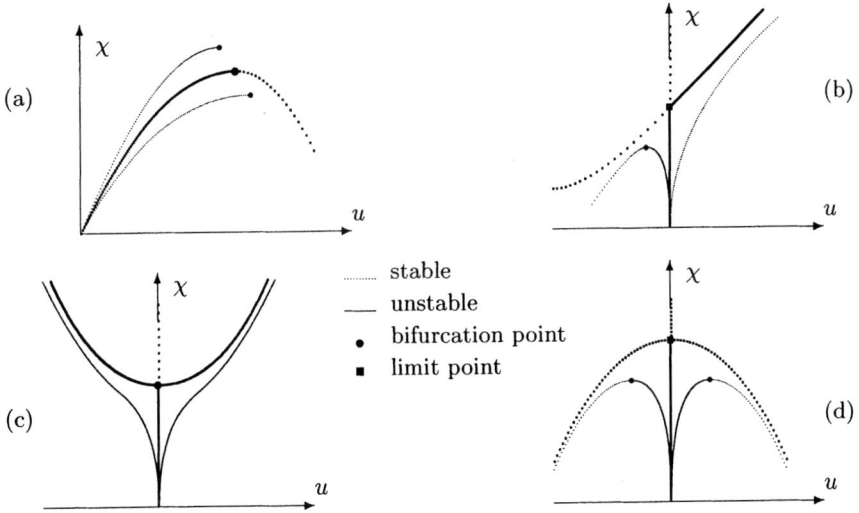

Figure 6.31: Classification of singular points and the effect of initial imperfections: (a) limit point, (b) asymmetric bifurcation, (c) stable symmetric bifurcation, (d) unstable symmetric bifurcation.

To know if these states are stable or not, however, we need to have an expression for the potential energy. Again, consider small changes in the total potential due to small changes in displacement (with χ fixed)

$$\delta\Pi(u,\chi) = \{\frac{\partial\Pi}{\partial u}\}\{\delta u\} + \frac{1}{2}\{\delta u\}^T[\frac{\partial^2\Pi}{\partial u\partial u}]\{\delta u\} + \frac{1}{6}\{\delta u\}^T[[\frac{\partial^3\Pi}{\partial u\partial u\partial u}]\{\delta u\}]\{\delta u\}$$

$$+ \frac{1}{24}\{\delta u\}^T[\{\delta u\}^T[\frac{\partial^4\Pi}{\partial u\partial u\partial u\partial u}]\{\delta u\}]\{\delta u\} + \cdots$$

The first term is zero because of equilibrium, and the second term is zero because it is a singular point. The energy change is therefore determined by the next higher terms. Reference [3] shows that these can be represented as

$$\delta\Pi(u,\chi) = \frac{1}{6}B_1\Delta\bar{s}^3 + \frac{1}{24}B_5\Delta\bar{s}^4 + \cdots \tag{6.18}$$

where

$$B_5 = \{\phi\}_1^T\left[\{\phi\}_1^T[\frac{\partial^3\mathcal{F}}{\partial u\partial u\partial u}]\{\phi\}_1\right]\{\phi\}_1 - \sum_{m=2}\frac{3}{\lambda_m}\left((\{\phi\}_1^T[\frac{\partial^2\mathcal{F}}{\partial u\partial u}]\{\phi\}_1)\{\phi\}_m\right)^2$$

We are now in a position to classify each of the singular points. It must be said, however, that the above formulas are only useful for a general classification scheme but are not of much practical use because of the difficulty in determining the various derivatives during a general finite element analysis.

For a limit point, $\Delta\chi = 0$, giving

$$B_1\Delta\bar{s}^2 + B_4\Delta^2\chi = 0, \qquad \Delta\bar{s} = \pm\sqrt{-\frac{B_4}{B_1}\Delta^2\chi}, \qquad \delta\Pi(u,\chi) = \frac{1}{6}B_1\Delta\bar{s}^3 + \cdots$$

There are two solutions symmetrically placed about the limit point. If B_1 is positive, the energy change is negative for a negative $\Delta\bar{s}$. Alternatively, if B_1 is negative, the energy change is negative for a positive $\Delta\bar{s}$. In either case, there is a direction in which negative energy results and hence a limit point is unstable.

For a bifurcation point $B_4 = 0$ giving

$$B_1\Delta\bar{s}^2 + 2B_2\Delta\bar{s}\Delta\chi + B_3\Delta\chi^2 = 0$$

There are two cases of this depending on the value of B_1. For an asymmetric bifurcation, $B_1 \neq 0$ leading to

$$\Delta\bar{s} = \left(-\frac{B_2}{B_1} \pm \sqrt{(\frac{B_2}{B_1})^2 - \frac{B_3}{B_1}}\right)\Delta\chi$$

There are two solutions, the one corresponding to the plus sign is the fundamental path. The energy change is the same as for the limit point and hence an asymmetric bifurcation point is unstable.

For a symmetric bifurcation, $B_1 = 0$ leading to

$$\left(2B_2\Delta\bar{s} + B_3\Delta\chi\right)\Delta\chi = 0 \qquad \Longrightarrow \qquad \Delta\chi = 0, \qquad \Delta\bar{s} = -\frac{B_3}{2B_2}\Delta\chi = Z\Delta\chi$$

Again there are two solutions, the first of which is the bifurcated path. The displacement increment for this path is

$$\{\Delta u_b\} = \{\phi\}_1 \Delta \bar{s}$$

For the fundamental path,

$$\{\Delta u_f\} = \Big(Z\{\phi\}_1 + \{v\}\Big)\Delta \chi$$

The stability is governed by the sign of B_5.

Example 6.22: Analyze the stability of the asymmetric truss structure shown in Figure 6.32

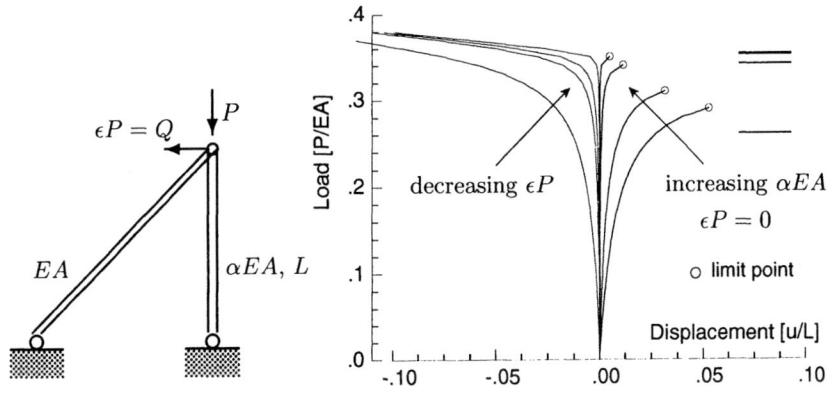

Figure 6.32: Example of a nonsymmetric bifurcation.

Unlike the previous example, we allow the load application point to have two degrees of freedom. As a result, this becomes an interesting little example because it can exhibit both a limit point and a bifurcation.

First consider when then is no component of horizontal force. The behavior is that of a limit point — the end of the plot in each case in approximately the limit point. The next equilibrium position is a snap-through all the way to the mirror image. As the stiffness of the vertical member is increased, the limit point moves to the vertical position. In fact, if the vertical member is rigid, we will not get any displacement to the right at all.

When the vertical member is quite stiff ($\alpha = 10^4$), a small horizontal component of force will induce the bifurcation to the left. This bifurcation is stable.

It is interesting to note that a linear buckling analysis of this problem gives the single buckling load of

$$\frac{P}{EA} = \frac{\alpha}{1 + \alpha 2\sqrt{2}}$$

(Actually, the second buckling load in infinite.) For α values of 1, 10, 100, this gives P/EA values of 0.26, 0.34, 0.35, respectively, shown as the horizontal lines in Figure 6.32.

This example highlights again, that many phenomena of stability can be truly observed only as limiting process of changing a loading parameter or a geometric parameter.

Discussion

Path-following schemes [20] are the computational implementation of the static methods; Reference [24] gives an excellent application and discussion (with many references) of current generalized path-following procedures. Severe difficulties can be encountered with limit points where the load/deflection curve becomes horizontal. The *arc-length* methods were introduced to overcome these difficulties. The essence of the arc-length method is that the load parameter becomes a variable just like the displacement variables; these $N + 1$ unknowns are solved by using N equilibrium equations and a constraint equation. Various forms of constraint equations can be used, a good discussion of some of the simpler ones is given in Reference [19] and a comprehensive survey is given in Reference [80].

Implementing the arc-length method requires a level of programming sophistication beyond the level directed by this book. As an alternative, the dynamic approach accepts that the structural behavior is dynamic in the vicinity of a critical point and the next section lays out some of the issues involved.

6.6   Dynamic View of Static Instabilities

As indicated in the introduction, a common intuitive notion of stability asks the question: if the structure is slightly perturbed, what happens to the ensuing dynamics? That is, if the structure returns to its current state, then it is stable; otherwise it is unstable and a dynamic process ensues. The purpose of this section is to pursue the implications of this simple notion of stability within the context of modern computational methods. One of the early papers taking this view is Reference [39] and a more recent paper is Reference [59].

The dynamic view considers instability to be synonymous with motion and large displacements, and therefore requires a fully nonlinear dynamic analysis capability. This section reviews some of the underlying theory, and its implementation as part of a finite element formulation. We use the example of a rectangular plate to illustrate the main features of the dynamic view — a range of other problems including frames and shells where the dynamic view was used can be found in Reference [81]. We first look at a simply supported plate because there are analytical solutions available for comparisons and for adding further insight into the study.

Monitoring the Spectral Behavior

As shown earlier, the discretized form of the equations of motion are

$$[\,M\,]\{\ddot{u}\} + [\,C\,]\{\dot{u}\} = \{P\} - \{F(u)\} \tag{6.19}$$

where $[\,M\,]$ and $[\,C\,]$ are the mass and damping matrices, respectively; $\{P\}$ is the total applied load vector, and $\{F\}$ is the vector of nodal forces. As shown in Section 5.8, this decomposition leads to the two equations

$$\epsilon^0 : \qquad [\,M\,]\{\ddot{u}_o\} + [\,C\,]\{\dot{u}_o\} = \{P_o\} - \{F(u_o)\} \tag{6.20}$$

$$\epsilon^1 : \qquad [\,M\,]\{\ddot{\xi}\} + [\,C\,]\{\dot{\xi}\} + [K_T]\{\xi\} = \{Q\} \tag{6.21}$$

It must be emphasized that this decomposition is only conceptual; in the discussions and results that follow, Equation (6.19) is treated as fully dynamic and fully nonlinear and used to generate all the responses. We see from Equation (6.21) that the response due to the ping (at a given load $\{P_o\}$) is that of a linear system with constant stiffness $[K_T]$. With changing deformation, governed by Equation (6.19), the tangent stiffness changes and clearly monitoring the spectral content of the free vibration response to $\{Q\}$ will then give information about the current tangent stiffness and hence stability.

Thus, a key ingredient of the dynamic approach is to monitor the spectral behavior of $[K_T]$. This can be done by imposing a ping, and doing a Fourier analysis on the response. This is conceptually appealing and has the significant advantage that it can be implemented with an explicit solver for the nonlinear dynamics. Unfortunately, it is too computationally intensive for use as a continuous monitor of the system. When the tangent stiffness is available (as when using an implicit solver), a more expedient method is to do an undamped vibration eigenanalysis — we will then refer to the eigenvalues as $\mu \to \lambda = \omega^2$, which are real only. It should be pointed out, however, that this is generally not effective for follower-force type problems because most codes use an approximate symmetric tangent stiffness matrix. Actually, as demonstrated in Chapter 3, for incremental schemes using Newton-Raphson iterations, an accurate tangent stiffness matrix is not essential as verified by the success of the various modified Newton-Raphson methods. What this means is that for a given level of discretization, the tangent stiffness matrix as used in an implicit solver may give imprecise estimates of the vibration eigenvalues; it may, for example, indicate a negative eigenvalue even though the system is actually stable. Of course, we expect an accurate stiffness matrix in the limit of a fine mesh discretization, but short of that, and consistent with our dynamic view, the structure can be pinged to assess the true stability state of the structure. Thus the monitoring is a combination of a vibration eigenanalysis at regular intervals plus selective use of ping. We elaborate more on this later.

For the present analysis, NonStaD was modified in three ways. First, it was changed to allow two independent load histories; one corresponds to the slowly varying primary loading, and the other to the ping loading. Second, the implicit module was modified to give a vibration eigenanalysis at regular stages of the loading. Vector iteration [7, 22] was initially implemented for the eigenanalysis because it is very efficient at determining the single lowest eigenmode; it became clear, however, that we needed to monitor many of the lowest eigenvalues simultaneously and so the subspace iteration [7, 22] method was also implemented. The third significant modification was in the spatial design of ping. Three options are implemented: a single point load ping, a random (in space) ping, and a designer ping. The second of these was used to simulate ambient disturbances. As will be shown later, the designer ping is related to the vibration eigenvectors that become unstable.

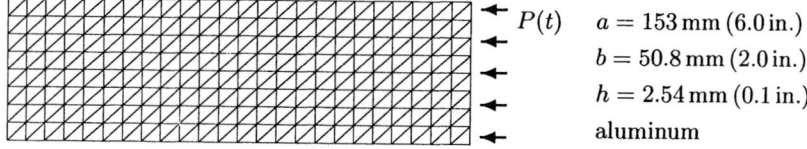

$$P(t) \qquad a = 153\,\mathrm{mm}\ (6.0\,\mathrm{in.})$$
$$b = 50.8\,\mathrm{mm}\ (2.0\,\mathrm{in.})$$
$$h = 2.54\,\mathrm{mm}\ (0.1\,\mathrm{in.})$$
aluminum

Figure 6.33: Dimensions of the simply supported plate and a typical mesh.

Simply Supported Plate

Consider a simply supported plate with the geometry shown in Figure 6.33. This was given a load/unload cycle that took it past its first bifurcation. The load and some responses are shown in Figure 6.34(a), and the deforming shape is shown in Figure 6.34(b); from these it is clear the plate has buckled to a $(3, 1)$ mode. The load history was chosen to give a dwell past the bifurcation load, during which time the ping $Q(t)$ was applied. The first four vibration eigenvalues are shown in Figure 6.35(a) and the corresponding mode shapes are in Figure 6.36(b). The loading portion of these histories can be divided into three stages: motion along the fundamental path, transition to the post-buckling stage, and the post-buckling behavior. We will now analyze each of these portions in detail and try to make connections with results from earlier in this chapter.

I: Motion Along the Fundamental Path

Referring to Figure 6.35, as the load (time) increases, all the eigenvalues decrease and one eventually goes to zero. It is worth pointing out that at the instant when $\lambda = 0$, the deformation state of the plate (as inferred from Figure 6.34) is still that of only the membrane compression. During this loading stage, the modes are intersecting; linear buckling theory can be used to explain this interesting result. Note, as shown in Figure 6.35(b), that if the load is continued way past the first critical load that the modes continue to change their intersections. Thus,

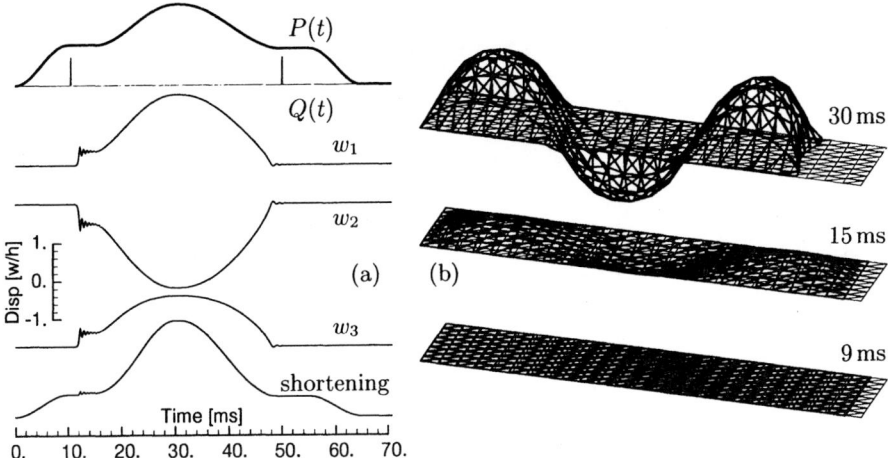

Figure 6.34: Deformation history of the simply supported [3.0:1] plate. (a) Displacements and loads as a function of time. (b) Exaggerated ($\times 4$) deformed shapes.

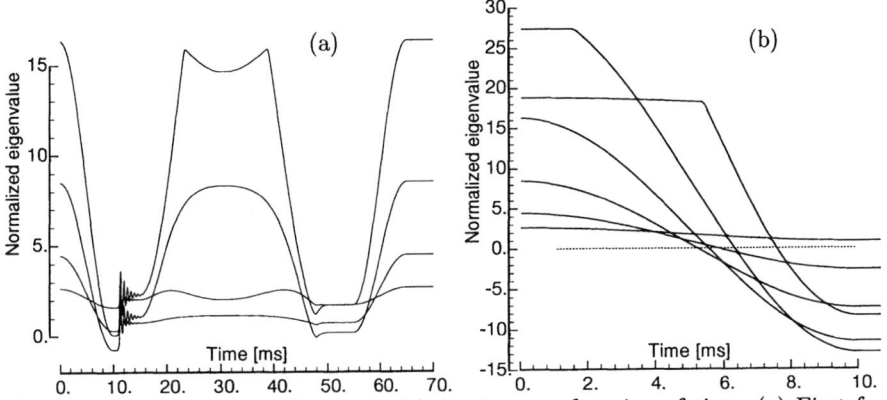

Figure 6.35: Monitoring the spectral behavior as a function of time. (a) First four vibration eigenvalues over the whole range of time. (b) First six vibration eigenvalues when the maximum load level is $P = 1.7P_c$.

at these high loads, it is not necessarily the first mode that became critical that will be the dominant mode.

The variation of the vibration mode shapes as the time (load) increases is shown in Figure 6.36; the sequence of intersecting mode shapes corresponds to that shown in Figure 6.22.

On further increase of load, the eigenvalue becomes negative giving an imaginary frequency and the plate is then unstable. An agent is necessary to cause it to deflect out of the plane. The agent could be the eventual accumulated round-off error as we will see shortly, or it could be the active use of a ping. It must be kept in mind that at this stage, ping has two slightly different roles to play: one is as tester of stability (in case the eigenvalues are not sufficiently

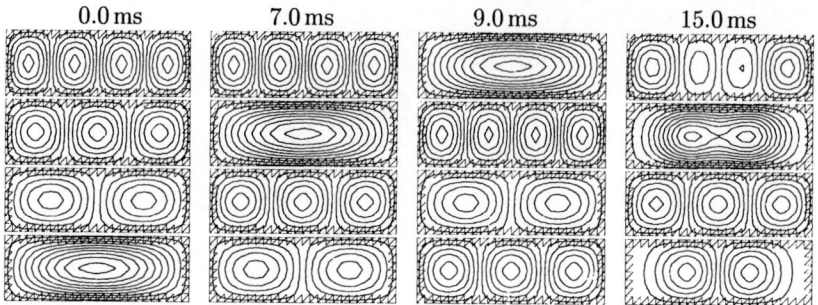

Figure 6.36: First four vibration mode shapes at selected times.

precise), and the other is as agent to "push" an unstable structure toward the new configuration.

II: Transition to Post-buckling Behavior

The transition to post-buckling behavior is a dynamic event. We will do a modal superposition analysis of the perturbed motion due to ping in order to get a closer look at this latter role.

The associated undamped vibration eigenvalue problem for the perturbed motion is given from Chapter 4 as

$$[K_T]\{\phi\} - \lambda[M]\{\phi\} = 0, \qquad \lambda = \omega^2, \qquad \{\phi\}_i^T[M]\{\phi\}_j = \delta_{ij}$$

We can represent the dynamic response due to the ping using the following modal superposition

$$\{\xi(t)\} = \sum_m \{\phi\}_m \eta_m(t)$$

where $\eta_m(t)$ are the principal coordinates obtained by solving the uncoupled equations

$$\ddot{\eta}_m + 2\zeta\omega_m\dot{\eta}_m + \omega_m^2\eta_m = \{\phi\}_m^T\{Q\}$$

For a ping-like $\{Q\}$ (short duration pulse), all $\eta_m(t)$ behave like a damped oscillator and eventually tend to zero. However, at a static singular point (limit or bifurcation), we have $\omega_1 = 0$, giving

$$\ddot{\eta}_1 = \{\phi\}_1^T\{Q\}$$

This is an unconstrained motion increasing almost linearly in time (since $\{Q\}$ is of short duration). Therefore, all other things being equal, for a ping with arbitrary spatial distribution, we expect (after a short time) the shape of the deforming structure to be dominated by the first vibration mode shape at that load level.

Example 6.23: Investigate the effect of ping magnitude on the transition to post-buckling.

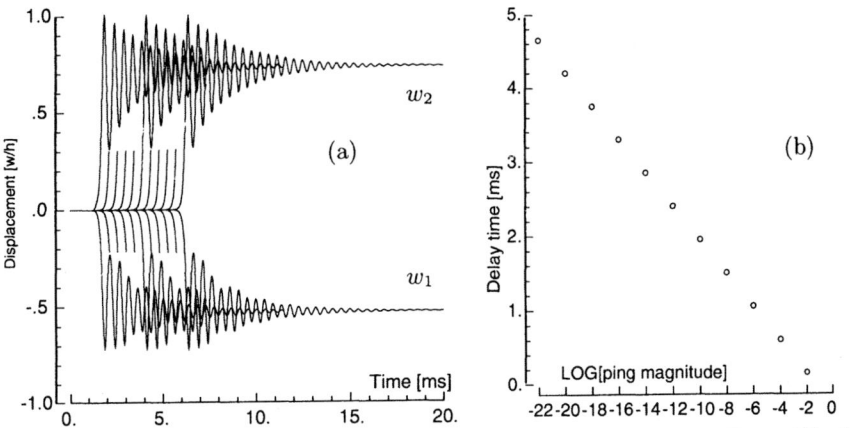

Figure 6.37: Effect of ping magnitude on the plate response when P is 10% above critical. (a) Out-of-plane responses. (b) Delay time for initiation.

In this example, we let the history of $P(t)$ be a smooth ramp up to the maximum and then held constant; $Q(t)$ is a sine-squared function that lasts for about $300\,\mu s$. In this case, the ping is a single load applied at the center of the plate. When the maximum P load level is below the critical, the application of Q only causes a vibration that eventually dies down and the plate returns to its original state. Changing the magnitude of ping causes a change in magnitude of the response, but otherwise the traces are self-similar. This is in line with the ping response being linear.

Figure 6.37(a) shows the response of the plate to a ping loading when the maximum P load level is 10% above the critical and for different peak magnitudes of Q. We see there is a transient response that eventually moves the plate to a new deformed configuration. Changing the magnitude of ping does not cause a change in magnitude of the response; in each case, the same final position is achieved but the time to initiate the transition is different. The plots shown are for ping magnitudes varying by 20 orders of magnitude. Note that once this motion initiates, it is as violent irrespective of the magnitude of ping that initiated it.

To explain the delay-time behavior, we need to realize that the effect of ping is to put the plate in motion with an initial velocity; this velocity is directly proportional to the magnitude of ping. Because P exceeds the critical value, the solution is of the form

$$w(t) \propto V e^{\alpha t} \propto Q_o e^{\alpha t} = \beta Q_o e^{\alpha t}$$

where α depends on the amount of $P > P_c$, V is the initial velocity, and β is some proportionality constant. This is an exponential increasing function. We now ask at what time the displacement reaches a certain value w_c. That is,

$$w_c = \beta Q_o e^{\alpha t_c} \qquad \text{or} \qquad t_c = \frac{1}{\alpha} \log(\frac{w_c}{\beta Q_o}) \propto \log(Q_o)$$

This predicts that the threshold time is linear with the log of ping amplitude. The plot of the actual data in Figure 6.37(b) shows remarkable linearity.

Phase-plane plots for the out-of-plane motions initiated by ping are shown in Figure 6.38(a) where the quarter points are monitored. The oscillations bear a strong resemblance to the nonlinear vibrations of a system with a nonsymmetric return force of Chapter 5. The plate eventually stabilizes in a new deformed configuration with all eigenvalues positive as seen in Figure 6.35.

The deformed shape is shown in Figure 6.34. This $(3, 1)$ shape is the same as the first mode predicted by the linear buckling analysis in Figure 6.22, and the same as the lowest vibration mode in Figure 6.36.

We conclude that once the structure is unstable $(\text{Im}[\mu] < 0$ or $\lambda < 0)$ that motion will eventually occur; the only effect of ping is to control when it begins. We are motivated to efficiently initiate this transition because we test for stability after each significant load stage. This control can be increased by enhancing the dominance of the first mode. This is achieved by letting the ping be designed such that

$$\{Q\} = \alpha[\,M\,]\{\phi\}_1 \tag{6.22}$$

where α is a proportionality factor. Then

$$\ddot{\eta}_m + 2\zeta\omega_m\dot{\eta}_m + \omega_m^2\eta_m = \{\phi\}_m^T\{Q\} = \alpha\{\phi\}_m^T[\,M\,]\{\phi\}_1 = 0\,, \qquad m \neq 1$$

Thus, the motions of the higher modes are initially quiescent. The designer ping of Equation (6.22) is implemented in NonStaD. Note that in many cases, multiple modes can go through zero almost simultaneously. For exploring those cases, NonStaD has actually implemented a ping designed as

$$\{Q\} = \alpha_1[\,M\,]\{\phi\}_1 + \alpha_2[\,M\,]\{\phi\}_2$$

In implementing this, each eigenvector is normalized so that the largest component is unity.

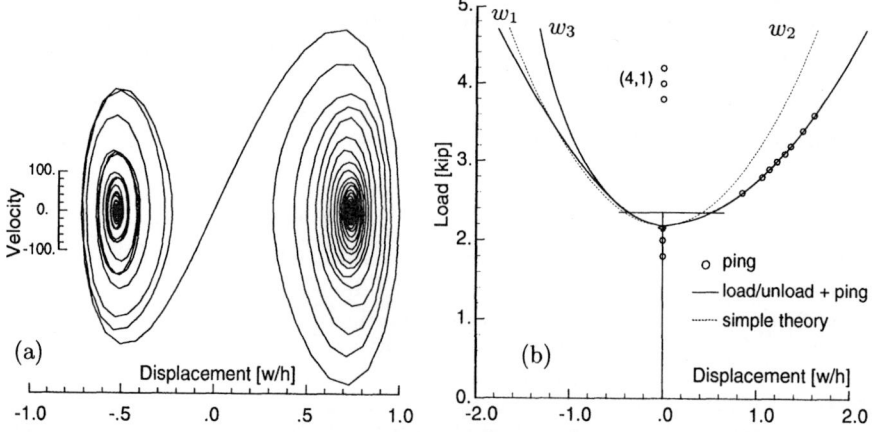

Figure 6.38: Post-buckling behavior. (a) Phase-plane plot. (b) Final equilibrium positions.

III: Post-Buckling Behavior

One way to use the dynamic approach is to ramp directly to the maximum load and then apply ping. Figure 6.38 shows results for the final out-of-plane equilibrium positions where a range of P values were used. This gives an idea of the post-buckling behavior. In this regard, ping can be thought of as a load imperfection, but unlike static imperfections, it is short lived and therefore the structure returns to its perfect state. Consequently, the results of the analysis are always that of the perfect structure. Note that if the load level is high enough, then other equilibrium positions (such as the $(4, 1)$ mode) may be found. Many examples using this approach are documented in Reference [81] for a variety of structures.

A more controlled dynamic procedure is to load the structure to just above critical, ping, and allow to settle. Then the load is slowly ramped up to the maximum level. The results for doing this are shown in Figure 6.38(b) as the thin continuous line. This line goes through the separate ping results. An advantage of this approach is that, on unloading, the initial portion of the post-buckling behavior can be obtained.

Example 6.24: Estimate the post-buckling behavior for a simply supported plate.

An approximate post-buckling analysis of the plate represents the out-of-plane displacements in the perturbation form

$$w(x, y) \approx \sum_{n=1,3,\cdots}^{N} \epsilon^n w_n(x, y), \qquad \epsilon \equiv \sqrt{\frac{P - P_c}{P_c}}$$

Using a perturbation procedure similar to that used in Chapter 5 and well documented in Reference [69], the first term approximation can be shown to be

$$w_1 = \sqrt{\frac{16 P_c}{Ehb} \frac{(n\pi/a)^2}{(n\pi/a)^4 + (m\pi/b)^4}}$$

This leads to the load deflection relation

$$P = P_c \left[1 + \alpha w_1^2 \right], \qquad \alpha = \frac{Ehb}{16} \left[(\frac{n\pi}{a})^4 + (\frac{m\pi}{b})^4 \right] (\frac{a}{n\pi})^2$$

This is shown in Figure 6.38 as the simple theory and indicates the bifurcation to be a stable symmetric bifurcation.

Discussion

Returning to Figures 6.34 and 6.35, we are now in a better position to interpret these histories and survey them from a dynamic perspective.

As the load increases, the vibration eigenvalues decrease with some eventually becoming negative. The plate is then unstable. Using ping as agent, the plate was set in motion in a controlled fashion and a new stable equilibrium position was found with a $(3, 1)$ deformed shape.

On further increase of load, the eigenvalues increase indicating a stiffening of the structure. This could not continue indefinitely, because the second mode is already indicating a downward trend before the load has peaked, and this could eventually become negative and unstable

On decreasing the load from its peak value, the eigenvalues again decrease with one becoming zero and again positive almost immediately. The plate has gone through a smooth transition from having out-of-plane deflections to being flat. A ping with an antisymmetric $(2, 1)$ spatial distribution confirms that the plate has found a new equilibrium shape. This transition is less violent than the bifurcation during loading. The structure stiffens on further decrease of the load.

In line with our intuitive notion of stability, the dynamic view treats all stability problems as dynamic events and thus there is no essential difference (or complication) between bifurcations, limit points, and phenomena such as mode jumping. In a sense, it recreates a situation quite close to an experiment where "what will happen will happen" and post-buckled states not easily attained by path-following methods are happened upon. The dynamic view is enhanced by implementing the following two aspects: monitoring the vibration eigenvalues to detect when a singularity is encountered, and using ping as agent to dynamically move the structure toward the new state. Both of these contribute a deeper insight into the static instability problem as well as giving some control over the loading process.

Monitoring the eigenvalues can also have a predictive aspect in the sense that it can show a trend toward zero, or if an eigenvalue goes close to zero. Both of these would indicate that if conditions changed slightly, an instability could arise.

There are quite a number of issues yet to be explored. First among them is the situation when multiple modes become unstable — the question of the uniqueness of the new found equilibrium state arises. The design of ping sends the structure in a certain direction, but this may bear little relation to the final rested state. Similar issues arise with an asymmetric bifurcation point. Both of these require that we address the question of stability of motion in the large [38, 44], which is done in the next chapter.

6.7 Mode Jumping

We finish this chapter with a discussion of a phenomenon called *mode jumping* or sometimes *secondary buckling*. The phenomenon is intimately associated with the interaction of two buckling modes and therefore can be very sensitive to the precise geometry and boundary conditions. This phenomenon received a good deal of attention after Stein [70] reported for his plate experiments that the changes of buckle pattern "occurred in a violent manner and were observed to go from 5 to 6 to 7 to 8 buckles." Our objective here is to use the dynamic view to navigate through a mode jump occurring in a plate. It is difficult to analyze within the context of plates, so we begin by developing some simple models through which we can explore some of its aspects.

Simple Models for Mode Jumping

Reference [72] and, more recently, Reference [61] show that the post-buckling behavior of plates is governed by cubic and quartic systems. This has been used as justification for neglecting higher-order terms in a Galerkin-type analysis. We use it as justification allowing the use of simple models through which we can explore, in a tractable way, some aspects of the mode jumping phenomenon.

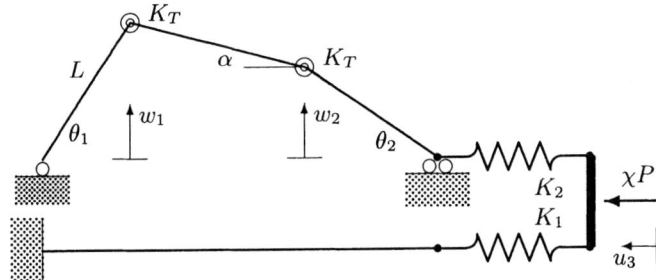

Figure 6.39: Three-degree-of-freedom simple model for a plate.

The essential deformations for the buckling of plates is an in-plane compression and a nonlinear out-of-plane flexure. Mode jumping is associated with the interaction of two modes, hence we need at least two flexural degrees-of-freedom. The simple model shown in Figure 6.39 (which is a modified version of that introduced by Stein [70]) is made of rigid links each of length L with torsional springs K_T representing the flexural stiffness. It also has axial springs associated with the shortening u_3. As we will see, the spring K_2 will introduce a nonlinearity into the system even for small deflections w_1 and w_2.

Consider the torsional springs first. With the small deflection assumption, the angles are given by

$$\theta_1 \approx \frac{w_1}{L}, \qquad \theta_2 \approx \frac{w_2}{L}, \qquad \alpha \approx \frac{w_1 - w_2}{L}$$

The twists of the spring are then

$$\theta_1 + \alpha \approx \frac{2w_1 - w_2}{L}, \qquad \theta_2 - \alpha \approx \frac{w_1 - 2w_2}{L}$$

The strain energy is

$$U_T = \tfrac{1}{2}K_T(\frac{2w_1 - w_2}{L})^2 + \tfrac{1}{2}K_T(\frac{2w_2 - w_1}{L})^2 = \frac{K_T}{2L^2}\left[5w_1^2 - 8w_1w_2 + 5w_2^2\right]$$

Introduce new generalized coordinates defined as

$$x_1 \equiv \tfrac{1}{2}(w_1 + w_2), \quad x_2 \equiv \tfrac{1}{2}(w_1 - w_2) \quad \text{or} \quad w_1 = \tfrac{1}{2}(x_1 + x_2), \quad w_2 = \tfrac{1}{2}(x_1 - x_2)$$

and $x_3 \equiv u_3$, where x_1 and x_2 are the amplitudes of the symmetric and anti-symmetric deformation modes. After substitution, we get

$$U_T = \frac{K_T}{4L^2}\left[x_1^2 + 9x_2^2\right]$$

This result is interesting for two reasons. First, the energies of the two modes are uncoupled and we can utilize this to add springs without affecting the coupling. Second, the anti-symmetric mode has a larger coefficient and hence we expect the symmetric mode to occur first.

The end shortening of the links is computed as

$$\Delta = L[3 - \cos\theta_1 - \cos\alpha - \cos\theta_2] \approx \frac{1}{2L}[w_1^2 + 8w_2^2 + (w_1 - w_2)^2] = \frac{1}{4L}[x_1^2 + 3x_2^2]$$

The strain energy of the axial springs is therefore

$$
\begin{aligned}
U_S &= \tfrac{1}{2}K_1 u_3^2 + \tfrac{1}{2}K_2(u_3 - \Delta)^2 \\
&= \tfrac{1}{2}(K_1 + K_2)u_3^2 + K_2(x_1^2 + 3x_2^2)^2/32L^2 - K_2(x_1^2 + 3x_2^2)u_3/4L
\end{aligned}
$$

We see the nonlinear contribution of the K_2 spring in the second and third terms. What is also interesting to observe is that this spring couples the in-plane and out-of-plane deflections.

We can write the total potential for the problem as

$$
\Pi = U + V = \frac{K_T}{4L^2}\left[x_1^2 + 9x_2^2\right] + \tfrac{1}{2}(K_1 + K_2)u_3^2 \\
- \frac{K_2}{4L}\left[x_1^2 + 3x_2^2\right]u_3 + \frac{K_2}{32L^2}\left[x_1^2 + 3x_2^2\right]^2 - \chi P u_3 \quad (6.23)
$$

This is the general structure of the potential function. This model can exhibit a bifurcation, but it cannot exhibit a mode jump. We can utilize the uncoupling of the modes to modify the potential in various ways. Consider a nonlinear axial and a nonlinear torsional spring attached at the center of the middle link, then we would just add to the potentials

$$U_A = \tfrac{1}{2}\alpha_1 x_1^2 + \tfrac{1}{4}\alpha_2 x_1^4 + \cdots, \qquad U_T = \tfrac{1}{2}\beta_1 x_2^2 + \tfrac{1}{4}\beta_2 x_2^4 + \cdots \quad (6.24)$$

respectively. Note that both α_2 and β_2 can be either positive or negative. This gives us a mechanism to change the parameters of the system without affecting the mechanics of the problem.

Rather than perform a parametric study using the general form of the potential — an exhaustive example of such a study for a plate on a nonlinear foundation is given in Reference [17] — we find it more to the point to compare the above to a system known to exhibit mode jumping.

Example 6.25: Show the connection between Allman's problem and the general potential.

Allman [3], as part of a discussion of mode jumping in plates, introduced an idealized problem with chosen parameters that make the manipulations less cumbersome. The potential energy (modified slightly) of the system is

$$\Pi = K\left[\tfrac{1}{2}(x_1^2 + 4x_2^2 + x_3^2) - \tfrac{1}{2}(x_1^2 + 3x_2^2)x_3 + \tfrac{1}{4}(x_1^4 + 5x_1^2x_2^2 + 6x_2^4)\right] - \chi P x_3$$

We can have a combination of Equations (6.23) and (6.24) coincide with Allman's by making the following associations:

$$K_1 + K_2 = K, \quad P = K, \quad \alpha_1 = K - \frac{K_T}{2L^2}, \quad \alpha_2 = \frac{11}{12}$$

$$\frac{K_2}{L} = 2K, \quad L = \frac{3}{10}, \quad \beta_1 = 4K - \frac{9K_T}{2L^2}, \quad \beta_2 = -\frac{3}{2}$$

and letting $u_3 \leftrightarrow x_3$. The main point of this is that, in terms of the physical model with nonlinear springs, in order to have mode jumping we need a softening mechanism associated with the antisymmetric mode in the large post-buckling region; that is, β_2 must be negative.

Example 6.26: Sketch the equilibrium paths for Allman's problem.

The equilibrium equations are

$$\mathcal{F}_1 = \frac{\partial \Pi}{\partial x_1} = x_1\left[1 + x_1^2 + \tfrac{5}{2}x_2^2 - x_3\right] = 0$$

$$\mathcal{F}_2 = \frac{\partial \Pi}{\partial x_2} = x_2\left[4 + \tfrac{5}{2}x_1^2 + 6x_2^2 - 3x_3\right] = 0$$

$$\mathcal{F}_3 = \frac{\partial \Pi}{\partial x_3} = \left[-\tfrac{1}{2}(x_1^2 + 3x_2^2) + x_3\right] - \chi = 0$$

There are four equilibrium paths. The first three are obtained by setting $x_1 = 0$, $x_2 = 0$; $x_2 = 0$; and $x_1 = 0$; respectively. The fourth path is obtained by setting the first two bracketed terms to zero. This results in

$$
\begin{array}{llll}
\text{I:} & x_1 = 0 & x_2 = 0 & x_3 = \chi \\
\text{II:} & x_1 = \sqrt{2}\sqrt{\chi - 1} & x_2 = 0 & x_3 = 2\chi - 1 \\
\text{III:} & x_1 = 0 & x_2 = \sqrt{(6\chi - 8)/3} & x_3 = 4\chi - 4 \\
\text{IV:} & x_1 = \sqrt{6\chi - 10} & x_2 = \sqrt{4 - 2\chi} & x_3 = \chi + 1
\end{array}
$$

These four equilibrium paths are shown plotted in Figure 6.40 where unstable segments are indicated as dashed lines. Note that these paths do not exist for all positive loads; for example, the second path exists only for $\chi \geq 1$.

We are also interested in distinguishing the stable and unstable portions of the paths. We get this stability information by looking at the spectral properties of

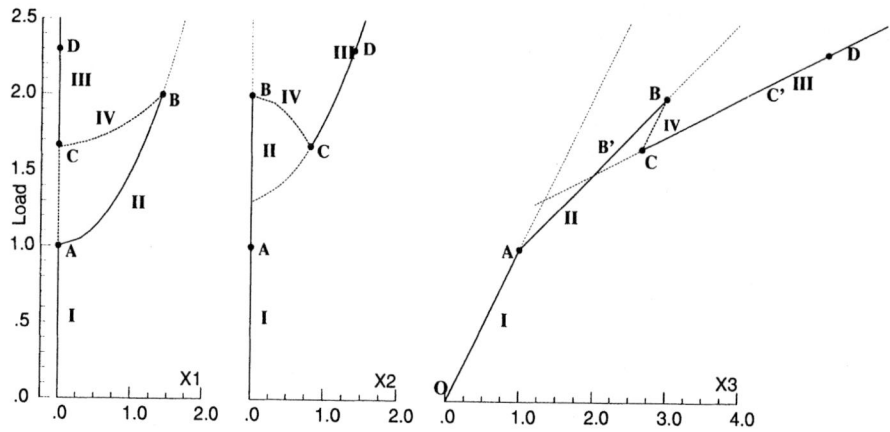

Figure 6.40: Equilibrium paths for Allman's problem.

the tangent stiffness matrix. This matrix is given by

$$[K_T] \equiv [\frac{\partial \mathcal{F}}{\partial x}] = \begin{bmatrix} 1 + 3x_1^2 + \frac{5}{2}x_2^2 - x_3 & 5x_1x_2 & -x_1 \\ 5x_1x_2 & 4 + \frac{5}{2}x_1^2 + 18x_2^2 - 3x_3 & -3x_2 \\ -x_1 & -3x_2 & 1 \end{bmatrix}$$

For convenience, let the mass be specified as $[\,M\,] = [K_T(x_i = 0)]$ or

$$[\,M\,] = \begin{bmatrix} 1 & 0 & 0 \\ 0 & 4 & 0 \\ 0 & 0 & 1 \end{bmatrix}$$

The eigenvalue problem to be solved for each equilibrium path is

$$[K_T]\{\phi\} - \lambda[\,M\,]\{\phi\} = 0, \qquad \lambda \equiv \omega^2$$

For Path I, for example, we get

$$\begin{bmatrix} 1 - \chi - \lambda & 0 & 0 \\ 0 & 4 - 3\chi - 4\lambda & 0 \\ 0 & 0 & 1 - \lambda \end{bmatrix} \begin{Bmatrix} \phi_1 \\ \phi_2 \\ \phi_3 \end{Bmatrix} = 0$$

This leads to the three eigenvalues

$$\lambda_1 = 1 - \chi, \qquad \lambda_2 = 1 - \tfrac{3}{4}\chi, \qquad \lambda_3 = 1$$

As the load increases to $\chi = 1$, λ_1 goes to zero. Above this load, it has a negative value resulting in a complex ω_1; consequently, the path is unstable above $\chi = 1$.

We summarize the behavior of all of the eigenvalues as

I: $\lambda_1 = 1 - \chi,$ $\lambda_2 = 1 - \tfrac{3}{4}\chi,$ $\lambda_3 = 1$

II: $\lambda_1 = \tfrac{1}{4}(2 - \chi),$ $\lambda_{2,3} = \tfrac{1}{2}(4\chi - 3) \mp \tfrac{1}{2}\sqrt{16\chi^2 - 32\chi + 17}$

III: $\lambda_1 = \chi - \tfrac{5}{3},$ $\lambda_{2,3} = \tfrac{1}{2}(6\chi - 7) \mp \tfrac{1}{2}\sqrt{3(12\chi^2 - 30\chi + 19)}$

These eigenvalues are shown plotted in Figure 6.41. Stable portions of Path II and Path III do not intersect in Figure 6.40. They are, in fact, connected by Path IV

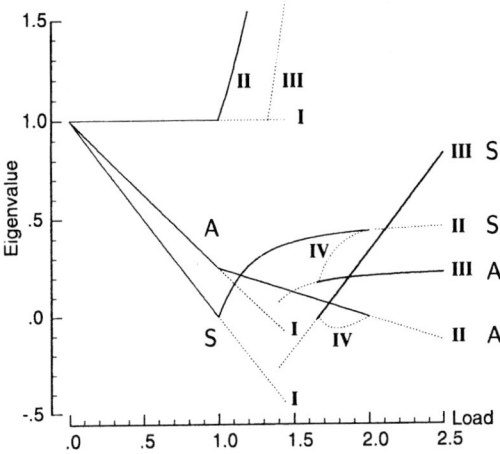

Figure 6.41: Eigenvalues as a function of load for Allman's problem. S=symmetric mode, A=antisymmetric mode.

whose solution exists only in the narrow region $\frac{5}{3} \le \chi \le 2$ and the eigenvalue λ_1 is always negative as shown in Figure 6.41. Consequently, Path IV is an unstable path.

Example 6.27: Trace a load/unload cycle for Allman's problem.

We can now trace a load/unload cycle for this simple model. With reference to Figure 6.40, we begin loading at O and increase up to A. During this stage $x_1 = 0$, $x_2 = 0$, $x_3 = \chi$. Point A is singular with $\{\phi\}_1^T = \{1, 0, 0\}$. This shifts the solution onto the stable Path II. The load can continue increasing until point B. During this stage the eigenvalue of the symmetric mode increases, but that of the antisymmetric mode decreases.

Point B is another singular point and has $\{\phi\}_1^T = \{0, 1, 0\}$. However, this is an unstable region. If we keep the load χ constant at a slightly higher value, then the only stable equilibrium point to be found is point C' on Path III. This involves a large displacement or "jump" resulting in the new configuration $x_1 = 0$, $x_2 = \sqrt{4/3}$, $x_3 = 4$. At this stage, the load can continue to increase indefinitely past point D and all eigenvalues increase.

On unloading along Path III, point C' is not special, and we can decrease to point C with the state $x_1 = 0$, $x_2 = 0$, $x_3 = 8/3$. This point is singular with $\{\phi\}_1^T = \{1, 0, 0\}$. Again, this is an unstable region, and if we keep the load constant (at a slightly lower value), then the only stable equilibrium point to be found is point B' on Path II. This involves a large displacement (but not as large as BC') resulting in the new configuration $x_1 = \sqrt{4/3}$, $x_2 = 0$, $x_3 = 7/3$.

The load can now be decreased to A where the symmetric mode becomes unstable and we get a change to Path I. We proceed to complete unloading with all eigenvalues increasing.

The cycle just completed is not conservative. That is, there is an energy loss corresponding to the quadrilateral $B'BC'C$. This loss manifests itself as kinetic energy during the mode jumping process.

Secondary Buckling of Plates

A common agreement is that the secondary buckling phenomenon is intimately associated with the interaction of two buckling modes and therefore can be very sensitive to the precise geometry and boundary conditions [33, 62]. We use a plate similar to that of the previous section and shown in Figure 6.33, but with two significant changes. The first is that it has clamped boundary conditions on the short edge; the second is that the plate aspect ratio is changed slightly to [3.4:1]. Both of these contribute to enhancing the interaction of buckling modes.

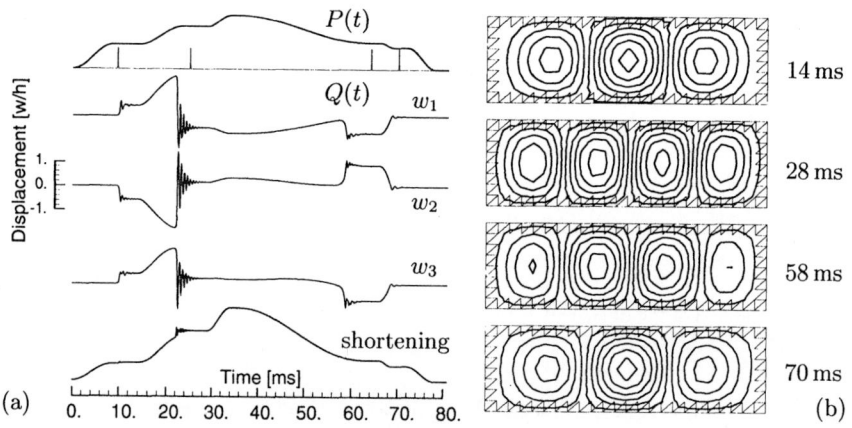

Figure 6.42: Deformation history of the [3.4:1] plate. (a) Displacements and loads as a function of time. (b) Contours of deformed shapes.

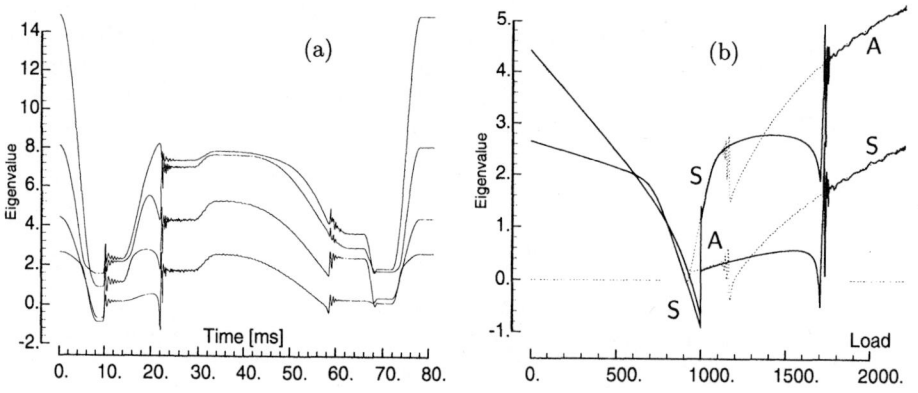

Figure 6.43: Monitoring the spectral behavior of the [3.4:1] plate. (a) First four vibration eigenvalues as a function of time over the full time range. (b) First two vibrations eigenvalues as a function of load; dashed lines are for unloading.

Figure 6.42 shows the load and some displacement histories for a complete load/unload cycle. The load history is such that after each significant stage it re-

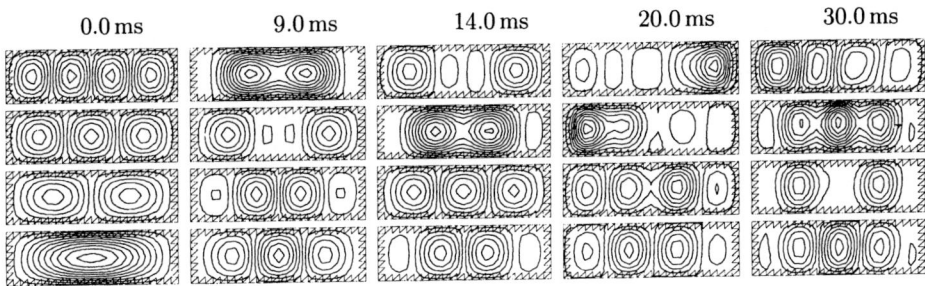

Figure 6.44: Sequence of vibration mode shapes.

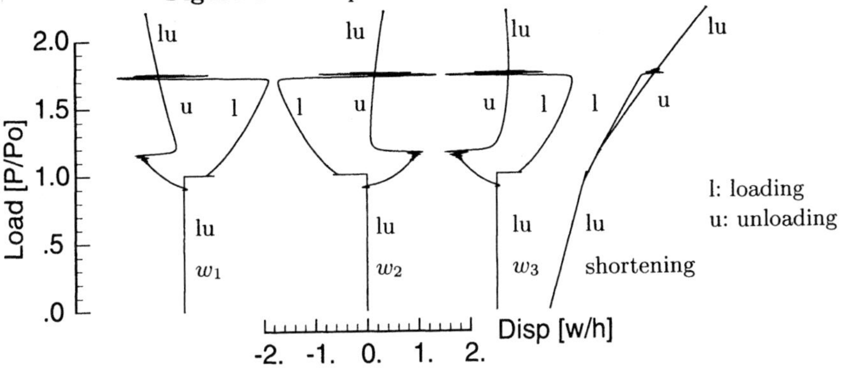

Figure 6.45: Load deformation plots.

mains constant until most of the dynamics associated with ping has died out, and Figure 6.42 shows deflection contours at some of these significant times. Clearly there are drastic changes in deformed shape. Figure 6.43 shows the eigenvalue histories and Figure 6.44 the vibration mode shapes. The latter figure shows a complex evolution over time; note that because of the clamped boundary condition, the vibration mode shapes are not as "crisp" and well defined as for the simply supported case.

For the first 10 ms the behavior seems about the same as in Figure 6.35 for the simply/supported plate. On closer look, however, we see some differences. First note that two modes become unstable almost simultaneously and that this mode pair is quite separated from the other modes. A second difference is that, after the bifurcation, the lowest mode has significantly less stiffness than for the simply supported case.

On further increase of load, first there is a stiffening, followed by a rapid loss of stiffness, which occurs at about 22 ms. Figure 6.42 shows that after the dynamics has settled down, the deformed shaped has changed significantly from a symmetric $(3,1)$ shape to an antisymmetric $(4,1)$ shape. This singular point encountered is so sensitive that the application of ping was not necessary to cause the transition. This is the mode jump phenomenon.

Between 30 ms and 34 ms, the load was then increased to show the stability of this new state — the four lower eigenvalues increased. On unloading, at a time of about 59 ms, the plate goes through another mode jump, this time from a $(4, 1)$ to a $(3, 1)$ shape as shown in Figure 6.42. This is not at the same load level as the first mode jump, as can be seen more clearly in the load/deflection plots of Figure 6.45 and the load/eigenvalue plots of Figure 6.43(b). After this jump, the unloading path is identical to the first post-buckling path (it is noted that for some computer runs the sense of the deflection switches). The transition through the first bifurcation is similar as for the simply supported plate.

The load/end-shortening plot of Figure 6.45 shows a hysteresis loop in the post-buckling region. This nonconservative behavior indicates an energy loss, energy that was dissipated during the dynamic mode jumping event.

Discussion of Plate Mode Jumps

We do not expect the simple model to precisely describe mode jumping in plates, but it should capture some of the essential features. We will now retrace the results of Figures 6.42 through 6.44 to see to what extent the simple model can explain what was observed.

At zero load, the vibration mode shapes have the familiar sequence for a rectangular plate. As the load increases, the sequence changes, and it is the $(3, 1)$ and $(4, 1)$ modes that become unstable. Both modes have comparable eigenvalues. After the transition, which occurs at a load higher than the critical load, these two modes interchange positions. The behavior of this first bifurcation is quite similar to that illustrated by the simple model.

As the load is increased, the $(4, 1)$ mode is unchanged, but there is a strong interaction between the $(3, 1)$ and $(1, 1)$ modes. All eigenvalues increase. The second and third modes peak first and only after they are definitely decreasing does the lowest mode peak and then decrease. It appears that it is the complicated large deflection of the plate that leads to the effective softening of the symmetric mode. As shown earlier in this chapter and in Chapter 3, in the FEM formulation, the tangent stiffness matrix is constructed of two parts as

$$[K_T] = [K_E] + [K_G(\sigma_{xx}, \sigma_{yy}, \sigma_{xy})]$$

where the geometric stiffness contribution depends on the current state of the membrane stresses. Figure 6.46 shows the stress contours during this loading stage and Figure 6.47 shows the distributions along the centerline. As the load changes, the shape of the contours remain essentially the same, but the range of the numbers change considerably. Clearly, there is not a uniform stress as was the case leading up to the first bifurcation. Indeed, there is also a significant compressive σ_{yy} stress. We identify the significant σ_{xx} stress as the softening mechanism anticipated by the simple model. Referring back to Figure 6.14 we see that a symmetric axial stress affects the symmetric vibration mode more than the antisymmetric one.

Figure 6.46: Contours of stress at time $t = 20$ ms.

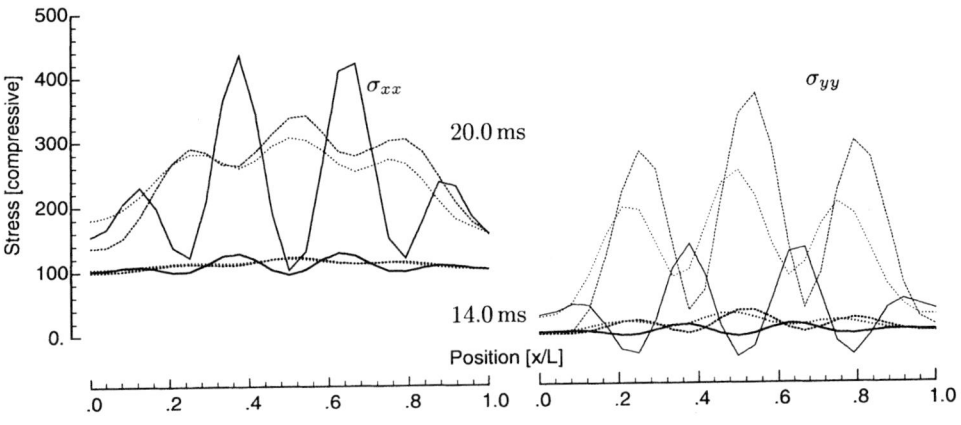

Figure 6.47: Stress distribution at times $t = 14$ ms and $t = 20$ ms.

Note that small changes in the boundary conditions can cause large changes in the σ_{xx} and σ_{yy} stresses, and, consequently, large changes in the post-buckling behavior. There is not much discussion in the literature about the role of this stress, although mention should be made of Reference [79].

Once the mode jump occurs, the symmetric mode is the lowest mode as also occurs in the simple model. On unloading, the second jump occurs a lower load just as in the simple model. The main difference between the simple model and plate is that, for the plate, it is the symmetric mode that causes the jumping.

Problems

6.1 Consider a cantilever beam as loaded in Figure 6.8.
- Plot the shear force and bending moment diagrams.
- Show that when the beam is loaded by an axial force and concentrated

moment applied at the free end the deflection shape is

$$v(x) = \frac{M_o L^2}{EI} \left[\frac{1 - \cos kx}{k^2 \cos kL} \right]$$

- If the cantilever beam has a tensile axial force applied, show that

$$v(L) = \frac{Q}{k^3 EI} \left[\frac{\sinh kL - kL \cosh kL}{\cosh kL} \right]$$

- Show that in the limit as P becomes very large that the deflection goes to zero.
- Recover the uncoupled solution for each of these situations.

[Reference[32], pp. 162]

6.2 Consider the buckling of a clamped/clamped beam.
- Show that the characteristic equation is $2 - 2 \cos kL - kL \sin kL = 0$.
- Show that one set of solutions is given by

$$P_{cr} = 4n^2 \pi^2 \frac{EI}{L^2}, \qquad v(x) = \left(\cos \frac{2n\pi}{L} x - 1 \right)$$

- Show that the lowest critical load of the second set of solutions is given by: $P_{cr} = 8.18\pi^2 EI/L^2$.
- Confirm these results using an FEM analysis.   [Reference[77], pp. 54]

6.3 A shaft is supported in two bearings and has two applied axial loads as shown.
- If $P_2 = 0$, what value of P_1 will make the shaft buckle?
- If both P_1 and P_2 are loads of $30EI/L^2$, what is the maximum distance a allowable without causing the shaft to buckle?
- Confirm these results using an FEM eigenanalysis.

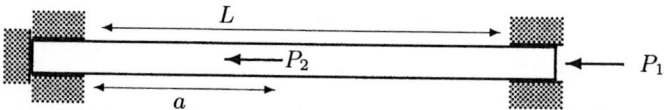

6.4 Consider a cantilever beam with a rigid "handle" of length a welded to the free end. The handle is oriented along the length of the beam with its tip closest to the fixed end of the beam.
- Show that if a force is applied to the handle and pointing toward the fixed end of the beam that the characteristic equation for the buckling load is $kL \tan kL = L/a$.
- If the force is reversed so that the beam is in tension, show that buckling occurs when $kL \tanh kL = L/a$. This is an example of buckling even though the beam is in tension.
- Confirm these results using an FEM eigenanalysis. Vary the length a.

[Reference[85], pp. 56]

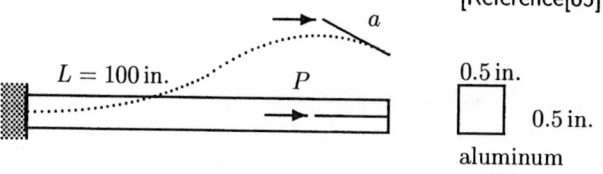

6.5 In reference to a cantilever beam supported at its end by a spring of stiffness K, an interesting special case arises when the spring has the special values $\alpha^* = 2K^2/L$.

- Show that the critical loads and the corresponding mode shapes are given by

$$P_{cr} = n^2\pi^2\frac{EI}{L^2}, \qquad \alpha = \frac{2\pi^2 EI}{L^3}, \qquad v(x) = c_2\left[\sin\frac{n\pi}{L}x \pm n\pi\frac{x}{2L}\right]$$

6.6 Consider the buckling of beams when the axial force varies.

- Show that the governing equation is

$$\frac{\partial^2}{\partial x^2}\left[EI\frac{\partial^2 v}{\partial x^2}\right] - \frac{\partial}{\partial x}\left[\bar{F_o}\frac{\partial v}{\partial x}\right] = 0$$

- Determine the critical load for a simply supported column loaded under self-weight.
- Confirm these results using an FEM analysis. [Reference[56], pp. 13]

6.7 For the truss shown:

- Determine the relationship between critical load and the angle θ.
- What orientation has a minimum buckling load?
- Confirm the results using an FEM eigenanalysis.

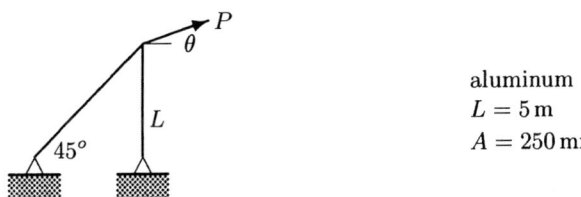

aluminum
$L = 5\,\mathrm{m}$
$A = 250\,\mathrm{mm}^2$

6.8 For the truss structure shown, and using an FEM analysis:

- Determine the buckling load(s) using different values of $\alpha = 10^0$, 10^2, 10^4, and different values of $\epsilon = \pm0.1, \pm0.01, \pm0.001$.
- Compare the results with the given theory,
- Do a nonlinear deflection analysis and compare.

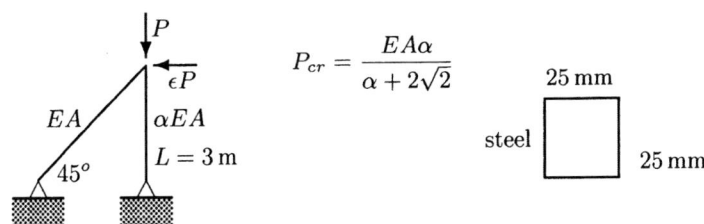

$$P_{cr} = \frac{EA\alpha}{\alpha + 2\sqrt{2}}$$

25 mm

steel

25 mm

6.9 Consider the folded plate structure of Figure 6.21.

- Do an FEM eigenanalysis to investigate the effect of different boundary conditions.
- Compare the results from those using a simple analysis where side plates are treated as simply supported.

7
Dynamic Stability

We considered stability in Chapter 6 in a statics context; clearly, however, it is in truth a dynamic phenomenon since we are talking of structural behavior changing. The purpose of this chapter is to analyze some of the relevant concepts from a fully dynamic perspective.

We again invoke our intuitive notion that perturbations of a dynamic system (which are always present in real situations) cause the system to move about its (dynamic) equilibrium configuration. If this movement becomes excessive, we talk of an instability of the motion. Some situations where the stability of the motion arises are aeroelastic flutter, whirling of shafts, rotating saw blades and computer disks, belt drives, galloping of power lines, and control of structures.

We also have the idea of a dynamic equilibrium position, although the system is in motion, it is stable in the sense that any disturbances eventually die down and the system returns to that dynamic state. This is illustrated in Figure 7.1 for a cantilever beam with a follower force: the phase plane-plots show that motions initiated either inside or outside the limit cycle will tend eventually to the limit cycle.

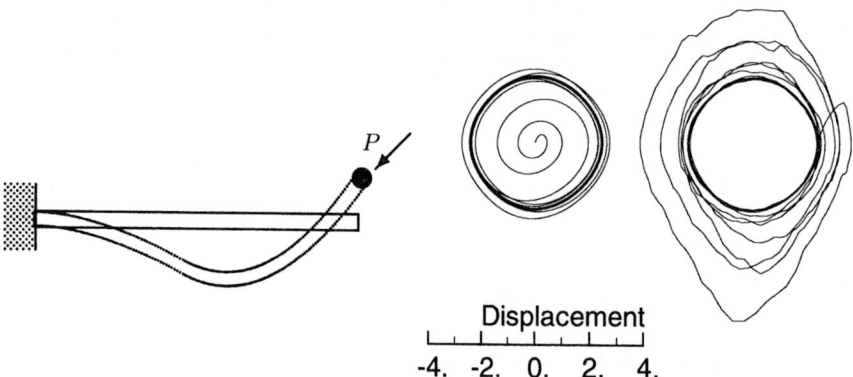

Figure 7.1: Beam with a follower force and the response for a load above critical. Phase-plane plots showing the achievement of a limit cycle from inside (left) and from outside (right).

7.1 Some Preliminary Ideas

The definition of stability chosen must be appropriate to the types of phenomena we wish to distinguish. There are two dynamic situations of primary interest to us:
- The free motion after an initial disturbance.
- The forced response.

In the first, we would like to have the dynamic response (e.g., displacement, velocity) "die down to zero asymptotically," that is, return (eventually) to its undisturbed position. In the second, we want "bounded response for bounded input." To clarify both of these concepts, we will look at a few simple structural systems. Reference [64] attempts to give an overview of stability under dynamic conditions.

Static Instabilities Revisited

Consider the simple pinned structure shown in Figure 7.2, which we have already looked at in Chapter 6. The initially vertical bar is assisted in remaining vertical by the action of the horizontal spring; the spring is unstretched when the bar is vertical. Let the bar be very stiff and massless; from a linear small deflection analysis we conclude that the displacement is only horizontal, that is, $v = 0$.

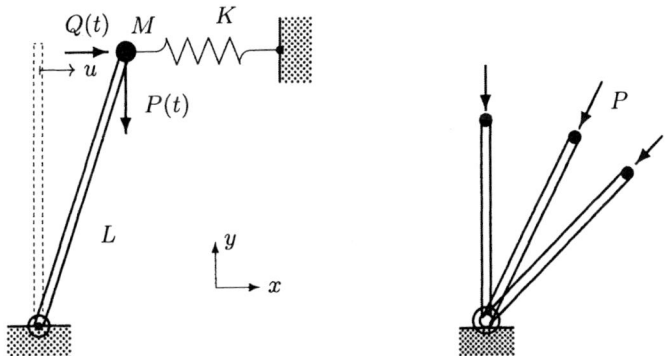

Figure 7.2: Disturbed equilibrium state of a pinned bar. Example of a follower force.

Look at dynamic equilibrium based on the deformed configuration. Consider the situation when the bar has already displaced by an amount u as shown in Figure 7.2. Summing the moments about the base gives

$$-QL - Pu + KuL + C\dot{u}L = -ML^2\ddot{u}/L$$

where a velocity-dependent resistance has also been added. This can be put in the usual form of our equations of motion

$$[K - P/L]u + [\,C\,]\dot{u} + [\,M\,]\ddot{u} = Q \tag{7.1}$$

We will now consider a couple of special cases. In each case, we treat $P(t)$ as being quasi-static (i.e., it alone will not cause inertia effects).

I: Free Vibration

Here we let $Q(t) = 0$ after some initial nonzero value and seek free vibration solutions of the form $u(t) = \hat{u}e^{i\mu t}$. On substituting this into the equation of motion get

$$\left[[K - P/L] + [i\mu C - \mu^2 M] \right] \hat{u} = 0$$

This can be true only if the term inside the large bracket is zero. This leads to

$$\mu = \frac{iC}{2M} \pm \sqrt{\frac{K - P/L}{M} - (\frac{C}{2M})^2}$$

The response is therefore

$$u(t) = \hat{u}e^{i\mu t} = \hat{u}e^{-(C/2M)t}e^{\pm i\sqrt{(K-P/L)/M-(C/2M)^2}t}$$

This solution is affected by the relative value of P and KL.

As long as the damping is nonzero and $P < KL$, we always have a bounded response. When the damping is zero, we get oscillatory behavior since

$$\mu = \pm\sqrt{\frac{K - P/L}{M}}, \qquad u(t) = \hat{u}e^{i\mu t} = \hat{u}e^{\pm i\sqrt{[(K-P/L)/M]}t}$$

This motion will persist for all time, that is, it does not asymptotically decay to zero. We consider this an instability, although it is stable in the sense of starts near, stays near. When $P > KL$, then

$$u(t) = \hat{u}e^{-(C/2M)t}e^{\pm i\sqrt{-(P/L-K)/M-(C/2M)^2}t} = \hat{u}e^{-(C/2M)t}e^{\pm\sqrt{(P/L-K)/M+(C/2M)^2}t}$$

and one of the solutions is monotonically increasing. The solution is therefore unstable. This is an example of the type of instability covered in Chapter 6. The complete picture as the load is changed is shown in Figure 7.3, where μ goes from being purely real to being purely imaginary.

In the undamped special case when $P = KL$, we get

$$\mu = +\frac{iC}{2M} \pm \frac{iC}{2M} = +\frac{iC}{M}, 0$$

This has the two responses

$$u(t) = \hat{u}e^{-(C/M)t} \qquad \text{and} \qquad u(t) = \hat{u} = \text{constant}$$

The velocity and acceleration go to zero but the mass does not return to its original position; this leads us to conclude that it is unstable. Let us look closer at the free motion of this system. The equation of motion is

$$C\dot{u} + M\ddot{u} = 0$$

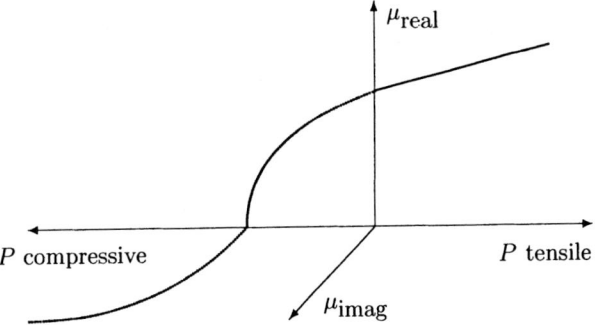

Figure 7.3: Undamped free vibration frequency as the load parameter P is changed.

This is integrated to give the total solution as

$$u(t) = \frac{1}{C}A + Be^{-(C/M)t}$$

For an initial velocity V_o when $u = 0$

$$u(t) = \frac{V_o M}{C}(1 - e^{-(C/M)t}), \qquad \dot{u}(t) = V_o e^{-(C/M)t}$$

Again, the original configuration is unstable because the bar does not return to its original position when it comes to rest. If the damping is very small, then by a Taylor series expansion we get

$$u(t) \approx \frac{V_o M}{C}[1 - (1 - (C/M)t \cdots)] = V_o t$$

which increases without bound. Actually, the damping is only affecting the distance the bar travels before it comes to rest.

II: Forced Vibration

Consider a forced vibration where $Q(t) = \hat{Q}e^{i\omega t}$ and again assume solutions $u(t) = \hat{u}e^{i\omega t}$, then

$$\left[[K - P/L] + [i\omega C - \omega^2 M]\right]\hat{u} = \hat{Q}$$

Rearrange this as

$$\hat{u} = \frac{[K - P/L - \omega^2 M] - [i\omega C]}{[K - P/L - \omega^2 M]^2 + [\omega C]^2}\hat{Q}$$

As long as the damping is nonzero, the denominator is nonzero, and we have a bounded response. However, for a given excitation $\hat{Q}$, the response could be arbitrarily large depending on the value of damping; hence, it may be desirable to set the stability criterion in terms of some threshold value of $\hat{u}/\hat{Q}$.

When the damping is zero, it is possible to get an unbounded response. This occurs when the denominator is zero

$$K - P/L - \omega^2 M = 0 \qquad \text{or} \qquad \omega_c = \sqrt{\frac{K - P/L}{M}}$$

From this, we consider resonance with no damping as unstable under forced frequency vibration. The relation between frequency and load is similar to the free vibration case except that here ω_c is the forcing frequency and the critical case is where it coincides with the natural frequency of free vibration.

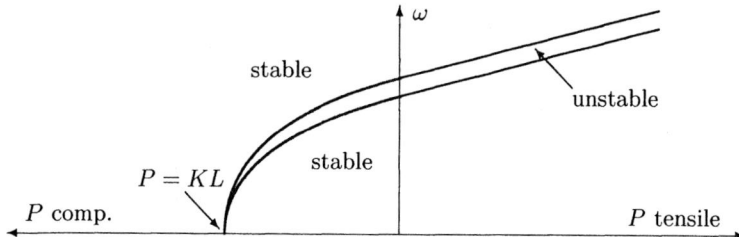

Figure 7.4: Locus of unstable points exceeding a threshold value under forced frequency response.

To continue on this point (but with damping): any combination of (positive) ω with (positive or negative) P gives a bounded solution. It is only in the vicinity of $\omega \approx \omega_c$ that the solution can exceed a threshold value and be unstable. This is shown plotted in Figure 7.4. We also see that static instability (buckling) is part of the continuum of forced frequency instability points. Actually, in looking at the figure and thinking in terms of a modal analysis, we can talk of the stability of the mode; that is, the figure shows one of the modes of a multi-degree-of-freedom system.

In the special case when $P = K/L$, even if there is damping, then

$$\hat{u} = \frac{[-\omega^2 M] - [i\omega C]}{[\omega^2 M]^2 + [\omega C]^2} \hat{Q} = \frac{-1}{\omega[\omega M - iC]} \hat{Q}$$

and most points are stable. Only in the limit as ω goes to zero does the response go to infinity and we say the system is unstable. That is, the system is stabilized by the inertia effects. Indeed, for all $P > KL$ and any forcing frequency, the system is stabilized by the dynamics. The static case, of course, does not depend on any of the dynamic parameters (mass or damping) and hence cannot be stabilized.

Follower Force and Self-Excited Oscillations

Consider the interesting special case when the constant load P always remains pointing along the axis of the rod as shown in Figure 7.2(b). Although the magnitude remains constant, the resolved components change as the rod rotates, and hence this is an example of a changing force. Taking moments about the base shows that this force does not appear in the equations of motion and we have

$$[K]u + [\,C\,]\dot{u} + [\,M\,]\ddot{u} = Q$$

Hence, there does not appear to be an opportunity for a static instability. That is, if Q is short lived, then after some vibration $u(t)$ will go to zero and the system will be back at its original state.

Intuition says, however, that there must be some sort of instability. Clearly, the equilibrium method is inadequate for this type of problem. We will later give the correct answer to this problem, but as a preliminary, we must refine our mathematical definition of stability. The aspect that characterizes the follower force problem is that the force is deformation dependent; that is, it changes as the displacement changes. The next example gives a simple illustration of the stability of such systems.

There are many dynamical systems in which a steady energy source is converted into oscillatory motion. Some examples are fluid flow around a structure, computer-controlled vibration systems, systems with chemical or biochemical reactions. What these have in common is that the input to the system is affected by the system response.

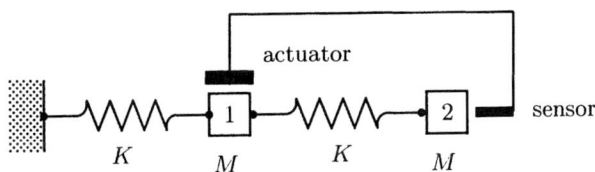

Figure 7.5: Simple system with feedback loop.

As a simple example, consider the spring/mass system shown in Figure 7.5 where an input is given to Mass 1 depending on the position of Mass 2. The equations of motion are

$$K \begin{bmatrix} 2 & -1 \\ -1 & 1 \end{bmatrix} \begin{Bmatrix} u_1 \\ u_2 \end{Bmatrix} + G \begin{bmatrix} 0 & 1 \\ 0 & 0 \end{bmatrix} \begin{Bmatrix} u_1 \\ u_2 \end{Bmatrix} + M \begin{bmatrix} 2 & 0 \\ 0 & 1 \end{bmatrix} \begin{Bmatrix} \ddot{u}_1 \\ \ddot{u}_2 \end{Bmatrix} = \begin{Bmatrix} Q_1 \\ Q_2 \end{Bmatrix}$$

where G is a "gain" parameter that could be either positive or negative. Note that the system is linear, and therefore the tangent stiffness matrix is

$$[K_T] = K \begin{bmatrix} 2 & -1 + G/K \\ -1 & 1 \end{bmatrix}$$

which is nonsymmetric. This nonsymmetry will affect the eigenvalues of the system.

As in the single-degree-of-freedom system, a stability analysis for free vibration begins by assuming solutions of the form

$$\left\{ \begin{matrix} u_1 \\ u_2 \end{matrix} \right\} (t) = \left\{ \begin{matrix} \hat{u}_1 \\ \hat{u}_2 \end{matrix} \right\} e^{i\mu t}$$

Substituting this into the governing equation gives the eigenvalue system

$$\left[K \begin{bmatrix} 2 & -1 \\ -1 & 1 \end{bmatrix} + G \begin{bmatrix} 0 & 1 \\ 0 & 0 \end{bmatrix} - \mu^2 \begin{bmatrix} 2 & 0 \\ 0 & 1 \end{bmatrix} \right] \left\{ \begin{matrix} \hat{u}_1 \\ \hat{u}_2 \end{matrix} \right\} = \left\{ \begin{matrix} 0 \\ 0 \end{matrix} \right\}$$

which leads to the characteristic equation

$$2\mu^4 - 4\omega_o^2\mu^2 + \omega_o^4 + \omega_o^2 G/M = 0, \qquad \omega_o \equiv \sqrt{K/M}$$

This is quadratic in μ^2, hence there are four roots appearing as $\pm$ pairs.

First look at the possibility of a static instability — this occurs when $\mu = 0$ giving

$$\omega_o^4 + \omega_o^2 G/M = 0 \qquad \text{or} \qquad G = -\omega_o^2 M = -K$$

Because the value of the gain is real only, we conclude that a static instability is possible. Note that for this value of G we have $\det[K_T] = 0$.

There are also other values of μ, which we now look at. The roots are given by

$$\mu^2 = 2\omega_o^2 \pm \omega_o^2 \sqrt{3 - G/K}$$

If μ is complex, then

$$e^{i\mu t} = e^{i(\mu_R + i\mu_I)t} = e^{-\mu_I t} e^{i\mu_R t}$$

and we want the imaginary part of μ to be greater than zero for stability. Now

$$\mu^2 = (\mu_R + i\mu_I)^2 = \mu_R^2 - \mu_I^2 + 2i\mu_R\mu_I$$

therefore we want $\text{Im}\,[\mu^2] > 0$ for stability. The imaginary component of μ^2 comes from the square-root term only, hence we ask if the radical can be negative. That is, for instability,

$$3 - G/K < 0 \qquad \text{or} \qquad G > 3K$$

Again, since the value of the gain is real only we conclude that an instability is possible.

This is an instability quite different from the static instability we just discussed (and was the focus of Chapter 6) because it does not occur when $\mu = 0$; at the point of instability $\mu = \sqrt{2}\omega_o$ and hence there is a vibration. That is, there is an oscillation occurring as the instability manifests itself — this is why it is called a *dynamic instability*. Furthermore, for this value of G, we have $\det[K_T] = 4K$, which is certainly not zero or even negative. Finally, it is also worth noting

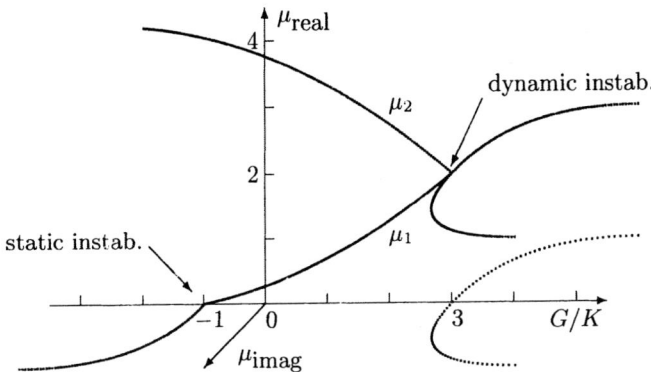

Figure 7.6: Roots as a function of the gain parameter G.

that the dynamic instability occurs when two modes coalesce, whereas the static instability occurs for a single mode.

The complete behavior is shown in Figure 7.6. At the static instability, we have both $\mu_R = 0$ and $\mu_I = 0$; however, at the dynamic instability we have $\mu_R > 0$ and only $\mu_I = 0$. Physically, the manifestations of a static and dynamic instability are quite different; however, we can cover both of them with the single criterion that for stability

$$\mu_I > 0$$

irrespective of the value of μ_R. This observation will be the backbone of our formal analysis of the stability of the motion.

Dynamic Stability of Discrete Systems

With the realization that all of our continuous systems can be reduced to discrete form, we will now consider general dynamic systems, both linear and nonlinear, presented in discrete form. The results and conclusions will also apply to continuous systems. In the linear description of systems, instability is equivalent to "blowing up." For nonlinear systems, "blowing up" is only one manifestation of an instability; there can be a large displacement (but not infinite) where the system comes to rest at a new equilibrium position. As we will see, there can also be *limit cycles* where the system remains in motion close to the (original) equilibrium point but it cannot remain arbitrarily close to it.

To carry on the discussion of stability as phrased in Chapter 6, our notion of stability is rooted in the question: If the system is slightly disturbed from its (dynamic) equilibrium position, what would happen? We see that this is fundamentally a question about the dynamics of the system.

The answers to the question are summarized in Figure 7.7, which is the dynamic companion to Figure 6.1. This figure shows a primary path (heavy dashed lines) and two examples of disturbed paths. In the first, illustrated by the segment A, the slightly disturbed motion returns to the primary path and we say

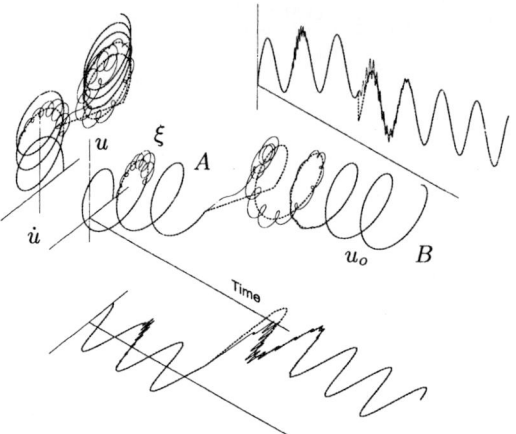

Figure 7.7: Some disturbed trajectories.

the system is stable. In the second, illustrated by the segment B, the primary path itself is changed. We say the system was unstable even though the ensuing dynamics (about the changed path) eventually die down indicating a sort of new stable motion.

As we did before, let us conceive of the total response and applied load as made up of two parts

$$\{u(t)\} = \{u_o(t)\} + \epsilon\{\xi(t)\}, \qquad \{P(t)\} = \{P_o(t)\} + \epsilon\{Q(t)\}$$

That is, there is the primary response $\{u_o\}$, which is due to $\{P_o\}$, and the smaller perturbation response $\{\xi\}$, which is due to the ping load $\{Q\}$. This leads (as in Chapter 6) to the separate equations

$$\epsilon^0: \qquad [\,M\,]\{\ddot{u}_o\} + [\,C\,]\{\dot{u}_o\} = \{P_o\} - \{F(u_o)\}$$

$$\epsilon^1: \qquad [\,M\,]\{\ddot{\xi}\} + [\,C\,]\{\dot{\xi}\} + [K_T]\{\xi\} = \{Q\}$$

The change of stiffness of the perturbed system is governed by the first equation but unlike our treatment of it in Chapter 6, here the equation remains fully dynamic.

The analysis of the ping free vibration response will give us the desired information about the stability of the system. Because the stiffness may change in time, in general, we are dealing with a nonautonomous system. We now look at a few particular cases.

I: Autonomous Systems

When the matrix $[K_T]$ is a constant, the stability criterion is relatively easy to construct. That is, starting with

$$[\,M\,]\{\ddot{\xi}\} + [\,C\,]\{\dot{\xi}\} + [K_T]\{\xi\} = \{Q\}, \qquad [K_T] = \text{constant}$$

we look for undamped free vibration solutions of the form

$$\{\xi(t)\} = \{c\}e^{i\mu t}$$

Substitute to get

$$\left[[K] - \mu^2[M]\right]\{c\}e^{i\mu t} = 0$$

This has to be true for all time, hence we must have

$$\left[[K] - \mu^2[M]\right]\{c\} = 0$$

We can have nontrivial solutions only if the determinant is zero. This leads to a characteristic equation to determine the eigenvalues μ_i and eigenvectors $\{c\}_i$. There are N roots (real or complex) some of which may be repeated.

The general solution is written as a combination of

$$\{c\}_1 e^{i\mu_1 t}, \quad \{c\}_2 e^{i\mu_2 t}, \quad \cdots, \quad \{c\}_N e^{i\mu_N t}$$

and as a result, the fundamental matrix can be written as

$$[\Phi(t)] = \left[\{c\}_1 e^{i\mu_1 t}, \{c\}_2 e^{i\mu_2 t}, \cdots, \{c\}_N e^{i\mu_N t}\right]$$

We state our stability criterion in terms of the properties of the eigenvalues μ_i. For the system to be asymptotically stable, we have

$$\text{Im}\,[\mu] > 0$$

If this is not true for any one of the roots, then the system is unstable.

Example 7.1: A two-degree-of-freedom system is described by

$$\begin{bmatrix} 4 & -2 \\ -2 & 6 \end{bmatrix} \begin{Bmatrix} u_1 \\ u_2 \end{Bmatrix} + G \begin{bmatrix} 0 & 1 \\ 0 & 0 \end{bmatrix} \begin{Bmatrix} u_1 \\ u_2 \end{Bmatrix} + \begin{bmatrix} 1 & 0 \\ 0 & 2a \end{bmatrix} \begin{Bmatrix} \ddot{u}_1 \\ \ddot{u}_2 \end{Bmatrix} = \begin{Bmatrix} P_1 \\ P_2 \end{Bmatrix}$$

where G is a gain parameter. Investigate the stability of the system as a function of the second mass $2a$.

First note that the tangent stiffness matrix is

$$[K_T] = \begin{bmatrix} 4 & -2+G \\ -2 & 6 \end{bmatrix}$$

and is nonsymmetric. This nonsymmetry will affect the eigenvalues of the system. A stability analysis for free vibration begins by assuming solutions of the form

$$\begin{Bmatrix} u_1 \\ u_2 \end{Bmatrix} (t) = \begin{Bmatrix} \hat{u}_1 \\ \hat{u}_2 \end{Bmatrix} e^{i\mu t}$$

Substituting this into the governing equation gives the eigenvalue system

$$\left[\begin{bmatrix} 4 & -2 \\ -2 & 6 \end{bmatrix} + G \begin{bmatrix} 0 & 1 \\ 0 & 0 \end{bmatrix} - \mu^2 \begin{bmatrix} 1 & 0 \\ 0 & 2a \end{bmatrix}\right] \begin{Bmatrix} \hat{u}_1 \\ \hat{u}_2 \end{Bmatrix} = \begin{Bmatrix} 0 \\ 0 \end{Bmatrix}$$

which leads to the characteristic equation

$$a\mu^4 - (3 + 4a)\mu^2 + 10 + G = 0$$

This is quadratic in μ^2, hence there are four roots appearing as $\pm$ pairs.

First look at the possibility of a static instability — this occurs when $\mu = 0$ or

$$10 + G = 0 \qquad \text{or} \qquad G = -10$$

Because the value of the gain is real only, we conclude that a static instability is possible. This is true irrespective of the value of the masses.

Now examine the system for a dynamic instability, that is, we consider cases where $\mu \neq 0$ but could have a negative imaginary part. The roots are given by

$$2\mu^2 = (3 + 4a) \pm \sqrt{9 - 4aG - 16a + 16a^2}$$

We want the imaginary part of μ to be greater than zero. Since

$$\mu^2 = (\mu_R + i\mu_I)^2 = \mu_R^2 - \mu_I^2 + 2i\mu_R\mu_I$$

we want $\text{Im}\,[\mu^2] > 0$. The imaginary component comes from the square-root term, hence we ask if the radical can be negative. That is,

$$9 - 4aG - 16a + 16a^2 < 0$$

The stability boundary is

$$G = \frac{9}{4a} - 4 + 4a \qquad \text{or} \qquad 8a = G + 4 \pm \sqrt{G^2 + 8G - 20}$$

The plot of $G(a)$ is parabolic with a minimum at $a = 0.75$, $G = 2$. If the second mass is zero $(a = 0)$ or infinite $(a = \infty)$, then an infinite gain is required to cause a dynamic instability. On the other hand, if $G < 0.75$, then irrespective of the second mass, there is no dynamic instability.

II: Nonautonomous Systems with Periodic Coefficients

We start with the system

$$[\,M\,]\{\ddot{\xi}\} + [\,C\,]\{\dot{\xi}\} + [K_T]\{\xi\} = \{Q\}, \qquad [K_T] = [K_T(\sin \omega t)]$$

We saw in Section 4.6 that the general solution is constructed from solutions of the form

$$\{c\}_1 e^{i\lambda_1 t} p_1(t), \qquad \{c\}_2 e^{i\lambda_2 t} p_2(t), \qquad \cdots, \qquad \{c\}_N e^{i\lambda_N t} p_N(t), \qquad \mu_i = e^{i\lambda_i T}$$

where $p_i(t)$ are Mathieu functions. The stability of the system is governed by the behavior of λ_i, which in turn is governed by μ_i. We concentrate on the roots μ_i.

Example 7.2: Determine the stability of the second-order system

$$\ddot{u} + [\alpha + \beta \cos t]\, u = 0$$

As was done in Section 4.6, we can arrange our equation of motion as a Mathieu equation in the form

$$\frac{d}{dt}\left\{\begin{matrix} u_1 \\ u_2 \end{matrix}\right\} = \begin{bmatrix} 0 & 1 \\ -\alpha - \beta\cos t & 0 \end{bmatrix} \left\{\begin{matrix} u_1 \\ u_2 \end{matrix}\right\}$$

The roots must be obtained from the quadratic form

$$\mu^2 - b\mu + c = 0$$

giving

$$\mu_1, \mu_2 = \tfrac{1}{2}b \pm \tfrac{1}{2}\sqrt{b^2 - 4c}$$

Using the result of Equation (4.18), and the fact that the trace of the $[A(t)]$ matrix is zero, we get

$$W(T) = \mu_1\mu_2 = \exp\left(\int_0^T \mathrm{Tr}[\ A\ ]\,ds\right) = e^0 = 1$$

This sets the value of $c = 1$. Replace $b \longrightarrow \phi(\alpha, \beta)$ giving the general form of the roots as

$$\mu_1, \mu_2 = \tfrac{1}{2}\phi(\alpha,\beta) \pm \tfrac{1}{2}\sqrt{\phi(\alpha,\beta)^2 - 4}$$

Although $\phi(\alpha, \beta)$ is not known explicitly, we can conclude the following [38]:

- If $\phi > 2$, μ_i are real hence λ_i are complex, and there are two unbounded solutions.
- If $\phi = 2$, $\mu_1 = \mu_2 = 1$, $\lambda = 0$, and there is one periodic solution (T) and one unbounded solution.
- If $-2 < \phi < 2$, μ_i are complex with $\mu_2 = \mathrm{conjg}(\mu_1)$, $|\mu_i| = 1$, and there are two bounded solutions.
- If $\phi = -2$, $\mu_1 = \mu_2 = -1$, $\lambda T = \pi$, and there is one periodic solution $(2T)$ and one unbounded solution.
- If $\phi < -2$ there are two unbounded solutions.

We conclude that the boundaries

$$\phi(\alpha, \beta) = \pm 2$$

separate regions where unbounded solutions exist from regions where all solutions are bounded. We also find that these are curves on which solutions of periodicities 2π or 4π occur. If we find, by whatever method, parameter values for which T and $2T$ periodic solutions exist, then we have also found the boundaries of the stability regions.

We will show an application of this when we consider the pulsating compression of a beam.

III: Nonlinear Systems

The full study of the stability of nonlinear systems is very difficult, extensive, and has a substantial literature of specialized techniques and solutions. Consistent with our contention that the nonlinearities of interest in structural dynamics are typically small perturbations from the linear system, we will use the results obtained from our linear analysis to understand, at least locally, the stability of the nonlinear systems. The final sections of this chapter considers some other aspects of the stability of nonlinear systems.

Let the nonlinear system have a stationary point $\{u_o(t)\}$, which we call the origin. Assuming $\{F\}$ is smooth, we can expand it in a Taylor's series about the origin to give

$$[M]\{\ddot{\xi}\} + [C]\{\dot{\xi}\} + [K_T]\{\xi\} = \{Q\}$$

This is our variational equation, but this form is also called the linearized version of the nonlinear equation. It is understood that an analysis based on this linearized equation is expected to give reliable results only in the neighborhood of the origin.

Ziegler [85] discusses the connection between nonlinear systems and their linearized form. The essential result is that if, in the linearized system, all the roots of the characteristic equation are such that $\text{Im}\,[\mu] > 0$, the equilibrium of the actual (nonlinear) system is stable. Furthermore, if in the linearized system, at least one of the roots of the characteristic equation is such that $\text{Im}\,[\mu] < 0$, the equilibrium of the actual (nonlinear) system is unstable. The results from the linear analysis are now directly applicable.

Because there can be many equilibrium points, this approach is applied at the relevant ones.

Example 7.3:   Analyze the stability of the free vibration of a pendulum taking into account its nonlinearity due to large rotations.

The equation of motion of the pendulum is given by

$$ML\ddot{\theta} + C\dot{\theta} + Mg\sin\theta = 0$$

The body force and tangent stiffness are, respectively,

$$\mathcal{F}_s = Mg\sin\theta, \qquad K_T = \frac{\partial F}{\partial \theta} = Mg\cos\theta$$

Hence the linearized equation of motion is

$$ML\ddot{\xi} + C\dot{\xi} + [Mg\cos\theta]\xi = 0 \qquad \text{or} \qquad \ddot{\xi} + c\dot{\xi} + [\omega^2\cos\theta]\xi = 0$$

with $c = C/ML$ and $\omega_o^2 = g/L$. (Note that θ is considered constant.) This system has many equilibrium points ($\mathcal{F}_s = 0$ when $\dot{\theta} = 0$ and $\ddot{\theta} = 0$) at

$$\theta = \pm j\pi, \qquad j = 0, 1, 2, \ldots$$

These points correspond to multiplicities of the origin and the upright position.

The characteristic equation, in the vicinity of $\theta = 0$, is given by

$$-\mu^2 + i\mu c + \omega_o^2 = 0 \qquad \text{or} \qquad \mu = \tfrac{1}{2}ic \pm \sqrt{\omega_o^2 - c^2/4}$$

The system is stable. This result is not unexpected since the linearized equation is identical to a simple oscillator and hence our analysis was no different than that of the introductory discussion.

To emphasize that we are, indeed, considering the nonlinear system, let us now look at stability in the vicinity of the point $\theta = \pi$. That is, we consider the stability of the equilibrium state when the pendulum is vertical. The tangent stiffness changes sign and the characteristic equation becomes

$$-\mu^2 + i\mu c - \omega_o^2 = 0 \qquad \text{or} \qquad \mu = \tfrac{1}{2}ic \pm i\sqrt{\omega_o^2 + c^2/4}$$

Because one of the roots has a negative imaginary part, the equilibrium is unstable. This is a result in agreement with experience.

Lyapunov Stability

There are a number of definitions of stability (see References[26, 38, 44] for further discussions) of which just three are shown in Figure 7.8 — the usefulness of a particular definition can be assessed only in particular applications. In the following, it is helpful to think of the motion as a point in configuration space, some trajectories of which are shown in Figure 7.7, and we adopt the terminology of Reference [26].

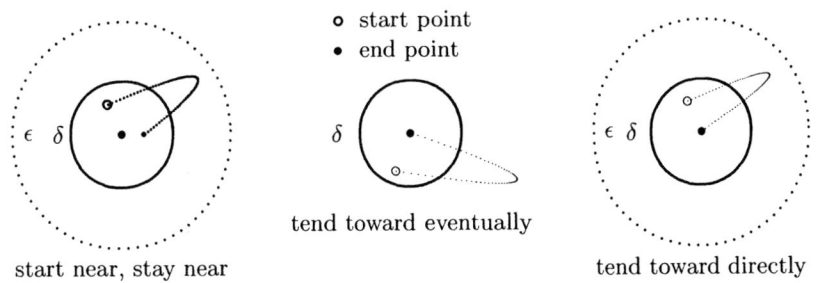

start near, stay near   tend toward eventually   tend toward directly

Figure 7.8: Three definitions of stability.

A point is Lyapunov stable if points which start nearby stay nearby or in short "start near, stay near." A point is quasi-asymptotically stable if nearby points tend toward it or, in short, "tend toward eventually." Note that this states what happens in the limit as time goes to infinity — in the meantime the solution can go all over the place. If a point is both Lyapunov and quasi-stable, then it is asymptotically stable or, in short, "tends toward directly." Each of these are illustrated in Figure 7.8 in the case where the point is a stationary point, although we need not restrict it in this way.

The last of these definitions is the one we will adopt. Mathematically, we state that a point is asymptotically stable if for all $\epsilon > 0$ (the region within which we restrict acceptable behavior) there exists a $\delta > 0$ (the region of possible initial conditions) such that, if $|x - x^*| < \delta$, then $|u(x,t) - u(x^*,t)| < \epsilon$ and that $|u(x,t) - u(x^*,t)| \longrightarrow 0$ as $t \longrightarrow \infty$. Necessarily, we have that $\delta \leq \epsilon$.

The inclusion of time in our definition has a subtle consequence. Consider a situation where, after a disturbance, the solution comes back to the undisturbed path but there is a phase lag, then our definition would still consider it unstable. When phase lags are not important, we say the solution is *Poincare* or *orbitally stable*. For autonomous systems where the governing equations do not depend explicitly on time, Poincare stability implies Lyapunov stability.

7.2   Follower Forces and Dynamic Instability

The introductory discussion of this chapter indicated that a follower force does not cause an instability. This seems counterintuitive and therefore we take that question up now in more detail. Strictly, this is a statical problem in that the applied loads are quasi-static; however, to analyze the stability we will consider the dynamic response to a small disturbance.

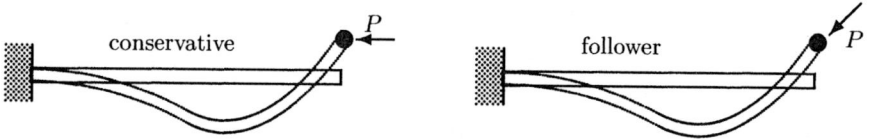

Figure 7.9: Beam with conservative and follower forces.

Governing Equations

Consider a column or beam with a quasi-static axial load P that is treated as a parameter. There are two situations that can prevail. The first is when the load P always remains horizontal. In this case P is a conservative load and hence we can write a potential for it. The second is when it remains tangential to the tip; while the magnitude remains essentially constant, the fact that it changes its orientation means it is a changing force and hence we cannot write a potential.

Assume small deflections; then from Section 6.2, the equation of motion in both cases is

$$EI\frac{\partial^4 v}{\partial x^4} - \bar{F}_o\frac{\partial^2 v}{\partial x^2} + \rho A\frac{\partial^2 v}{\partial t^2} = q$$

with $\bar{F}_o$ being the tensile axial force. Note that, since $\bar{F}_o$ is assumed essentially constant, this is a linear equation with constant coefficients. The spectral form of this is obtained by seeking solutions of the form $\hat{v}e^{i\mu t}$ and leads to

$$EI\frac{d^4\hat{v}}{dx^4} - \bar{F}_o\frac{d^2\hat{v}}{dx^2} - \rho A\mu^2\hat{v} = \hat{q}$$

This is an ordinary differential equation with constant coefficients, hence seek solutions of the form $\hat{v}(x) = v_o e^{-ikx}$, which leads to

$$\hat{v}(x) = c_1\cos k_1 x + c_2\sin k_1 x + c_3\cosh k_2 x + c_4\sinh k_2 x$$

where the spectrum relations are

$$k_1 = \sqrt{\frac{P}{2EI} + \sqrt{\frac{\rho A \mu^2}{EI} + (\frac{P}{2EI})^2}}, \qquad k_2 = i\sqrt{-\frac{P}{2EI} + \sqrt{\frac{\rho A \mu^2}{EI} + (\frac{P}{2EI})^2}}$$

In both of our cases the boundary conditions at $x = 0$ are

$$\hat{v} = 0 = c_1 + c_3, \qquad \frac{d\hat{v}}{dx} = 0 = k_1 c_2 + k_2 c_4$$

These allow the solution to be written as

$$\hat{v}(x) = c_1[\cos k_1 x - \cosh k_2 x] + c_2[\sin k_1 x - \frac{k_1}{k_2}\sinh k_2 x]$$

At the loaded end, we will have natural boundary conditions in terms of the shear force and bending moments

$$\hat{M} = EI\frac{d^2\hat{v}}{dx^2}, \qquad \hat{V} = -EI\frac{d^3\hat{v}}{dx^3} + \bar{F}_o\frac{d\hat{v}}{dx}$$

These are specified differently for the two cases of interest.

Stability Behavior

When the load is not a follower load, we have that both $\hat{M}$ and $\hat{V}$ are zero

$$\hat{M} = EI\frac{d^2\hat{v}}{dx^2} = 0, \qquad \hat{V} = -EI\frac{d^3\hat{v}}{dx^3} + \bar{F}_o\frac{d\hat{v}}{dx} = 0$$

giving

$$0 = c_1[-k_1^2 \cos k_1 L - k_2^2 \cosh k_2 L] + c_2[-k_1^2 \sin k_1 L - \frac{k_1}{k_2}k_2^2 \sinh k_2 L]$$

and

$$0 = c_1[k_1^3 \sin k_1 L - k_2^3 \sinh k_2 L] + c_2[-k_1^3 \cos k_1 L - \frac{k_1}{k_2}k_2^3 \cosh k_2 L]$$

These give a system of homogeneous equations for the two unknowns c_1 and c_3. The determinant of the system, after simplification, is

$$\Delta(P, \mu) = \frac{2\rho A}{EI}\mu^2(1 + CC_h) + \frac{P^2}{EI}CC_h - \sqrt{\frac{\mu^2}{EI}\frac{P}{EI}}SS_h = 0$$

where $C \equiv \cos k_1 L$, $C_h \equiv \cosh k_2 L$, and so on.

The roots (those values of μ for a given value of P that make $\Delta = 0$) are shown in Figure 7.10(a). Each mode goes to zero as the axial load is increased.

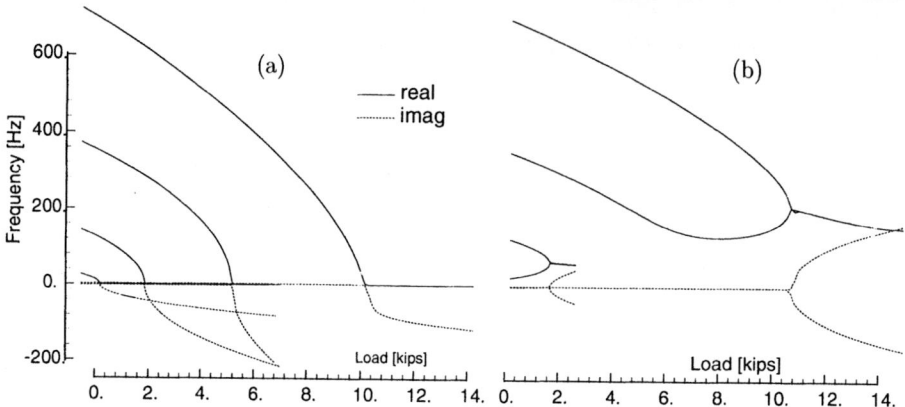

Figure 7.10: Roots for the continuous beam. (a) Conservative force. (b) Follower force.

This behavior is the same as illustrated in our introductory example. Thus there are only static instabilities.

When the force is a follower force, the moment is zero, but there is a nonzero component of the shear. That is, we have for the natural boundary conditions

$$\hat{M} = EI\frac{d^2\hat{v}}{dx^2} = 0, \qquad \hat{V} = -EI\frac{d^3\hat{v}}{dx^3} + \bar{F}_o\frac{d\hat{v}}{dx} = \bar{F}_o\frac{d\hat{v}}{dx} \quad \text{or} \quad -EI\frac{d^3\hat{v}}{dx^3} = 0$$

This leads to the determinental equation

$$\Delta(P,\mu) = \frac{2\rho A}{EI}\mu^2(1 + CC_h) + \frac{P^2}{EI} + \sqrt{\frac{\mu^2}{EI}\frac{P}{EI}}SS_h = 0$$

The roots are shown in Figure 7.10(b).

The behavior is very different from that of the previous case. First note that there are no static instabilities. That is, if $|\mu| = 0$, we can only get $\Delta = 0$ when $P = 0$. Furthermore, we see that pairs of roots coalesce at a nonzero real part and then produce positive and negative imaginary parts. Thus the instability occurs while the beam is vibrating. This is called *dynamic instability* or *flutter*. It is interesting to compare this result with that shown in Figure 3.10(b), which indicated that no instability would occur at any load for a perfect axial loading of a beam. The dynamic analysis has shown that there is, indeed, an instability load.

This is an important example to complement our static examples of instability. In the static case, we associate an instability with a loss of stiffness (also illustrated by the frequency going to zero) but in the dynamic case there is still stiffness.

Example 7.4: Use a ping test to demonstrate the presence of a dynamic instability for a beam axially loaded with a follower force.

Consider a cantilever beam of dimensions $[254 \times 25 \times 2.5\,\text{mm}^2]$ $(10 \times 1.0 \times 0.1\,\text{in.}^3)$ with an axial P applied load as a follower load. We will apply the ping load as a transverse point load of approximately 4 ms duration. The beam is modeled with 10 elements and damping used to diminish the vibrations within a reasonable time. The procedure applied the axial load quasi-statically, wait until the transients settled down, then apply the ping.

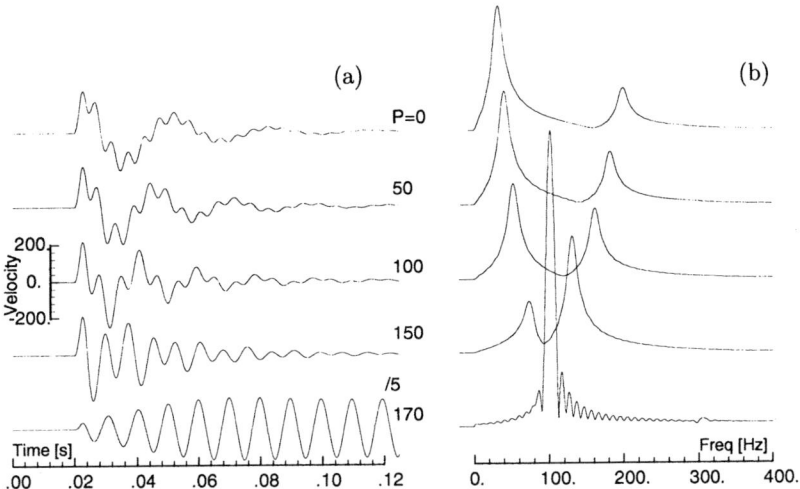

Figure 7.11: A ping test at different values of axial load. (a) Tip velocity responses. (b) Frequency response functions.

The results are shown in Figure 7.11 for different values of the axial load. Clearly, beyond a certain axial load value (approximately $P = 170$), the system is unstable. What is most interesting, however, is that this can be anticipated by looking at the frequency response functions — the instability occurs close to where two modes coalesce. It is interesting to compare this result with that shown in Figure 5.33, which indicated both modes going to zero. Here, one decreases while the other increases until there is a coalescence.

The linear theory predicts infinite displacements at the critical load. What actually happens is that the beam goes into large but finite amplitude oscillations, the magnitude of which depends on the magnitude of P.

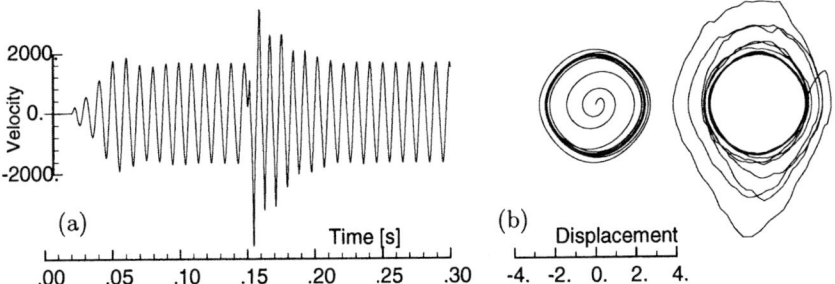

Figure 7.12: Response for a follower axial load above critical. (a) Time trace showing when second ping was applied. (b) Phase plane plots showing the achievement of a limit cycle.

As we showed earlier, the response to the ping (after $\{Q\}$ goes to zero) is that of a linear, free vibration of a system with a tangent stiffness $[K_T]$. The spectral analysis of the ping response will give information about the tangent stiffness $[K_T]$. It is worth noting that the FEM solution in this case used the explicit formulation, and hence did not directly deal with the tangent stiffness matrix. But clearly ping is extracting the desired information. Monitoring the spectral content as done in Chapter 6 would be to no avail here, since (for expediency reasons) the tangent stiffness is symmetricized and hence does not give complex eigenvalues.

Our stability analysis predicts that we get an exponential increase in displacement and velocity above the critical load. This does not occur in Figure 7.11, which shows an increase but only by a factor of about five. To further investigate this, a second ping was applied as indicated in Figure 7.12 at a time of $t = 0.15\,\text{s}$. The phase-plane plots are for the responses up to the ping (left) and from the ping onwards (right). The first plot exhibits a spiral outwards but achieves a limiting orbit. The second plot exhibits an excursion away from the orbit but then returns to it. The limiting orbit is a *limit cycle* and it is stable because the motion returns to it after the application of ping. It is worth noting that the strength of the second ping was two orders of magnitude larger than the first.

Example 7.5:   Replace the axially loaded continuous beam by a discrete version.

We explore the effect of discretization by looking at the two-mode discrete system shown in Figure 7.13 where the column is modeled as two rigid bars of length L, two torsional springs of stiffness K_1 and K_2, and two masses M_1 and M_2. The two degrees of freedom are the rotations of the bars. We also take the opportunity to illustrate the use of virtual work in establishing the generalized applied loads.

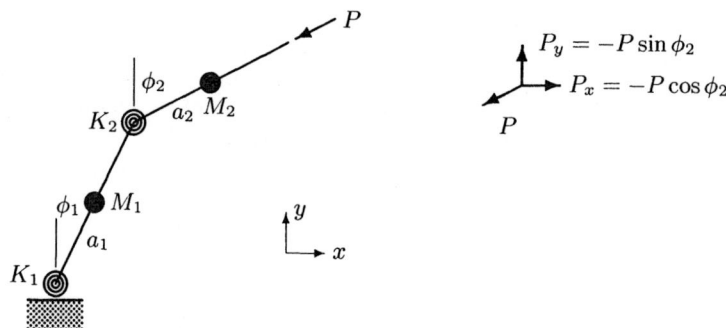

Figure 7.13: Discretized column.

The strain energy is confined to the two springs and is given by

$$U = \tfrac{1}{2}K_1\phi_1^2 + \tfrac{1}{2}K_2(\phi_2 - \phi_1)^2$$

The stiffness matrix is obtained as

$$K_{ij} = \frac{\partial^2 U}{\partial\phi_i \partial\phi_j} \qquad \text{giving} \qquad [\,K\,] = \begin{bmatrix} K_1 + K_2 & -K_2 \\ -K_2 & K_2 \end{bmatrix}$$

The kinetic energy is confined to just the masses. The position of these masses are

$$x_1 = a_1 \sin \phi_1 \approx a_1 \phi_1$$
$$y_1 = a_1 \cos \phi_1 \approx a_1 [1 - \tfrac{1}{2}\phi_1^2]$$
$$x_2 = L \sin \phi_1 + a_2 \sin \phi_2 \approx L\phi_1 + a_2\phi_2$$
$$y_1 = L \cos \phi_1 + a_2 \cos \phi_2 \approx L[1 - \tfrac{1}{2}\phi_1^2] + a_2[1 - \tfrac{1}{2}\phi_2^2]$$

were we have used the small angle approximations. This leads to the velocities

$$\dot{x}_1 = a_1\dot{\phi}_1, \qquad\qquad \dot{y}_1 = -a_1\phi_1\dot{\phi}_1 \approx 0$$
$$\dot{x}_2 = L\dot{\phi}_1 + a_2\dot{\phi}_2, \qquad \dot{y}_2 = -L\phi_1\dot{\phi}_1 - a_2\phi_2\dot{\phi}_2 \approx 0$$

The kinetic energy is therefore

$$
\begin{aligned}
T &= \tfrac{1}{2}M_1(\dot{x}_1^2 + \dot{y}_1^2) + \tfrac{1}{2}M_2(\dot{x}_2^2 + \dot{y}_2^2) \\
&\approx \tfrac{1}{2}M_1 a_1^2 \dot{\phi}_1^2 + \tfrac{1}{2}M_2[L^2\dot{\phi}_1^2 + 2a_2 L\dot{\phi}_1\dot{\phi}_2 + a_2^2\dot{\phi}_1^2]
\end{aligned}
$$

and the mass matrix is obtained as

$$M_{ij} = \frac{\partial^2 T}{\partial \dot{\phi}_i \partial \dot{\phi}_j} \qquad \text{giving} \qquad [\,M\,] = \begin{bmatrix} M_1 a_1^2 + M_2 L^2 & M_2 a_2 L \\ M_2 a_2 L & M_2 a_2^2 \end{bmatrix}$$

The overall inertia of the system depends on the particular location of the masses.

We will consider the special case when $a_1 = a_2 = L/2$, $M_1 = M_2 = M$, and $K_1 = K_2 = K$, then the equations of motion are

$$K \begin{bmatrix} 2 & -1 \\ -1 & 1 \end{bmatrix} \left\{ \begin{array}{c} \phi_1 \\ \phi_2 \end{array} \right\} + \frac{ML^2}{4} \begin{bmatrix} 5 & 2 \\ 2 & 1 \end{bmatrix} \left\{ \begin{array}{c} \ddot{\phi}_1 \\ \ddot{\phi}_2 \end{array} \right\} = \left\{ \begin{array}{c} P_1 \\ P_2 \end{array} \right\}$$

It remains to consider the applied loads.

For the conservative, noncirculatory system, the work done can be expressed in terms of a potential as

$$V = -P \times \text{vertical deflection} = -P[L - L\cos\phi_1 + L - L\cos\phi_2] \approx -\tfrac{1}{2}PL[\phi_1^2 + \phi_2^2]$$

This gives the generalized forces as

$$P_1 = -\frac{\partial V}{\partial \phi_1} = +PL\phi_1, \qquad P_2 = -\frac{\partial V}{\partial \phi_2} = +PL\phi_2$$

Note that these are deformation dependent and hence the equations of motion become

$$K \begin{bmatrix} 2 & -1 \\ -1 & 1 \end{bmatrix} \left\{ \begin{array}{c} \phi_1 \\ \phi_2 \end{array} \right\} - PL \begin{bmatrix} 1 & 0 \\ 0 & 1 \end{bmatrix} \left\{ \begin{array}{c} \phi_1 \\ \phi_2 \end{array} \right\} + \frac{ML^2}{4} \begin{bmatrix} 5 & 2 \\ 2 & 1 \end{bmatrix} \left\{ \begin{array}{c} \ddot{\phi}_1 \\ \ddot{\phi}_2 \end{array} \right\} = \left\{ \begin{array}{c} 0 \\ 0 \end{array} \right\}$$

The axial load influences the system as a symmetric stiffness matrix.

Now consider the follower force situation. Since angles are small, we have for the two components of force

$$P_x = -P \sin \phi_2 \approx -P\phi_2, \qquad P_y = -P \cos \phi_2 \approx -P$$

We see that while P_y is constant, P_x changes. We could treat the vertical component as before, but, since the horizontal component depends on the rotation ϕ_2, it does not have a potential and therefore must be treated differently. We will establish the generalized forces for our system by looking at the virtual work done by both components.

The virtual horizontal displacement is

$$\delta u = \delta x_p = \delta[L\sin\phi_1 + L\sin\phi_2] = L\cos\phi_1\delta\phi_1 + L\cos\phi_2\delta\phi_2 \approx L\delta\phi_1 + L\delta\phi_2$$

The virtual work of the horizontal component is then

$$\delta W_x = P_x\delta u = -P\phi_2 L[\delta\phi_1 + \delta\phi_2] = PL[-\phi_2\delta\phi_1 - \phi_2\delta\phi_2]$$

The virtual vertical displacement is

$$\delta v = \delta y_p = \delta[L(\cos\phi_1 - 1) + L(\cos\phi_2 - 1)] \approx -L\phi_1\delta\phi_1 - \phi_2 L\delta\phi_2$$

The virtual work of the vertical component is then

$$\delta W_y = P_x\delta u = -P[-L\phi_1\delta\phi_1 - L\phi_2\delta\phi_2] = PL[\phi_1\delta\phi_1 + \phi_2\delta\phi_2]$$

The total virtual work done by both components is

$$\delta W = \delta W_x + \delta W_y = PL[\phi_1\delta\phi_1 - \phi_2\delta\phi_1]$$

The generalized forces associated with the degrees of freedom ϕ_1 and ϕ_2 give the virtual work

$$\delta W = P_1\delta\phi_1 + P_2\delta\phi_2$$

From which we conclude that $P_2 = 0$ and $P_1 = PL[\phi_1 - \phi_2]$. Again, the applied loads are deformation dependent giving the equations of motion

$$K\begin{bmatrix} 2 & -1 \\ -1 & 1 \end{bmatrix}\begin{Bmatrix} \phi_1 \\ \phi_2 \end{Bmatrix} - PL\begin{bmatrix} 1 & -1 \\ 0 & 0 \end{bmatrix}\begin{Bmatrix} \phi_1 \\ \phi_2 \end{Bmatrix} + \frac{ML^2}{4}\begin{bmatrix} 5 & 2 \\ 2 & 1 \end{bmatrix}\begin{Bmatrix} \ddot{\phi}_1 \\ \ddot{\phi}_2 \end{Bmatrix} = \begin{Bmatrix} 0 \\ 0 \end{Bmatrix}$$

Note that the contribution of the PL term is that of a nonsymmetric stiffness matrix.

Example 7.6: Compare the stability of the discrete form of the conservative and follower force problem.

The stability analysis for free vibration begins by assuming solutions of the form

$$\begin{Bmatrix} \phi_1 \\ \phi_2 \end{Bmatrix}(t) = \begin{Bmatrix} \hat{\phi}_1 \\ \hat{\phi}_2 \end{Bmatrix}e^{i\mu t}$$

Substituting this into the governing equation for the conservative system gives the eigenvalue problem

$$\left[K\begin{bmatrix} 2 & -1 \\ -1 & 1 \end{bmatrix} - PL\begin{bmatrix} 1 & 0 \\ 0 & 1 \end{bmatrix} - \frac{ML^2}{4}\mu^2\begin{bmatrix} 5 & 2 \\ 2 & 1 \end{bmatrix}\right]\begin{Bmatrix} \hat{\phi}_1 \\ \hat{\phi}_2 \end{Bmatrix} = \begin{Bmatrix} 0 \\ 0 \end{Bmatrix}$$

which leads to the characteristic equation

$$\alpha^4 + \alpha^2[6PL - 11K] + [P^2L^2 - 3KPL + K^2] = 0, \qquad \alpha \equiv \mu\sqrt{\frac{ML^2}{4}}$$

This is quadratic in α^2, hence there are four roots appearing as $\pm$ pairs.

First look at the possibility of a static instability — this occurs when $\alpha = \mu = 0$ or

$$[P^2 L^2 - 3KPL + K^2] = 0$$

This gives

$$PL = \tfrac{1}{2}(3 - \sqrt{5})K = 0.38K , \quad \tfrac{1}{2}(3 + \sqrt{5})K = 2.62K ,$$

Because the loads are real only, we conclude that static instability is possible. Note that this does not depend on the particulars of the mass distribution, since it occurs when the inertia is zero.

Now examine the system for a dynamic instability; that is, we consider cases where $\mu \neq 0$ but could have a negative imaginary part. The roots are given by

$$2\alpha^2 = -(6PL - 11K) \pm \sqrt{(6PL - 11K)^2 - 4(P^2 L^2 - 3KPL + K^2)}$$

We want the imaginary part of μ (and hence α) to be greater than zero, hence, we want $\mathrm{Im}\,[\mu^2] > 0$. The imaginary component comes from the square-root term, hence we ask if the radical can be negative. Expanding and collecting terms gives

$$32P^2 L^2 - 120PLK + 117K^2 < 0$$

This is zero when

$$PL = \frac{60K \pm i12K}{32}$$

Because a complex load is required, we conclude that a dynamic instability is not possible.

The full results are shown in Figure 7.12(a), which illustrates both roots going to zero before the imaginary terms are generated.

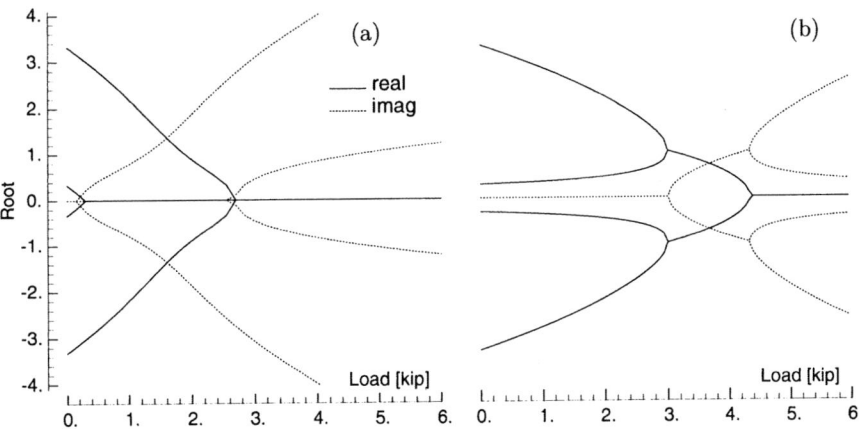

Figure 7.14: Roots for the discrete form of the loaded beam/column problem. (a) Conservative. (b) Follower.

We now consider the follower force problem. The stability analysis parallels that of the conservative system and leads to the characteristic equation

$$\alpha^4 + \alpha^2[3PL - 11K] + [K^2] = 0 , \quad \alpha^2 \equiv \mu^2 \sqrt{\frac{ML^2}{4}}$$

Again, this is quadratic in α^2, hence there are four roots appearing as $\pm$ pairs.

First look at the possibility of a static instability — this occurs when $\alpha = \mu = 0$. This is not possible unless $K = 0$. This is in agreement with the conclusion in the introductory section, but it also applies to the two DoF system.

Now examine the system for a dynamic instability, that is, we consider cases where $\mu \neq 0$ but have a negative imaginary component. The roots are given by

$$2\alpha^2 = -(3PL - 11K) \pm \sqrt{(3PL - 11K)^2 - 4K^2}$$

We ask if the radical can be negative. Expanding and collecting terms gives

$$9P^2L^2 - 66PLK + 117K^2 < 0$$

This is zero when

$$PL = \frac{33k \pm 6K}{9} = 3K, \quad 4.33K$$

There are two loads at which the radical can be zero, hence we can get an imaginary component of μ for a nonnegative real part.

The full results are shown in Figure 7.12(b), which illustrates both modes coalescing and generating both positive and negative imaginary parts. Thus, the instability occurs while the system is vibrating.

7.3   Aeroelasticity and Flutter

Aeroelasticity is the study of the structural dynamics when the loadings are aerodynamic in origin. To make the example concrete, we will look at the flutter of aircraft wings, but the analysis is also applicable to other structural systems (bridges, buildings, cars) under wind and similar loading.

Structural Modeling

The shape of the wing profoundly influences the aerodynamic loads, but it has little effect on the structural modeling. Therefore, we will replace the 3-D wing with a Ritz model approximation with a minimum of degrees of freedom.

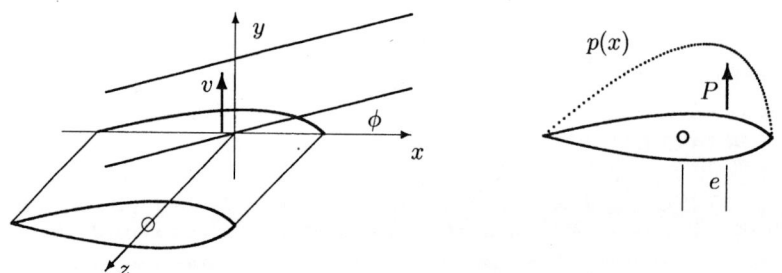

Figure 7.15: Wing under aerodynamic loading.

For a complex structure modeled using FEM with many (N) degrees of freedom $\{u\}$, we have for the total strain energy

$$U = \tfrac{1}{2}\{u\}^T [\,K\,]\{u\}$$

Let us represent the degrees of freedom with a reduced system given by

$$\{u\} = [\,\Psi\,]\{\eta\}$$

where $\{\eta\}$ is only of size $M << N$. The strain energy is then

$$U = \tfrac{1}{2}\{\eta\}^T [\,\Psi\,]^T [\,K\,][\,\Psi\,]\{\eta\} = \tfrac{1}{2}\{\eta\}^T [\,\tilde{K}\,]\{\eta\}, \quad [\,\tilde{K}\,] \equiv [\,\Psi\,]^T [\,K\,][\,\Psi\,]$$

The matrix $[\,\tilde{K}\,]$ is the $[M \times M]$ reduced stiffness matrix and the matrix $[\,\Psi\,]$ is the transformation matrix that imposes the kinematic restrictions. In a similar manner, we can establish the mass matrix as

$$[\,\tilde{M}\,] = [\,\Psi\,]^T [\,M\,][\,\Psi\,]$$

The virtual work of the applied loads is

$$\delta W = \{P\}^T \{\delta u\} = \{P\}^T [\,\Psi\,]\{\delta\eta\} = \{\tilde{P}\}\{\delta\eta\}, \quad \{\tilde{P}\} \equiv \{P\}^T [\,\Psi\,]$$

The reduced applied loads vector $\{\tilde{P}\}$ is of size $[M \times 1]$. The system of equations becomes

$$[\,\tilde{K}\,]\{\eta\} + [\,\tilde{C}\,]\{\dot{\eta}\} + [\,\tilde{M}\,]\{\ddot{\eta}\} = \{\tilde{P}\} = \{P\}^T [\,\Psi\,]$$

In this way, a conventional finite element model can be used to establish the mass, stiffness, and load matrices, and the above then used to reduce it to a smaller size for stability analysis.

The question remains as to what to use as the reduction functions $[\,\Psi\,]$. Actually, if an eigenanalysis is performed, then the lowest eigenmodes can be used as the reduction vectors. This, in fact, is basically what happens in the subspace iteration method [22]. The resulting stiffness and mass matrices are then diagonal. At first sight, it might appear that, if the eigenvectors are used as the reduction functions, this will lead to a diagonal system with no coupling and hence no possibility of a dynamic instability. Remember, however, that the effective loads will also be transformed and these will lead to the coupling.

Example 7.7: Replace the wing with a reduced model.

We saw in the follower force problems that at least two modes are needed in order to get a dynamic instability. Hence, let us replace the deflection of the wing with a vertical deflection and a rotation as shown in Figure 7.15. This gives

$$\bar{u}(x,y,z) = -x + x\cos\bar{\phi} - y\sin\bar{\phi} \approx -y\bar{\phi} = -y[\tfrac{z}{L}]\phi$$

$$\bar{v}(x,y,z) = -y + x\sin\bar{\phi} + y\cos\bar{\phi} + \bar{v} \approx \bar{v} + x\bar{\phi} = [\tfrac{z^2}{L^2}]v + x[\tfrac{z}{L}]\phi$$

where $[\frac{z}{L}]$, $[\frac{z^2}{L^2}]$, and $[\frac{z}{L}]$ play the role of Ritz functions. The velocities are given by

$$\dot{\tilde{u}} = -y\dot{\phi}[\frac{z}{L}], \qquad \dot{\tilde{v}} = \dot{v}[\frac{z^2}{L^2}] + x\dot{\phi}[\frac{z}{L}]$$

leading to the kinetic energy

$$T = \frac{1}{2}\iiint \rho[\dot{\tilde{v}}^2 + \dot{\tilde{u}}^2]\,dxdydz = \frac{1}{2}\iiint \rho[\dot{v}^2[\frac{z^2}{L^2}]^2 + 2\dot{v}\dot{\phi}x[\frac{z}{L}]^3 + \dot{\phi}^2[x^2+y^2][\frac{z}{L}]^2]\,dxdydz$$

After performing the integrations, we can represent the kinetic energy as

$$T = \frac{1}{2}M_B\dot{v}^2 + M_{BT}\dot{v}\dot{\phi} + \frac{1}{2}M_T\dot{\phi}^2$$

Thus the mass matrix is given by

$$[\,M\,] = \begin{bmatrix} M_B & M_{BT} \\ M_{BT} & M_T \end{bmatrix}$$

We can establish the stiffness matrix in a similar manner. Let the strain energy ultimately have the form

$$U = \frac{1}{2}K_B v^2 + K_{BT}v\phi + \frac{1}{2}K_T\phi^2$$

giving the stiffness matrix as

$$[\,K\,] = \begin{bmatrix} K_B & K_{BT} \\ K_{BT} & K_T \end{bmatrix}$$

The equations of motion are

$$\begin{bmatrix} K_B & K_{BT} \\ K_{BT} & K_T \end{bmatrix}\begin{Bmatrix} v \\ \phi \end{Bmatrix} + \begin{bmatrix} M_B & M_{BT} \\ M_{BT} & M_T \end{bmatrix}\begin{Bmatrix} \ddot{v} \\ \ddot{\phi} \end{Bmatrix} = \begin{Bmatrix} \tilde{P} \\ \tilde{T} \end{Bmatrix}$$

where $\tilde{P}$ is the resultant vertical load and $\tilde{T}$ is the resultant torque.

Aerodynamic Loading

Let the aircraft be in steady flight with velocity V so that the lift just balances the weight. Now consider the additional lift due to rotation ϕ of the wing. This is given by

$$P = \frac{1}{2}\rho_a V^2(2\pi)S\phi = CV^2\phi$$

where C is a group coefficient that depends on the surface area S, the airfoil shape, and the air density ρ_a. This acts at the aerodynamic center, which is a distance ec from the elastic (or shear) center. Thus, the torque acting is

$$T = Pec = \frac{1}{2}\rho V^2(2\pi)Sec\phi = CV^2ec\phi$$

Both of these loads are deformation dependent through the twist ϕ. We can represent them in matrix form as

$$\begin{Bmatrix} P \\ T \end{Bmatrix} = CV^2\begin{bmatrix} 0 & 1 \\ 0 & ec \end{bmatrix}\begin{Bmatrix} v \\ \phi \end{Bmatrix}$$

from which it is seen that the aerodynamic loading contribution is that of an unsymmetric stiffness matrix (since it is of the form $[\,K\,]\{u\}$). This will lead to the possibility of having complex roots.

There are also unsteady aerodynamic effects. Because the airfoil itself has a velocity (at the aerodynamic center) of $\dot{v} + ec\dot{\phi}$, the effective angle of attack is really

$$\phi_{\text{eff}} = \phi - \frac{\dot{v} + ec\dot{\phi}}{V}$$

The additional terms contribute to the loads as

$$\left\{ \begin{matrix} P \\ T \end{matrix} \right\} = -\frac{CV^2}{V} \begin{bmatrix} 1 & 0 \\ 0 & ec \end{bmatrix} \left\{ \begin{matrix} \dot{v} \\ \dot{\phi} \end{matrix} \right\}$$

This has the form of a symmetric damping matrix.

The full equations of motion become

$$\begin{bmatrix} K_B & K_{BT} \\ K_{BT} & K_T \end{bmatrix} \left\{ \begin{matrix} v \\ \phi \end{matrix} \right\} - CV^2 \begin{bmatrix} 0 & 1 \\ 0 & ec \end{bmatrix} \left\{ \begin{matrix} v \\ \phi \end{matrix} \right\}$$

$$+ CV \begin{bmatrix} 1 & 0 \\ 0 & ec \end{bmatrix} \left\{ \begin{matrix} \dot{v} \\ \dot{\phi} \end{matrix} \right\} + \begin{bmatrix} M_B & M_{BT} \\ M_{BT} & M_T \end{bmatrix} \left\{ \begin{matrix} \ddot{v} \\ \ddot{\phi} \end{matrix} \right\} = \left\{ \begin{matrix} 0 \\ 0 \end{matrix} \right\}$$

The CV term appears like a damping term and hence we would conclude it only contributes to the stability of the motion. It must be kept in mind, however, that ec can be negative.

Stability Analysis

As a preliminary to a stability analysis, introduce the following normalizations:

$$s^2 = \frac{K_T}{K_B}, \quad r^2 = \frac{M_T}{M_B}, \quad \delta = \frac{M_{BT}}{\sqrt{M_B M_T}}, \quad \tau = t\sqrt{\frac{K_B}{M_B}}, \quad \bar{V}^2 = \frac{cV^2}{K_B}, \quad \gamma = \sqrt{\frac{C}{M_B}}$$

Furthermore, without loss of generality, we can take $K_{BT} = 0$. These allow us to rewrite the equations of motion as

$$\begin{bmatrix} 1 & 0 \\ 0 & s^2 \end{bmatrix} \left\{ \begin{matrix} v \\ \phi \end{matrix} \right\} - \bar{V}^2 \begin{bmatrix} 0 & 1 \\ 0 & e \end{bmatrix} \left\{ \begin{matrix} v \\ \phi \end{matrix} \right\} + \gamma\bar{V} \begin{bmatrix} 1 & 0 \\ 0 & e \end{bmatrix} \left\{ \begin{matrix} v' \\ \phi' \end{matrix} \right\} + \begin{bmatrix} 1 & \delta r \\ \delta r & r^2 \end{bmatrix} \left\{ \begin{matrix} v'' \\ \phi'' \end{matrix} \right\} = \left\{ \begin{matrix} 0 \\ 0 \end{matrix} \right\}$$

where the prime indicates differentiation with respect to normalized time τ.

A stability analysis begins by assuming solutions of the form

$$\left\{ \begin{matrix} v \\ \phi \end{matrix} \right\} (\tau) = \left\{ \begin{matrix} \hat{v} \\ \hat{\phi} \end{matrix} \right\} e^{i\mu\tau}$$

Substituting this into the equations of motion gives the eigenvalue system

$$\begin{bmatrix} 1 + \gamma\bar{V}i\mu - \mu^2 & -\bar{V}^2 - \delta r\mu^2 \\ -\delta r\mu^2 & s^2 - \bar{V}^2 e + \gamma\bar{V}ei\mu - r^2\mu^2 \end{bmatrix} \left\{ \begin{matrix} \hat{v} \\ \hat{\phi} \end{matrix} \right\} = \left\{ \begin{matrix} 0 \\ 0 \end{matrix} \right\}$$

Multiplying out leads to the characteristic equation

$$[r^2(1 - \delta^2)]\mu^4 - [\gamma\bar{V}(r^2 + e)]i\mu^3 - [(r^2 + s^2) + \bar{V}^2(\delta r - e + \gamma^2 e)]\mu^2$$
$$+ [\gamma\bar{V}(s^2 + e - \bar{V}^2 e)]i\mu + [s^2 - \bar{V}^2 e] = 0$$

This is a fourth-order polynomial in μ and would require numerical methods for its solution. We therefore look at some special cases.

I: Divergence

First see if the system can have a static instability. This occurs when $\mu = 0$, giving

$$s^2 - \bar{V}^2 e = 0 \qquad \text{or} \qquad \bar{V} = \sqrt{s^2/e}$$

Restoring the original variables

$$V = \sqrt{\frac{K_T}{Cec}}$$

This is called the *divergence velocity*. If the aerodynamic center coincides with the shear center so that $ec = 0$, then the divergence velocity goes to infinity. This was the case with the Spitfire fighter plane of WWII.

II: No Mass Coupling, $\delta = 0$

When $\delta = 0$, there is no mass coupling, that is, the mass center and the shear center coincide, and the characteristic equation can be factored as

$$[\mu^2 - \gamma\bar{V}i\mu - 1][r^2\mu^2 - \gamma\bar{V}ei\mu - (s^2 - \bar{V}^2 e)] = 0$$

The roots are

$$\mu_1 = \tfrac{1}{2}i\gamma\bar{V} \pm \sqrt{1 - \tfrac{1}{4}\gamma^2\bar{V}^2}$$

$$\mu_2 = \frac{i\gamma\bar{V}e}{2r^2} \pm \sqrt{\left[\frac{s^2 - \bar{V}^2 e}{r^2}\right] - \left[\frac{\gamma\bar{V}e}{2r^2}\right]^2} \qquad (7.2)$$

Note from the second mode that the system is unstable if $e < 0$. The first mode is bending while the second is torsion. These roots are shown plotted against velocity in Figure 7.16(a). The fact that they intersect means that in the coupled case there is a possibility of getting a dynamic instability.

Let $\gamma = 0$, and restoring to the original variables, gives

$$\mu_1 = \omega_b$$

$$\mu_2 = \sqrt{\frac{s^2 - V^2 e}{r^2}} = \sqrt{\frac{\omega_T^2}{\omega_B^2} - \frac{CV^2 e}{M_T^2}}$$

The second mode exhibits divergence.

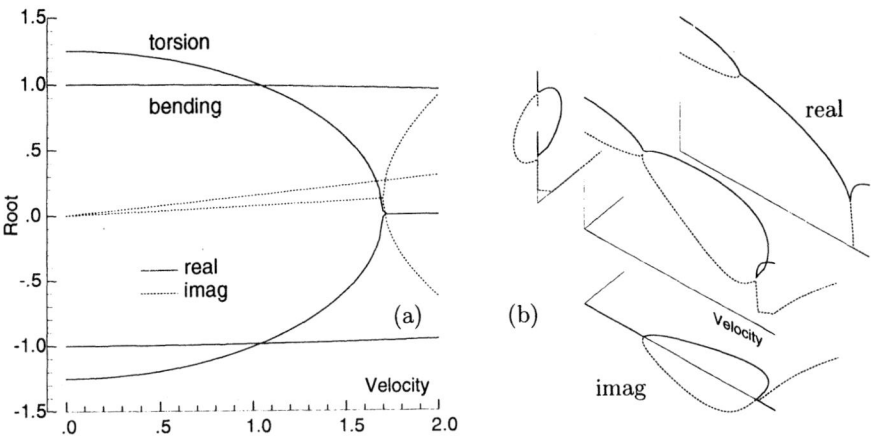

Figure 7.16: Dependence of roots on velocity. (a) Uncoupled case. (b) General case.

III: No Damping, $\gamma = 0$

When $\gamma = 0$, there are no unsteady aerodynamic effects, and the characteristic equation becomes

$$[r^2(1 - \delta^2)]\mu^4 - [(r^2 + s^2) + \bar{V}^2(\delta r - e)]\mu^2 + [s^2 - \bar{V}^2 e] = 0$$

This is quadratic in μ^2, which is easily solved.

These roots are shown plotted against velocity in Figure 7.16(b). There is indeed a region of dynamic instability, but this requires $\delta < 0$.

Stall Flutter

When a structure is in a steady stream of air or water, a vortex sheet is formed behind the structure with the vortices flowing off at a definite periodicity. The periodicity depends on the shape and dimensions of the structure as well as the velocity of the stream. For example, Figure 7.17 shows vortex sheets forming around a cylinder and the number of vortices flowing off per second is given by

$$2\pi\omega = f = \frac{0.22V}{D} \text{ Hz}$$

where D is the diameter, V is the velocity of the stream, and the coefficient 0.22 (Strouhall number) depends on the particular shape of the object. Reference [50] has a nice photograph of vortex shedding.

Because the vortices separate alternately from either side of the cylinder, the cylinder experiences a periodic force perpendicular to the flow direction. This force can be approximated by

$$P(t) = C_L \rho V^2 S \sin \omega t$$

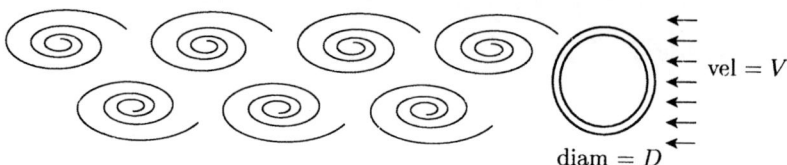

Figure 7.17: Vortex shedding off a cylinder.

The coefficient C_L depends on the particular structure; the poorer the streamlining of the structure, the larger is C_L and hence the amplitude of the force. For example, flow around an airplane wing at small angles of attack is quite smooth, so that the vortex sheet is very well defined and the force is small. Poorly streamlined structures such as suspension bridges or off-shore oil drilling rigs can experience very large forces.

The structure experiences a forced frequency oscillation; if the frequency is close to a natural frequency of the structure, then excessive deflections and hence damage can occur. This sort of excitation has been observed in large factory chimneys, and it is now considered to be the cause of the collapse of the Tacoma Narrows bridge [56].

This phenomenon is called *stall flutter* and as described is not essentially connected with aeroelasticity. But clearly, once the structure begins to oscillate, there will be additional aeroelastic effects thus complicating the analysis considerably.

Other aspects of wind, wave, and blast loadings are covered in a readable fashion in Reference [73].

7.4   Pulsating Compression of a Beam

In Chapter 4, we considered the vibration of a beam with a periodic axial loading. While developing the solution, we noted parameter regions where a solution did not exist. Using Floquet's theory, we are now in a position to look closer at these regions and establish aspects of their stability.

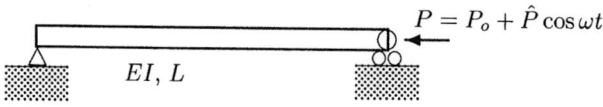

Figure 7.18: Beam with pulsating axial load.

Equation of Motion

Because nothing in the derivation of the governing equation in Chapter 6 required P to be constant, we can write

$$EI\frac{\partial^4 v}{\partial x^4} + P(t)\frac{\partial^2 v}{\partial x^2} + \rho A\frac{\partial^2 v}{\partial t^2} = 0$$

This gives rise to a case where the coefficients are variable; solutions to these equations are very difficult to find. To simplify matters, we look at the case when the load varies harmonically as

$$P(t) = P_o + \hat{P}\cos\omega t$$

Thus the coefficients are periodic but even this is very difficult to solve. To further simplify matters, we will consider the case of simply supported boundary conditions and use a type of Fourier series expansion.

At any instant in time, we can imagine the shape to be represented by a Fourier series

$$v(x,t) = \sum_m C_m(t)\sin(\frac{m\pi x}{L})$$

In this, we have considered the Fourier coefficients, $C_m(t)$, to be functions of time. Furthermore, each term individually satisfies the pinned boundary conditions. After substitution, we get for the time varying coefficients

$$(\frac{m\pi}{L})^4 EIC_m - (\frac{m\pi}{L})^2 PC_m + \rho A\ddot{C}_m = 0$$

This can be rearranged as

$$\ddot{C}_m + \omega_m^2[(1-\frac{P_o}{P_m}) - \frac{\hat{P}}{P_m}\cos\omega t]C_m = 0, \quad \omega_m \equiv (\frac{m\pi}{L})^2\sqrt{\frac{EI}{\rho A}}, \quad P_m \equiv (\frac{m\pi}{L})^2 EI$$

where ω_m are the natural frequencies of the beam while P_m are the static buckling loads of the beam. Note that when $\hat{P}$ is zero, we have free vibration solutions of the form $Ae^{i\mu t}$ if

$$-\mu^2 + \omega_m^2(1 - \frac{P_o}{P_m}) = 0 \quad \text{or} \quad \mu = \omega_m\sqrt{1 - \frac{P_o}{P_m}}$$

Thus, as long as $P_o < P_m$, we get a stable solution. We also see the effect of axial load on the resonance frequencies; this was already covered in Chapter 6.

Transition Curves and Their Approximation

As was done in Chapter 4, we arrange our equation of motion as a Mathieu equation in the form

$$u'' + [\alpha + \beta\cos\tau]u = 0, \quad \alpha \equiv \frac{\omega_m^2}{\omega^2}(1 - \frac{P_o}{P_m}), \quad \beta \equiv (-)\frac{\omega_m^2}{\omega^2}\frac{\hat{P}}{P_m}, \quad u \equiv C_m$$

where a new time variable was introduced as $\tau \equiv \omega t$. We concluded that the boundaries

$$\phi(\alpha, \beta) = \pm 2$$

separate regions where unbounded solutions exist from regions where all solutions are bounded. We also found that these are curves on which solutions of periodicities 2π or 4π occur. We will get an approximation for these curves by way of perturbation methods.

Consider β to be small and let

$$\alpha = \alpha(\beta) \approx \alpha_0 + \alpha_1\beta + \alpha_2\beta^2 + \cdots, \qquad u(\tau) \approx u_0(\tau) + u_1(\tau)\beta + u_2(\tau)\beta^2 + \cdots$$

Substitute these into the governing equation and group equal powers of β to get

$$\beta^0: \qquad u_0'' + \alpha_0 u_0 = 0$$
$$\beta^1: \qquad u_1'' + \alpha_0 u_1 = -\alpha_1 u_0 - u_0 \cos\tau$$
$$\beta^2: \qquad u_2'' + \alpha_0 u_2 = -\alpha_2 u_0 - \alpha_1 u_1 - u_1 \cos\tau$$

and so on. The solution to the first equation is

$$u_0(t) = A \cos\sqrt{\alpha_0}\,\tau + B \sin\sqrt{\alpha_0}\,\tau$$

We are interested in solutions that are periodic, having the same or multiple periods of the forcing term. Specifically, we want it with period 2π and 4π as concluded from the transition curve analysis. We achieve this with

$$\sqrt{\alpha_0} = 0, 1, 2, 3, \cdots, n, \qquad \sqrt{\alpha_0} = \tfrac{1}{2}, 1\tfrac{1}{2}, 2\tfrac{1}{2}, 3\tfrac{1}{2}, \cdots, (n+\tfrac{1}{2})$$

Both sequences can be combined into the single sequence

$$\alpha_0 = \tfrac{1}{4}n^2, \qquad n = 0, 1, 2, 3, \cdots$$

In the following, we will look at the cases of $n = 0$ and $n = 1$.

For $n = 0$, we have that $\alpha_0 = 0$ giving the sequence

$$u_0'' = 0 \qquad \Rightarrow \qquad u_0 = a_0 + a_1\tau$$

We take $a_1 = 0$ since otherwise that would destroy the periodicity. The second equation becomes

$$u_1'' = -\alpha_1 a_0 - a_0 \cos\tau \qquad \Rightarrow \qquad u_1 = -\alpha_1 a_0 \tfrac{1}{2}t^2 + b_1 t + b_0 - a_0 \cos\tau$$

from which we conclude that $\alpha_1 = 0$ and $b_1 = 0$. The third equation becomes

$$u_2'' = -\alpha_2 a_0 - (b_0 + a_0 \cos\tau)\cos\tau = -\alpha_2 a_0 - \tfrac{1}{2}a_0(1 + \cos 2\tau) - b_0 \cos\tau$$

We conclude that

$$-\alpha_2 a_0 - \tfrac{1}{2}a_0 = 0 \qquad \text{or} \qquad \alpha_2 = -\tfrac{1}{2}$$

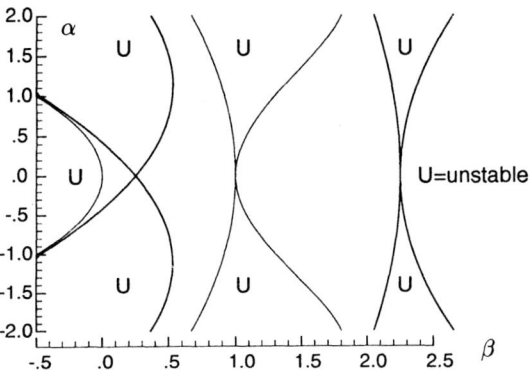

Figure 7.19: Transition curves for Mathieu's equation.

Our expression for the transition curve is

$$\alpha = -\tfrac{1}{2}\beta^2$$

This is shown plotted in Figure 7.19 as the leftmost curve.

For $n = 1$, we have that $\alpha_0 = \tfrac{1}{4}$ giving the sequence

$$u_0'' + \tfrac{1}{4}u_0 = 0 \qquad \Rightarrow \qquad u_0 = a_0 \cos \tfrac{1}{2}\tau + a_1 \sin \tfrac{1}{2}\tau$$

The second equation becomes

$$u_1'' + \tfrac{1}{4}u_1 = -a_0(\alpha_1 + \tfrac{1}{2})\cos \tfrac{1}{2}\tau - a_1(\alpha_1 - \tfrac{1}{2})\sin \tfrac{1}{2}\tau - \tfrac{1}{2}a_0 \cos \tfrac{3}{2}\tau - \tfrac{1}{2}a_1 \sin \tfrac{3}{2}\tau$$

There are periodic solutions (of 4π) only if

$$a_1 = 0, \qquad \alpha_1 = -\tfrac{1}{2} \qquad \text{or} \qquad a_0 = 0, \qquad \alpha_1 = \tfrac{1}{2}$$

Our expression for the transition curve is

$$\alpha = \tfrac{1}{4} - \tfrac{1}{2}\beta \qquad \text{and} \qquad \alpha = \tfrac{1}{4} + \tfrac{1}{2}\beta$$

These are also shown plotted in Figure 7.19.

We can proceed in this way to determine the other transition curves. The next are

$$\alpha = 1 - \tfrac{1}{12}\beta^2 \qquad \text{and} \qquad \alpha = 1 + \tfrac{5}{12}\beta^2$$

Thereafter, they are almost vertical appearing at $\alpha = \tfrac{1}{4}n^2$. These plots are also called Strutt's diagram.

The portion of this graph between $0 < \alpha < 1$ corresponds to Figure 4.34. We now know that the center wedge is an unstable region.

Stability of the Beam

It is seen from Figure 7.19 that the stable regions are mostly below the line $\alpha = \beta$, which corresponds to

$$(1 - \frac{P_o}{P_m}) = \frac{\hat{P}}{P_m} \qquad \text{or} \qquad P_o + \hat{P} = P_m$$

The only major deviation from this is in the region of $\alpha = 1/4$. The region $\alpha < \frac{1}{4} - \frac{1}{2}\beta^2$ leads to

$$\frac{\omega^2}{\omega_m^2} > 4(1 - \frac{P_o}{P_m}) + 2\frac{\hat{P}}{P_m}$$

while the region $\alpha > \frac{1}{4} + \frac{1}{2}\beta^2$ leads to

$$\frac{\omega^2}{\omega_m^2} < 4(1 - \frac{P_o}{P_m}) - 2\frac{\hat{P}}{P_m}$$

These two regions are shown plotted in Figure 7.20. Note the inversion with respect to frequency. In fact, all of the higher transition curves (which are infinitesimally thin vertical lines) map into the lower left region.

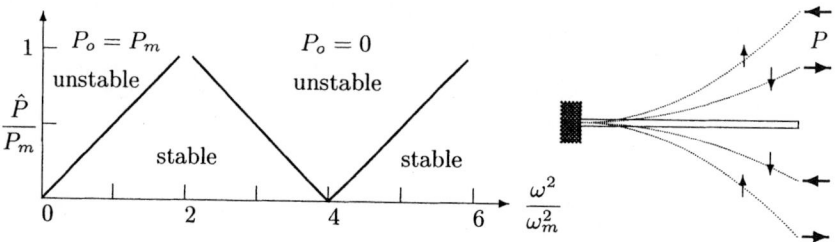

Figure 7.20: Stability regions.

It is thus clear that the frequency region near

$$\omega^2 \approx 4\omega_m^2 (1 - \frac{P_o}{P_m}) \qquad \text{or} \qquad \omega \approx 2\bar{\omega}_m$$

where $\bar{\omega}_m$ is the natural frequency of vibration of the beam with axial compression, is to be avoided. This is easily justified by looking at the inset figure in Figure 7.20 for a cantilever beam under axial varying load. On the vertical excursion, the load is compressive and adds to the motion; when the motion reverses, the load pulls thus further aiding the motion. A similar effect occurs on the lower side. We see there are four load reversals for two displacement reversals.

Example 7.8: Investigate the stability of a cantilever beam to an oscillating axial load.

We will just look at the effect of the fluctuating load and ignore the static component P_o. The results for various frequencies and amplitudes are shown plotted

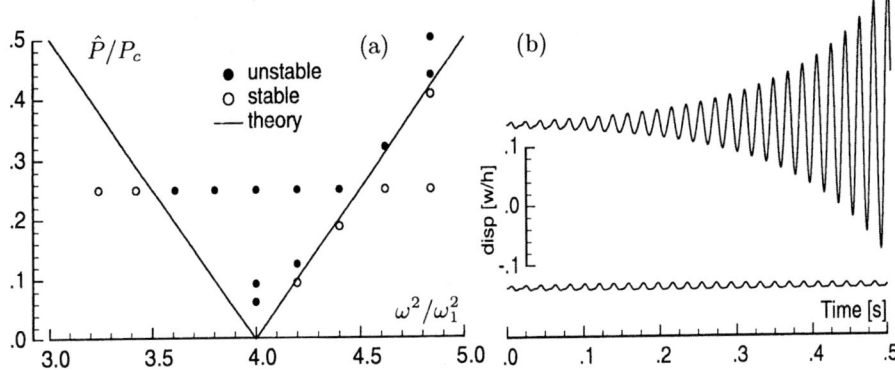

Figure 7.21: Cantilever beam under axial compression. (a) Stable and unstable points for various combinations of frequency and amplitude. (b) Displacement histories for just above and just below critical.

in Figure 7.21(a). Clearly, there is a boundary between the stable and unstable regions that depends on the combination of frequency and amplitude. Figure 7.21(b) shows the early responses for two amplitudes just straddling the boundary when $(\omega/\omega_1)^2 = 4.4$. For the unstable cases, the transverse displacement continues to rise to the order of $L/2$ and then numerical instability sets in. The rate of growth is sensitive to the amount of damping. Also, a small transverse load imperfection ($\times 10^{-3}$ that of the axial) was used as agent to manifest the instability in a reasonable time period.

To plot the theory, we realize that we can do an analysis similar to that for the simply supported beam but the parameters P_m and ω_m we use are from the FEM analysis.

Example 7.9: Investigate the stability of a model of a bridge subjected to oscillating vertical loads.

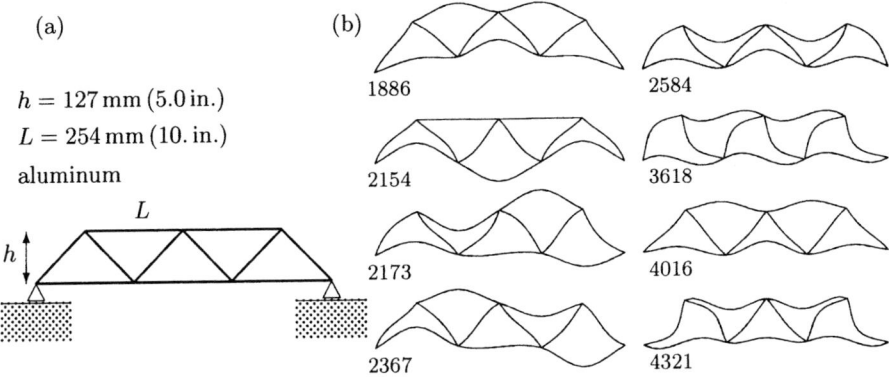

(a)

$h = 127\,\text{mm}\,(5.0\,\text{in.})$
$L = 254\,\text{mm}\,(10.\,\text{in.})$
aluminum

(b)

1886   2584

2154   3618

2173   4016

2367   4321

Figure 7.22: Bridge under oscillating vertical loads. (a) Geometry. (b) Vibration mode shapes with frequencies in [r/s].

The dimensions of the bridge are shown in Figure 7.22(a). The loading is considered to be that of vertical loads applied at each of the joints. Thus some of the members are in compression (top ones) while others are in tension (bottom ones). As the loads oscillate, we suspect there might be a parametric instability in the compressed members.

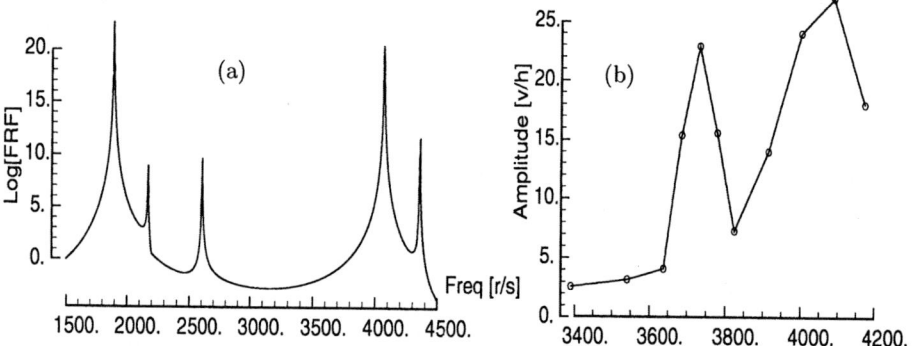

Figure 7.23: Forced response of bridge. (a) Linear frequency response function (FRF). (b) Nonlinear response showing instability at twice the fundamental frequency.

Shown in Figure 7.22(b) are the first eight vibration mode shapes. These suggest that it would not be possible to distinguish parametric instability from the large response of one of the resonance frequencies. In looking at a complex structure under external loading, the first thing to do is to determine which modes are actually excited. From Figure 7.23(a) it is quite clear that only a subset of the modes are excited. Furthermore, the modes excited do not have the same magnitudes. Note that the particular frequency scan shown uses all the degrees of freedom and these are referenced to the first frequency result (which becomes zero on the log plot). Thus, modes 1 and 7 are strongly excited, modes 2, 3, 5, and 8 are mildly excited, modes 4 and 6 do not register at all. Consequently, the somewhat complex sequence of mode shapes reduces to a discussion of modes 1 and 7 being excited by this particular loading.

As illustrated for the beam, we expect a parametric behavior for each of the modes and this appears roughly at twice the fundamental frequency. Thus we expect that if the structure is excited at a frequency close to $1886 \times 2 = 3772\,r/s$ we will see a significant change in the response. From Figure 7.23(a) we see that this does not coincide with any forced resonance frequencies of the bridge.

Figure 7.23(b) shows the results of exciting the bridge at frequencies close to $2\omega_1$. The model had 10 elements per member to ensure that correct local behavior was obtained. Clearly there is significant responses close to $2\omega_1$. The measure of amplitude used is the RMS amplitude.

7.5   Stability of Forced Vibration of Plates

Consider a rectangular plate vibrating out-of-plane; if the edges do not slide, then a membrane load is generated (if the displacements are large). We are interested in the effect this membrane load has on the dynamic response and stability of the plate. This is a difficult problem in its general form; to make it analytically tractable we will invoke some simplifications. We take the governing equation as

$$D\nabla^2\nabla^2 w - \bar{N}_{xx}\frac{\partial^2 w}{\partial x^2} - 2\bar{N}_{xy}\frac{\partial^2 w}{\partial x\partial y} - \bar{N}_{yy}\frac{\partial^2 w}{\partial y^2} + \eta h\frac{\partial w}{\partial t} + \rho h\frac{\partial^2 w}{\partial t^2} = q(x,y,t)$$

which is valid for somewhat moderate deflections. Also, we will assume all boundary conditions are simply-supported.

Reduced Equation of Motion

Because the boundary conditions are simply supported, we can assume a solution of the form

$$w(x,y,t) = C_{nm}(t)\sin(\frac{n\pi x}{a})\sin(\frac{m\pi y}{b})$$

which is a typical term in a Fourier series expansion of the deflection of a plate of size $[a \times b]$. Substituting into the governing equation and neglecting the membrane shear contribution (because of symmetry of the boundary conditions) leads to

$$D[(\frac{n\pi}{a})^2 + (\frac{m\pi}{b})^2]^2 C_{mn} + [N_{xx}(\frac{n\pi}{a})^2 + N_{yy}(\frac{m\pi}{b})^2]C_{mn} + \eta h\dot{C}_{nm} + \rho h\ddot{C}_{nm} = q_{nm}$$

The membrane forces are functions of the out-of-plane deflections; we will estimate their effect based on the assumption that the deflections are not very large. The deflected length of the middle line in the x-direction is

$$L_x = \int dL_x = \int \sqrt{1+(\frac{\partial w}{\partial x})^2}dx \approx \int [1+\tfrac{1}{2}(\frac{\partial w}{\partial x})^2]dx$$

$$= \int [1+(\frac{n\pi}{2a})^2 C_{nm}^2(1+\cos(\frac{2n\pi x}{a}))]dx = [1+(\frac{n\pi}{2a})^2 C_{nm}^2]a$$

An approximation for the strain is therefore

$$\epsilon_{xx} \approx \frac{L_x - a}{a} \approx (\frac{n\pi}{2a})^2 C_{nm}^2$$

Similarly, for the other strain

$$\epsilon_{yy} \approx \frac{L_y - b}{b} \approx (\frac{m\pi}{2b})^2 C_{nm}^2$$

Although the out-of-plane deflection of the plate alternates in sign, the axial strains are positive quantities. Using Hooke's law under plane stress conditions, we get

$$\bar{N}_{xx} = \sigma_{xx}h = \frac{Eh}{1-\nu^2}[\epsilon_{xx} + \nu\epsilon_{yy}] = \frac{Eh}{1-\nu^2}[(\frac{n\pi}{2a})^2 + \nu(\frac{m\pi}{2b})^2]C_{nm}^2$$

$$\bar{N}_{yy} = \sigma_{yy}h = \frac{Eh}{1-\nu^2}[\epsilon_{yy} + \nu\epsilon_{xx}] = \frac{Eh}{1-\nu^2}[(\frac{m\pi}{2b})^2 + \nu(\frac{n\pi}{2a})^2]C_{nm}^2$$

The approximation of the contribution of the membrane forces to the governing equation is now

$$\frac{Eh}{4(1-\nu^2)}\left[(\frac{n\pi}{a})^4 + 2\nu(\frac{n\pi}{a})^2(\frac{m\pi}{b})^2 + (\frac{m\pi}{b})^4\right]C_{nm}^2$$

It clearly depends on the square of the deflection.

We can now write the transformed equation of motion as

$$\ddot{C}_{nm} + \xi\dot{C}_{nm} + \omega_{nm}^2[1 + \alpha_{nm}C_{nm}^2]C_{nm} = \tilde{q}_{nm}(t)$$

The parameters introduced are

$$\omega_{nm}^2 \equiv \frac{D}{\rho h}\left[(\frac{n\pi}{a})^2 + (\frac{m\pi}{b})^2\right]^2, \quad \alpha_{nm} \equiv \frac{3}{h^2}\frac{\left[(\frac{n\pi}{a})^4 + 2\nu(\frac{n\pi}{a})^2(\frac{m\pi}{b})^2 + (\frac{m\pi}{b})^4\right]}{\left[(\frac{n\pi}{a})^2 + (\frac{m\pi}{b})^2\right]^2}$$

and $\xi = \eta/\rho$. The nonlinear contribution is approximately

$$\alpha_{nm}C_{nm}^2 \approx 3\left(\frac{C_{nm}}{h}\right)^2$$

We therefore expect the nonlinear effects to become significant when the deflections are on the order of the plate thickness.

We have thus reduced the problem to the study of a single-degree-of-freedom system with a symmetric return force. Consequently, we can draw on some of the results from Chapter 5 to predict the response.

Nonlinear Response

Consider when the loading is sinusoidal of the form $\tilde{q} = \tilde{P}\cos\omega t$, then the equation of motion for a typical mode can be written as

$$\ddot{u} + \eta\dot{u} + \omega_o^2[1 + \bar{\alpha}u^2]u = \tilde{q} = \tilde{P}\cos\omega t$$

We have already established in Chapter 5 that the solution can be approximated as

$$u(t) \approx A\cos\omega t + B\sin\omega t \qquad (7.3)$$

where the amplitudes are obtained from

$$\eta\omega A - B[(\omega_o^2 - \omega^2) + \tfrac{3}{4}\omega_o^2\alpha(A^2 + B^2)] = 0$$
$$\eta\omega B + A[(\omega_o^2 - \omega^2) + \tfrac{3}{4}\omega_o^2\alpha(A^2 + B^2)] = \tilde{P} \qquad (7.4)$$

The magnitude of these can be obtained from

$$\eta^2\omega^2 C^2 + C^2[(\omega_o^2 - \omega^2) + \tfrac{3}{4}\omega_o^2\alpha C^2]^2 = \tilde{P}^2, \qquad C^2 \equiv A^2 + B^2$$

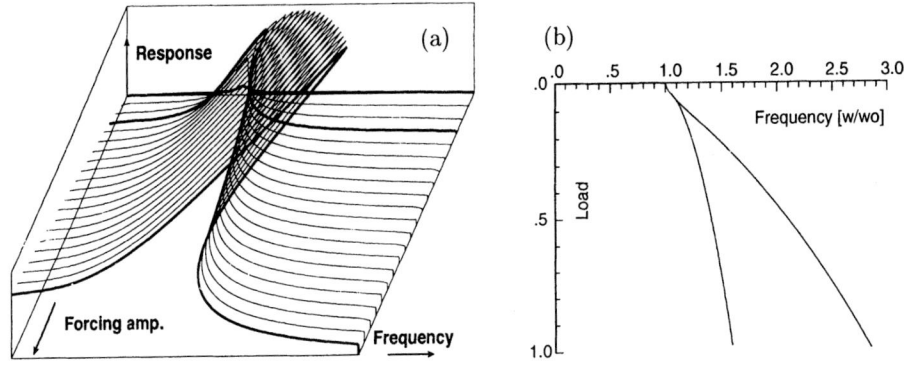

Figure 7.24: Forced frequency behavior of a plate. (a) Response amplitude showing a fold. (b) Projection of the fold onto the $w - P$ plane showing a cusp.

Figure 7.24(a) shows the response amplitude as a function of the forcing frequency w and the forcing amplitude P. For small excitation amplitudes the response is similar to a linear damped oscillator. For higher excitation amplitudes, the surface has a "fold" — inside the fold there are three possible amplitudes and outside it there is only one. Clearly a change of solution character will occur for parameters inside the fold.

The boundary of the fold as projected onto the $w - P$ plane is shown in Figure 7.24(b). It has a cusp, which is why this singularity is known as a *cusp catastrophe* [38, 75]. The equation of the boundary is obtained from the condition that

$$\left. \frac{\partial w^2}{\partial A^2} \right|_{P=\text{constant}} = 0$$

This leads to the quadratic equation for C^2

$$[3(\tfrac{3}{4}w_o^2\alpha)]C^4 + [4(w_o^2 - w^2)\tfrac{3}{4}w_o^2\alpha]C^2 + [\eta^2 w^2 + (w_o^2 - w^2)^2] = 0$$

This can then be used in Equation (7.3) to determine the excitation load. When the damping is negligible, this is easily factored to give

$$C = A = \sqrt{\frac{(w^2 - w_o^2)}{3\tfrac{3}{4}w_o^2\alpha}}, \quad \sqrt{\frac{(w^2 - w_o^2)}{\tfrac{3}{4}w_o^2\alpha}}$$

We can also rearrange this as

$$w^2 = w_o^2 + \tfrac{3}{4}w_o^2\alpha A^2, \quad w_o^2 + 3\tfrac{3}{4}w_o^2\alpha A^2$$

We will now obtain similar results but using a different approach.

Forced Response as a Mathieu Equation

Write the equation of motion in the convenient form

$$\ddot{u} + \eta\dot{u} + \omega_o^2[1 + \alpha u^2]u = \tilde{q}(t) \quad \text{or} \quad \ddot{u} + \eta\dot{u} = -F + P$$

Our idea of stability is what happens the response due to a small disturbance. Therefore, as shown in the introduction, we will decompose the total response into two parts: a zero solution and a variational solution according to

$$\{u(t)\} = \{u(t)\}_o + \epsilon\{\xi(t)\}$$

This lead to the two equations

$$\epsilon^0: \qquad \ddot{u}_o + \eta\dot{u}_o + \omega_o^2[1 + \alpha u_o^2]u_o = \tilde{q}$$

$$\epsilon^1: \qquad \ddot{\xi} + \eta\dot{\xi} + \omega_o^2[1 + 3\alpha u_o^2]\xi = Q$$

We can use the solution of Equations (7.3) and (7.4) to estimate the tangent stiffness in the perturbation equation.

Neglecting damping gives $B = 0$ and put the estimate for u_o into the variational equation to get

$$\ddot{\xi} + \omega_o^2[1 + 3\epsilon\alpha A^2(\tfrac{1}{2} + \tfrac{1}{2}\cos 2\omega t)]\xi = Q, \qquad A[(\omega_o^2 - \omega^2) + \tfrac{3}{4}\omega_o^2\alpha A^2] = \tilde{P}$$

It is interesting that the transverse loading results in a linear equation with periodic coefficients. We have already established the stability of such systems in the previous section and we now use these results. We reduce the system to the standard form

$$\xi'' + [\bar{\alpha} + \bar{\beta}\cos\tau]\xi = 0$$

by the substitutions

$$\tau = 2\omega t, \qquad \bar{\alpha} = \omega_o^2[2 + 3\epsilon A^2]/8, \qquad \bar{\beta} = 3\omega_o^2\epsilon\alpha A^2/8$$

Because $\epsilon \ll 1$, the stability boundaries are near $\bar{\alpha} = \tfrac{1}{4}$. For positive ϵ, then for stability

$$\bar{\alpha} < \tfrac{1}{4} - \tfrac{1}{2}\bar{\beta}, \qquad \bar{\alpha} < \tfrac{1}{4} + \tfrac{1}{2}\bar{\beta}$$

In terms of the original variables

$$\omega^2 < \omega_o^2 + \tfrac{3}{4}\omega_o^2\alpha A^2, \qquad \omega^2 > \omega_o^2 + 3\tfrac{3}{4}\omega_o^2\alpha A^2$$

This is the region bounded by the cusp equations.

Example 7.10: Illustrate how the plate instability manifests itself under a forced frequency situation.

This is an example of how, for nonlinear systems, different responses are obtained depending on how the state was arrived at.

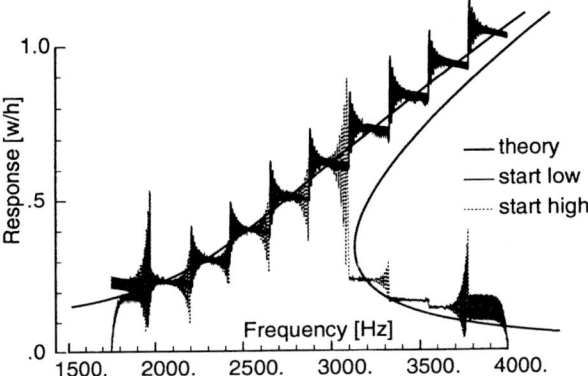

Figure 7.25: Forced frequency response of a plate.

A plate, similar to that used in Chapter 6 with aspect ratio [3 : 1], was excited with a uniform pressure whose frequency changed in steps. It is worth noting that the stepped procedure gives more consistent results than using a continuously increasing frequency. The test was run twice: first, the frequency scan began low and was then increased; second, the frequency scan began high and was then decreased. The results are shown in Figure 7.25 and compared to the theory just derived. The jump in the amplitude is apparent.

The actual responses are sinusoidal in time so the plot is for the amplitude obtained as

$$C = \sqrt{w^2 + \dot{w}^2/\omega^2}$$

After each change of frequency there are dynamic transients that require time to settle down. This is reminiscent of the behavior of plates under membrane loading. Indeed, in this instance, the frequency is playing a similar role to that of the axial load.

In computing the comparison theory, the nonlinear contribution was determined as

$$\alpha_{nm} C_{nm}^2 = 2.6 \left(\frac{C_{nm}}{h} \right)^2$$

The load level was chosen so that the magnitude of the response was on the order of the plate thickness.

7.6   Stability of Motion in the Large

So far we have been mainly concentrating on motions local to the equilibrium point. We end this chapter with a brief consideration of situations where the motion is large and we are interested in establishing the domain of attraction of the equilibrium points. (The domain of attraction is the largest set of points such that their trajectories converge to the region.) Some of the ideas come from References [26, 38, 43, 49]

Motion in the Large

Let us return to a variation of a truss problem shown Figure 7.26. We know from before that this can exhibit a snap-through buckling resulting in a quite large displacement.

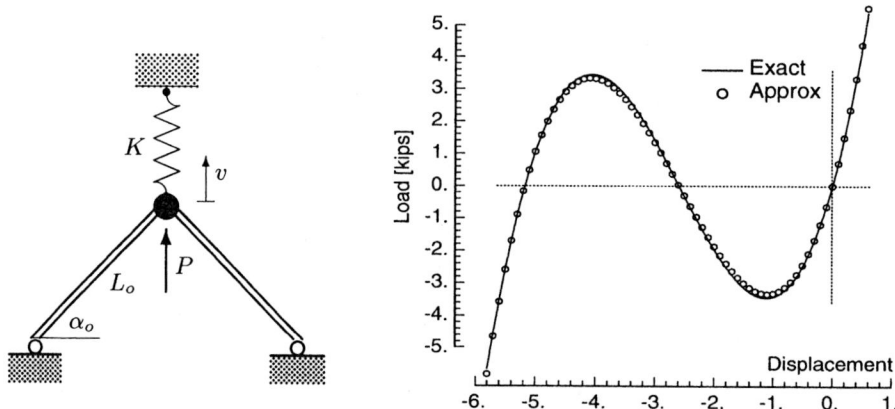

Figure 7.26: Pinned truss.

We showed in Chapter 6 that the dynamic equilibrium equations are given by

$$M\ddot{v} + C\dot{v} = P - 2EA\left[\sin\alpha_o + \frac{v}{L_o}\right]\left[1 - \frac{1}{\sqrt{1 + 2\dfrac{v}{L_o}\sin\alpha_o + (\dfrac{v}{L_o})^2}}\right] - Kv$$

The member force term is too complicated for our present purpose, so we will approximate it with a polynomial. Noting that the three zero-load points occur when v is 0, $-L_o \sin\alpha_o$, and $-2L_o \sin\alpha_o$, respectively, then an approximation is

$$M\ddot{v} + C\dot{v} = P - EA\sin^3\alpha_o\left[\bar{v}(1+\bar{v})(2+\bar{v})\right] - Kv, \qquad \bar{v} \equiv \frac{v}{L_o \sin\alpha_o}$$

A comparison of the approximate and true static behavior of the truss is shown in Figure 7.26. Clearly the approximation is good over the whole range of loads.

Consider the free motion of the system when there is no damping; we will write it as

$$M\ddot{v} = -F(v)$$

Noting that

$$\ddot{v} = \frac{d\dot{v}}{dt} = \frac{d\dot{v}}{dv}\frac{dv}{dt} = \frac{1}{2}\frac{d\dot{v}^2}{dv}$$

Then we can obtain a first integration of our equation as

$$M\tfrac{1}{2}\dot{v}^2 = -\int F\,dv + \text{constant}$$

The first term is the kinetic energy, and the second is the strain energy of the elastic forces, so rewrite as

$$\tfrac{1}{2}M\dot{v}^2 + \tfrac{1}{2}U(v) = E = \text{constant}$$

where E is the total energy. In the phase plane, this represents integral curves with E as the parameter. In the particular case we are considering here, the energy relation is

$$\tfrac{1}{2}M\dot{v}^2 + \beta\left[\bar{v}^2 + \bar{v}^3 + \tfrac{1}{4}\bar{v}^4\right] L_o \sin\alpha_o + \tfrac{1}{2}Kv^2 = E, \quad \beta \equiv EA\sin^3\alpha_o, \quad \bar{v} \equiv \frac{v}{L_o \sin\alpha_o}$$

We can solve for the velocity as a function of displacement as

$$\dot{v} = \pm\sqrt{\left[E - \beta[\bar{v}^2 + \bar{v}^3 + \tfrac{1}{4}\bar{v}^4]L_o \sin\alpha_o - \tfrac{1}{2}Kv^2\right]2/M}$$

The contours are shown in Figure 7.27. The role of the linear spring K is simply to make the two minima different.

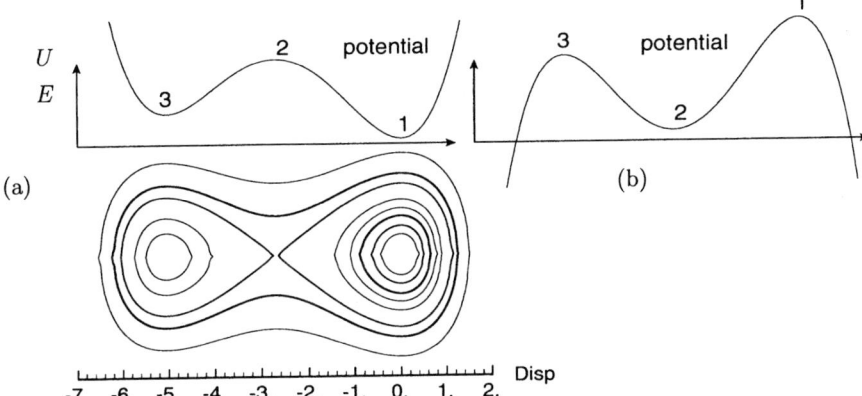

Figure 7.27: Energy contours. (a) Stable in the large. (b) Unstable in the large.

We are now in a position to discuss the motion in the large and not just local to the equilibrium points. Points corresponding to $\partial U/\partial v = 0$ are equilibrium points (since the force is zero); there are three equilibrium points in Figure 7.27(a). These points are remote from each other and are referred to as *isolated equilibrium points*. For $E < U_1$, no motion is possible. For $U_1 < E < U_2$, there is periodic motion about point 1 and point 1 behaves as a center. This is possible only if the initial displacements put it in the vicinity of the center. Consider a motion local to point 3. For energy $E < U_3$, no motion is possible but for $U_3 < E < U_2$, there is periodic motion and the point behaves as a center.

Now consider motion local to point 2. This is a saddle point, and for any energy $E < U_2$ the motion will be about one of the other equilibrium points. For energy $E > U_2$, the motion is periodic but the trajectories enclose both centers and the saddle point. This motion is qualitatively different from that about either center. Clearly, if damping is added to the system, then the motions become spirals to one of the centers.

It is typical of conservative systems that a closed trajectory encloses an odd number of equilibrium points [49], with the number of centers exceeding the number of saddle points by one.

A final point about terminology. The closeness or remoteness of the equilibrium points is to be measured against the level of excitation. In Figure 7.27(a), point 2 is unstable in the small; that is, a small disturbance will cause a large displacement. But if points 1 and 3 are relatively close, then point 2 is stable in the large; that is, a large disturbance will not cause an excessively large displacement because it goes in an orbit around the three equilibrium points. According to Reference [44], this is called "stable for practical purposes." Conversely, in Figure 7.27(b), point 2 is stable in the small (because a small disturbance will only cause a small displacement) but is unstable in the large. This is called "unstable for practical purposes."

Limit Cycles

Closed trajectories, as we just saw, imply periodic motion for conservative systems and they enclose centers and saddle points. Closed trajectories can also occur in nonconservative systems, but they must be such that over a single cycle there is no net energy change. This implies that energy is dissipated over some of the cycle while energy is imparted over the remainder. These trajectories are called *limit cycles*.

Limit cycles are equilibrium motions (in contrast to equilibrium points where the system is at rest) where the system is performing a periodic motion. Neighboring paths are not closed but spiral into (stable) or away (unstable) from the limit cycle. That is why, in discussing the stability of these systems, we refer to their *orbital stability* instead of their equilibrium points. It is worth pointing out that for a conservative system the amplitude of the motion depends on the energy imparted to the system — this is easily seen in Figure 7.27; the amplitude of a limit cycle, however, depends on the system parameters alone. This is also why we concluded that the follower force behavior of Figure 7.12 was that of a limit cycle.

Linear systems with constant coefficients cannot exhibit limit cycles. One of the early examples studied was by Rayleigh in connection with the oscillation of a violin string, which led to the equation for the string displacement

$$\ddot{v} + \epsilon[\tfrac{1}{3}\dot{v}^2 - 1]\dot{v} + v = 0$$

The nonlinearity is associated with the "damping" term, which comes from the friction between the string and the bow — the bow acts as the external source of energy. Before discussing this equation further, we will put it into the standard form of a *Van der Pol equation* by differentiation and making the substitution $\dot{v} \to u$ to give

$$\ddot{u} + \eta[u^2 - 1]\dot{u} + u = 0$$

The sign of the damping contribution depends on the relative magnitude of the displacement: for $|u| < 1$, it is negative and the system receives energy, while for $|u| > 1$, it is positive and the system dissipates energy.

Example 7.11: Analyze the stability of the origin of the Van der Pol equation. Introduce the new variables $x_1 = \dot{u}$ and $x_2 = u$, then the equations become

$$\dot{x}_1 = -\eta[x_2^2 - 1]x_1 - x_2$$
$$\dot{x}_2 = x_1$$

The origin $x_1 = 0$, $x_2 = 0$ is an equilibrium point, and we are interested in its stability. The linearized system of equations near the origin is

$$\{\dot{\xi}\} = [\, A\,]\{\xi\}, \qquad [\, A\,] = [\frac{\partial \mathcal{F}}{\partial x}] = \begin{bmatrix} -\eta[x_2^2 - 1] & -2x_1x_2 - 1 \\ 1 & 0 \end{bmatrix} = \begin{bmatrix} \eta & -1 \\ 1 & 0 \end{bmatrix}$$

As usual for stability analysis, we assume solutions of the form $\{\xi\} = \{c\}e^{i\mu t}$. This leads to the eigenvalue problem

$$\det \begin{bmatrix} \eta - i\mu & -1 \\ 1 & -i\mu \end{bmatrix} = 0 \qquad \text{or} \qquad \mu^2 + i\eta\mu - 1 = 0$$

This gives the two roots

$$\mu_{1,2} = -\tfrac{1}{2}i\eta \pm \sqrt{1 - \tfrac{1}{2}\eta^2}$$

For $\eta > 0$, this gives a negative imaginary component irrespective of the magnitude of η and hence the origin is unstable. On the other hand, for $\eta < 0$, this gives a positive imaginary component again irrespective of the magnitude of η and hence the origin is stable.

Thus for $\eta > 0$, any motion initiated in the neighborhood of the origin will leave the origin. The very difficult problem to solve is to determine where it goes, and in particular, if it finds another equilibrium point. We will consider this further using the Lyapunov method.

Example 7.12: Use numerical methods to show the large motion behavior of the Van der Pol equation.

Figure 7.28 shows phase-plane plots where the damping parameter is $\eta = 1.0$ and the origin is unstable. The figures correspond to initial conditions inside and outside the limit cycle. Clearly all points, irrespective of where they begin, end up on the limit cycle. Hence the cycle acts as an attractor for trajectories and we say that it is stable.

As η is made smaller, the cycle takes on more of a circular shape. When η is reversed, the limit cycle still exists but it is unstable. That is, points inside it spiral to the origin, points outside it spiral to infinity.

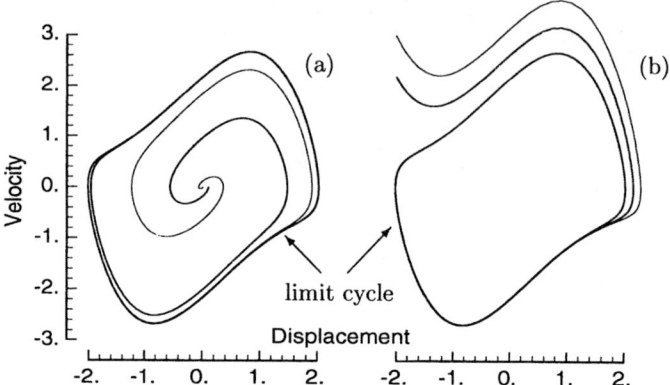

Figure 7.28: Phase plane plots for the Van der Pol equation. (a) Small initial conditions spiral out. (b) Large initial conditions spiral in.

Lyapunov Method

Consider the nonlinear system

$$M\ddot{u} + C\dot{u} + F(u) = 0$$

Assume the mass is pulled away from the natural length of the spring by a large displacement and then released. Will the resulting motion be stable? It is very difficult to answer this using our definitions of stability because the general solution is not available. The linear method cannot be used because the motion starts outside the linear range. However, examination of the system energy can tell us a lot about the motion pattern. There is a correspondence between our notion of stability and mechanical energy:

$$
\begin{array}{ccl}
\text{zero energy} & \Longleftrightarrow & \text{equilibrium point} \\
\text{asymptotic stability} & \Longleftrightarrow & \text{convergence of energy to zero} \\
\text{instability} & \Longleftrightarrow & \text{growth of energy}
\end{array}
$$

Note that we are dealing with scalar functions and energy indirectly reflects a measure of the state vector (strain energy $\to u$, kinetic energy $\to \dot{u}$).

The basic procedure of the Lyapunov method is to generate a scalar "energy-like" function for the dynamic system and examine the time variation of that scalar function. Thus conclusions may be drawn on the stability of systems without requiring explicit knowledge of the solutions. Lyapunov functions (which we indicate as V) are like energy functions in that, if the total energy of a real system is continuously dissipated, then the system must eventually settle down to an equilibrium point. They are used to prove that a stationary point is stable. The reason these functions are useful (when an appropriate one has been found) can be seen in Figure 7.29 which shows that V decreases along trajectories.

The Lyapunov theory was mainly developed for the stability of nonlinear systems with respect to their initial conditions.

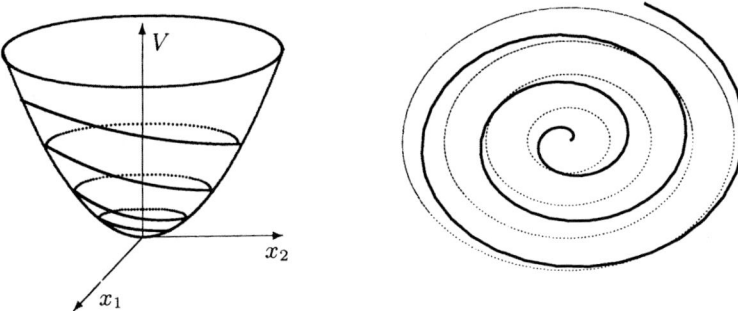

Figure 7.29: Lyapunov function showing decreases along trajectories.

Let the N degree-of-freedom system we are interested in be described by the $2N$ first order autonomous differential equations

$$\{\dot{x}\} = \{\mathcal{F}(x)\}$$

Furthermore, let the origin of the phase space be an equilibrium point where we have that $\{\mathcal{F}(0)\} = \{0\}$. We are interested in knowing if the motion tends to zero as time increases indefinitely.

Consider a set of closed curves (e.g., ellipses but not necessarily), these are our reference curves and they must be nonintersecting and increasing away from the origin. A trajectory that approaches the origin must penetrate the reference curves from the exterior to the interior; that is, the Lyapunov function must decrease along the trajectory. We have for a time derivative along a trajectory

$$\dot{V} = \frac{dV}{dt} = \sum_i \frac{\partial V}{\partial x_i}\dot{x}_i = \sum_i \frac{\partial V}{\partial x_i}\mathcal{F}_i$$

If we can find a V with the following properties:

- $V(x)$ is positive definite;
- $\dot{V}$ is negative semi-definite,

then the equilibrium point is (Lyapunov) stable. If, furthermore, $\dot{V}$ is negative definite, then the equilibrium point is asymptotically stable. This then is the objective of the Lyapunov method of stability analysis.

Suppose we are to determine the stability of the stationary points $x_i = a_i$. First we change the coordinates so that the stationary point is at the origin, that is, set $y_i = x_i - a_i$ so that we have the new system $\dot{y}_i = g_i(y)$ with $g_i(0) = 0$. Then we find a neighborhood of $y_i = 0$ for which the function $V(y)$ has the above properties.

Example 7.13: Use the Lyapunov method to discuss the free vibration stability of the linear oscillator.

We will start with the equation in the simple form

$$Mü + Cü + Ku = 0 \qquad \text{or} \qquad ü + \eta \dot{u} + \omega_o^2 u = 0$$

and convert it to the two first-order equations

$$\dot{x}_1 = x_2$$
$$\dot{x}_2 = -\eta x_2 - \omega_o^2 x_1$$

where we have used $x_1 = u$, $x_2 = \dot{u}$. The stationary point is at $x_1 = 0$, $x_2 = 0$.

According to Reference [26], finding a Lyapunov function is a matter of luck or judgment, hard work or a stroke of genius; we will consider the energies of the system and hope that leads somewhere. The strain and kinetic energies are, respectively,

$$U = \tfrac{1}{2} K x_1^2, \qquad T = \tfrac{1}{2} M x_2^2$$

so we will try

$$V = a x_1^2 + c x_2^2$$

where a and c are undetermined constants except that they must be positive (so that V be positive definite). Differentiating V with respect to time gives

$$\dot{V}(x_1, x_2) = [2ax_1]\dot{x}_1 + [2cx_2]\dot{x}_2 = [2ax_1]x_2 + [2cx_2][-\eta x_2 - \omega_o^2 x_1]$$

Regroup as

$$\dot{V}(x_1, x_2) = [2a - 2c\omega_o^2]x_1 x_2 - 2c\eta x_2^2$$

Because c is positive, then for positive η the second term is negative for $|x_2| > 0$. We can ensure that the whole left-hand side is negative by forcing the coefficient of $x_1 x_2$ to be zero. That is, choose $c = a/\omega_o^2$ to get

$$V = a[x_1^2 + \frac{1}{\omega_o^2}x_2^2], \qquad \dot{V}(x_1, x_2) = -2\frac{a}{\omega_o^2}\eta x_2^2$$

We conclude that so long as $\eta > 0$ and $x_2^2 > 0$ that the oscillator is asymptotically stable.

For the Lyapunov function as is, however, if $x_2 = 0$ for $x_1 > 0$, then $\dot{V}$ is only negative semi-definite and we are not sure about the asymptotic stability — it is Lyapunov stable but not necessarily asymptotically stable. Lyapunov functions provide sufficiency conditions on the stability behavior; if, however, for a particular choice of V the conditions on $\dot{V}$ are not met, then no conclusions about stability can be drawn and another function must be sought. We will now try

$$V = a x_1^2 + b x_1 x_2 + c x_2^2$$

Following the procedure as above we get

$$V = a\left[x_1^2 + \frac{1}{(\eta^2 + \omega_o^2)}[x_2^2 + 2\eta x_1 x_2]\right], \qquad \dot{V}(x_1, x_2) = -\frac{2a\eta\omega_o^2}{\omega_o^2 + \eta^2}x_1^2$$

Therefore, as long as $x_1 > 0$ we have $\dot{V} < 0$ and conclude that the system is asymptotically stable. Furthermore, since the Lyapunov function is valid in the whole $x_1 - x_2$ plane, we conclude that this system is globally stable, that is, it is asymptotically stable for any initial conditions. Global stability implies that the origin is the only equilibrium point of the system.

We draw these conclusion even though the Lyapunov function has no immediately obvious physical meaning for the additional term. It is useful to realize that the Lyapunov function need not have any physical meaning but is really a geometric construct. In reference to Figure 7.29, what we actually did was construct ellipses slightly rotated about the origin so that the trajectories (which are perpendicular to the $x_2 = 0$ axis) intersect the ellipses with a normal component.

Example 7.14: Use the Lyapunov method to discuss the free vibration stability of the nonlinear oscillator with a symmetric return force.

We will start with the equation in the simple form

$$\ddot{u} + \eta \dot{u} + \omega_o^2(1 + \alpha u^2)u = 0$$

and convert it to the two first-order equations

$$\begin{aligned}
\dot{x}_1 &= x_2 \\
\dot{x}_2 &= -\eta x_2 - \omega_o^2 x_1 - \epsilon x_1^3
\end{aligned}$$

where we have used $x_1 = u$, $x_2 = \dot{u}$, and $\epsilon \equiv \omega_o^2 \alpha$. The stationary point is at $x_1 = 0$, $x_2 = 0$.

Following on from the previous example, we will try for our Lyapunov function

$$V = ax_1^2 + bx_1 x_2 + cx_2^2 + dx_1^4$$

where all coefficients are to be positive constants for V to be positive definite. Differentiating V with respect to time gives

$$\begin{aligned}
\dot{V}(x_1, x_2) &= [2ax_1 + bx_2 + 4dx_1^3]\dot{x}_1 + [bx_1 + 2cx_2]\dot{x}_2 \\
&= [2ax_1 + bx_2 + 4dx_1^3]x_2 + [bx_1 + 2cx_2][-\eta x_2 - \omega_o^2 x_1 - \epsilon x_1^3]
\end{aligned}$$

Regroup as

$$\dot{V} = [-b\omega_o^2]x_1^2 + [-b\epsilon]x_1^4 + [2a - b\eta - 2c\omega_o^2]x_1 x_2 + [4d - 2c\epsilon]x_1^3 x_2 + [b - 2c\eta]x_2^2$$

We can ensure that the whole left-hand side is negative by forcing the coefficients of $x_1 x_2$ and $x_1^3 x_2$ to be zero. That is, choose $c = a/(\eta^2 + \omega_o^2)$, $d = c\epsilon/2 = a\alpha/2$, $b = 2c\eta$, and let $a = 1$ to get

$$V = x_1^2 + \frac{1}{(\eta^2 + \omega_o^2)}\left[2\eta x_1 x_2 + x_2^2 + \tfrac{1}{2}\alpha \omega_o^2 x_1^4\right], \quad \dot{V} = -\frac{2\eta}{(\eta^2 + \omega_o^2)}\left[x_1^2 + \alpha \omega_o^2 x_1^4\right]$$

Again, we conclude that so long as $\eta > 0$ that the oscillator is asymptotically stable.

But suppose that α is negative (which corresponds to the pendulum), then d becomes negative and we do not have a positive definite function for $x_1 \geq \sqrt{2/\alpha}$ (with very small η). We have just delimited the range of our knowledge of the stability of the system. This does not prove that the system is unstable outside this region only that we do not know its stability.

Example 7.15: Use the Lyapunov method to discuss the stability of the Van der Pol oscillator.

We will start with the equation in the standard form

$$\ddot{u} - \eta[u^2 - 1]\dot{u} + u = 0$$

and convert it to two first-order equations. However, instead of using $x_1 = u$, $x_2 = \dot{u}$ as before, we will avail of a trick [26] often used for converting equations of the form

$$\ddot{u} + f(u)\dot{u} + g(u) = 0$$

That is, introduce $F(u) = \int f(u)\,du$ because it has the property

$$\frac{dF}{dt} = \frac{dF}{du}\frac{du}{dt} = f(u)\dot{u}$$

Now define $x_2 = \dot{u} + F(u)$, then $\dot{x}_2 = \ddot{u} + f(u)\dot{u} = -g(u)$. This gives the two equations

$$\begin{aligned}
\dot{x}_1 &= x_2 - F(x_1) \\
\dot{x}_2 &= -g(x_1)
\end{aligned}$$

In our particular case, this leads to

$$\begin{aligned}
\dot{x}_1 &= x_2 + \eta[\tfrac{1}{3}x_1^3 - x_1] \\
\dot{x}_2 &= -x_1
\end{aligned}$$

In the transformed coordinates, the stationary point is at $x_1 = 0$, $x_2 = 0$.

In comparison to our two previous examples, we have that $\omega_o = 1$, hence we will try a Lyapunov function of the form

$$V = \tfrac{1}{2}x_1^2 + \tfrac{1}{2}x_2^2$$

to get

$$\dot{V}(x_1, x_2) = [x_1]\dot{x}_1 + [x_2]\dot{x}_2 = x_1[x_2 + \eta[\tfrac{1}{3}x_1^3 - x_1]] + x_2[-x_1] = \eta x_1^2[\tfrac{1}{3}x_1^2 - 1]$$

For $\eta > 0$, then $\dot{V} < 0$ when $x_1^2 < 3$. The largest domain that lies entirely within the stable region is therefore

$$x_1^2 + x_2^2 < 3$$

To get an interpretation of this, we will revert back to the original coordinates. That is,

$$u^2 + \left[\dot{u} + F(u)\right]^2 < 3 \qquad \text{or} \qquad \dot{u}^2 + 2F(u)\dot{u} + F^2(u) + u^2 - 3 < 0$$

We can solve for the velocity/displacement relation along the borderline to be

$$\dot{u} = -F(u) \pm \sqrt{3 - u^2} = \eta[\tfrac{1}{3}u^3 - u] \pm \sqrt{3 - u^2}$$

This is shown plotted in Figure 7.30 for different values of η. Note that at $u = 0$ and $u = \pm\sqrt{3}$, $\dot{u}$ is independent of η and given by $\pm\sqrt{3}$, 0, respectively.

The figure, of course, resembles the limit cycles already computed in Figure 7.28. For $\eta > 0$ the inside of the cycle is stable and points outside are unstable, whereas for $\eta < 0$ the reverse is true.

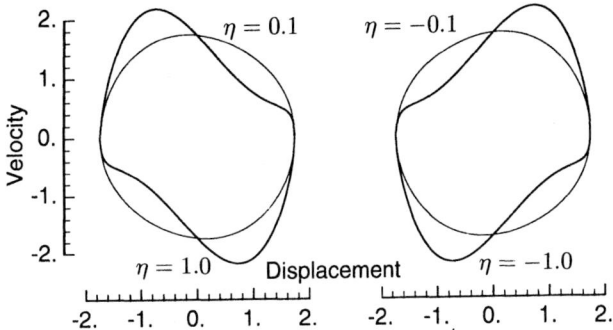

Figure 7.30: Stability boundary for Van der Pol's equation. The near circular limit cycles are for the smaller η.

7.7 Stability Under Transient Loading

We close this chapter with a look at the stability behavior of structures under transient loading. There are two situations of interest. First is impulse or short term loading, and the second is long-term loading such as a suddenly applied load, which then remains constant. The first could be associated with impact-type loading and is covered in good detail in Reference [27]; the second is the rapid application of a nominally quasi-static load. These two may also be considered as limiting cases of blast loading as shown in Figure 7.31; a readable introduction to blast loading is given in Reference [73]. Aspects of these loadings have already been considered in other sections, but this section will attempt to bring the considerations together.

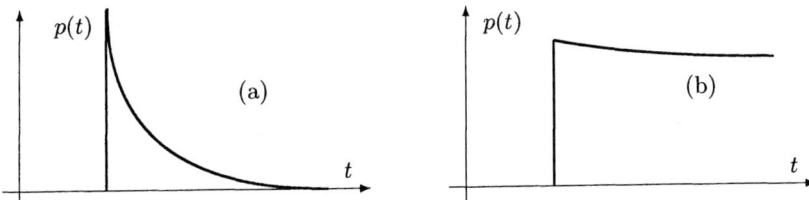

Figure 7.31: Limiting situations of blast loading. (a) Long time scale, blast appears of short duration and impulse like. (b) Short time scale, blast appears of long duration and steady like.

Impulse Loading

A way of viewing these problems is to realize that once the impulse is over the structure is now in free motion. This makes the problem amenable to the Lyapunov method where we view the conditions at the end of the impulse as a set of initial conditions.

The problem then reduces to the question of whether the energy of the impulse is sufficient to put the system outside the stable region.

Example 7.16: Illustrate the stability response of the truss of Figure 7.26 due to impulse loading.

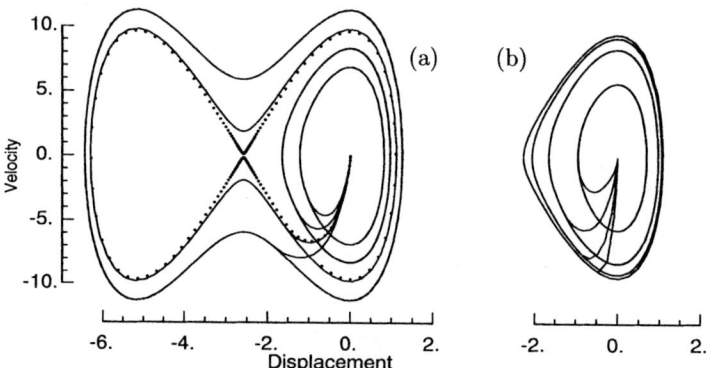

Figure 7.32: Phase-plane response for the truss. (a) Effect of increasing impulse for constant duration. (b) Effect of pulse duration for constant impulse.

For simplicity we remove the vertical spring. The impulse is taken as a triangular force history, $I = \frac{1}{2}P_o\Delta T$. Figure 7.32(a) shows the phase-plane responses for different strengths of impulse; P_o is changed but ΔT is constant. As expected, the effect of increasing the impulse is to send the system into orbit with increasing radius. Note that the end of the impulse coincides with where the trajectory touches the orbit. Values of impulse sufficient to put the system on an outer energy contour need not correspond to maximum displacement. That is, no matter how the system gets onto this outer contour, it is unstable. Thus an impulse that gives a positive initial velocity (so that the truss gets higher) could also cause an instability.

Figure 7.32(b) shows the effect of pulse duration for constant impulse; both P_o and ΔT are changed so that $I = \frac{1}{2}P_o\Delta T$ is constant. The shorter duration pulse causes a larger initial velocity (and thus kinetic energy) and is more likely to send the system to the outer orbit. This is understandable, since the reverse situation, where the duration gets larger and larger, limits to a quasi-static case with diminishing amplitude and hence should not cause an instability.

To utilize this result for the general case, we perform a static analysis to determine the critical load and hence critical strain energy. The critical kinetic energy is then obtained as

$$U_c = \frac{1}{2}\sum_m \{u\}_m^T [k_m]\{u\}_m \quad \Longleftrightarrow \quad \frac{1}{2}\{\dot{u}\}[M]\{\dot{u}\} = T_c = U_c$$

Now imagine the load is applied impulsively over a very short duration such that there is only kinetic energy with very little straining. Thus, the initial velocities are related to the impulse by

$$\int \{P\}\,dt = [\,M\,]\{\dot{u}\} \qquad \text{or} \qquad \{\dot{u}\} = [M^{-1}]\int \{P\}\,dt$$

Finally, if we approximate the impulse as a triangular history and a space distribution that is proportional to the static case, that is, $\chi\{P\}$, then the velocities are estimated as

$$\{\dot{u}\} = \tfrac{1}{2}\chi[M^{-1}]\{P\}\Delta T$$

The criterion is

$$\tfrac{1}{2}\{\dot{u}\}[\,M\,]\{\dot{u}\} < T_c = U_c$$

Corresponding to different expected durations ΔT we will have different load scalings χ.

Long-Term Loading

There are three situations that are of interest to us here. These are broadly divided into their quasi-static stability behavior. Clearly, if a system can become unstable under a quasi-static load, then it will also become unstable when it is applied dynamically; the question we are interested in, however, is if it becomes unstable at a lower dynamically applied load.

I: Stiff Systems

By stiff systems, we mean those which exhibit increasing stiffness with increasing deformation; these are globally stable when loads are applied quasi-statically. We take some of the ideas from Reference [43].

Consider the system

$$\{\dot{x}\} = [F(x,t)]$$

If the solution can be extended or computed for all t, then we say it is *defined in the future*. On the other hand, if at some time T we get $|x| \to \infty$, then the solution is said to have a *finite escape time*. In this latter case, the solutions are unbounded. The boundedness of solutions is a kind of stability described as *Lagrange stability*. We are interested in determining conditions under which solutions are defined in the future (i.e., are bounded for finite times) and this will lead to some statements about its stability.

To help focus ideas, consider the nonlinear system

$$M\ddot{u} + C(u,\dot{u},t)\dot{u} + F(u) = P(t) \qquad \text{or} \qquad \ddot{u} + c(u,\dot{u},t)\dot{u} + f(u) = p(t)$$

where $P(t)$ is the forcing term. Assume the damping term is always positive and that the restoring force is such that

$$G(u) = \int_0^u f(u)\,du \to +\infty \qquad \text{as} \qquad |u| \to \infty$$

That is, away from the origin, $f(u)$ behaves like a stiffening spring. Rewrite the system in the equivalent form

$$\dot{x}_1 = x_2, \qquad \dot{x}_2 = -f(x_1) - c(x_1, x_2, t)x_2 + p(t)$$

Let the Lyapunov function be

$$V = \tfrac{1}{2}x_2^2 + G(x_1)$$

The derivative becomes

$$\dot{V} = x_2\dot{x}_2 + \frac{dG}{dx_1}\dot{x}_1 = -cx_2^2 + px_2$$

The first term on the right-hand side is always negative and so we have the inequalities

$$\dot{V} \le px_2 \le |p||x_2| \le |p||2V - G|^{1/2} \le |p|\sqrt{2}|V|^{1/2}$$

We are interested in positive V solutions, so consider the differential equation

$$\dot{v} = g(t)v^{1/2} \qquad \text{or} \qquad v^{1/2} = \tfrac{1}{2}\int g(t)\,dt$$

It does not have a finite escape time and possibly becomes unbounded only as $t \to \infty$. Applying this result to the inequality relation, we therefore conclude that there are no solutions with finite escape times, that all the solutions are defined in the future.

The final step is to impose that

$$\int_0^\infty |p(t)|\,dt < \infty$$

which corresponds to forcing terms that die out sufficiently fast as time increases. The solutions are then bounded and the system is Lagrange stable. Note that this restriction rules out periodic forcing terms.

As another example, consider the linear system with time varying coefficients

$$M\ddot{u} + P(t)\dot{u} + [A^2 + Q(t)]u = 0 \qquad \text{or} \qquad \ddot{u} + p(t)\dot{u} + [a^2 + q(t)]u = 0$$

where both $P(t)$ and $Q(t)$ are forcing terms. Rewrite the system in the equivalent form

$$\dot{x}_1 = x_2, \qquad \dot{x}_2 = -p(t)x_2 - [a^2 + q(t)]x_1$$

Let the Lyapunov function be

$$V = \tfrac{1}{2}x_2^2 + \tfrac{1}{2}a^2x_1^2$$

The derivative becomes

$$\dot{V} = x_2\dot{x}_2 + a^2x_1\dot{x}_1 = -p(t)x_2^2 - q(t)x_1x_2$$

The first term on the right-hand side is always negative and so we have the inequalities

$$\dot{V} \le -qx_1x_2 \le -|qx_1x_2| + \frac{|q|}{|2a|}[|x_2| + |ax_1|]^2 \le \frac{|q|}{|2a|}[x_2^2 + a^2x_1^2] \le \frac{|q|}{|2a|}V$$

This is of the form that we can conclude that there are no solutions with finite escape times, and that all solutions are defined in the future. If we now impose that

$$\int_0^\infty |q(t)|\, dt < \infty$$

then the solutions are bounded and the system is Lagrange stable. Again, we are ruling out periodic forcing terms.

Example 7.17: Illustrate the distinction between Lagrange stability and Lyapunov stability.

Consider the linear system [67]

$$\ddot{u} + [2 + e^t]\dot{u} + u = p(t)$$

Let $p(t)$ be impulsive so that after a time it is zero and hence its integral is bounded. Since the damping is always positive, we conclude there is Lagrange stability.

At the end of the impulse action, the system is in free motion. Let the initial conditions for this be $u(0) = 2$, $\dot{u}(0) = -1$, then the solution is

$$u(t) = 1 + e^{-t}$$

As time increases, this tends to $u = 1$ and not the origin and hence it is not Lyapunov stable.

II: Bifurcational Systems

What we mean by bifurcational systems are those systems which have a primary loading path that does not cause a change in the bifurcated degrees of freedom — Euler buckling of columns and flat plates are examples.

To help focus ideas, we will concentrate on the simply supported beam shown in Figure 7.33. The linearized governing equation for the motion (at the instant after the load reaches its maximum value P_o) is

$$EI\frac{\partial^4 v}{\partial x^4} + P_o\frac{\partial^2 v}{\partial x^2} + \rho A\frac{\partial^2 v}{\partial t^2} = 0$$

The total potential for this problem is

$$\Pi = \tfrac{1}{2}EI\int_L(\frac{\partial^2 v}{\partial x^2})^2 dx - \tfrac{1}{2}P_o\int_L(\frac{\partial v}{\partial x})^2 dx$$

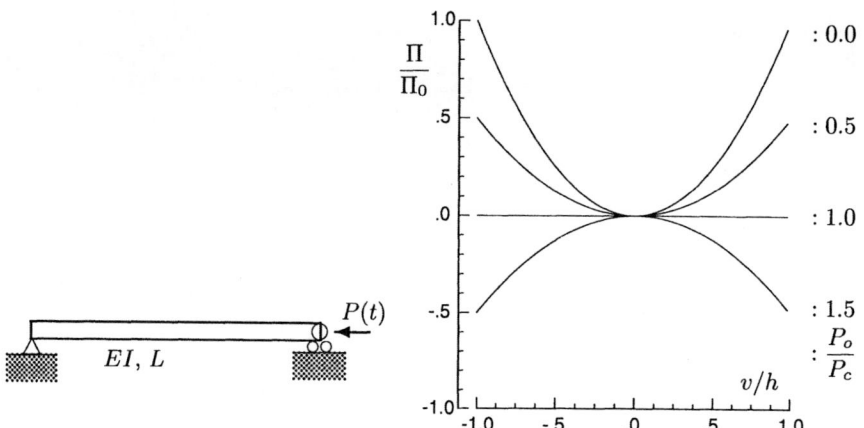

Figure 7.33: Potential energy for a simply supported beam under different values of axial loading.

To get an idea about the motion in the large, let the deflected shape have the representation $v(x) = \hat{v}\sin(\pi x/L)$, then the potential becomes

$$\Pi = \frac{\pi^2}{4}\left[\frac{EI\pi^2}{L^3} - \frac{P_o}{L}\right]\hat{v}^2$$

This is shown plotted in Figure 7.33 for different values of P_o. It is clear that the system becomes unstable only when $P_o \geq EI\pi^2/L^2$, which is the Euler buckling load. We therefore conclude that for this type of system, the dynamics associated with the application of the applied load do not affect the stability behavior.

In a similar manner, if we let the deflected shape of the simply supported plate have the representation $w(x,y) = \hat{w}\sin(\pi x/a)\sin(\pi y/b)$, then the potential becomes

$$\Pi = ab\left[\frac{D}{8}\left[(\frac{n\pi}{a})^2 + (\frac{m\pi}{b})^2\right]^2 + \bar{N}_{xx}(\frac{n\pi^2}{a})^2\right]\hat{w}^2$$

The expression inside the brackets is the buckling loads for plate. Thus for particular a, b and n, m this potential has the same shape as in Figure 7.33.

Example 7.18: Investigate the stability of a plate under the action of a rapidly applied axial force.

We will use the plate investigated in Chapter 6. In that chapter, we were careful to apply the load "slowly" so that excessive dynamics were not induced — this was judged simply by there not being too many oscillations once the maximum load was obtained. We now wish to apply the load "fast."

Loadings are fast when they are applied on a time scale comparable to the time it takes a wave to travel a significant dimension of the structure. In the present case, the axially applied load generates a longitudinal wave that travels at the speed of $c_o = \sqrt{E/\rho}$. For the aluminum plate of our example, this corresponds to a total travel time (down the plate and back) of $60\,\mu s$. A load history with this rise time was constructed.

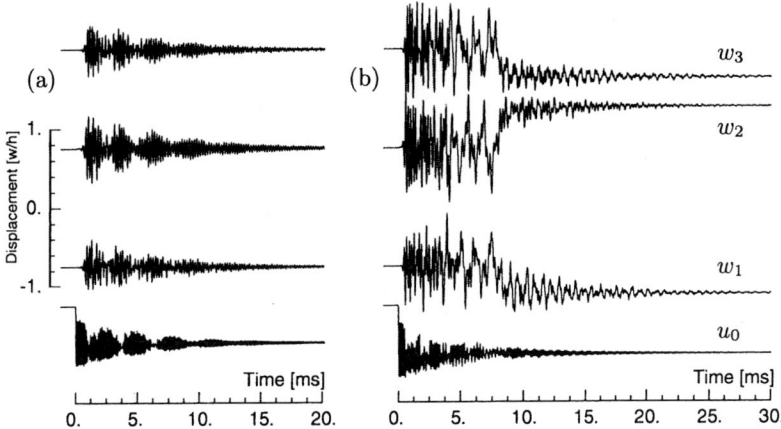

Figure 7.34: Response of a simply supported plate to dynamically applied axial loading. (a) Load is 10% below static critical. (b) Load is 10% above static critical.

The deflections of the quarter points (w_i) and end-shortening (u_0) are shown in Figure 7.34 for cases when the load is 10% below the static critical value and 10% above. In both cases, a transverse load of 0.001 times the axial load was used as an imperfection to induce the instability. In comparison to Chapter 6, we see a great deal of activity during the initial stage of the response; indeed for the second case, the oscillation swings exceed the final equilibrium displacements. Nonetheless, when equilibrium is achieved, the load below the static critical value has caused negligible transverse deflection; the load above static critical causes the plate to deflect into the expected [3,1] mode shape.

III: Limit-Point Systems

In contrast to the bifurcational systems, the limit-point systems show a change of deformation in the degrees of freedom along the primary loading path. Snap-through buckling of a shallow arch is an example, but so also is the buckling of shells. Background material can be found in References [35, 39, 40, 64, 65, 66].

To help focus ideas, consider the arch of Figure 7.35 modeled as a truss similar to Figure 7.26 (without the vertical spring). The governing equation for the motion (at the instant after the load reaches its maximum value P_o) is

$$M\ddot{v} + K[1 + \beta v]v = P_o$$

where we have approximated the truss stiffness with just two terms. The total potential for this problem is

$$\Pi = \tfrac{1}{2}K[v^2 + \tfrac{2}{3}\beta v^3] - P_o v$$

The static instability occurs when $v = -1/2\beta$, so we will normalize the potential using $v_c \equiv |1/2\beta|$ to give

$$\frac{\Pi}{\Pi_o} = \tfrac{1}{2}(\frac{v}{v_c})^2 + \tfrac{1}{6}(\frac{v}{v_c})^3 - \tilde{P}\frac{v}{v_c}, \qquad \Pi_o \equiv \frac{Kv_c}{2\beta}, \qquad \tilde{P} \equiv \frac{P_o 2\beta}{K}$$

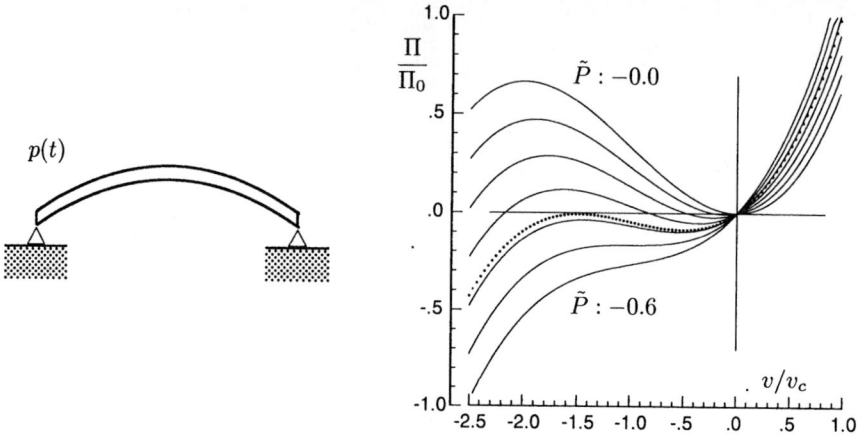

Figure 7.35: Arch under distributed load.

This is shown plotted in Figure 7.35 for different values of P_o.

There are two equilibrium points for each load value given by

$$\frac{\partial \Pi}{\partial v} = 0 \quad \Longrightarrow \quad v = \frac{-1 \pm \sqrt{1 + 4\beta P_o/K}}{2\beta}$$

One is stable, and the other is unstable. The static instability load corresponds to where both equilibrium points coincide giving $P_o = -K/4\beta$ and $v = -1/2\beta$.

Consider the scenario where the load is applied up to the value P_o but the deflection is constrained to be zero. Let the value of load coincide with the dotted line in Figure 7.35. Now release the constraint. Just at the instant of release, the system is at the $(0,0)$ configuration, which is not an equilibrium configuration, and motion ensues. If there is no dissipation mechanism, the motion goes all the way to the unstable equilibrium point and back again to the $(0,0)$ point repeatedly. Now repeat the scenario but with a slightly higher load, then the motion will go past the unstable equilibrium point and there is an unstable response. We therefore identify the dotted value of load as the critical load, and clearly this is less than the static critical load.

We obtain the critical value of dynamic load from the conditions

$$\Pi = 0, \quad \frac{\partial \Pi}{\partial v} = 0 \quad \Longrightarrow \quad P_d = -3K/16\beta, \quad v = -3/4\beta$$

Thus the dynamic collapse load is 0.75 times that of the static case.

Example 7.19: Investigate the stability of an arch under the action of a rapidly applied transverse load.

Dynamic loading is when a vibration mode is excited and a very rapid loading is when very many modes are excited. Under zero loading the first four vibration frequencies of the arch are 95, 153, 232, 309 Hz, respectively, with the middle two being symmetric modes. Based on the first mode, any loading whose rise time is

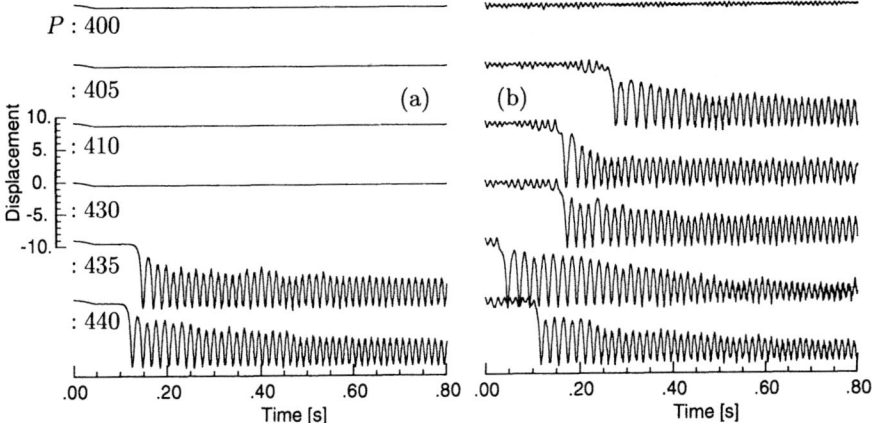

Figure 7.36: Center response of an arch. (a) Quasi-static loading. (b) Dynamic loading.

greater than $T = 1/f = 10\,\text{ms}$ will give a quasi-static loading. Figure 7.36 show the results when the rise time is 50 ms and 200 μs. There is a difference in the results with the dynamically applied load causing snap-through at a lower load. However, the difference (about 8%) is not as significant as predicted by the simple analysis. This was also reported for shells in Reference [64].

Problems

7.1 Consider a cantilever beam with an axially applied load and let the transverse deflection be written as

$$v(x) = (\frac{x}{L})^2 \left[3 - 2(\frac{x}{L})\right] v_L + (\frac{x}{L})^2 \left[-1 + (\frac{x}{L})\right] L\phi_L$$

$$\equiv g_3(x)v_L + g_4(x)\phi_L$$

where v_L, ϕ_L are the tip deflection and rotation, respectively.
• Show that $g_3(x)$, $g_4(x)$ are suitable Ritz functions.
• Use these Ritz functions to establish the elastic stiffness, mass, and geometric stiffness matrices for the cantilever beam.
• Write down the equations of motion.

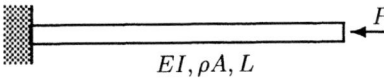

$$EI, \rho A, L$$

7.2 For the system of the previous problem,
• Determine the generalized force vector $\{P\}$ when the axial load remains horizontal.
• Investigate the stability of this system.
• Make a sketch of the behavior of the roots as a function of P.

7.3 For the system of the previous problem
• Determine the generalized force vector $\{P\}$ when the axial load is a

follower load.
- Investigate the stability of this system.
- Make a sketch of the behavior of the roots as a function of P.
- Add damping to the system and investigate its effect on the stability.

7.4 A two-degree-of-freedom system is described by

$$\begin{bmatrix} 4 & -2 \\ -2 & 6 \end{bmatrix} \begin{Bmatrix} u_1 \\ u_2 \end{Bmatrix} + G \begin{bmatrix} 0 & 1 \\ 0 & 0 \end{bmatrix} \begin{Bmatrix} u_1 \\ u_2 \end{Bmatrix} + \begin{bmatrix} 1 & 0 \\ 0 & 2 \end{bmatrix} \begin{Bmatrix} \ddot{u}_1 \\ \ddot{u}_2 \end{Bmatrix} = \begin{Bmatrix} P_1 \\ P_2 \end{Bmatrix}$$

where G is a gain parameter.
- Investigate the eigenvalues as a function of the gain.
- Is a static instability possible ?
- Is a dynamic instability possible ?

7.5 Continuing discussion of the discretized version of the follower force problem:
- Add a form of proportional damping to system.
- Investigate the effect of the damping on the stability.

7.6 Consider an aircraft wing as a cantilevered thin plate.
- Use the Ritz method to replace the equations of motion for the plate with a discretized $[2 \times 2]$ system dominated by flexure and torsion.
- Assume there is a pressure distribution of the form $p = p_o + p_1 x$ where x is in the chord direction. Replace this pressure with an equivalent torque and force.

7.7 For the system of the previous problem, assume that the pressure has the same distribution but that it is linearly dependent on $\dfrac{\partial w}{\partial x}$.
- Replace this pressure with an equivalent torque and force.
- Investigate the stability of this system.

7.8 Treat the launch of a rocket as a follower force problem.
- Investigate the effect of the location of a concentrated payload.

7.9 The equation $\ddot{u} + u - \frac{1}{6}u^3 = 0$ has an approximate solution $a \cos \omega t$ where $\omega^2 = 1 - \frac{1}{8}a^2$, $a \ll 1$.
- Show that the solution is unstable.

— Reference [38], p. 266

7.10 The equation $\ddot{u} + \alpha u + \epsilon u^3 = \epsilon \gamma \cos \omega t$ has the exact subharmonic solution $u = (4\gamma)^{1/3} \cos \frac{1}{3}\omega t$ when $\omega^2 = 9(\alpha + 3\epsilon \gamma^{2/3}/4^{1/3})$.
- Show that the solution is stable.

— Reference [38], p. 265

7.11 Consider the system of equations $\dot{x}_1 = x_2 + \alpha x_1 [x_1^2 + x_2^2 - 1]^\gamma$, $\dot{x}_2 = -x_2 + \beta x_2 [x_1^2 + x_2^2 - 1]^\gamma$,
- Show that $\alpha = -1$, $\beta = -1$, $\gamma = 1$ gives a stable limit cycle.
- Show that $\alpha = +1$, $\beta = +1$, $\gamma = 1$ gives an unstable limit cycle.
- Show that $\alpha = -1$, $\beta = -1$, $\gamma = 2$ gives a semi-stable limit cycle.

— Reference [67], p. 35

7.12 Compare the following two Lyapunov functions for testing the stability of the pendulum: $V_1 = (1 - \cos x_1) + \frac{1}{2}x_2^2$, $V_2 = 2(1 - \cos x_1) + \frac{1}{2}x_1^2 + x_1 x_2 + x_2^2$.

— Reference [67], p. 67

Afterword

> Thus, for a complete investigation of dynamical systems, we require not only a computer and the direct integration methods. These provide no more than an ideal computer laboratory in which an arbitrary number of experiments can be performed, yielding an immense data flow. We require, in addition, certain principles according to which the data may be evaluated and displayed, thus giving an insight into the astonishing variety of responses of dynamical systems.
>
> J. Argyris and H-P. Mlejnek [4]

Nonlinear problems present special difficulties for a book in that they pose a great challenge to have sufficient representative results. A book example by its nature is a single realization — a single geometry, material, or load case; multiple test cases and examples are just not feasible if the book is to remain reasonable in size. But engineers, being introduced to something new, need to see other examples as well as variations on the given examples. Consequently, a new way must be found to present the results of exploring nonlinear problems.

As computer modeling becomes easier to use and faster to run, this opens up the possibility of understanding problems by visualizing the results. Whereas computer programs once sufficed to provide numbers — discrete solutions of "stress at a point" and the like — now complete simulations, which present global behavior, trends, and patterns, can be explored and sensitivity studies analyzing the relative importance of geometric and material parameters can be evaluated. Being able to observe phenomena and zoom in on significant parameters, making judgments concerning those that are significant and those that are not, will greatly enhance the depth of understanding.

The situation becomes one of how to present and interpret information, what trends to look for, what conclusions can be drawn. The purpose of this discussion is to explore (in very broad terms) the make-up of such a computer laboratory, to share some experiences of a prototype version called QED, and to place the content of the current book within this context.

QED: A Computer Laboratory

As envisioned, the simulation program runs simultaneously with reading the text and becomes a resource to be interacted with and tested against as the examples are explored. For instance, in the stress analysis of a plate with a hole introduced in Chapter 2, some logical questions to ask are:

- What if the hole is bigger or smaller?
- What if there are two holes instead of one?
- What if the hole is in a shell?
- What if the loading is dynamic?
- What if the clamped boundary has some elasticity?

These questions are too cumbersome and expensive to answer in a single text but are very appropriate for a simulation program.

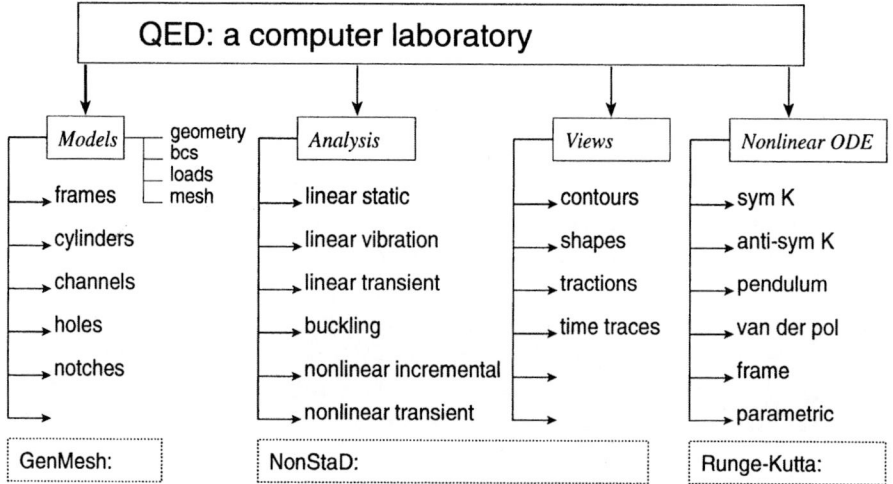

Figure A.1: Overview of a simulation program called QED.

The intent of QED, then, is to provide an interactive simulation environment for studying and understanding a variety of problems in nonlinear structural mechanics — a laboratory for exploration and accumulation of experiences. Figure A.1 shows a possible schematic of the functional parts of the program. Its design is such that it isolates the user from having to cope with the underlying enabling programs (indicated in dashed boxes) and presents each problem in terms of a limited (but richly adaptable) number of choices and combinations.

The process of finite element analysis can be broken down into three separate stages. These are presented as independent modules in QED. The pre-processing stage allows the model geometry to be defined, the boundary conditions imposed, the loads applied, and the mesh generated. In the second stage, the analytical solution is obtained. Choices as to the type of solution required and the parameters best suited to guide the procedure are made. In the post-processing stage, results are displayed in a variety of ways. Contour plots of nodal results, the deformed shape, free body diagrams, and time history traces are available. The essence of many nonlinear phenomena is captured in single differential equations. A gallery of standard some nonlinear equations are also presented.

I: Creating Model Geometries

The design philosophy for the model building is to make each module very specific but flexible through parameterization. This is important as it helps to fix focus on the significant aspects of behavior. Each problem has a template of properties, therefore the user need only focus on what they want to change.

The model geometries presented by QED are chosen to encompass a range of interesting shapes, loading patterns, and boundary conditions. The models are separated into categories based upon their geometric configurations. Within each module parameters can be changed to "morph" the shape from one form to another. For instance, the "Frames" module includes the option to define a multispan beam, a plane frame, and a space frame as shown in Figure A.2. These share common geometric characteristics and are thus classified in a single module. There are corresponding variations of the loads.

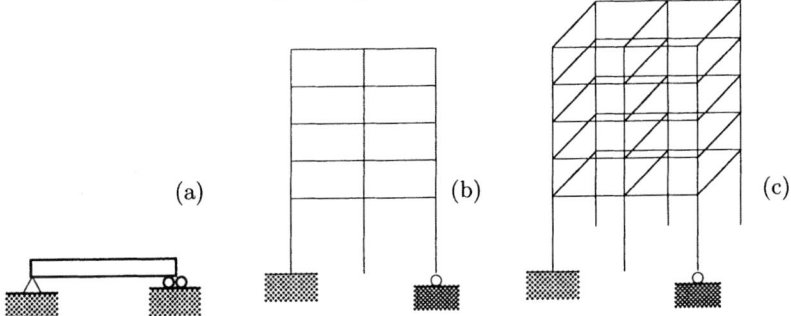

Figure A.2: Progression of associated geometries. (a) Beam. (b) Plane frame. (c) Space frame.

II: Analyzing the Problem

Irrespective of how the model was created, the same collection of analysis procedures can be applied and all loads are considered to be applied dynamically. Each has a set of parameters associated with the solution algorithm. Thus the "nonlinear incremental" option includes parameters for the time step and convergence tolerance for Newton-Raphson iterations. Generally, the algorithmic issues addressed in the text are presented as parameters. Any load can be considered to be applied dynamically or statically.

III: Viewing the Results

In the post-processing stage, a variety of schemes are provided for viewing the results. The same collection of post-processed information can be interrogated for all models, but knowing which is best suited to a particular problem is very valuable to a deep understanding of a problem. For example, for buckling and wave propagation problems, each mode or snapshot can be displayed and investigated separately as if an instance of a static problem. But time traces (of velocity or stress at a point) is additionally needed for the wave propagation problem but not the buckling problem.

IV: Gallery of Nonlinear Equations

The collection could range from the symmetric return force to Van der Pol equation to parametric excitation. The typical equation is of the form

$$M\ddot{u} + C(u)\dot{u} + K(u,t) = P(t)$$

with $C(u)$ and $K(u,t)$ containing many parameters that are adjustable. The loads can be a combination of ramp, forced frequency, multiple pings, and uniform.

The results are presented as trajectories, component histories, phase-plane plots, Poincare plots, energy and isocline contours, and Fourier transforms. At any stage, an immediate comparison can be made with the linearized model.

V: Enabling Programs

In terms of philosophy, any FEM program could be used for the underlying computations and any graphically oriented 4GL can be used for programming the user interface. Reference [81] is a nice illustration of using MatLab as the host to Ansys for the computations.

NonStaD is the finite element analysis program used as a test bed implementing the major features discussed in this book. Its operation under QED is by way of driver or script files. That is, QED creates the script files to execute NonStaD in batch mode.

The program GenMesh is used to create structure datafiles for use by NonStaD. Its main capabilities involve generating meshes and performing executive functions such as merging meshes and adding material properties and boundary conditions. Here also, QED creates the script files which run GenMesh in batch mode.

The Role of the Book

The role of the book is to provide the principles by which dynamical systems are explored, it is a guide book; the role of the computer is to make many examples readily available. Having a flexible sophisticated simulation running on the computer to augment/counterpoint the text examples would profoundly affect the engineers' perception of both theory and computational methods. The interplay of both would establish an interesting and exciting dynamic.

References

[1] **Abramowitz, M. and Stegun, I. A.**, *Handbook of Mathematical Functions*, Dover, New York, 1965.

[2] **Allman, D.J.**, "Evaluation of the Constant Strain Triangle with Drilling Rotations," *International Journal for Numerical Methods in Engineering*, **26**, pp. 2645–2655, 1988.

[3] **Allman, D.J.**, "On the General Theory of the Stability of Equilibrium of Discrete Conservative Systems," *Aeronautical Journal*, **27**, pp. 29–35, 1989.

[4] **Argyris, J.H. and Mlejnek, H-P.**, *Dynamics of Structures*, Elsevier, Amsterdam, 1991.

[5] **Argyris, J.**, "An Excursion into Large Rotations," *Computer Methods in Applied Mechanics and Engineering*, **32**, pp. 85–155, 1982.

[6] **Baker, G.L. and Gollub, J.P.**, *Chaotic Dynamamics: an Introduction*, Cambridge University Press, Cambridge, 1990.

[7] **Bathe, K.-J.**, *Finite Element Procedures in Engineering Analysis*, Prentice-Hall, Englewood Cliffs, NJ, 1982, 2/E 1995.

[8] **Batoz, J-L., Bathe, K-J. and Ho, L-W.**, "A Study of Three-Node Triangular Plate Bending Elements," *International Journal for Numerical Methods in Engineering*, **15**, pp. 1771–1812, 1980.

[9] **Batoz, J-L.**, "An Explicit Formulation for an Efficient Triangular Plate-Bending Element," *International Journal for Numerical Methods in Engineering*, 1982.

[10] **Belytschko, T., Schwer, L., and Klein, M.J.**, "Large Displacement, Transient Analysis of Space Frames," *International Journal for Numerical Methods in Engineering*, **11**, pp. 65–84, 1977.

[11] **Belytschko, T. and Hsieh, B.J.**, "Non-Linear Transient Finite Element Analysis with Convected Coordinates," *International Journal for Numerical Methods in Engineering*, **7**, pp. 255–271, 1973.

[12] **Belytschko, T.**, An Overview of Semidiscretization and Time Integration Procedures, In *Computaional Methods for Transient Analysis*, T. Belytschko and T.J.R. Hughes, editors, pp. 1–65, Elsevier, 1993.

[13] **Bergan, P.G. and Felippa, C.A.**, "A Triangular Membrane Element with Rotational Degrees of Freedom," *Computer Methods in Applied Mechanics and Engineering*, **50**, pp. 25–69, 1985.

[14] **Bergan, P.G. and Felippa, C.A.**, Efficient Implementation of a Triangular Membrane Element with Drilling Freedoms, In *FEM Methods*, pp. 128–152, 1987.

[15] **Boley, B.A. and Weiner, J.H.**, *Theory of Thermal Stress*, Wiley & Sons, New York, 1960.

[16] **Brigham, E.O.**, *The Fast Fourier Transform*, Prentice-Hall, Englewood Cliffs, New Jersey, 1973.

[17] **Chen, Y-C.**, "Stability and Bifurcation of Finite Deformations of Elastic Cylindrical Membranes — Part I: Stability Analysis," *Naturwissenchaften*, **34**(14), pp. 1735–1749, 1997.

[18] **Cook, R.D., Malkus, D.S. and Plesha, M.E.**, *Concepts and Applications of Finite Element Analysis, 3/E*, Wiley & Sons, New York, 1989.

[19] **Crisfield, M.A.**, *Nonlinear Finite Element Analysis of Solids and Structures*, **Vol 1**: *Essentials*, Wiley & Sons, New York, 1991.

[20] **Crisfield, M.A.**, *Nonlinear Finite Element Analysis of Solids and Structures,* **Vol 2:** *Advanced Topics*, Wiley & Sons, New York, 1997.

[21] **Doyle, J.F., and Sun, C-T.**, *Theory of Elasticity*, A&AE 553 Course Notes, Purdue University, 1999.

[22] **Doyle, J.F.**, *Static and Dynamic Analysis of Structures*, Kluwer, The Netherlands, 1991.

[23] **Doyle, J.F.**, *Wave Propagation in Structures, 2/E*, Springer-Verlag, New York, 1997.

[24] **Eriksson,A., Pacoste, C. and Zdunek, A.**, "Numerical Analysis of Complex Instability Behavior using Incremental-Iterative Strategies," *Computer Methods in Applied Mechanics and Engineering*, **179**, pp. 265–305, 1999.

[25] **Galvanetto, U. and Crisfield, M.A.**, "An Energy-Conserving Co-Rotational Procedure for the Dynamics of Planar Beam Structures," *International Journal for Numerical Methods in Engineering*, **39**, pp. 2265–2282, 1992.

[26] **Glendinning, P.**, *Stability, Instability and Chaos*, Cambridge University Press, Cambridge, UK, 1994.

[27] **Goldsmith, W.**, *Impact*, Edward Arnold, London, 1960.

[28] **Gordon, J.E.**, *Structures, or Why Things Don't Fall Down*, Penguin, London, 1978.

[29] **Graff, K.F.**, *Wave Motion in Elastic Solids*, Ohio State University Press, Columbus, 1975.

[30] **Guz, A.N., Makhort, F.G., Gushcha, O.I. and Lebedev, V.K.**, "Theory of Wave Propagation in an Elastic Isotropic Body with Initial Deformations," *Soviet Applied Mechanics*, **6**, pp. 1308–1313, 1970.

[31] **Hamilton, W.R.**, *The Mathematical Papers of Sir W.R. Hamilton*, Cambridge University Press, Cambridge, 1940.

[32] **Hoff, N.J.**, *The Analysis of Structures*, Wiley & Sons, New York, 1956.

[33] **Holder, E.J. and Schaeffer, D.**, "Boundary Conditions and Mode Jumping in the von Karman Equations," *SIAM Journal of Mathematical Analysis*, **15**(3), pp. 446–458, 1984.

[34] **Hsiao, K-M.**, "Nonlinear Analysis of General Shell Structures by Flat Triangular Shell Element," *Computers & Structures*, **25**(5), pp. 665–675, 1986.

[35] **Hsu, C.S.**, "Stability of Shallow Arches against Snap-through under Timewise Step Loads," *Journal of Applied Mechanics*, **35**, pp. 31–39, 1968.

[36] **Jeyachandrabose, C., Kirkhope, J. and Babu, C.R.**, "An Alternative Explicit Formulation for the DKT Plate-Bending Element," *International Journal for Numerical Methods in Engineering*, **21**, pp. 1289–1293, 1985.

[37] **Jones, R.M.**, *Mechanics of Composite Materials*, McGraw-Hill, New York, 1975.

[38] **Jordon, D.W. and Smith, P.**, *Nonlinear Ordinary Differentia Equations, 2/E*, Clarendon Press, Oxford, 1987.

[39] **Kleiber, M., Kotula, W., and Saran, M.**, "Numerical Analysis of Dynamic Quasi-Bifurcation," *Engineering Computations*, 4, pp. 48–52, 1987.

[40] **Kounadis, A.N. and Raftoyiannis, J.**, "Dynamic Stability Criteria of Nonlinear Elastic Damped/Undamped Systems under Step Loading," *AIAA Journal*, **28**(7), pp. 1217–1223, 1990.

[41] **Kuhl, D. and Crisfield, M.A.**, "Energy-Cnserving and Decaying Algorithms in Non-Linear Structural Dynamics," *International Journal for Numerical Methods in Engineering*, **45**, pp. 569–599, 1999.

[42] **Lau, J.H.**, "Large Deflections of Beams with Combined Loads," *Engineering Mechanics, ASCE*, **108**, pp. 180–185, 1982.

[43] **La Salle, J. and Lefschetz, S.**, *Stability by Liapunov's Direct Method*, Academic Press, New York, 1961.

[44] **Leipholz, H.**, *Stability Theory, an Introduction to the Stability of Dynamic Systems and Rigid Bodies, 2/E*, John Wiley & Sons and B.G. Teubner, Stuttgart, 1987.

[45] **Leissa, A.W.**, *Vibration of Plates*, NASA SP-160, 1969.

[46] **Leissa, A.W.**, *Vibration of Shells*, NASA SP-288, 1973.

[47] **Markus, S.**, *Mechanics of Vibrations of Cylindrical Shells*, Elsevier, New York, 1988.

[48] **McConnell, K.G. and Riley, W.F.**, Force-Pressure-Motion Measuring Transducers, In *Handbook on Experimental Mechanics*, A.S. Kobayashi, editor, pp. 79–118, Society for Experimental Mechanics, 1993.

[49] **Meirovitch, L.**, *Elements of Vibration Analysis*, McGraw-Hill, 1986.

[50] **Milne-Thomson, L.M.**, *Theoretical Hydrodynamics, 4/E*, Macmillan, New York, 1960.

[51] **Naghdi, P.M. and Berry, J.G.**, "On the Equations of Motion of Cylindrical Shells," *Journal of Applied Mechanics*, **21**(2), pp. 160–166, 1964.

[52] **Nour-Omid, B. and Rankin, C.C.**, "Finite Rotation Analysis and Consistent Linearization using Projectors," *Computer Methods in Applied Mechanics and Engineering*, **93**(1), pp. 353–384, 1991.

[53] **Novozhilov, V.V.**, *Foundations of the Nonlinear Theory of Elasticity*, Graylock Press, Rochester, N.Y., 1953.

[54] **Pacoste, C. and Eriksson, A.**, "Beam Elements in Instability Problems," *Computer Methods in Applied Mechanics and Engineering*, **144**, pp. 163–197, 1997.

[55] **Pacoste, C.**, "Co-rotational Flat Facet Triangular Elements for Shell Instability Analyses," *Computer Methods in Applied Mechanics and Engineering*, **156**, pp. 75–110, 1998.

[56] **Panovko, Y.G. and Gubanova, I.I.**, *Stability and Oscillations of Elastic Systems: Paradoxes, Fallacies, and New Concepts*, Consultants Bureau, New York, 1965.

[57] **Peng, X. and Crisfield, M.A.**, "A Consistent Co-Rotational Formulation for Shells using the Constant Stress/Constant Moment Triangle," *International Journal for Numerical Methods in Engineering*, **35**, pp. 1829–1847, 1992.

[58] **Reissner, E.**, "Stress and Displacement of Shallow Spherical Shells," *Journal of Mathematical Physics*, **25**(1), pp. 80–85, 1946.

[59] **Riks, E., Rankin, C.C., and Brogan, F.A.**, "On the Solution of Mode Jumping Phenomena in Thin-walled Shell Structures," *Computational Methods in Applied Mechanics and Engineering*, **136**, pp. 59–92, 1996.

[60] **Saaty, T.L., and Bram, J.**, *Nonlinear Mathematics*, Dover, New York, 1981.

[61] **Salerno, G. and Casciaro, R.**, "Mode Jumping and Attractive Paths in Multimode Elastic Buckling," *International Journal for Numerical Methods in Engineering*, **40**, pp. 833–861, 1997.

[62] **Schaeffer, D. and Golubitsky, M.**, "Boundary Conditions and Mode Jumping in the Buckling of a Rectangular Plate," *Communications in Mathematical Physics*, **69**, pp. 209–306, 1979.

[63] **Shames, I.H. and Dym, C.L.**, *Energy and Finite Element Methods in Structural Analysis*, Hemisphere, Washington, 1985.

[64] **Simitses, G.J., and Sheinman, I.**, "Static and Dynamic Buckling of Pressure Loaded Ring-stiffened Cylindrical Shells," *Journal of Ship Research*, **27**(2), pp. 113–120, 1983.

[65] **Simitses, G.J.**, "On the Dynamic Buckling of Shallow Spherical Shells," *Journal of Aplied Mechanics*, **41**, pp. 299–300, 1974.

[66] **Simitses, G.J.**, "Suddenly Loaded Structural Configurations," *Journal of Engineering Mechanics, ASCE*, **110**(9), pp. 1320–1334, 1984.

[67] **Slotine, J-J.E. and Li, W.**, *Applied Nonlinear Control*, Prentice Hall, New Jersey, 1991.

[68] **Southwell, R.V. and Skan, S.W.**, "On the Stability under Shearing Forces of a Flat Elastic Strip," *Proc. Royal Society*, **A 105**, pp. 582–606, 1924.

[69] **Stein, M.L.**, *Loads and Deformations of Buckled Rectangular Plates*, NASA Technical Report R-40, 1959.

[70] **Stein, M.L.**, *The Phenomenon of Change in Buckle Patern in Elastic Structures*, NASA Technical Report R-39, 1959.

[71] **Stricklin, J.A and Haisler, E.E. and Tisdale, P.R. and Gunderson, R.**, "A Rapidly Converging Triangular Plate Element," *AAIA Journal*, **7**(1), pp. 180–181, 1969.

[72] **Suchy, H., Troger, H., and Weiss, R.**, "A Numerical Study of Mode Jumping of Rectangular Plates," *ZAMM*, **65**(2), pp. 71–78, 1985.

[73] **Tedesco, J.W., McDougal, W.G., and Allen Ross, C.**, *Structural Dynamics: Theory and Applications*, Addison-Wesley, Menlo Park, CA, 1999.

[74] **Thompson, J.M.T. and Hunt, G.W.**, *A General Theory for Elastic Stability*, Wiley & Sons, London, 1973.

[75] **Thompson, J.M.T. and Hunt, G.W.**, *Elastic Stability*, Wiley & Sons, London, 1993.

[76] **Thomson, W.T.**, *Theory of Vibrations with Applications*, Prentice-Hall, New Jersey, 1981.

[77] **Timoshenko, S.P. and Gere, J.M.**, *Theory of Elastic Stability*, McGraw-Hill, New York, 1988.

[78] **Timoshenko, S.P. and Woinowsky-Krieger, S.**, *Theory of Plates and Shells*, McGraw-Hill, New York, 1968.

[79] **Uemura, M. and Byon, O-IL**, "Secondary Buckling of a Flat Plate under Uniaxial Compression. Part I: Theoretical Analysis of Simply Supported Flat Plate," *International Journal of Nonlinear Mechanics*, **12**, pp. 355–370, 1977.

[80] **Ulrich, K.**, *State of the Art in Numerical Methods for Continuation and Bifurcation Problems with Applications in Continuum Mechanics — A Survey and Comparative Study*, Laboratorio Nacional de Comutacao, Brazil, 1988.

[81] **Webster, E.M**, *A Dynamic View of Static Instabilities*, Ph.D. Thesis, Purdue University, 1999.

[82] **Whittaker, E.T. and Watson, G.N.**, *A Course of Modern Analysis*, Cambridge University Press, Cambridge, 1927.

[83] **Xu, X-P., and Needleman, A.**, "Numerical Simulations of Fast Crack Growth in Brittle Solids," *Journal of Mechanics and Physics of Solids*, **42**(9), pp. 1397–1434, 1994.

[84] **Yourgrau, W. and Mandelstam, S.**, *Variational Principles in Dynamics and Quantum Theory*, Dover, New York, 1979.

[85] **Ziegler, H.**, *Principles of Structural Stability*, Ginn and Company, Massachusetts, 1968.

Index

R.A. Layton, **Principles of Analytical System Dynamics**

C.V. Madhusudana, **Thermal Contact Conductance**

D.P. Miannay, **Fracture Mechanics**

D.K. Miu, **Mechatronics: Electromechanics and Contromechanics**

D. Post, B. Han, and P. Ifju, **High Sensitivity Moiré: Experimental Analysis for Mechanics and Materials**

F.P. Rimrott, **Introductory Attitude Dynamics**

S.S. Sadhal, P.S. Ayyaswamy, and J.N. Chung, **Transport Phenomena with Drops and Bubbles**

A.A. Shabana, **Theory of Vibration: An Introduction, 2nd ed.**

A.A. Shabana, **Theory of Vibration: Discrete and Continuous Systems, 2nd ed.**